Transition State Modeling
for Catalysis

ACS SYMPOSIUM SERIES **721**

Transition State Modeling for Catalysis

Donald G. Truhlar, EDITOR
University of Minnesota

Keiji Morokuma, EDITOR
Emory University

Developed from a symposium sponsored by the Division
of Computers in Chemistry at the 215th National Meeting
of the American Chemical Society,
Dallas, Texas,
March 29–April 2, 1998

American Chemical Society, Washington, DC

Library of Congress Cataloging-in-Publication Data

Transition state modeling for catalysis: developed from a symposium sponsored by the Division of Computers in Chemistry at the 215th National Meeting of the American Chemical Society, Dallas, Texas, March 29–April 2, 1998 Donald G. Truhlar, editor, Keiji Morokuma, editor.

p.   cm.—(ACS symposium series, ISSN 0097–6156 ; 721)

Includes bibliographical references and indexes.

ISBN 0-8412-3610-0 (cloth)

1. Catalysis—Computer simulation—Congresses. I. Truhlar, Donald G., 1944– . II. Morokuma, K. (Keiji), 1934– . III. American Chemical Society. Computers in Chemistry of Division. IV. American Chemical Society. Meeting (215ᵗʰ : 1998 : Dallas, Tex.) V. Series.

QD505.T735   1999
541.3′.95′0113—dc21

98-53092
CIP

Advisory Board

ACS Symposium Series

Foreword

The ACS Symposium Series was first published in 1974 to provide a mechanism for publishing symposia quickly in book form. The purpose of the series is to publish timely, comprehensive books developed from ACS sponsored symposia based on current scientific research. Occasionally, books are developed from symposia sponsored by other organizations when the topic is of keen interest to the chemistry audience.

Before agreeing to publish a book, the proposed table of contents is reviewed for appropriate and comprehensive coverage and for interest to the audience. Some papers may be excluded in order to better focus the book; others may be added to provide comprehensiveness. When appropriate, overview or introductory chapters are added. Drafts of chapters are peer-reviewed prior to final acceptance or rejection, and manuscripts are prepared in camera-ready format.

As a rule, only original research papers and original review papers are included in the volumes. Verbatim reproductions of previously published papers are not accepted.

ACS Books Department

Contents

METALS AND METALLOIDS AS CATALYSTS

Enzymatic Reactions

Preface

This volume contains contributions from the speakers at the Symposium on "Transition State Catalysis in Computational Chemistry," sponsored by the Division of Computers in Chemistry at the National Meeting of the American Chemical Society in Dallas, March 30–April 2, 1998. Most of the papers are by invited speakers. The European community, which is very strong in the area, is well represented in this international symposium.

The symposium was unique, and the volume is unique in the same way. This is because it includes both homogeneous and heterogeneous catalysis, including organometallic catalysts, biological systems, zeolites and metal oxides, and metal surfaces. Furthermore, the emphasis is on actual *transition state modeling*, in contrast to the much more widely studied topic of precursor binding.

Computational chemistry is making explosive inroads on many areas of practical concern, and it is becoming an important tool for catalyst design. We believe that this symposium and volume catch the cutting-edge field of transition state modeling for catalysis in its infancy. These are the seminal studies that will set a tone for the next decade and beyond. We hope the book will be of great interest to a broad group of industrial and academic chemists worldwide.

All contributions to the volume received peer review, coordinated by the editors. Peer review for the three chapters for which one or both editors is a coauthor was arranged by Anne Wilson of ACS Books. We would like to take this opportunity to thank Anne Wilson for this and also for making our interactions with the publisher smooth and enjoyable at every step of the planning and production of this volume.

The editors were also the symposium organizers. We were very pleased to be assisted in organizing the conference by the following International Advisory Board of experts in the field: Gernot Frenking, Gabor Náray-Szabó, Vern L. Schramm, Rutger A. van Santen, and Arieh Warshel.

We would like to close by thanking the authors for adhering remarkably well to deadlines and thanking the referees for providing speedy review. Copies of the proceedings papers were due 2.5 weeks before the symposium, and the book was completed only 4 weeks after the conclusion of the symposium.

DONALD G. TRUHLAR
Department of Chemistry
University of Minnesota
Minneapolis, MN 55455–0431

KEIJI MOROKUMA
Department of Chemistry
Emory University
Atlanta, GA 30322

Dedicated to

Eiko, Jane, Sara, and Stephanie

INTRODUCTION

Chapter 1

Quantum Catalysis: The Modeling of Catalytic Transition States

Michael B. Hall[1], Peter Margl[2], Gabor Náray-Szabó[3], Vern L. Schramm[4], Donald G. Truhlar[5], Rutger A. van Santen[6], Arieh Warshel[7], and Jerry L. Whitten[8]

[1]Department of Chemistry, Texas A&M University, College Station, TX 77843–3255
[2]Department of Chemistry, University of Calgary, 2500 University Drive N.W., Calgary, Alberta T2N 1N4, Canada
[3]Department of Theoretical Chemistry, Loránd Eötvös University, Pázmány Péter st. 2, H-1117 Budapest, Hungary
[4]Department of Biochemistry, Albert Einstein College of Medicine, 1300 Morris Park Avenue, Bronx, NY 10461–1602
[5]Department of Chemistry, University of Minnesota, 207 Pleasant Street S.E., Minneapolis, MN 55455
[6]Eindhoven University of Technology, P.O. Box 513, 5600 MB Eindhoven, Netherlands
[7]Department of Chemistry, University of Southern California, Los Angeles, CA 90089–1062
[8]Physical and Mathematical Sciences, North Carolina State University, Box 8201, Raleigh, NC 27695–8201

We present an introduction to the computational modeling of transition states for catalytic reactions. We consider both homogeneous catalysis and heterogeneous catalysis, including organometallic catalysts, enzymes, zeolites and metal oxides, and metal surfaces. We summarize successes, promising approaches, and problems. We attempt to delineate the key issues and summarize the current status of our understanding of these issues. Topics covered include basis sets, classical trajectories, cluster calculations, combined quantum-mechanical/molecular-mechanical (QM/MM) methods, density functional theory, electrostatics, empirical valence bond theory, free energies of activation, frictional effects and nonequilibrium solvation, kinetic isotope effects, localized orbitals at surfaces, the reliability of correlated electronic structure calculations, the role of d orbitals in transition metals, transition state geometries, and tunneling.

Computational chemistry has achieved great strides in recent years, and it has become a strong partner with experimental methods, displacing purely analytic theory almost completely. The essential element in computational chemistry is the use of very general theoretical frameworks that can be applied broadly across a number of fields. Examples are molecular orbital theory, molecular mechanics, transition state theory, classical trajectories, and the Wentzel-Kramers-Brillouin method. As the techniques of these fields are developed, they can be applied across broad ranges of problems inchemical structure and reactivity. When this pack of techniques is used to attack a herd of problems, it is only natural that attention is first focussed on the easiest ones. As these begin to succumb, computational chemistry sets its sights on more recalcitrant targets. Thus computational chemistry has been very successful for calculating structures and energies of small molecules, structures and energies of large molecules, and properties of simple reactions of these molecules. An emerging competency is the treatment of complex reactions involving catalysts. A critical element in this endeavor is the ability to directly model the transition state.

Systems of interest range from heterogeneous catalysis, through transition metal complexes, both classical and organometallic, to bioinorganic chemistry of enzymes and enzyme mimics. Often a reasonably accurate structure and a semi-quantitative relative energy for possible transition states can be helpful in pointing the way to the correct reaction mechanism. Knowledge of the mechanism is important as it directs the thinking about how to modify the reaction under study or how to create a new reaction. In many cases experimental work will provide an approximate relative free energy for the key or rate-determining transition state, but ascertaining its structure from experimental data is much more difficult. Here, theory can help. In fact conventional quantum chemistry is arguably at its best when determining structures.

Catalysts are molecular devices that are able to perform complicated chemical manipulations on individual molecules. Although they are technologically attractive entities, engineering them for a given task is a challenge that is only slowly coming within reach. Whereas it is possible to efficiently design and build macroscopic machines using classical mechanics, it is quite another matter to do so with individual atoms and molecules, which behave according to the laws of quantum mechanics. Thus quantum mechanical modeling methods play a prominent role in catalyst modeling and computational catalyst design.

An appropriate model of a catalytic system is, in most cases, quite complicated and consists of many atoms. Owing to their relative mathematical simplicity molecular mechanics and classical trajectory studies may be applied with explicit consideration of all components; however full models of practically important systems are beyond the scope of quantum mechanical methods. A more and more popular approach to quantum mechanical modeling is partitioning of the catalytic system (substrate interactions with an enzyme, zeolite pore, solid surface, or solvent) into various regions (*1–8*). The active site or central machinery (C) is surrounded by a polarizable environment (P) which is embedded in a non-polarizable region (N) and solvent (S). Some regions may overlap and others may be dropped completely from the model, depending on the degree of sophistication of the treatment and their effect on the process. A very great variety of implementations are possible, differing in details, and we will elaborate on some variations at appropriate points in this article.

The minimum-size active site model (C) should be composed of those atoms that undergo chemical changes during catalysis, i.e., participate in the bond(s) that will be cleaved or formed during the process. For example, in enzyme catalysis by serine proteases, which hydrolyze peptide bonds, C must contain at least the =N–C=O moiety of the peptide bond, the –OH group of the attacking serine, and the proton acceptor imidazole ring of the nearby histidine. For systems where electronic excitation or electron transfer plays a role (e.g., heme proteins) the minimum-size active site may contain several dozen atoms. Reactions at metal surfaces require considering a large number of metal atoms in order to converge the work function and the tendency to create electron-hole pairs. Reactions in high-dielectric-constant solvents, such as water, require considering a large number of solvent molecules in order to obtain converged electrostatic energies.

Transition State Theory

The rate constants for reactions in the condensed phase can be written (9)

$$k = \kappa \frac{k_B T}{h} \exp[-\Delta G^{\ddagger,0}(T)/RT] \qquad (1)$$

where $G^{\ddagger,0}$ is the standard-state free energy of activation that reflects the probability of reaching the transition state, R is the molar Boltzmann constant, and κ is the transmission coefficient (10) which contains dynamical effects. $G^{\ddagger,0}(T)$ reflects a compromise of quasiequilibrium energetic and entropic effects; κ reflects the probability of a trajectory recrossing the transition state, and it may include effects variously labeled as dynamic or nonequilibrium effects or friction. The transmission coefficient must also include quantum effects on the reaction coordinate motion, which is otherwise intrinsically classical in transition state theory (11). In conventional transition state theory (12), $G^{\ddagger,0}(T)$ is calculated at the saddle point of the Born-Oppenheimer potential energy surface (PES) for the reaction, whereas in variational transition state theory it is calculated at the maximum of the free energy of activation profile (5,9,12,13), which is the free energy of activation as a function of a reaction coordinate (RC). Models for transition states of catalytic reactions should focus on determining $G^{\ddagger,0}$ and κ.

If barriers are low enough and/or temperatures are high enough, dynamics calculations can be directly used to simulate a reaction in real time and thereby focus on kinetic, thermodynamic, and mechanistic issues. However, this is not feasible if the reaction of interest has a high barrier compared to $k_B T$, where k_B is Boltzmann's constant and T is temperature, unless one employs infrequent event sampling and/or a constraint on the reaction coordinate. In such a case, one can show that it is still possible to obtain statistical information about a given reaction along a reaction coordinate by using umbrella sampling or by enforcing a constrained reaction coordinate and employing generalized normal mode analysis, thermodynamic integration, or free energy perturbation theory where the incremental charge in the reaction coordinate is the perturbation (1–3,14–28). In this approach, one selects (or "distinguishes") a specific RC that most conveniently leads the system through an area of interest, for instance the transition state region. The chemical system is then led slowly along the RC, thereby mapping out the regions adjacent to the RC. This slow mapping yields two types of information. First, the dynamical scanning process can escape from local catchment basins and find reaction pathways unforeseen by the operator (29,30). This is an especially important feature if one is trying to explore as yet uncharted territory. Several recent papers show that this approach works very well for finding lowest-energy pathways in catalytic systems like the Ziegler-Natta metallocene type homogeneous catalysts (29,30). Second, by umbrella sampling, generalized normal mode analysis, integrating the effective forces along the reaction coordinate, or free energy perturbation theory, one arrives at the free energy of activation profile mentioned below Eq. (1), that is, the free energy of activation (5,9,12,13) required to move the system from one cut through the RC to another or from reactants to a particular point along the RC. This technique is a very efficient tool to calculate activation free energies and enthalpies in molecular dynamics.

In order to accelerate the rate of a chemical reaction the catalyst must find a way to interact with the substrate(s) so that the resulting free energy of activation barrier or reaction friction becomes lower than that of the original process in the gas phase or in solution. There is no universal theory of the origin of catalytic rate acceleration; however, in light of the considerations of the previous paragraph, we may attempt to classify the effects into energetic, entropic, or frictional/dynamical Since energetic stabilization takes place in most cases by non-covalent interactions, it is plausible that electrostatic effects and the participation of ionic and polar species are of utmost importance in catalysis. Electrostatics plays a central role in all of chemistry, and catalytic reactions are no exception. Catalysts may produce a strong electrostatic field around a substrate, and an appropriate model should therefore contain the environment of C, including polar or charged groups and solvent. As mentioned above, aqueous solutions are especially complicated since water is a liquid with a large dielectric constant; thus it has quite a strong influence on molecular transformations of polar species or with polar or charged catalysts. In some cases, transition-metal catalysis may be electrostatic in nature, but in most cases it can be attributed to changing overlap between interacting atomic or molecular orbitals; thus, the effect should typically be described in terms of quantum mechanics. There is evidence that electrostatic catalysis is important for a number of pericyclic reactions (31), and we may speculate that potentially similar effects arise for rearrangements in water or other polar solvents where the initial state of the reaction involves less polar species than the transition state. In such cases metal or other small ions stabilize the transition state by electrostatic interactions. However, for some transition-metal ions, orbital interactions may overweigh electrostatics; thus homogeneous catalysis involving these does not necessarily obey the above rule. For heterogeneous catalysis taking place at interfaces, the arrangement and properties of participating atoms or molecules may differ strongly from the bulk; specific interactions take place here. There are at least two systems, crystal surfaces and zeolite pores, where the electrostatic potential changes extremely rapidly, providing a large electrostatic field difference between the gas phase and the bulk. This may vary in the range of 5 to 20 V/nm and strongly polarize reactant bonds which may result in their fission or enhancement of attack by certain reagents (32). It has been postulated that the catalytic effect of zeolites is mainly due to the strong electrostatic field that may emerge within the pores (33), although this is an incomplete explanation, and one must consider the specific Lewis or Brønsted acid sites that may be present. Electrostatic catalysis in zeolites may be attributed to two main effects. One is the stabilization of cations by the large negative molecular electrostatic potential inside zeolite pores. This effect favors, for example, the formation of protonated species, and thus it may result in stabilization of the corresponding transition states. The other effect is related to the high electrostatic field inside pores that polarizes covalent bonds and promotes them for fission and stabilizes ion pairs.

Entropy effects may play an important role for a number of catalytic processes by immobilizing the reacting partners and thus reducing translational, rotational, or, to a lesser extent, vibrational entropy in the transition state. It is an old hypothesis that enzymes should be capable of efficiently catalyzing reactions with unfavorable entropies of activation by acting as "entropy traps;" this means that the binding energy of the enzyme is used to freeze out rotational and translational degrees of freedom by converting them to confined vibrations in the process of forming the activated complex. These effects may be smaller than previously thought since enzyme molecules are quite flexible, e.g., an examination of the entropic contribution to the rate acceleration of serine proteases indicate that this is a small effect (1).

Frictional effects are usually assumed to be less important than equilibrium free energy effects, but are not always negligible (5,34). As discussed elsewhere (35), "friction" and "nonequilibrium solvation" provide two different ways of looking at the effects that decrease the transmission factor. Nonequilibrium solvation may play a significant role in determining the temperature dependence of reaction rates.

Many catalytic processes involve general acid or base catalysis which can provide an alternative reaction path with lower energetic barriers and different free energies. The extent of barrier reduction of a proton transfer step is determined by the difference between the proton affinities of the donor and acceptor, which are intrinsic properties of the reactants and catalyst. Catalyst interactions with the reactant may induce changes in its proton affinity or in its state of protonation in the catalyst-substrate precursor complex. In addition, a step of the given reaction that involves proton transfer may be modified by participation of an acidic or basic residue or ligand of the catalyst. In this way the free energy of the proton transfer step is reduced and the reaction is accelerated. On the other hand, fixation of a general acid or base by the catalyst backbone or framework near the active site provides a gross entropy gain as compared to water solution. One common motif is that aqueous solvent may serve as a catalyst by serving as a bridge that both donates and accepts protons (36,37).

In addition to these general issues that play a prominent role in all types of catalytic processes, there are many more specific questions to be considered in each kind of catalysis, as discussed in the next section.

Catalytic Systems

Homogeneous catalysis. Transition metal reactions and catalysis have been the subject of several recent reviews (38–40). Driven to a large extent by improvements in computer technology and algorithms, high-level quantum chemistry has recently become powerful enough for realistic modeling of homogeneous organometallic catalytic systems on the femto- to picosecond time scale and the sub-nanometer length scale. Molecular orbital calculations employing density functional theory (41) (especially with generalized-gradient-approximation density functionals) sometimes allow one to predict chemical activation parameters to within a few kcal/mol. This is already enough to decide whether a catalyst is worthwhile so that expensive synthetic screening experiments can be directed to better use. These powers can also be harnessed to find or explain the actual mechanism of a reaction by investigating the topology of the PES associated with the catalytic system at hand. In the mid-1980s, a novel approach (42,44) to *ab initio* PES mapping was conceived by Car and Parrinello (CP). In that approach, the PES is mapped by simulating the classical dynamics of nuclear motion at finite temperature. Although this in itself is only the time-honored "molecular dynamics" or "classical trajectory" approach that has been used since the 1950s to describe chemical dynamics, the CP method places an emphasis on systematic methods for using atomic forces that are calculated directly from *ab initio* electronic structure calculations on the chemical system at hand. Thus the method is sometimes called a direct dynamics approach. In the CP approach the calculation of forces is fast since the wavefunctions themselves are treated as dynamical variables which are propagated through time in conjunction with their attached nuclei. Nowadays, it is not uncommon to calculate the dynamics of systems with more than 100 atoms over several tens of picoseconds. For the chemist who is interested in predicting activation parameters and reaction pathways, this approach bears several advantages compared to more traditional methods which restrict themselves to locating only a few stationary points on the potential surface.

The application of *ab initio* molecular dynamics to calculate free energy of activation profiles for catalytic reactions is a recent development (29,30) and bears much promise, as it does not rely on any analytical approximations to the potential surface. This becomes especially significant when the potential surfaces that one is trying to describe are very complex, so that obtaining an analytical expression for them is difficult. However, the original CP approach suffers from the limitations of using classical mechanics for the nuclear motion, and this can lead to severe quantitative errors in some cases (45). Direct dynamics methods that include quantum effects such as tunneling and zero point energy are also under active development and should see more applicability to catalytic systems in the future (*45–48*).

Calculations on large catalyst systems have been dramatically speeded up by the use of quantum-mechanical/molecular-mechanical (QM/MM) partitioning (*1,3,8,49*) of the catalyst system in CP simulations. This is an example of the general C/P/N/S partitioning techniques mentioned above. Here, parts of the simulated system that do not involve any bond making or breaking processes are described by a molecular mechanics force field, whereas the chemically active catalyst site is still described by high-level density functional theory. QM and MM parts of the system interact through van der Waals and Coulomb forces. Enormous time savings are achieved when these techniques are compared to a pure QM dynamics calculation because MM calculations are typically 10,000 times faster than analogous QM calculations. This workable and very efficient model has recently been demonstrated for the Brookhart ethylene polymerization catalyst (*50*).

Further developments in dynamical reaction modeling will undoubtedly arise soon. For example one will see QM and QM/MM CP simulations that account not only for the immediate reactants, but also for a whole reaction ensemble, including solvent and "spectator" species. Other approaches to treating the solvent quantum mechanically are also under investigation (*51,52*). Dual-level techniques involving high-order and low-order quantum mechanical methods (*53–56*), rather than mixing quantum mechanics and molecular mechanics, may also be useful.

The above emphasis on the *interface* of electronic structure (the ultimate source of the PES) and dynamics does not mean that the problems with the electronic structure *per se* are fully solved for transition metal catalysis. *Au contraire*, as anyone who has done calculations knows, there seem to be more pitfalls and pathological problems with transition metals than one encounters with lighter elements. When one uses effective core potentials to avoid explicit inclusion of core-electron basis functions, one finds the counter-intuitive result that the heavier third-transition-row metals can be treated more accurately than the first-transition-row metals. The reason for this unexpected behavior is the larger relative spatial extent of the $5d$ orbitals compared to the $3d$. Thus, third-transition-row metals form stronger, more covalent bonds with most ligands due to the large overlap provided by the $5d$ orbitals. Thus, for geometry optimizations of third-transition-row systems with monovalent (H, R, Cl) and coordinate covalent (NR_3, PR_3, CO) ligands, hybrid density functional methods such as B3LYP (*57*) and the standard second-order Møller-Plesset perturbation method (*58*) (MP2) for treating electron correlation usually yield similar and reasonable geometries (*59*). Even Hartree-Fock-Roothaan single-determinant self-consistent field methods (*58*) (HFR) can be adequate for geometries of many systems.

The choice of basis set provides an especially difficult problem for modeling catalytic reactions involving transition metals. Many of the standard basis sets in both commercial and public domain codes have serious inadequacies. Sometimes the d space is missing an essential diffuse d function, or the outer $(n + 1)s$ and $(n + 1)p$ functions are treated inadequately (*60,61*). This is a dangerous situation for new practitioners in the field as they may have some understanding of standard light-atom basis sets but be unaware that the transition-metal basis set, which is coupled to a reasonable ligand basis set, may have serious deficiencies. Thus, if one calculates unsystematic behavior, such as in a survey of structural trends in electronically similar molecules, and is unaware of the basis set inadequacy, one can attribute the observed behavior to faults in methodology or—even worse—to new and unexpected chemistry.

Although the relative energies one calculates are often not as accurate as the structures, they may be accurate enough for trends. Higher-order electron-correlation methods (*62*) such as MP4, QCISD, CCSD, and especially CCSD(T), can provide more reliable energies. When one moves from monovalent ligands to divalent ligands such as O, additional problems arise because of the stronger near-degeneracy situations one encounters. In such cases, HFR geometries may no longer be adequate and one must use at least the B3LYP or MP2 methods. Although the calculated structures may be realistic, the energies may be unreliable, especially if these divalent ligands participate in the reaction under study. Second-transition-row metals with monovalent and coordinate covalent ligands fall into the same category. When second-transition-row metals are bonded to multivalent ligands, MP2 energies are inadequate even for trends, and HFR makes unacceptably large errors in the geometry (*63*). The situation for first-transition-row metals is dramatically worse. Here, even for monovalent and coordinate covalent ligands, MP2 may have unexpected pathological problems in determining the structure, and higher-order electron-correlation methods such as MP3 and MP4 may oscillate strongly and provide unacceptably large errors in relative energy. Geometries from B3LYP and relative energies from QCISD, CCSD, and CCSD(T) are usually adequate and more consistent than those from MP2, MP3, or MP4 (*64*). The most unsatisfactory situation occurs for first-transition-row elements with multivalent ligands such as found in metal oxo or nitrosyl complexes (65). Here the near-degeneracy problem is so great that MP2 is totally inadequate even for the geometry; thus one usually resorts to B3LYP or similar methods for the geometry in such cases and to CCSD or CCSD(T) for energies, although even these methods can fail. Thus, as in so many other branches of computational chemistry, the treatment of systems with near-degeneracy effects remains an impediment to progress. All these problems become much worse for transition states than for reactants, and they may become even more critical when one considers metal clusters.

Zeolites. Transition states of elementary reaction steps in zeolite catalysis pose yet another set of specialized problems. The discovery that carbenium (*66*) and carbonium (*67*) ions in zeolite catalysis are usually transition states and not stable ions, as proposed in homogeneous acids, is due to applied quantum chemical research on protonation by zeolites. Early work, especially by Kazansky (*68,69*), showed that the low dielectric constant of zeolites makes charge separation difficult. Charge separation occurs when protons are generated by dissociation from their bound state in the zeolite. Protonated species become stabilized by the strong electrostatic attraction between their positive charge and the negativity charged zeolite wall. This has found early confirmation in quantum chemical studies of ammonium formation by protonation of ammonia (*70*). Proton transfer can only occur when ammonium binds back to the zeolite by directing two or three of its positively charged hydrogen atoms to the negatively charged oxygen atoms. The relatively high activation energies for proton-induced reactions stem essentially from the high zeolite deprotonation energies (~ 300 kcal/mol).

A major issue in quantum chemical calculations on zeolites is the validity of cluster models that are commonly used. Of course there is also the question of the level and kind of quantum chemical method that is adequate. Models and methods should be of about the same sophistication. For a model that is too small even the most precise *ab initio* quantum mechanical calculations may lead to erroneous results. On the other hand, low-level calculations on appropriately large models may provide artifacts as well. A minimum requirement that cluster models have to satisfy is that they are neutral. Because the deprotonation energy of a cluster tends to oscillate and slowly converges with size, the question of which cluster size is adequate to obtain representative results for the zeolite situation becomes paramount. As cluster models evolve, various C/P/N/S partitionings (in the classification introduced above) may be identified as particularly appropriate.

Geometry relaxation of cluster geometries in reactants as well as transition states is usually essential. A main effect of embedding of a cluster in the zeolite lattice is to reduce full relaxation of the cluster slightly (71) and to affect transition state energies by the non-equivalence of the oxygen atoms around the protonation site (72). Long-range electrostatic effects appear to be minor for geometry relaxation in embedded clusters. A shortcoming of the cluster approach is the absence of any steric constraints on the size of reaction intermediates or transition states due to limitations induced by the size of the zeolite micropores. Preliminary results (73) indicate that such special constraints can be incorporated by suitable embedding approaches and, dependent on the size of reaction systems, they can induce dramatic changes compared to results obtained with non-embedded clusters.

A major issue with lattice approaches such as CP techniques (42–44) when used to analyze the effects of zeolite cavity size is the non-reliability of current density functional techniques (41) to properly compute van der Waals interactions. Despite these and other limitations, the impact of quantum chemistry on the fundamental understanding of zeolite catalysis has been significant. As mentioned above, the proper formulation of reaction energy diagrams was first accomplished theoretically and is now gradually being confirmed by experimental studies (74). One reason for the success of computational methods applied to this kind of catalysis is the compensation of the forces controlling the covalent hydroxyl bond strength and the counteracting stabilization of the electrostatic attraction between protonated species and zeolite wall. Embedding techniques have reached the stage where quantitative agreement between measured and experimental protonation energy of ammonia has become possible (75). Because of the well understood geometry of zeolites and their accessibility to detailed spectroscopic probing, theoretical study of catalysis in these materials will remain very fruitful to deepen our understanding of catalysis as well as theoretical approaches.

Metal and metal oxide surfaces. Electronic structure methods aimed at a molecular-level description of the chemisorption and reaction of adsorbates on metal surface pose yet another set of problems. Some of these problems are also relevant to catalysis on metal oxide surfaces. Applications of these methods are relevant to transition metal heterogeneous catalysis (76). Questions of interest include:
- chemisorption energetics of molecules and molecular fragments
- adsorbate structure and spectra as a function of surface site
- interaction of coadsorbed species
- dissociative chemisorption
- surface reactions (heats of reaction and activation energies)
- nonequilibrium geometries, mechanistic issues, and potential energy surfaces
- electronic and geometric effects on trapping and activation

First-principles theory is needed to unravel complexities associated with such surface phenomena and to provide a framework for the interpretation of experiments.

The difficulty in treating metal and oxide surfaces using first-principles theory is that there are conflicting demands on the theory. At the solid surface, the treatment must be accurate enough to describe surface-adsorbate bonds and energy changes accompanying chemical reactions (generally, this is most readily achieved if the system is small), while for metals and oxides a large number of atoms is required to describe either conduction and charge transfer processes or the Madelung potential (in this case methods for treating large symmetric systems are most appropriate). Embedding methods (another example of the C/P/N/S approach discussed above) seek to balance the accuracy versus size aspects of the problem.

Many theoretical studies and calculations, varying considerably in quality, have been reported for adsorbate/metal systems (76). Most of these have been performed on a portion of an ideal metal lattice, and only a few allow local reconstruction of the substrate. In cluster studies of metals, one of the fundamental questions is the convergence of the adsorption energy and other properties with respect to cluster size. Many calculations have shown that even very small clusters of fewer than ten atoms allow strong bonds to develop with adsorbates. Highly ionic adsorbates are more problematic since charge transfer should occur from a much larger number of metal atoms than for covalent or slightly ionic bonds for which the screening length for charge transfer is much shorter. Erratic behavior has been observed for calculated adsorption energies on small clusters. Although energies are sometimes seriously in error, other properties such as bond length or vibrational frequencies are often in rather good agreement with experiment. On the question of convergence with respect to cluster size, it was expected that gradually enlarging the cluster would lead to convergence of the adsorption energy. This goal has proved to be elusive. There are different interpretations of the convergence situation: (1) a realistic approximation of the metal band structure and assurance that orbitals of the correct symmetry in the appropriate energy range are available to interact with an adsorbate requires *a priori* a large cluster, and (2) a cluster boundary near the adsorption site can obviously distort the adsorbate-surface bonding.

Although embedding and periodic methods may be required for *high* accuracy in the treatment of metals, there are many examples in which small cluster models have provided *useful* accurate descriptions of adsorbate energetics, vibrations, and core level shifts. When clusters reach a certain size, there apparently are compensating effects that allow many such problems to be treated, and the use of different references states of a cluster has provided a practical way to improve the accuracy of adsorption energy calculations. Slab calculations, which traditionally have provided the most powerful way of treating adsorbate overlayers and photoemission, have now been extended to surface and interstitial impurities. Clearly, in order to describe coadsorbed species and surface reactions, the theoretical model must include a sufficient number of surface atoms so as to eliminate artifacts associated with the boundary.

It has been possible, for quite a few years, to perform high-quality quantum chemistry calculations on the electronic structure of molecules. More recently, the capability of performing band calculations for metallic systems has also advanced. Early attempts at understanding adsorbate bonding to metal surfaces focused on the metal work function, the local density of states of the metal, orbital hybridization in the molecular fragments, and, for transition metals, the interaction of s, p, and d bands with the adsorbate. Although the electrons in the metal are delocalized, one of the early observations was that the strong interactions, typically on the order of 30–100 kcal/mol, between nonmetal adsorbates and the surface split out a set of orbitals that could be regarded as localized in the vicinity of the adsorption site.

Early calculations on copper and nickel revealed that the main contribution to bonding between adsorbates such as oxygen and hydrogen and the metal surface comes from the metal s and p orbitals, which have overlap interactions with the adsorbate that are much larger than those of the d orbitals. There is now little doubt that the s and p electrons provide the main contribution to the energetics when species with unsaturated valence are bonded to transition metal surfaces (*76*). For this reason, one can often replace the d electrons of atoms by effective core potentials and still obtain good energetics (*77,78*). Although the s and p contributions dominate in the first-row transition metals, this does not mean that d electron contributions are unimportant, since, quite to the contrary, they are responsible for the interesting differences in transition metal behavior. One of the reasons the d electron participation in bonding is sometimes obscure is because d orbitals bond indirectly in most instances by mixing with the metal s and p orbitals and with the adsorbate, analogous to atomic orbital hybridization, but with more extensive s, p

delocalization. The explicit participation of d electrons in bonding modulates the s and p interactions and gives rise to the special properties of transition metal systems *(76)*. A simple hybridization picture that describes the reactivity of transition metals such as Ni, Pt, Cu, and Au has been reported *(79)*. The adsorbate orbitals are viewed as first interacting with the metal s band to produce a bonding state. Such interactions do not vary much for different transition metals. The resulting bonding state then interacts with the metal d states to produce bonding and antibonding states; the net interaction depends on the coupling matrix elements with the renormalized state and the extent to which the d band is filled. The authors explain the variation in transition metal reactivity in terms of the location of the latter antibonding state *(79)*.

Other issues concerning the role of d electrons in transition metals lie in their differential participation in transition states for adsorbate reactions on surfaces *(76,80)*. Evidence from studies of methane dissociation on nickel points to the importance of Ni $3d$ electron contributions in enabling the C-H bond stretch. Likewise, when adsorbates move across transition metal surfaces, the d contributions to the bonding are found to differ significantly between atop sites and sites where there is bonding with multiple surface atoms. For higher-series transition metals such as Pd and Pt, there is an entirely new dimension to the participation of d electrons that derives from the reduced spacing between the $d^n s^1$ and the $d^{n+1} s^0$ atomic states. Studies reported in this volume of the reaction of adsorbed ethyl on Pt(100) to product ethylene coadsorbed with hydrogen reveal closely spaced electronic states that differ in d occupancy and overall electron spin. The complexities of processes with competing reactant states are just beginning to be appreciated.

Another class of transition states for catalysis at metal surfaces is that associated with photochemical reactions. A study describing the photoinduced dissociation of methane physisorbed on Pt(111) has been reported in which the key feature of the process is the formation of an electronically excited state characterized as a CH_4^- electron attachment complex that interacts with a positive image charge localized near the metal surface *(81)*. The reaction takes place by movement on the electronically-excited-state PES to produce a distorted methane geometry followed by decay to the ground state and methane dissociation.

Metal oxide surfaces pose special difficulties in treating the strong electrostatic effects and are not as well understood as metal surfaces.

Enzymes. Enzymes, like other catalysts, usually work by stabilizing the corresponding transition states *(1)*. Nevertheless, enzymatic reactions, like other types of catalysis, have their unique features. A reliable modeling of enzymatic reactions must capture the energetics and dynamics of the relevant transition states. This should allow one to determine which factors are crucial for efficient catalysis. Modeling enzyme transition states is especially challenging because of the large size and complexity of proteins. It is certainly not enough to obtain an accurate quantum mechanical description of the isolated substrate. In principle, one can model enzymatic reactions by QM/MM approaches using *ab initio* or semiempirical Hamiltonians, including the empirical valence bond (EVB) Hamiltonian *(1)*. The EVB method is also called molecular-mechanics/valence-bond (MMVB); it involves creating a $n \times n$ configuration interaction matrix (where n is the number of configurations, typically 2 to 6, but in some cases as large as 14) with MM diabatic energies on the diagonal and a semiempirical expression for the nonzero couplings. Enzyme transition states can be modeled by reasonable potential surfaces, and more refined treatments will emerge in the future. At this stage, it is already possible to explore key issues about the nature of the enzyme-substrate transition states. Such exploratory studies indicate that the transition states of enzymes are similar to the corresponding transition states in solution except that the enzyme is a "better" solvent since it has to undergo a smaller reorganization in order to reach its transition state configuration (1). For example, one may hypothesize that in enzymatic mechanisms where the polarity of the transition state is greater than that of the reactant, the major source of catalytic rate acceleration is electrostatic stabilization of the transition state by protein dipoles (6,82). At present, however, most methods do not allow one to sort out dynamical effects and non-equilibrium free energy contributions. Here we will examine what can be deduced from current simulation methods with the EVB approach about these effects and the nature of transition states of enzymes.

Before proceeding, though, it is important to define the relevant questions in a clear way. In particular, it is important to emphasize that the catalytic effect of an enzyme is defined as the ratio between the reaction rate in the presence of enzyme and the rate of a reference reaction, where the reference reaction is most conveniently taken as the same reaction in solution (1). The corresponding *difference* in activation free energies is easier to evaluate than are the absolute free energies of activation (1,2). Usually, electrostatic stabilization of the transition state charges provides the largest catalytic effect. This electrostatic effect is exerted by preoriented polar groups of the active site. Thus very often the most important feature of an enzymatic transition state is the arrangement of the surrounding environment and not the actual nature of the reacting fragments. This point should be explored further by *ab initio* quantum mechanical approaches and also correlated with experimental studies. Despite the obvious importance of the reduction of activation free energies it is also important to be able to explore other factors such as dynamical effects, non-equilibrium effects, and nuclear tunneling effects. Since the importance of such factors cannot be determined uniquely by experiments, it is essential to use simulation studies in analyzing those factors and in relating them to the observed rate constant.

The importance of frictional effects has sometimes been invoked for enzyme catalysis (83). In order for such effects to be catalytic, one must have a different frictional correction (or different κ) for the rate constant in the presence of the enzyme and in water. The transmission factor can be related to the time-dependent autocorrelation $C(t)$ of the difference between diabatic EVB or MMVB potential energy surfaces of the reactant and product or to the autocorrelation of the corresponding reactant and product environmental reaction fields (1). Thus, under the assumptions where this relation holds, comparing $C(t)$ for the enzymatic and solution reactions should tell us whether dynamical effects are important in enzyme catalysis. An EVB study of this issue (1) indicates that $C(t)$ is similar in enzymes and in solutions, and therefore frictional effects are not likely to provide a substantial catalytic contribution. In other words, the fluctuations of the environment at the transition state are not so different in enzymes and in solution. Frictional effects are inseparable from nonequilibrium solvation. Preliminary EVB studies have shown that nonequilibrium solvation effects are quite similar in enzymes and in the corresponding reaction in solution. The main difference is the reduction of the reorganization energy in enzyme active sites, which is not a dynamical effect but rather a factor that contributes to $G^{\ddagger,0}(T)$

A related issue is associated with quantum mechanical nuclear tunneling. This issue has been explored by path integral centroid calculations using EVB potential surfaces, which indicate that tunneling effects are similar in the enzyme and solution (48), but more complete studies are required. In particular multidimensional semiclassical tunneling methods show that nuclear tunneling can be a very sensitive function of small differences in effective potentials or tunneling path lengths (34,47,84). Experimental kinetic isotope effects (KIEs), which are discussed below, should provide a way to verify whether tunneling is important.

The relationship between transition state structure and KIEs is implicit in transition state theory (*11*), and it was presented in beautiful detail by Biegeleisen and Wolfsberg in a landmark paper in 1958 (*85*). It was clear from this work that direct information about transition state structure could be obtained from isotopic reaction rates in chemical systems. Systematic determination of kinetic isotope effects combined with bond energy/bond order vibrational analysis (BEBOVA) led to semiempirical transition state structures of several non-enzymatic reactions by the early 1970s. A Steenbock Symposium was held in 1976 on the use of kinetic isotope effects in enzymology, and a book of biological transition state applications was published in 1978 (*86,87*). A computer program, BEBOVIB, was written in 1977 to facilitate BEBOVA calculations for matching transition state structures to intrinsic kinetic isotope effects (*88*). These developments for understanding isotope effects for chemical reactions set the stage for applications to enzymes early in the 1980s.

Major problems in the application of KIEs to enzymology arise from the ability of enzymes to partition the progress along the reaction coordinate into several energetically discrete steps. The chemical step rarely provides the major energetic barrier in the conversion of substrates to products. Thus the kinetic isotope effects can be partially or completely obscured by slow steps associated with substrate binding, product dissociation, or protein conformation changes. In recent years, these problems have been euphemistically termed "kinetic complexity," and, when present, they cause the expression of fractional intrinsic isotope effects. The intrinsic KIEs require quantitation of each complicating step and correction to give the isotope effects associated with the chemical step. These problems were summarized by Northrup in 1981, and they have been systematically investigated and solved in the past 15 years (*89–91*). In 1998, practical solutions are available for the problems caused by almost any form of kinetic complexity, and a growing list of enzymes is available where complete families of intrinsic KIEs have been measured. In these cases, the isotope effects can be used to test models of enzymatic transition states. Recent efforts in transition state analysis have included complex reactions in which the substrate is also a protein undergoing covalent modification by the enzymatic protein catalyst (*92*).

Using the conventional transition state formalism of Beigeleisen and Wolfsberg, enzymatic transition state structures can be constructed systematically from the intrinsic KIEs. One possible procedure is as follows. First one constructs a truncated structure that includes at least all atoms within two bonds of the reaction center. Normal vibrational modes are generated for the reactant and transition state structures, and from these the isotopic partition functions are determined to yield the KIEs for each isotopically labeled position. From the truncated structure, a full transition state molecule is created by adding back the cutoff atoms and re-optimizing the structure while holding fixed the geometry defined by the KIEs (*93*). The analysis provides a full molecular model of the transition state, constrained by the experimental values of the isotope effects. Single-point calculations using GAUSSIAN94 (*94*) lead to the electronic wave function for the transition state. Wave functions for the substrate and transition state are compared to identify the characteristics of the transition state. It should be kept in mind that the accuracy of this procedure is limited by the accuracy of conventional transition state theory, which assumes that the transition state structure and force constants are independent of isotopic substitution and which neglects tunneling. In many cases the structures and force constants of the variational transition states are expected to be different for different isotopic versions of the reaction (*95*). The critical configurations along tunneling paths may also be isotope dependent (*96*), and if tunneling is important, it is essential to include reliable multidimensional tunneling approximations when calculating the KIEs (*97,98*). Light-atom transfers, such as proton and hydride transfer are especially susceptible to isotope-dependent transition state locations, recrossing effects, and tunneling corrections (99).

What are we learning about enzymatic transition states? The most revealing studies compare the transition states for the uncatalyzed solution reactions to the same reactions catalyzed by enzymes. In most cases, the transition states for the enzymatic reactions occur earlier along the reaction coordinate and demonstrate bond distortions remote from the reaction center. The presence of enzymatic activators which increase the reaction rate are also capable of changing the transition state structure. Substrates that undergo attack by nucleophiles show earlier transition states when better nucleophiles are participating. When kinetic complexity has obscured the intrinsic chemical KIEs, appropriate mutations have been made to slow the reaction rate and make the intrinsic KIEs observable. In some cases mutational changes have been demonstrated to alter the transition state structure. Finally, enzymatic transition state structure is sufficiently specific to provide information that guides the synthesis of transition state inhibitor analogues. These powerful inhibitors have now been synthesized for at least four enzymes which have been analyzed by transition state methods. Much of this work has been summarized in recent reviews (91,100).

The conjunction of methods for measurement of enzymatic kinetic isotope effects, determination of their intrinsic values, and the use of this information to constrain the limits of computationally determined transition states has opened a new era of understanding enzymatic catalysis. The approach requires no information about the structure of the catalyst and can provide both comparative and direct information about the transition state. It is anticipated that this approach will have an active future in understanding enzymatic catalysis and in the design of transition state inhibitors.

Outlook

What does the future hold for computational modeling of catalytic transition states? First, we need more reliable, affordable electronic structure methods for predicting saddle point structures, not only in the gas phase, but at gas-solid interfaces, in zeolites, and in solution. Furthermore experience with more accurate dynamical methods (17,101) shows that there are often important corrections to treatment based on classical dynamics and conventional transition state theory, so we will often need to apply more complete dynamical theories. The transition state concept will be broadened and stretched in the process, but it is safe to assume that it will continue to be the most fruitful theoretical concept in the currently emerging field of computation-driven catalyst design.

Acknowledgment. This research was supported in part by the National Science Foundation and the National Institutes of Health.

Literature Cited

(1)   Warshel, A. *Computer Modeling of Chemical Reactions in Enzymes and Solutions*; John Wiley & Sons: New York, 1991.

(2)   Åqvist, J.; Warshel, A. *Chem. Rev.* **1993**, *93*, 2523.

(3)   Field, M. J. In *Computer Simulation of Biomolecular Systems*; van Gunsteren, W. F., Weiner, P. K., Wilkinson, A. J., Eds.; ESCOM: Leiden, 1993; p. 82.

(4)   *Electronic Properties of Solids Using Cluster Methods*; Kaplan, T. A., Mahanti, S. D., Eds.; Plenum: New York, 1995.

(5)   *Solvent Effects and Chemical Reactivity*; Tapia, O., Bertrán, J., Eds.; Kluwer: Dordrecht, 1996.

(6)   *Computational Approaches to Biochemical Reactivity;* Náray-Szabó, G., Warshel A., Eds.; Kluwer: Dordrecht, 1997.
(7)   Náray-Szabó G. *Ann Repts. Prog. Chem.* **1998**, *94*, in press.
(8)   *Combined Quantum Mechanical and Molecular Mechanical Methods*; Gao, J., Thompson, M. A., Eds.; American Chemical Society: Washington; in press.
(9)   Kreevoy, M. M.; Truhlar, D. G. In *Investigation of Rates and Mechanisms of Reaction* (*Techniques of Chemistry*, 4th ed., Vol. 6); Bernasconi, C. F., Ed.; John Wiley & Sons; New York, 1986; Part 1, p. 13.
(10)   Anderson, J. B. *J. Chem. Phys.* **1973**, *58* 4684.
(11)   Glasstone, S.; Laidler, K. J.; Eyring, H. *The Theory of Rate Processes*; McGraw-Hill: New York, 1941.
(12)   Tucker, S. C.; Truhlar, D. G. In *New Theoretical Methods for Understanding Organic Reactions*; Bertrán, J. Csizmadia, I. G., Eds.; Kluwer: Dordrecht, 1989; p. 291.
(13)   Garrett, B. C.; Truhlar, D. G. *J. Chem. Phys.* **1979**, *70*, 1593.
(14)   Bennett, C. H. *ACS Symp. Ser.* **1977**, *46*, 63.
(15)   McCammon, J. A.; Karplus, M. *Annu. Rev. Phys. Chem.* **1980**, *31*, 29.
(16)   Rebertus, D. W.; Berne, B. J.; Chandler, D. *J. Chem. Phys.* **1979**, *71*, 2975.
(17)   Chandler, D. W.; Berne, B. J. *J. Chem. Phys.* **1979**, *71*, 5386.
(18)   Isaacson, A. D.; Truhlar, D. G. *J. Chem. Phys.* **1982**, *76*, 1380.
(19)   Northrup, S. H.; Pear, M. R.; Lee, C.-Y.; McCammon, J. A. Karplus, M. *Proc. Natl. Acad. Sci. U.S.A.* **1982**, *79*, 4035.
(20)   Chandresekhar, J.; Smith, S. F.; Jorgensen, W. L. *J. Am. Chem. Soc.* **1984**, *107*, 154.
(21)   Madura, J. D.; Jorgensen, W. L. *J. Am. Chem. Soc.* **1986**, *108*, 2517.
(22)   Hwang, J.-K.; King, G.; Creighton, S.; Warshel, A. *J. Am. Chem. Soc.* **1988**, *110*, 5297.
(23)   Carter, E. A.; Ciccotti, G.; Hynes, J. T.; Kapral, R. *Chem. Phys. Lett.* **1989**, *156*, 472.
(24)   Wilson, M. A.; Chandler, D. *Chem. Phys.* **1990**, *149*, 11.
(25)   Ryckaert, J. P. In *Computer Simulation in Materials Science*; Meyer, M., Pontikis, V., Eds.; Kluwer: Dordrecht, 1991; p. 43.
(26)   Cicotti, G. *ibid.*; p. 119.
(27)   Jackels, C. F.; Gu, Z.; Truhlar, D. G. *J. Chem. Phys.* **1995**, *102*, 3188.
(28)   Villà, J.; Truhlar, D. G. Theor. Chem. Acc. **1997**, *97*, 317.
(29)   Woo, T. K.; Margl, P. M.; Bloechl, P. E.; Ziegler, T. *Organometallics* **1997**, *16*, 3454.
(30)   Margl, P. M.; Woo, T. K.; Bloechl, P. E.; Ziegler, T. *J. Am. Chem. Soc.* **1998**, *120*, 2174.
(31)   Jiao, H.; Schleyer, P. v. R. *J. Am. Chem. Soc.* **1995**, *117*, 11529.
(32)   Náray-Szabó G. In *Encyclopedia of Computational Chemistry*; Schleyer, P. v. R., Allinger, N. L., Clark, T., Gasteiger, J., Kollman, P., Schaefer, H. F. III, Eds.; Wiley: Chichester, 1998; in press.
(33)   Rabo, J. A. *Catal. Rev. Sci. Eng.* **1981**, *23*, 293.
(34)   McRae, R. P.; Schenter, G. K.; Garrett, B. C. *J. Chem. Soc. Faraday Trans.* **1997**, *93*, 997.
(35)   Truhlar, D. G.; Garrett, B. C.; Klippenstein, S. J. *J. Phys. Chem.* **1996**, *100*, 12771.
(36)   Yamabe, T.; Yamashita, K.; Kaminoyama, M.; Koizumi, M.; Tachibana, A.; Fukui, K. *J. Phys. Chem.* **1984**, *88*, 1459.
(37)   Nguyen, K. A.; Gordon, M. S.; Truhlar, D. G. *J. Am. Chem. Soc.* **1991**, *113*, 1596.
(38)   Musaev, D. G.; Morokuma, K. *Adv. Chem. Phys.* **1996**, *95*, 61.

16

(39) Siegbahn, P. E. M. *Adv. Chem. Phys.* **1996**, *93*, 333.
(40) Gordon, M. S.; Cundari, T. R. *Coord. Chem. Rev.* **1996**, *147*, 87.
(41) *Chemical Applications of Density Functional Theory*; Laird, B. B., Ross, R. B., Ziegler, T., Eds.; American Chemical Society: Washington, 1996.
(42) Car, R.; Parrinello, M. *Phys. Rev. Lett.* **1985**, *55*, 2471.
(43) Galli, G.; Parrinello, M. In *Computer Simulation in Materials Science*; Meyer, M., Pontikis, V., Eds.; Kluwer: Dordrecht, 1991; p. 283.
(44) Payne, M. C.; Teter, M. P.; Allan, D. C.; Arias, T. A.; Joannopoulos, J. D. *Rev. Mod. Phys.* **1992**, *64*, 1045.
(45) Espinosa-García, J.; Corchado, J. C.; Truhlar, D. G. *J. Am. Chem. Soc.* **1997**, *119*, 9891.
(46) Truhlar, D. G. In *The Reaction Path in Chemistry: Current Approaches and Perspectives*; Heidrich, D., Ed.; Kluwer: Dordrecht, 1995; p. 229.
(47) Chuang, J. C.; Cramer, C. J.; Truhlar, D. G. *Int. J. Quantum Chem.*, to be published.
(48) Hwang, J.-K.; Warshel, A. *J. Am. Chem. Soc.* **1996**, *118*, 11745.
(49) Singh, U.; Kollman, P.A. *J. Comp. Chem.* **1986**, *7*, 718.
(50) Woo, T. K.; Margl, P. M.; Bloechl, P. E.; Ziegler, T. *J. Phys. Chem. B* **1997**, *101*, 7877.
(51) Nadig, G.; Van Zant, L. C.; Dixon, S. L.; Merz, K. M. Jr., to be published.
(52) Wesolowski, T.; Muller, R. P.; Warshel, A. *J. Phys. Chem.* **1996**, *100*, 15444.
(53) Náray-Szabó, G.; Surjan, P. R. *Chem. Phys. Lett.* **1983**, *96*, 499.
(54) Svensson, M.; Humbel, S.; Morokuma, K. *J. Chem. Phys.* **1996**, *105*, 3654.
(55) Noland, M.; Coitiño, E. L.; Truhlar, D. G. *J. Phys. Chem. A* **1997**, *101*, 1193.
(56) Corchado, J. C.; Truhlar, D. G. In Ref. 8, in press.
(57) Stephens, P. J.; Devlin, F. J.; Ashvar, C. S.; Bak, K. L.; Taylor, P. R.; Frisch, M. J. In Ref. 41, p. 105.
(58) Hehre, W. J.; Radom, L.; Schleyer, P. v. R.; Pople, J. A. *Ab Initio Molecular Orbital Theory*; John Wiley: New York, 1986.
(59) Lin, Z.; Hall, M. B. *Coord. Chem. Rev.* **1994**, *135/136*, 845.
(60) Couty, M.; Hall, M. B. *J. Comput. Chem.* **1996**, *17*, 1359.
(61) Couty, M.; Bayse, C. A.; Jimenez-Catano, R.; Hall, M. B. *J. Phys. Chem.* **1996**, *100*, 13976.
(62) Raghavachari, K.; Anderson, J. B. *J. Phys. Chem.* **1996**, *100*, 12960.
(63) Pietsch, M. A.; Couty, M.; Hall, M. B. *J. Phys. Chem.* **1995**, *99*, 16315.
(64) Thomas, J. L. C.; Hall, M. B. *Organometallics* **1997**, *16*, 2318.
(65) Niu, S.; Hall, M. B. *J. Phys. Chem. A* **1997**, *101*, 1360.
(66) Senchenya, I. N.; Kazansky, V. B. *Catal. Lett.* **1991**, *8*, 317.
(67) Kramer, G. J.; van Santen, R. A.; Emeis, C. A.; Novak, A. *Nature* **1993**, *363*, 529.
(68) Kazansky, V. B. *Acc. Chem. Res.* **1991**, *24*, 379.
(69) Kazansky, V. B. *Kinet. Catal.* **1980**, *21*, 159.
(70) Teunissen, E.; van Santen, R. A.; Jansen, A. P. J.; van Duyneveldt, E. M. *J. Phys. Chem.* **1993**, *97*, 1993.
(71) Kramer, G. J.; van Santen, R. A. *J. Am. Chem. Soc.* **1995**, *117*, 1766.
(72) Greatbanks, S. P.; Sherwood, P.; Nillier, J. H.; Hall R. J.; Burton, N. A.; Gould, I. R. *Chem. Phys. Lett.* **1995**, *234*, 367.
(73) Sinclair, P.; van Santen, R. A.; in preparation.
(74) van Santen, R. A.; Kramer, G. J. *Chem. Rev.* **1995**, *95*, 637.
(75) Sauer, J.; Uglienga, P.; Garrone, E.; Saunders, V. R. *Chem. Rev.* **1995**, *94*, 2095.

(76) Whitten, J. L.; Yang, H. *Surf. Sci. Repts.* **1996**, *24*, 55.
(77) Panas, I.; Siegbahn, P.; Wahlgren, U. *Chem. Phys.* **1987**, *112*, 325.
(78) Pettersson, L. G. M.; Faxen, T. *Theor. Chim. Acta* **1993**, *85*, 345.
(79) Hammer, B.; Nørskov, J. K. *Nature* **1995**, *376*, 238.
(80) Yang, H.; Whitten, J. L. *J. Chem. Phys.* **1992**, *96*, 5529.
(81) Whitten, J. L. *Chem. Phys.* **1997**, *225*, 189.
(82) Warshel A.; Levitt, M. *J. Mol. Biol.* **1976**, *103*, 227.
(83) Careri, G.; Fasella, P.; Gratton, E. *Annu. Rev. Biophys. Bioeng.* **1979**, *8*, 69.
(84) Hu, W.-P.; Liu, Y.-P.; Truhlar, D. G. *J. Chem. Soc. Faraday Trans.* **1994**, *90*, 1715.
(85) Biegeleisen, J.; Wolfsberg, M. *Adv. Chem. Phys.* **1958**, *1*, 15.
(86) *Isotope Effects on Enzyme-catalyzed Reactions*; Cleland, W. W., O'Leary, M. H., Northrop, D. B., Eds.; Steenbock Symposium Series; University Park Press: Baltimore, 1977.
(87) *Transition States of Biochemical Processes*, Gandour, R. D., Schowen, R. L., Eds.; Plenum: New York, 1978.
(88) Sims, L. B.; Burton, G. W.; Lewis, D. E. BEBOVIB-IV: QCPE Program No. 337, *Quantum Chemistry Program Exchange Newsletter* **1977**, *58*, 20.
(89) Northrop, D. B. *Ann. Rev. Biochem.* **1981**, *50*, 103.
(90) Cleland, W. W. *Methods Enzymol.* **1982**, *87*, 625.
(91) Cleland, W. W. *Methods Enzymol.* **1995**, *249*, 341.
(92) Scheuring, J.; Berti, P. J.; Schramm, V. L. *Biochemistry* **1998**, *37*, 2748.
(93) Stewart, J. J. P. *J. Comput. Chem.* **1989**, *10*, 221.
(94) Frisch, M. J.; Trucks, G. W.; Schlegel, H. B.; Gill, P. M. W.; Johnson, B. G.; Robb, M. A.; Cheeseman, J. R.; Keith, T.; Petersson, G. A.; Montgomery, J. A.; Raghavachari, K.; Al-Laham, M. A.; Zakrzewski, V. G.; Ortiz, J. V.; Foresman, J. B.; Cioslowski, J.; Stefanov, B. B.; Nanayakkara, A.; Challacombe, M.; Peng, C. Y.; Ayala, P. Y.; Chen, W.; Wong, M. W.; Andres, J. L.; Replogle, E. S.; Gomperts, R.; Martin, R. L.; Fox, D. J.; Binkley, J. S.; Defrees, D. J.; Baker, J.; Stewart, J. P.; Head-Gordon, M.; Gonzalez, C.; Pople, J. A. GAUSSIAN94, *Revision C.2*. **1995**, Gaussian, Inc.: Pittsburgh, PA.
(95) Tucker, S. C.; Truhlar, D. G.; Garrett, B. C.; Isaacson, A. D. *J. Chem. Phys.* **1985**, *82*, 4102.
(96) Kreevoy, M. M.; Ostović, D.; Truhlar, D. G.; Garrett, B. C. *J. Phys. Chem.* **1986**, *90*, 3766.
(97) Garrett, B. C.; Truhlar, D. G.; Wagner, A. F.; Dunning, T. H., Jr. *J. Chem. Phys.* **1983**, *78*, 4400.
(98) Allison, T. C.; Truhlar, D. G. In *Modern Methods for Multidimensional Dynamics Calculations in Chemistry*; Thompson, D. L., Ed.; World Scientific: Singapore, 1998; in press.
(99) Truhlar, D. G.; Garrett, B. C. *Acc. Chem. Res.* **1980**, *13*, 440.
(100) Schramm, V. L. *Annu. Rev. Biochem.* **1998**, *67*, in press.
(101) *Dynamics of Molecules and Chemical Reactions*; Wyatt, R. E., Zhang, J. Z. H., Eds.; Marcel Dekker: New York, 1996.

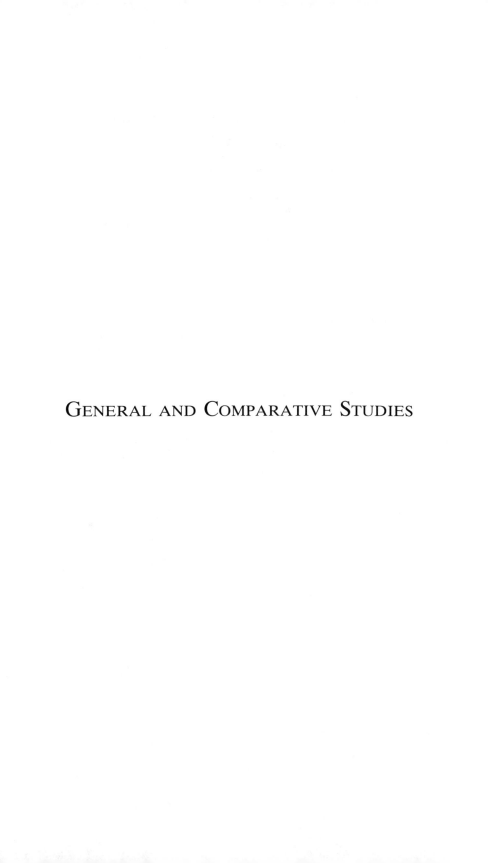

GENERAL AND COMPARATIVE STUDIES

Chapter 2

Performance of Density Functionals for Transition States

D. R. Salahub[2], S. Chrétien[1], A. Milet[1,3], and E. I. Proynov[1]

[1]Département de chimie, Université de Montréal, C.P. 6128, Succursale Centre-Ville, Montréal, Québec H3C 3J7, Canada
[2]Centre de Recherche en Calcul Appliqué CERCA, 5160, boul. Décarie, bureau 400, Montréal, Québec H3X 2H9, Canada

An overview of the performance of GGA, hybrid and LAP density functionals for transition states in a variety of benchmark systems is given. Examples, such as hydrogen abstraction and intramolecular proton transfer in malonaldehyde, illustrate the difficulties which even the most advanced density functionals meet in describing transition states. Comparisons are also reported for the oligomerization of acetylene $Fe(C_2H_2)_2^+ \longrightarrow Fe(C_4H_4)^+$.

Advances in computational techniques and the vast experience of the past decades have allowed an understanding of catalytic reactions at a very detailed level, involving full geometry optimization, characterization of transition states (TS) and reaction intermediates (1–3). The contemporary methods of density functional theory (DFT), especially with the latest progress in the development of efficient exchange-correlation (XC) functionals, appear rather successful in these cases, embodying a large amount of electron correlation at a much lower computational cost (1–3). However, in some cases the reaction path and especially the structure and the energy of the TS may present a stringent test for the existing approximate XC functionals. A typical example is the hydrogen abstraction reaction going through one of the seemingly simplest TS structures, a linear three-atomic complex. The correct reproduction of the energy barrier for this reaction turns out to be a very difficult task not only for approximate DFT methods but also for Hartree-Fock (HF) and most of the post-HF *ab initio* methods (4,5). In this chapter we shed some light on why this linear three-center TS structure is so difficult to describe.

Another big challenge for DFT and post-HF methods is the correct description of inter-molecular and intra-molecular proton transfer (PT) energetics. The energy barrier for the internal proton transfer (IPT) in malonaldehyde is particularly difficult

[3]Current address: Laboratoire de Chimie Quantique, Université Louis Pasteur, Strasbourg, France.

to describe quantitatively (*6*). We revisit this problem, presenting more evidence that the quality of the functionals is of crucial importance in DFT studies of PT processes.

A large class of organometallic reactions involving Pd, Fe and other transition metals appeared also as a very demanding task for theoretical treatment. Often detailed experimental data concerning the structure and the energetics of such reactions are not directly available and having a reliable theoretical treatment is the only alternative. In this chapter we discuss examples of such reactions having a complex TS structure and mechanism.

One of the main purposes of this chapter is to illustrate the performance of different XC functionals. Emphasis will be given to the recently developed kinetic-energy-density (τ)- dependent XC schemes BLAP, PLAP (*7,8*). The BLAP1 and PLAP1 (*7*) schemes involve the τ-dependent correlation functional LAP1, carefully synchronized with the Generalized Gradient Approximation (GGA) exchange functionals of Becke (*9*) and Perdew (*10*) respectively. The XC schemes BLAP3 and PLAP3 are constructed in a similar manner, employing the more recent LAP3 correlation functional (*8*). The latter is an extension of the LAP1 correlation functional to incorporate parallel-spin electron correlation beyond the Kohn-Sham (KS) exchange level. Other XC functionals implemented in the deMon-KS3 code are also used for comparison, such as the popular GGA schemes Becke exchange-Perdew correlation (BP86) and Perdew exchange-Perdew correlation (PP86) (*10*). Literature results of hybrid HF-DFT methods and post-HF methods for the cases studied are referred to whenever possible.

Computational Details

The KS DFT results discussed in this chapter were obtained using the latest release of the Linear Combination of Gaussian Type Orbitals-DFT code deMon-KS3 (*11–13*). For the hydrogen abstraction and the IPT in malonaldehyde we used a variety of orbital bases ranging from double-ζ plus polarization (DZVP) and triple-ζ plus polarization (TZVP2) (*14*), to the exhaustive basis set of Sadlej (*15*). Auxiliary basis sets used to fit the charge density and the XC potential were chosen from the deMon-KS basis library (*14*). For the Fe^+-based reactions, the orbital basis sets (*14,16*) were respectively (43321/4211*/311+) for iron, (721/51/1*) for carbon and (41/1*) for hydrogen. The charge density and the XC potential were fitted using the auxiliary basis set (5,5;5,5) for iron, (4,3;4,3) for carbon, and (3,1;3,1) for hydrogen.

Minima and transition states on the potential energy surfaces (PES) were determined without symmetry constraints, using the Broyden-Fletcher-Goldfarb-Shanno algorithm (*17*). During the optimization the norm of the gradient and the maximum atomic contribution to the gradient were minimized below 10^{-4} and the convergence criteria for the charge density and the energy were 10^{-6} and 10^{-7} respectively. The second derivatives of the energy were calculated by finite difference of the gradient using a displacement of 0.03 a.u.. The vibrational frequencies were obtained in the harmonic approximation.

Activation Barrier for Hydrogen abstraction

The hydrogen abstraction reaction:

$$H_2 + H \longrightarrow [\,H{-}H{-}H\,]^{\ddagger} \longrightarrow H + H_2$$

has received much attention in the past few years. This seemingly simple reaction turned out to be very difficult for an accurate theoretical description and is still a challenge to both *ab initio* and DFT methods. Pople et al. (*4*) performed exhaustive DFT validation tests on this reaction using various XC schemes and, surprisingly, a massive DFT failure was detected in reproducing accurately the classical activation barrier. One possible reason for this could be related to the self-interaction (SI) error inherent to most of the approximate DFT schemes used. Adding to a given XC scheme an orbital-dependent self-interaction correction (SIC) (*18*) in a perturbative fashion, Pople et al. obtained a shift of the activation barrier in the right direction. However, this improvement arises not only due to elimination of the SI error itself, but also due to other consequences of introducing orbital dependent terms in the XC functional and potential. The hybrid HF-KS-DFT method B3LYP for example has a substantially reduced SI error due to the fraction of exact exchange incorporated in it but, nevertheless, its activation energy estimate is not as good, as is seen from Table I. Concerning benchmarks for this reaction, very reliable estimates of the classical activation barrier exist due to highly accurate quantum Monte Carlo (QMC) (*19,20*) and variational calculations (*21*). Comparing these results with the experimental estimate of Schulz (*22*) one finds a difference of about 4.2 - 6.3 kJ/mol, after subtracting the finite-temperature and the zero-point energy (ZPE) contributions. This difference could be either due to imprecisions of the experimental estimate (as the authors themselves warn (*22*)), or might be an indication that some factors are still missing in the benchmark theoretical estimates.

Turning to the DFT side, a comparison between the results of BLYP and BP86 with BLAP3 and B3LYP indicates that the correlation energy plays a key role in this reaction (Table I): going from BLYP to B3LYP, i.e. changing just the exchange functional, leads to a smaller shift of the activation energy compared to the shift caused by changing the correlation functional alone, as comparing BLYP with BLAP3. On the other hand, comparing the B3LYP result with its predecessor HaHLYP(half HF exchange - half GGA exchange), or with its recent modification B3(H) (*23*), one sees that increasing the amount of exact HF exchange injected in the hybrid scheme also leads to a noticeable improvement (Table I).

That electron correlation plays a substantial role in this reaction follows also from the comparison between the HF result (with no electron correlation) of about 74.1 kJ/mol and the Local Spin Density (LSD) result using the Vosko-Wilk-Nusair (*24*) (VWN) approximation (with an overestimated correlation) of about -11.8 kJ/mol, compared to the exact estimate of 40.2 kJ/mol. It should be noted that these energy differences are not structure related as the TS geometry is reproduced reasonably well by most of the methods (except for LS-VWN which yields a somewhat larger elongation of the H-H distances in the TS).

Comparing the PLAP1 with the PLAP3 energy estimates it follows that the anti-parallel-spin component of the correlation energy is a key player here: in PLAP1 the parallel-spin correlation (beyond the KS exchange) is missing, but its energy estimate is very close to that of PLAP3.

Table I. Energy barrier (in kJ/mol) and transition state structure (distances in Ångstroms) for hydrogen abstraction.

method/basis set	$\Delta E^{\ddagger}$	r(H-H)	references
LS-VWN	-11.8	0.950	(4,5)
B-VWN	15.3; 19.3	0.927	(4,5)
BLYP	12.0; 12.4; 12.1	0.936; 0.935	(4,5),([a])
BP86	11.9	0.940	(4)
BP86(SIC)	46.2	0.931	(4)
BLAP3/6311++G(**)	30.5	0.936	
BLAP3/DZVP(41/1*)	31.0	0.937	
PLAP3/DZVP(41/1*)	32.3	0.938	
PLAP1/DZVP(41/1*)	32.7	0.938	
B1B95	31.7	0.929	([a])
VSXC	23.0	0.929	([a])
B3LYP	18.1; 17.8; 18.0	0.929	(5,23),([a])
B3(H)	30.0		(23)
HaHLYP	27.0		(25)
HF	73.9; 74.1	0.932	(4),([a])
MP2	55.2; 57.7	0.918	(4,5)
CCSD(T)	41.4	0.930	(4)
Variational	39.9 to 40.4	0.930	(21)
QMC(CA)	40.4	0.930	(19)
QMC(best)	40.2	0.930	(20)
Expt(at 300 K)	≈40.6		(22)
	ZPE[CCSD(T)] ≈ -3.5		(4)

[a] (Van Voorhis, T.; Scuseria, G.E. *J. Chem. Phys*, in press.)

The main question we address in this section is why the τ- and Laplacian-dependent XC schemes PLAP1, BLAP3 and PLAP3 (7,8), as well as some other τ-dependent functionals such as B1B95 and VSXC (Van Voorhis, T.; Scuseria, G.E. *J. Chem. Phys*, in press.) lead to a significant improvement compared to the other DFT methods used? A related question is why this TS (a simple linear triatomic) is so difficult to describe quantitatively? It is a system with a doubly degenerate ground state (GS) (two degenerate spin doublets) which would strictly require a MRCI type treatment from the CI point of view. Such situations are known to be difficult for DFT (Duarte, H.; Proynov, E.; Salahub, D.R. *J. Chem. Phys.* in press.) as one does not deal explicitly with correlated many-electron wavefunctions. Instead of describing the GS in many-electron configuration space, in KS-DFT the GS is described in real 3-D space in terms of electron density and spin density distributions and related quantities (pair density distribution, local response functions, etc.) A close inspection of some of these distributions reveals quite interesting features of the TS: while in H_2 the GS is a true singlet with zero spin density everywhere in space (covalent bond pair situation), the TS has partly localized magnetic moments on the atoms (about +0.52 atomic spin-up population on both terminal hydrogens and about −0.04 on the central atom), the interatomic distances being here about 0.2 Å longer than in H_2. The single spin-down electron interacts with the two spin-up electrons in a way to keep the latter as far apart as possible. In such a situation one may expect strong spin-spin fluctuations associated with the formation of an unstable 3-center bond. In a sense, one may view this TS as a "strongly correlated system" as one needs at least CCSD(T) to achieve a reasonably accurate description (Table I). In such systems the degree of spatial inhomogeneity is also very large. The way strong inhomogeneity is handled in the LAP XC schemes is different from that in the GGA functionals. The derivation of the nonlocal terms in the LAP correlation functional is not based on a series expansion about the homogeneous gas limit of E_c. In GGA the inhomogeneity is described by just one single variable, the reduced density gradient (10). Its real space distribution alone does not reflect real attributes of inhomogeneity in atoms and molecules, such as the shell structure of atoms, the location of bond pairs and lone pairs in molecules, etc. Moreover, the applicability of the GGA becomes questionable at high values of the reduced density gradient (10,26). In the LAP correlation functional the nonlocal terms include the kinetic energy density τ (7,8) in a form resembling the nonlocal variable used in the electron localization function approach (27). The latter has proven to be a fine tuned detector of inhomogeneity, reproducing correctly the shell structure of atoms and the location of electron pairs in molecules. Having τ as a basic nonlocal variable in XC functionals, as in the LAP XC schemes or in VSXC and B1B95 functionals (Table I) leads to a better reflection of strong inhomogeneity, and hence to better activation energy for the hydrogen abstraction. Increasing the fraction of exact HF exchange in the hybrid schemes (as in B3(H) and HaHLYP (23,25)) also leads to a considerable improvement in the same vein. Particularly concerning the LAP XC schemes, their correlation potential does not depend on τ, as it was derived in the gradientless limit of the LAP correlation energy expression (7,8). In spite of this, its long range behavior is very close to that of the LSD-SIC potential and compared to GGA correlation potentials provides a relatively better description of states with multireference character, as was shown recently by Moscardó et al. (28).

Internal Proton Transfer in Malonaldehyde

Malonaldehyde is a typical example of a system with an internal hydrogen bond that can participate in an IPT process. Similar processes may play a substantial role in some enzymatic reactions.

Systems with intra-molecular hydrogen bonds appear to represent a very demanding task for the current GGA and hybrid DFT methods (6,29–31). The asymmetric GS configuration of malonaldehyde corresponds to a symmetric double-minimum potential with a low barrier between the two equivalent minima (6). Use of the most accurate nonlocal XC functionals available is required to reproduce satisfactorily the structure of this compound, especially concerning the long O···H and O···O bond lengths (6,29) and the energy barrier for IPT (6,32). In a previous work (6) the TS structure and the IPT energetics in malonaldehyde were successfully described using the τ-dependent XC schemes BLAP1 and PLAP1 (7). Here we also report new results for the more recent LAP3 correlation functional (8). Table II contains some of the calculated structural characteristics of the ground and transition states of malonaldehyde and the energy barrier for IPT.

Table II. Energy barrier (in kJ/mol) and geometry (distances in Ångstroms) for the internal proton transfer in malonaldehyde.

method/basis set//geom.	$\Delta E^{\ddagger}$	GS r(H···O)	GS r(O···O)	TS r(H···O)	TS r(O···O)
PP86/DZVP[a]	8.8	1.568	2.520	1.223	2.400
B3LYP/6-311+G(3df,2p)[b]	12.6	1.606	2.530	1.208	2.367
BLAP1/DZVP[a]	13.8	1.684	2.591	1.214	2.381
BLAP1/TZVP2[a]	13.7	1.697	2.595	1.216	2.382
BLAP1/Sadlej[c]	13.4	1.674	2.582	1.213	2.379
PLAP1/DZVP[a]	18.8	1.731	2.626	1.222	2.394
PLAP1/TZVP2[a]	19.1	1.760	2.638	1.224	2.395
PLAP1/Sadlej[c]	18.6	1.733	2.623	1.220	2.389
BLAP3/TZVP2[c]	13.8	1.688	2.586	1.214	2.378
PLAP3/TZVP2[c]	19.0	1.758	2.637	1.224	2.394
UMP2/6-311+G(d,p)[b]	13.8	1.678	2.581	1.197	2.355
CCSD(T)/DZP// MP2/TZP[b]	18.8				
CCSD(T)/DZP//B3LYP/DZP[b]	17.2				
G2(U)[b]	18.3				
Expt[d]		1.68	≈ 2.58		

[a] ref. (6); [b] ref. (32); [c]present work; [d] ref. (33)

It is a difficult DFT task to accurately reproduce the weak bond lengths r(H···O) and r(O···O). The LAP XC schemes lead here to considerably improved values, compared to the GGA results. The TS structure governing the IPT in malonaldehyde has a C_{2v} geometry (Figure 1, Table II), the hydrogen-bond bridge O···H···O forming a kind of three-center hydrogen bond (6). We have carefully examined the influence of the grid and basis set size. A relatively good stability of the results upon increasing the basis set size was found at the TZVP2 level and beyond, especially concerning the IPT energy barrier (Table II). Similar basis set stability was found for the relative conformer energetics of glycine (6). This can be attributed to the fact that the basis sets used in deMon-KS3 are specifically optimized for DFT calculations with this code (14).

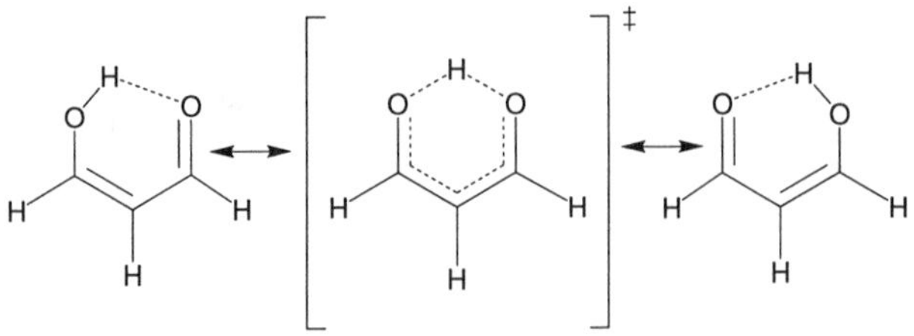

Figure 1. Internal proton transfer in malonaldehyde.

There is no experimental estimate reported for the barrier in malonaldehyde. Various high quality post-HF estimates range from 13 to 19 kJ/mol ($6,32$). A recent theoretical benchmark for this quantity is the composite G2(U) estimate of 18.3 kJ/mol of ref. (32). The best GGA value (PP86) of 8.8 kJ/mol (6) is too low. B3LYP yields about 12.6 kJ/mol (32). Some other realizations of the hybrid HF-DFT approach lead to a slight improvement over the B3LYP estimate (34). The τ-dependent XC schemes involving the GGA exchange of Becke, BLAP1 and BLAP3 yield an barrier between 13.4 and 13.8 kJ/mol, while the XC schemes PLAP1 and PLAP3 involving the GGA exchange of Perdew-Wang (86) give between 18.6 and 19.1 kJ/mol (6), very close to the G2 benchmark value of 18.3 kJ/mol. Similar to the hydrogen abstraction reaction, the IPT in malonaldehyde is a good example of how different methods and levels of theory can yield different answers, as far as the subtle TS energetics is concerned. Some of the possible reasons for the improvement achieved with the LAP XC schemes have already been mentioned above in relation to the hydrogen abstraction reaction. Improved results for other hydrogen-bonded systems have also been obtained with these functionals ($6,8,30$).

Acetylene oligomerization with Fe^+

Cyclotrimerization of acetylene in the gas phase induced by transition metal atoms (35–37), clusters (38–40) and surfaces (41–45) is being studied intensively, but the reaction mechanism is still unknown. In particular, Irion et al. (38–40) showed experimental evidence that when ethylene reacts with Fe_4^+, a product of stoechiometry $Fe_4(C_6H_6)^+$ is formed through dehydrogenation of ethylene, and benzene is released during the collision-induced dissociation experiment. In the proposed mechanism, it is acetylene rather than ethylene that is adsorbed on the Fe_4^+ cluster. In order to improve our knowledge of this reaction and to characterize some reaction intermediates, we initiated calculations, first, on a smaller system, the Fe^+-mediated oligomerization of acetylene in the gas phase (35). The immediate goal is to obtain equilibrium structures for the following complexes; $Fe(C_2H_2)^+$, $Fe(C_2H_2)_2^+$, $Fe(C_2H_2)_3^+$ and $Fe(C_6H_6)^+$ and the transition states between them. We want to determine what is going on when the $Fe(C_2H_2)_2^+$ complex is formed. Does the complex prefer to add a third acetylene or will it go towards the formation of a

$Fe(C_4H_4)^+$ complex (cyclobutadiene or metallacycle)? Another interesting question is whether a complex of the form $Fe(C_2H_2)_3^+$ will go towards the formation of benzene in a concerted mechanism or will it prefer to go through a cyclobutadiene or metallacycle complex, $Fe(C_2H_2)(C_4H_4)^+$ and then form benzene? We also want to understand why, in the experiment (35), no $Fe(C_4H_4)^+$ products are observed whereas the C_4H_4 unit seems to be an important reaction intermediate in the cyclotrimerization of acetylene on the Pd(111) surface (42-45) and on a U^+ cation (37).

Recently, DFT has been successfully used to describe the activation of C-C and C-H bonds in ethane by Fe^+ (47). The relative energies are in good agreement with experiment and the predicted mechanisms respect the available experimental information. This lends confidence that DFT is an appropriate tool. In this section, we focus on the TS connecting the $Fe(C_4H_4)^+$ complex in its metallacycle and cyclobutadiene forms. We will present and discuss the optimized structures and the relative energies for the $Fe(C_2H_2)_2^+$ and $Fe(C_4H_4)^+$ complexes. Full details on the entire reaction scheme will be published in due course.

In order to locate some minima on the PES, each geometry of Figure 2 has been fully optimized for different occupation numbers (multiplicities). In Table III we present the calculated relative energies (without ZPE correction) at the optimized structures of the $Fe(C_2H_2)_2^+$ and $Fe(C_4H_4)^+$ complexes. The Roman numbers in the column "initial structure" are related to Figure 2 and correspond to the starting structure for the geometry optimization.

In all tables N_s is the difference between the number of electrons with spin up and spin down. For each optimized structure we performed a vibrational analysis to characterize the nature of this extremum on the PES. If one or several negative frequencies were detected, we disturbed the geometry according to the vibrational modes and re-optimized the structure again. Hence, all the results in Table III correspond to minima on the PES. Since the structures **III** and **IV** were relatively high in energy with the PP86 functional, we did not optimize these structures with the PLAP3 and BLAP3 functionals. The geometry optimization of structure **II**, leads to the same minima as structure **I** which is why **II** does not appear in Table III. From the initial structure **I** with $N_s=3$, which corresponds to a saddle point, we obtain two different minima. The first one correspond to structure **II** in which one of the acetylenes rotates to get closer to the other one. The second is similar to structure **I** but the Fe atom is out of the plane defined by the carbon skeleton. The energies of these two structures are relatively close with the PP86 functional, and the inclusion of the LAP3 correlation (PLAP3) decreases further the energy difference from 9.5 to 0.8 kJ/mol. The exchange functional seems to have a small effect since the relative energy difference increases to 3.2 kJ/mol when BLAP3 is used instead of PLAP3. In general, all the isomers keep the same energetic order with all the functionals, except for the structures **V** and **VI** ($N_s=1$). PLAP3 gives almost equal energies for these, while PP86 and BLAP3 give structure **VI** lower in energy than structure **V** by about 35 and 20 kJ/mol respectively.

In Table IV we present the optimized structures for the $Fe(C_2H_2)_2^+$ and $Fe(C_4H_4)^+$ complexes for $N_s=1$ (Figure 3). The differences in the bond lengths computed by different functionals are smaller than 0.02 Å and smaller than 1 degree for the angles. In general, PLAP3 gives longer bonds than PP86 and BLAP3. The optimized structures with PP86 and BLAP3 are very similar, especially the Fe-C bonds. These observations are also applicable to all other optimized structures.

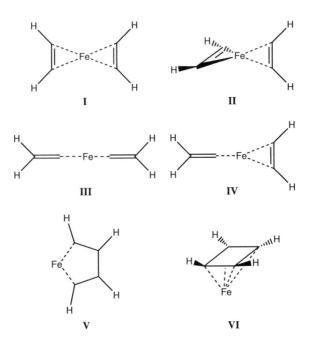

Figure 2. Starting points for the geometry optimizations.

For all the different values of N_s studied, the formation of a cycle (cyclobutadiene or metallacycle) from two separated acetylenes on an iron cation is thermodynamically favored (Table III). There are three possible transition states between the $Fe(C_2H_2)_2^+$ and $Fe(C_4H_4)^+$ complexes. For the present chapter, we present the results obtained for the TS which connects the two lowest energy structures for $N_s=1$ (Figure 3, Table V). The activation energy, for the reaction from the metallacycle (**V**) to the cyclobutadiene complex (**VI**) is 115.0 kJ/mol, 138.0 kJ/mol and 153.2 kJ/mol respectively, for the PP86, BLAP3 and PLAP3 functionals. The differences in the bond lengths between the TS structures optimized by these two functionals are smaller than 0.02 Å and smaller than 3 degrees for the dihedral angles. The TS structure is not symmetric as can be seen from Table V. The Fe atom is closer to C_1 than C_4 and the carbon skeleton is not planar. The negative frequencies (Table V) confirm that the optimized structures are transition states. The animation of the eigenmodes associated with the negative frequencies showed the connection between the two limiting structures indicating that the right TS has been found.

Table III. Calculated relative energies (in kJ/mol) between $Fe(C_2H_2)_2^+$ and $Fe(C_4H_4)^+$ complexes for various functionals.

Initial structure	N_s	PP86	PLAP3	BLAP3
	1	63.8	90.4	97.5
I	3^a	21.9	30.2	32.0
	3^b	31.4	31.0	35.2
	5	186.0	-	-
	1	118.3	-	-
III	3	129.3	-	-
	5	218.0	-	-
	1	74.8	-	-
IV	3	69.8	-	-
	5	186.5	-	-
	1	0.0	0.0	0.0
V	3	8.9	9.0	8.2
	5	77.4	-	-
VI	1	-35.3	1.4	-20.1
	3	39.1	65.9	44.9

^{a} optimized structure similar to **II**;
^{b} optimized structure similar to **I**

Table IV. Optimized geometries (distances in Ångstroms and angles in degrees) of the $Fe(C_4H_4)^+$ complexes ($N_s=1$) for various functionals.

	Metallacycle (**V**)			Cyclobutadiene (**VI**)		
	PP86	PLAP3	BLAP3	PP86	PLAP3	BLAP3
$r(FeC_1)$	1.806	1.809	1.793	1.975	1.985	1.979
$r(C_1C_2)$	1.444	1.429	1.431	1.478	1.468	1.465
$r(C_2C_3)$	1.402	1.399	1.392	1.478	1.468	1.465
$r(C_1H_1)$	1.108	1.091	1.090	1.095	1.079	1.078
$\angle(C_1FeC_4)$	86.9	87.0	86.5	43.9	43.4	43.4
$\angle(FeC_1C_2)$	114.6	114.1	114.9	63.9	63.1	63.1
$\angle(C_1C_2C_3)$	112.0	112.4	111.8	90.0	90.0	90.0
$\angle(H_1C_1C_2)$	121.5	121.8	121.3	134.9	134.9	134.8
$\angle(H_2C_2C_3)$	124.2	123.9	124.3	134.9	134.9	134.8
$\angle(FeC_1C_4C_2)$	177.8	178.0	178.6	66.1	66.4	66.5
$\angle(C_1C_2C_3C_4)$	0.1	0.1	0.1	0.0	0.0	0.0

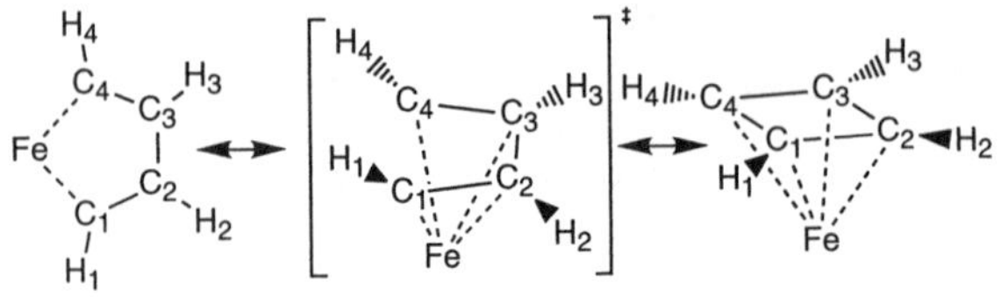

Figure 3. Optimized structure of the TS connecting the metallacycle and the cyclobutadiene form of the $Fe(C_4H_4)^+$ complex ($N_s=1$).

Table V. Optimized structure (distances in Ångstroms and angles in degrees) and frequencies (in cm^{-1}) for the TS connecting the two forms of the $Fe(C_4H_4)^+$ complex.

	PP86	PLAP3	BLAP3
$r(FeC_1)$	1.854	1.866	1.844
$r(FeC_4)$	1.796	1.796	1.784
$r(C_1C_2)$	1.424	1.417	1.415
$r(C_2C_3)$	1.441	1.428	1.425
$r(C_3C_4)$	1.447	1.440	1.436
$r(C_1C_4)$	2.066	2.055	2.063
$r(C_1H_1)$	1.100	1.084	1.083
$r(C_2H_2)$	1.098	1.083	1.081
$\angle(FeC_1C_4C_2)$	85.3	93.0	87.2
$\angle(C_1C_2C_3C_4)$	9.9	8.8	7.2
$\angle(H_1C_1C_4H_4)$	15.1	16.9	16.3
$\angle(H_2C_2C_3H_3)$	4.4	3.2	2.4
$\bar{\nu}$	-606.7	-566.8	-580.5

Conclusions

While contemporary DFT methods are quite successful in predicting GS structures and properties, transition states are more difficult to handle. Seemingly simple reactions like the hydrogen abstraction and the IPT in malonaldehyde turn out be very hard to describe, and the quality of the DFT results depends strongly on the quality of the XC functional used. Using the τ- and Laplacian-dependent correlation functional LAP3 leads to improvement in these difficult cases in spite of the approximations in the GGA exchange used. However a special synchronization between the two XC parts is required (7,8).

The optimized structures for the acetylene dimerization on Fe^+ are in good agreement for the various functionals used. No large geometry differences were detected and PLAP3 usually gives slightly longer bonds than PP86 and BLAP3. The relative energetic order is the same for all functionals used. For the TS energetics, PP86, PLAP3 and BLAP3 give reasonably similar energies, the overall spread between the two LAP functionals being of the order of 15 kJ/mol and the closer of these being some 25-35 kJ/mol away from the PP86 results. While this 20-50 kJ/mol uncertainty allows the general semi-quantitative features of the reaction to be described with confidence, obtaining greater accuracy will require further work – on the functionals, on DFT technology, and experimentally.

Another class of challenging reactions for DFT are ligand substitutions in bidentate Pd(II) complexes. Here different levels of theory, which appear *a priori* reasonable, often give different answers (*48*). Preliminary DFT results for one such reaction were reported at this meeting, their complete analysis will be presented elsewhere.

Acknowledgments

We thank S. Sirois and D. Wei for discussions about malonaldehyde, and H.Chermette for discussions and data for hydrogen abstraction. Financial support from NSERC (Canada) is gratefully acknowledged as is the provision of computing resources by the Services Informatiques de l'Université de Montréal.

Literature Cited

(*1*) Siegbahn, P.E.M. *Advances in Chemical Physics*, 1996 Vol. XCIII, pp 333.

(*2*) Schmid, R; Herrmann, W.A.; Frenking, G. *Organometallics* **1997**, *16*, 701.

(*3*) Musaev, D.G.; Mebel, A.M.; Morokuma, K. *J. Am. Chem. Soc.* **1994**, *116*, 10693.

(*4*) Johnson, B.G.; Gonzales, C.A.; Gill, P.M.W.; Pople J.A. *Chem. Phys. Lett.* **1994**, *221*, 100.

(*5*) Jursic, B.S. *Chem. Phys. Lett.* **1996**, *256*, 603.

(*6*) Sirois, S.; Proynov, E.I.; Nguyen, D.; Salahub, D.R. *J. Chem. Phys.* **1997**, *107*, 6770.

(*7*) Proynov, E.I.; Ruiz, E.; Vela, A.; Salahub, D.R. *Int. J. Quant. Chem. (Symp.)* **1995**, *29*, 61.

(*8*) Proynov, E.I.; Sirois, S.; Salahub, D.R. *Int. J. Quant. Chem.* **1997**, *64*, 427.

(*9*) Becke, A.D. *Phys. Rev. A* **1988**, *38*, 3098.

(*10*) Perdew, J.P. *Electronic structure of solids*, P. Ziesche, H. Eschrig, Academic Verlag, Berlin, 1991.

(*11*) St-Amant, A.; Salahub, D.R. *Chem. Phys. Lett.* **1990**, *169*, 387.

(*12*) St-Amant, A. Thesis, University of Montreal, **1992**.

(*13*) M.E. Casida, C.D. Daul, A. Goursot, A. Koester, L. Pettersson, E. Proynov, A. St-Amant, D.R. Salahub, H. Duarte, N. Godbout, J. Guan, C. Jamorski, M. Leboeuf, V. Malkin, O. Malkina, F. Sim, A. Vela, *deMon Software- deMon-KS3 Module* Université de Montréal, 1996.

(*14*) Godbout, N.; Salahub, D.R.; Andzelm, J.; Wimmer, E. *Can. J. Chem.* **1992**, *70*, 560.

(*15*) Sadlej, A.J. *Collec. Czech. Chem. Commun.* **1988**, *53*, 1995.

(*16*) Andzelm, J.; Radzio, E.; Salahub, D.R. *J. Comput. Chem.* **1985**, *6*, 520.

32

(17) H.B. Schlegel, In *Ab-Initio Methods in Quantum Chemistry-I* , Edited by K.P. Lawley, Wiley, New York, 1987; pp 459-500.

(18) Perdew, J.P.; Zunger, A. *Phys. Rev. B* **1981**, *23*, 5048.

(19) Ceperley, D.M.; Alder, B.J. *J. Chem. Phys.* **1984**, *81*, 5833.

(20) Diedrich, D.L.; Anderson, J.B. *Science* **1992**, *258*, 786.

(21) Liu, B. *J. Chem. Phys.* **1984**, *80*, 581.

(22) Schulz, W.R.; LeRoy, D.J. *J. Chem. Phys.* **1965**, *42*, 3869.

(23) Chermette, H.; Razafinjanahary, H.; Carrion, L. *J. Chem. Phys.* **1998**, *107*, 10643.

(24) Vosko, H.; Wilk, L.; Nusair, M. *Can. J. Phys.* **1980**, *58*, 1200.

(25) Durant, J.L. *Chem. Phys. Lett.* **1996**, *256*, 595.

(26) Dreizler, R.M.; Gross, E.K.U. in *Density Functional Theory. An Approach to the Quantum Many-Body Problem*, Springer-Verlag, Berlin, 1990.

(27) Becke, A.D.; Edgecombe, K.E. *J. Chem. Phys.* **1990**, *92*, 539.

(28) Moscardó, F.; Perez-jiménes, A.J. *Int. J. Quant. Chem.* **1998**, *67*, 143.

(29) Sim, F.; St-Amant, A.; Papai, I.; Salahub, D.R. *J. Am. Chem. Soc.* **1992**, *114*, 4391.

(30) Guo, H.; Sirois, S.; Proynov, E.I.; Salahub, D.R. In *Theoretical treatment of Hydrogen Bonding.* Hadzi, D. ed., Wiley, 1996; pp 49-74

(31) Nguyen, D.T.; Scheiner, A.C.; Andzelm, J.W.; Sirois, S.; Salahub, D.R.; Hagler, A.T. *J. Comput. Chem.* **1997**, *18*, 1609.

(32) Barone, V.; Adamo, C. *J. Chem. Phys.* **1996**, *105*, 11007.

(33) Baughcum, S.L.; Duerst, R.W.; Rowe, W.F.; Smith, Z.; Wilson, E.B.J. *J. Am. Chem. Soc.* **1981**, *103*, 6296.

(34) Adamo, C.; Barone, V. *J. Comput. Chem.* **1998**, *19*, 418.

(35) Schröder, D.; Sülzle, D.; Hrušák, J.; Böhme, D.K.; Schwarz, H. *Int. J. Mass Spectrom. Ion. Processes* **1991**, *110*, 145.

(36) Berg, C.; Kaiser, S.; Schindler, T.; Kronseder, C.; Neidner-Scatteburg, G.; Bondybey, V. E. *Chem. Phys. Lett.* **1994**, *231*, 139.

(37) Heinemann, C.; Cornehl H.H.; Schwarz H. *J. Organomet. Chem.* **1995**, *501*, 201.

(38) Schnabel, P.; Irion, M.P.; Weil, K.G. *J. Phys. Chem.* **1991**, *95*, 9688.

(39) Schnabel, P.; Irion, M.P.; Weil, K.G. *Chem. Phys. Lett.* **1992**, *190*, 255.

(40) Gehret, O.; Irion, M.P. *Chem. Phys. Lett.* **1996**, *254*, 379.

(41) Alter, W.; Borgmann, D.; Stadelmann, M.; Wörn, M.; Welder, G. *J. Am. Chem. Soc.* **1994**, *116*, 10041.

(42) Abdelrehim, I.M.; Thornburg, N.A.; Sloan, J.T.; Caldwell, T.E.; Land, D.P. *J. Am. Chem. Soc.* **1995**, *117*, 9509.

(43) Sesselmann, W.; Woratschek, B.; Ertl, G.; Küppers, J.; Haberland, H. *Surface Sci.* **1983**, *162*, 245.

(44) Patterson, C.H.; Mundenar, J.M.; Timbrell, P.Y.; Gellman, A.J.; Lambert, R.M. *Surface Sci.* **1989**, *208*, 93.

(45) Pacchioni, G.; Lambert, R.M. *Surface Sci.* **1994**, *304*, 208.

(46) Holthausen, M.C.; Fiedler, A.; Schwarz, H.; Koch, W. *J. Phys. Chem.* **1996**, *100*, 6236.

(47) Milet, A. Thesis, Université Louis Pasteur, Strasbourg, France, **1997**.

Chapter 3

Transition State Modeling of Asymmetric Epoxidation Catalysts

K. N. Houk, Jian Liu, and Thomas Strassner

Department of Chemistry and Biochemistry, University of California, Los Angeles, CA 90095-1569

The catalytic epoxidation of alkenes produces optically active epoxides from prochiral precursors. Reaction of peracids, dioxiranes, oxaziridines, and oxaziridinium cations with alkenes have been explored with density functional theory. Examples of catalytic reactions involving chiral ketones, and iminium cations, using Oxone as oxidant, have been modeled by quantum mechanical and force field methods. The geometries and multiplicities of the manganese-oxo intermediates involved in the Jacobsen asymmetric catalytic epoxidation using manganese-salen complexes have been explored.

A large number of catalytic and stoichiometric asymmetric epoxidations have been developed (1). What are the mechanisms of these reactions, and what is the origin of their high selectivities? It is the purpose of our theoretical studies to understand how these reactions work, to develop computational models to be used for the prediction of reactions of new substrates, and to design new catalysts. This paper describes the current status of our work.

Figure 1 shows examples of some stereoselective epoxidations using a variety of optically active epoxidizing agents, such as oxaziridines (2), peroxyacids (3-6), and dioxiranes (7-11). Peracids and hydroperoxides give low enantioselectivities. There are now many examples of dioxirane epoxidations producing optically active products, and several groups are working actively on catalytic versions of these reactions. Davis and coworkers have prepared oxaziridines which epoxidize alkenes with moderate stereospecificity (2). A few examples of chiral oxaziridinium compounds have been reported (12-17).

34

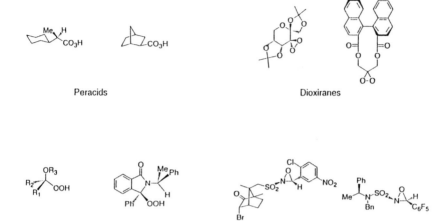

Peracids   Dioxiranes

Hydroperoxides   Oxaziridines

Figure 1. Examples of stereoselective epoxidations.

Dioxirane and oxaziridinium reactions have been made catalytic. Figure 2 outlines the general strategy for catalytic asymmetric epoxidations. For this cycle to be effective, the oxidation of the ketone, imine or iminium should be efficient, and in the case of imines and iminium, the oxidation should be stereoselective, or else diastereoisomeric oxaziridine derivatives will be formed.

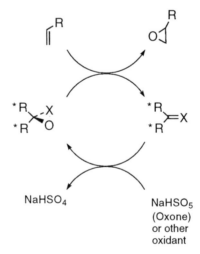

Figure 2. Catalytic cycle for epoxidation by dioxiranes (X=O), oxaziridine (X=NR), or oxaziridinium (X=NR$_2^+$).

Our initial work established the utility of B3LYP calculations for the prediction of transition states. In a collaboration with Dan Singleton at Texas A&M, we compared computed isotope effects for several different transition structures to experimental (*18*).

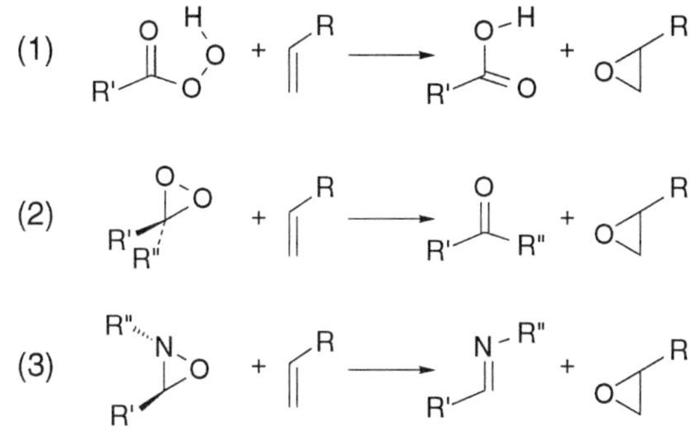

Figure 3. Prototype epoxidations.

Pioneering calculations by Bach and others (*19-24*) have provided insights into the nature of epoxidation transition states. Over the years, as increasingly accurate computational methods became available, models of epoxidation transition states have evolved. For the extensively discussed oxaziridine epoxidations, we have found that the spiro transition structures are always favored, but either synchronous or asynchronous transition structures can be formed (*25*). Figure 3 shows the reactions we have studied in detail.

Computational Methodology

Calculations reported here were conducted using the Gaussian94 program (*26*). Full geometry optimizations were carried out using density functional theory with the Becke3LYP functional and the 6-31G* basis set. Both restricted and unrestricted Becke3LYP/6-31G* were used for the transition structure searches for the parent systems with oxygen transfer from unsubstituted performic acid, dioxirane, peroxynitrous acid and oxaziridine to ethylene. Frequency calculations have been carried out for all the transition structures to ensure the presence of only one imaginary frequency corresponding to C-O bond forming and O-X (X= O, N) bond breaking. The optimized reactants were also checked by frequency calculations in order to confirm that they are minima. Charges cited in the discussions are Mulliken charges.

Transition Structures for the Epoxidation of the Parent Systems. Our previous results for the parent systems are summarized in Figure 4, which shows the transition states for epoxidations of ethylene by performic acid, dioxirane, peroxynitrous acid (*27*), and oxaziridine calculated by Becke3LYP/6-31G* method. The transition structures for the epoxidations by performic acid and by dioxirane are very similar: the breaking O-O bond lengths are 1.86 Å and 1.87 Å, respectively, and the two forming C-O bond lengths are 2.03 Å and 2.01 Å, respectively.

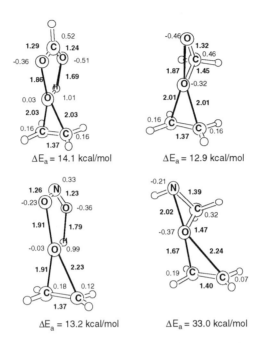

ΔE_a = 14.1 kcal/mol ΔE_a = 12.9 kcal/mol

ΔE_a = 13.2 kcal/mol ΔE_a = 33.0 kcal/mol

Figure 4. Transition structures (B3LYP/6-31G*) for epoxidations of ethylene by performic acid, dioxirane, peroxynitrous acid, and oxaziridine.

The activation energies for the epoxidations by performic acid and dioxirane by both Becke3LYP/6-31G* and MP2/6-31G* method are in reasonable agreement with experimental results in solution, which are 15-18 kcal/mol for performic acid (28) and 14.1 kcal/mol for dioxirane (29).

The epoxidation of ethylene by oxaziridine has been studied by different methods (19-25). Previous calculations using MP2/6-31G* yielded a synchronous transition structure (19-24).

Peroxynitrous acid is a very short-lived species which has not yet been observed to epoxidize alkenes (27). The activation energy for the oxygen transfer from peroxynitrous acid to ethylene is only 13.2 kcal/mol by Becke3LYP/6-31G* theory. It has an asynchronous transition structure, in between the synchronous peracid and the asynchronous oxaziridine in character.

The difference in energy between the spiro and planar geometries was assessed computationally using Becke3LYP/6-31G*. The preference for spiro increases as the length of the most fully formed CO bond decreases, from 5.1 kcal/mol with performic acid to 11.4 kcal/mol with oxaziridine. In all cases, there is a very large spiro preference.

There is obvious diradical character in the asynchronous transition structures of epoxidation by oxaziridine. The partial radical centers are on the N-terminus of the oxaziridine and one C-terminus of ethylene. Because of this, we also explored these reactions with unrestricted DFT theory (UBecke3LYP/6-31G*) for the parent system. We have demonstrated the suitability of UB3LYP calculations for reactions involving diradicals and have discussed elsewhere the potential problems in radical calculations (*30*). In the case of the oxaziridine transition state, the unrestricted solution is more stable than the restricted by 4.7 kcal/mol, and the spin densities are -0.68 and 0.53 at the N and C termini, respectively.

Transition structures for epoxidations of ethylene substituted by four different groups - methoxy, methyl, vinyl and cyano, ranging from electron-donating to electron-withdrawing - were located. We concentrate here on the reactions of dioxirane, which are characteristic of all the oxidants, and are most relevant to the observed catalytic processes.

For the oxygen transfer from dioxirane, all the transition structures are asynchronous, with the substituted groups on the carbon with the longer forming C-O bonds (Figure 5).

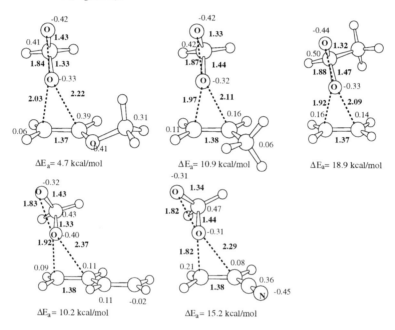

Figure 5. Transition structures (B3LYP/6-31G*) for epoxidations of substituted ethylenes by dioxirane and ethylene by methyldioxirane.

Unsymmetrical substitution such as that of methyl on the dioxirane causes the transition structure to become somewhat asynchronous. The methoxy, methyl and vinyl group all decrease the activation energies of the epoxidation reaction, while the electron-withdrawing cyano group increases the activation energy.

The transition structure for oxygen transfer from oxaziridine to ethylene is already very asynchronous for the parent system, and the four different substituents have little influence on the timing of the two forming C-O bonds (Figure 6). All four substituents decrease the activation energies as compared to the parent system, with the methoxy and vinyl groups stabilizing the transition structures the most. Instead of destabilizing the transition structures as in the performic acid and dioxirane system, the cyano group decreases the activation energy of epoxidation by oxaziridine. This reflects the significant diradical character of the transition structure for the oxaziridine system.

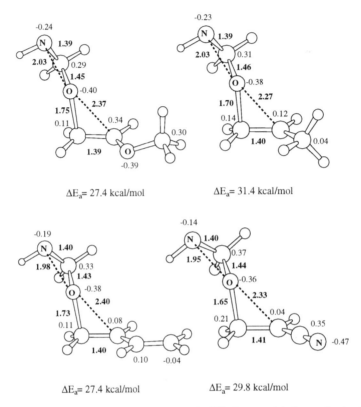

ΔE$_a$= 27.4 kcal/mol   ΔE$_a$= 31.4 kcal/mol

ΔE$_a$= 27.4 kcal/mol   ΔE$_a$= 29.8 kcal/mol

Figure 6. Transition structures (B3LYP/6-31G*) for epoxidations of substituted ethylenes by oxaziridine.

The partial charges are given next to the heavy atoms in the Figures. All the transition structures have no more than 0.35 charge transfer from alkene to the epoxidizing reagent, indicating diradical, rather than zwitterionic, character for the transition structures. The more asynchronous the transition structures are, the less the charge transfer, and the more diradical character. There is very little CO bond-breaking in the dioxirane and oxaziridine reactions.

Figure 7 summarizes the principal frontier orbital interactions occurring in the transition states. Relatively early transition states involve S$_N$2-like interactions have been described in detail by Bach (22). The alkene HOMO interacts in a stabilizing fashion with the OX σ* orbital, and a synchronous geometry is favored. When the leaving group is less able to stabilize a negative charge, the transition state becomes more advanced and S$_H$2 in character, which favors an asynchronous geometry. This involves interaction of the O lone pair with the alkene π* orbital, which is enhanced in an asynchronous geometry. Those two FMO interactions give a diradicaloid transition state when they are comparable in magnitude.

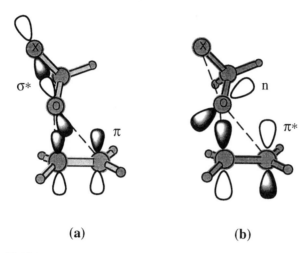

(a)   **(b)**

Figure 7. FMO interactions in the transition states of epoxidations: (a) alkene HOMO-oxidant LUMO and (b) oxidant HOMO-alkene LUMO.

     The trajectory predicted by this model follows that trajectory which is well-known for carbene cycloadditions. The spiro geometry is favored significantly, because the oxygen lone-pair HOMO interaction with the alkene LUMO is maximized in the spiro geometry but is eliminated in the planar transition state. The importance of this increases as the transition state becomes more advanced.

     This transition state model provides a good qualitative explanation for stereoselectivities which have been observed. Figure 8 shows the transition states of epoxidation of *trans*-stilbene by one of the Davis chiral oxaziridine reagents (*2*) and the Yang chiral binaphthyl dioxirane (*5*). The transition states were built by constraining the five atoms involved in bonding changes to the transition state optimal geometry determined by DFT calculation and minimizing all the other parts by MM2* using MacroModel 5.0 (*31*). The energy difference predicted between the transition states corresponding to experimentally favored and disfavored products are 4.2 kcal/mol for oxaziridine and 2.0 kcal/mol for dioxirane (Figure 8). This energy difference corresponds to a higher selectivity than that observed by experiment, but the models are nevertheless useful for the prediction of the sense of the observed selectivity.

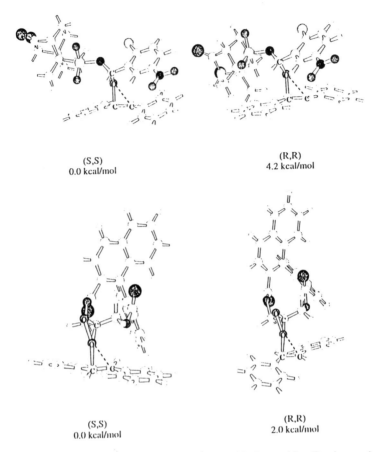

(S,S)
0.0 kcal/mol

(R,R)
4.2 kcal/mol

(S,S)
0.0 kcal/mol

(R,R)
2.0 kcal/mol

Figure 8. Force field models of asymmetric epoxidations with a Davis oxaziridine and a Yang chiral dioxirane.

The factors which control the stereoselectivity are shown in Figure 9. In the oxaziridine, experimental results can be rationalized by the model shown: on the left carbon of the alkene, the larger substituent prefers to be back, away from the protruding H of the oxaziridine. At the other terminus, π-stacking causes aromatic substituents to prefer the forward position. Alkyl or bulky groups will prefer to be back. Radical stabilization causes the substituents to be favored at one terminus in an unsymmetrical case; π-stacking as shown in the sketch and steric hindrance, involving groups on the reagent and on the alkene, dictate the position of groups in the transition state.

The qualitative model shown for dioxirane epoxidation also rationalizes experimental data obtained by Yang (21). Here the interaction of the larger dioxirane substituent with the alkene substituent dominates, and trans-alkenes place groups at S1 and L in spite of the fact that the large substituent on the left carbon would prefer to be back.

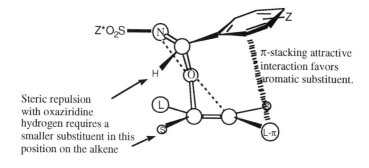

Steric repulsion
with oxaziridine
hydrogen requires a
smaller substituent in this
position on the alkene

π-stacking attractive
interaction favors
aromatic substituent.

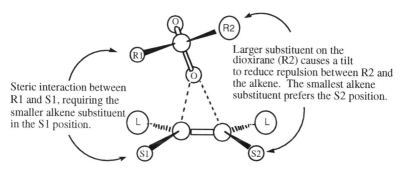

Steric interaction between
R1 and S1, requiring the
smaller alkene substituent
in the S1 position.

Larger substituent on the
dioxirane (R2) causes a tilt
to reduce repulsion between R2 and
the alkene. The smallest alkene
substituent prefers the S2 position.

Figure 9. Qualitative models to rationalize stereoselective epoxidations by oxaziridines and dioxiranes.

Catalytic epoxidations involving oxaziridinium intermediates are most promising. Iminium salts can be oxidized to oxaziridinium salts very easily in situ, and the oxaziridinium salts can epoxidize alkenes (12-17). Iminium salts are organic salts which can have good solubility in aqueous and organic solvents. Hence, iminium salts can act as catalysts for epoxidation reactions using oxidants such as Oxone.

The pioneering work by Hanquet and Lusinchi demonstrates a catalytic reaction involving an iminium salt. The iminium salt can be converted by Oxone to the corresponding oxaziridinium which effectively epoxidize alkenes (Figure 10) (12-15). In the investigation of the asymmetric epoxidation catalyzed by chiral iminium salts, two chiral iminium salts, **1** and **2**, have been reported (12-16). The asymmetric epoxidations of stilbene catalyzed by both chiral iminium salts give promising results of about 30% ee, and the epoxidation of 1-phenylcyclohexene catalyzed by **2** gives 71% ee. In a recent paper by Armstrong and co-workers (17), a series of pyrrolidine-derived iminium salts **3** of different substituted benzaldehyde have been prepared and used in the catalytic epoxidation. The electron-releasing substituent MeO- on phenyl completely deactivates the catalyst, while the electron-withdrawing group such as o-CF$_3$ or o-Cl increases the catalytic activity of the iminium salt.

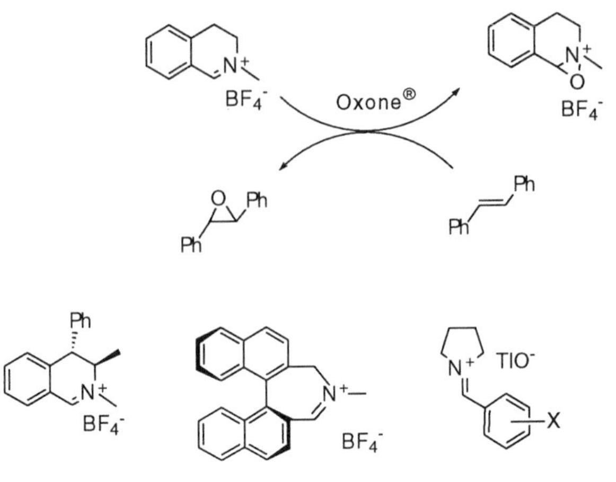

Figure 10. Iminium ion catalysts for epoxidations involving oxaziridinium ions.

B3LYP/6-31G* calculations predicts a symmetric spiro transition structure (TS1 in Figure 11) for the epoxidation of ethylene by oxaziridinium, similar to that obtained for performic acid or dioxirane. The activation energy for oxygen transfer from oxaziridinium to ethylene is only 4.9 kcal/mol, and the transition state is a very early one. The two forming epoxide C-O bonds are 2.6 Å, much longer than those of about 2.0 Å in the epoxidation by performic acid or dioxirane. The bond length of O-N bond has elongated from 1.44 Å to 1.57 Å in the transition state.

The influence of the methyl substituent on the transition structure has been investigated. Figure 11 also shows the transition states for the epoxidations of ethylene by N,N-dimethyl oxaziridinium (TS2), N-methyl oxaziridinium (TS3) and C-methyl oxaziridinium (TS4). The methyl substituents on oxaziridinium make the transition structure substantially later than TS1, but it is still quite early. The geometries of transition states still are spiro synchronous, even though TS3 and TS4 have unsymmetrical substituents. Since all these epoxidations by oxaziridiniums have early transition states, the methyl substituents do not have as big an influence on the synchronicity of the transition structure as those for dioxirane, performic acid and oxaziridine. The transition state TS3 and TS4 have negative activation energies, indicating that a complex between the cation and ethylene forms before the transition states.

Compared to the epoxidation by performic acid, dioxirane and oxaziridine, calculations predicts much lower activation energy for the epoxidation by oxaziridinium. Experimentally, epoxidations by oxaziridinium generated from 1 are faster than m-CPBA epoxidations, since a catalytic amount of iminium 1 can accelerate the reaction, similar to trifluoroacetone. The calculation most probably

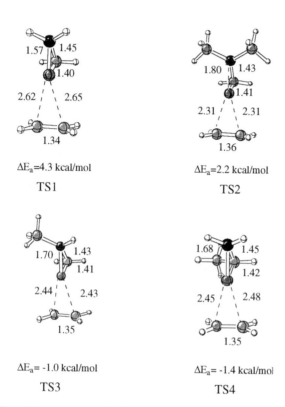

Figure 11. Transition structures for the epoxidations of ethylene by oxaziridinium (TS1), N,N-dimethyl oxaziridinium (TS2), N-methyl oxaziridinium (TS3), C-methyl oxaziridinium (TS4).

underestimated the activation energy, since the transition state was located in gas phase, and there is likely to be a large solvent effect for the cationic mechanism.

TS2 has been used to build transition state models for the study of epoxidations by chiral oxaziridiniums in the same way we have modeled epoxidations by oxaziridines and dioxiranes. We have tried to use force field modeling to elucidate the selectivity for the epoxidation of *trans*-stilbene catalyzed by iminium salt **1**. Since it was reported that oxaziridinium ring can only form on the side of the methyl substituent of iminium salt while the other side is blocked by the phenyl substituent, we used the oxaziridinium formed from iminium salt **1** for modeling. Figure 12 shows the modeled transition structures for the epoxidations of *trans*--stilbene by which lead to (R,R) and (S,S) epoxides.

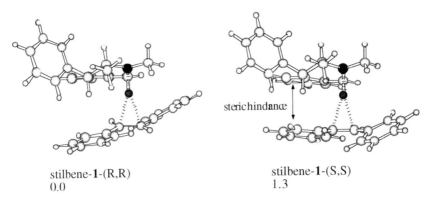

stilbene-**1**-(R,R)   stilbene-**1**-(S,S)
0.0   1.3

Figure 12. Modeled transition structures for the epoxidation catalyzed by iminium salt **1**. The relative MM2* energies are in kcal/mol.

The modeling predicts that the transition structure which leads to (R,R) is 1.3 kcal/mol lower in energy than stilbene-**1**-(S,S). The steric hindrance of the two phenyl rings in stilbene-**1**-(S,S) cause the higher energy of this model. The calculation is consistent with the experiment which gave 33% ee (R,R) selectivity. The top face of the iminium double bond is blocked by substituent groups. The bottom face of the iminium double bond is open to the formation of oxaziridinium.

Jacobsen Oxidations. We have followed avidly the development of manganese salen catalyzed epoxidation reactions and have developed models to understand the stereoselectivity (*32*). The reaction is summarized in Figure 13.

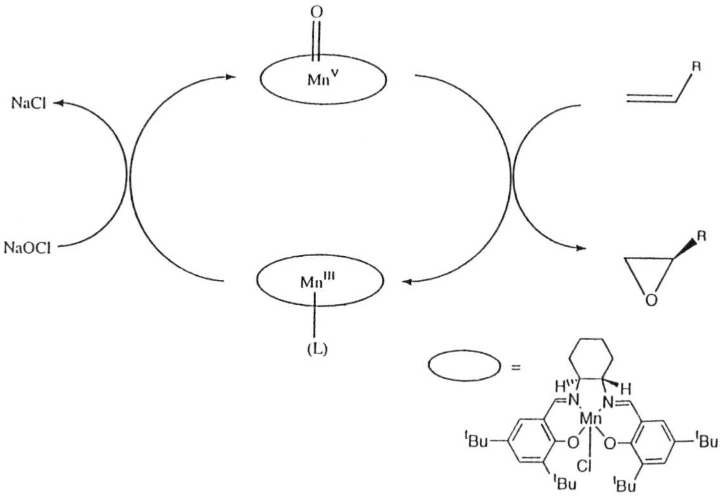

Figure 13. The Jacobsen asymmetric epoxidation.

There have been several suggestions about the mechanism of this reaction, including a concerted reaction, a stepwise process involving diradical intermediates, and a stepwise process involving a manganooxetane intermediate. A recent article by Jacobsen has summarized electronic effects on stereoselectivity and has used a stepwise/diradical model to interpret the results (34). Here, we concentrate on characterization of geometries and multiplicities of the metal-oxo intermediates, and consider the possibilities that two states of different multiplicity are involved in these reactions, analogous to the model recently proposed for cytochrome p-450 reactions (35).

Calculations were performed with B3LYP using the split valence double-zeta (DZ basis set, 3-21G, or a DZ plus polarization basis set, 6-31G*, for C, H, N, and O, together with double-zeta (DZ) (36) or triple zeta (TZ) (37) valence basis sets for manganese. The model systems used in these calculations are shown in Figure 14.

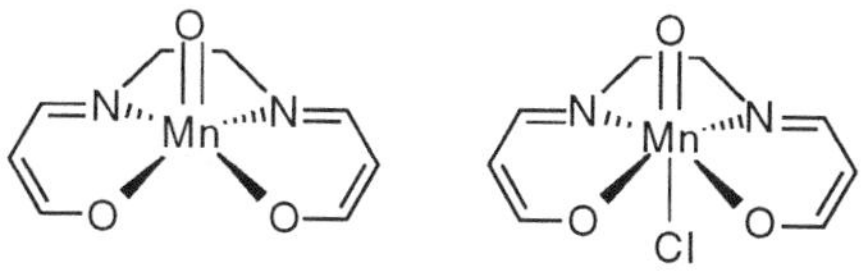

Figure 14. The models for Mn(V)-oxo complexes.

Calculations were performed on the structures in the oxidation states +III to compare to existing X-ray data and +V to make predictions about the structure of the actual catalyst. The manganese (III) compounds have d^4 and the manganese (V) compounds d^2 electronic configurations. For the d^2 electron configuration on an octahedral field, each electron will occupy one of the t_{2g}-orbitals with parallel spins.

It is well known that the Mn (III) favors an octahedral coordination sphere, while little is known about manganese (V) compounds.

The geometries of the Mn (III) complexes compare favorably with ample X-ray crystallographic data available in the literature. These results will be reported elsewhere (38). For the manganese (V) systems, calculations on oxo compounds lacking the Cl ligand give the geometries shown in Figure 15. Both calculations predict low- and high-spin to be close in energy. The best calculations predict the low-spin system to be lower in energy by 3.5 kcal/mol. By contrast, with the Cl ligand, the high-spin state is 10.5 kcal/mol more stable. Thus, with relatively strong ligands, the triplet is the ground state, and reactions are expected to be the characteristic of a diradical (35): the Jacobsen mechanism (34) is fully consistent with this.

By removal of the ligand, or perhaps by reducing to a less nucleophilic ligand, there may be a change of the spin state from a triplet species to a singlet spin state. This may be related to the fact that the experimental results show that for different ligands (39) or ligand substitutents (34), there might be different reaction pathways and enantioselectivities (39). The nature of the ligand might play a role for the stereoselectivity of the reaction.

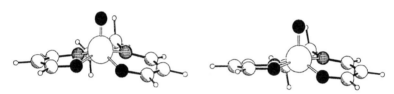

Figure 15. The low-spin (singlet) and high-spin (triplet) forms of the Mn (V) model system.

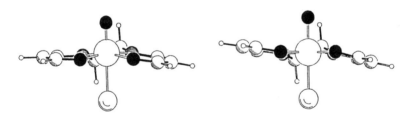

Figure 16. Manganese (V) model compounds calculated by the manganese triple-zeta basis/Becke3LYP/6-31G(d).

The multiplicity of the oxo-complex is important for the course of the reaction. A high-spin complex is more likely to react via a stepwise mechanism which involves a radical intermediate than a low-spin complex which would more easily react via a concerted pathway. On the other hand, the multiplicity of the Mn (III) catalyst is quintet, and so the correlation may be much more complex than indicated in this simple picture. These findings make a recently proposed (2+2) intermediate unlikely, but would very well agree with a radical intermediate in the transition state. We are currently exploring possible transition state structures.

Acknowledgments. We thank Kevin R. Condroski and Nicholas C. DeMello for initial results and helpful discussions. We are grateful to the 'Deutsche Forschungsgemeinschaft' for a fellowship (T.S.) and to the National Institute of General Medical Sciences, National Institutes of Health for financial support of this research.

Literature Cited.

(1) Jacobsen, E. N. In *Catalytic Asymmetric Synthesis*; Ojima, I., Ed.; VCH: New York, **1993**; Chapter 4.2.

(2) Davis, F. A.; Sheppard, A. C. *Tetrahedron* **1989**, *45*, 5703-5742, and references therein.

(3) Ewins, R. C.; Henbest, H. B.; McKervey, M. A. *Chem. Commun.* **1967**, 1085.

(4) Monlanari, F.; Moretti, J. ; Torre, G. *Gazz. Chim. Ital.* **1974**, *104*, 7.

(5) Pirkle, W. H.; Rinaldi, P. L. *J. Org. Chem.* **1977**, *42*, 2080-2082.

(6) For a claim of planar transition state, see Rebek, Jr., J.; Marshall, L.; McMairy, J.; Wolak, R. *J. Org. Chem.* **1986**, *51*, 1649.

(7) Curci, R.; D'Accolti, L.; Fiorentino, M.; Rosa, A. *Tetrahedron Lett.* **1995**, *36*, 5831.

(8) Murray, R. W.; Singh, M.; Williams, B. L.; Moncrieff, H. M. *Tetrahedron Lett.* **1996**, *36*, 2437.

(9) Yang, D.; Yip, Y.-C.; Tang, M.-W.; Wong, M.-K; Zheng, J.-H.; Cheung, K.-K. *J. Am. Chem. Soc.* **1996**, *118*, 491-492.

(10) Tu, Y.; Wang, Z.-X.; Shi, Y. *J. Am. Chem. Soc.* **1996**, *118*, 9806.

(11) Adam, W.; Smerz. A. K. *J. Org. Chem.* **1996**, *61*, 3506.

(12) Hanquet, G.; Lusinchi, X.; Milliet, P. *Tetrahedron Lett.* **1988**, *29*, 3941.

(13) Hanquet, G.; Lusinchi, X. *Tetrahedron Lett.* **1993**, *24*, 5299.

(14) Bohe, L.; Hanquet, G.; Lusinchi, M.; Lusinchi, X. *Tetrahedron Lett.* **1993**, *34*, 7271.

(15) Lusinchi, X.; Hanquet, G. *Tetrahedron* **1997**, *53*, 13727.

(16) Aggarwal, V. K.; Wang, M. F. *Chem. Commun.* **1996**, 191.

(17) Armstrong, A.; Ahmed, G.; Garnett, I.; Goacolou, K. *SYNLETT* **1997**, 1075.

(18) Singleton, D. A.; Merrigan, S. R.; Liu, J.; Houk, K. N. *J. Am. Chem. Soc.* **1997**, *119*, 3385-3386.

(19) Bach, R. D.; Coddens, B. A.; McDouall, J. J. W.; Schlegel, H. B.; Davis, F. A. *J. Org. Chem.* **1990**, *55*, 3325-3330.

(20) Bach, R. D.; Andres, J. L.; Davis, F. A. *J. Org. Chem.* **1992**, *57*, 613-618.

(21) Bach, R. D.; Winter, J. E.; McDouall, J. J. W. *J. Am. Chem. Soc.* **1995**, *117*, 8586-8593, and references therein.

(22) Bach, R. D.; Glukhovtsev, M. N. Gonzalez, C.; Marquez, M.; Estevez, C. M.; Baboul, A. G.; Schlegel, H. B. *J. Phy. Chem.* **1997**, *101*, 6092.

(23) Yamabe, S.; Kondou, C.; Minato, T. *J. Org. Chem.* **1996**, *61*, 616-620.

(24) Bach, R. D.; Canepa, C.; Winter, J. E.; Blanchette, P. E. *J. Org. Chem.* **1997**, *62*, 5191.

(25) Houk, K. N.; Liu, J.; DeMello, N. C.; Condroski, K. R. *J. Am. Chem. Soc.* **1997**, *119*, 10153.

(26) Gaussian 94 (Revision A.1), Frisch, M. J.; Trucks, G. W.; Schlegel, H. B.; Gill, P. M. W.; Johnson, B. G.; Robb, M. A.; Cheeseman, J. R.; Keith, T. A.; Peterson, G. A.; Montgomery, J. A.; Raghavachari, K.; Al-Laham, M. A.; Zakrzewski, V. G.; Oriz, J. V.; Foresman, J. B.; Cioslowski, J.; Stefanov, B. B.; Nanayakkara, A.; Challacombe, M.; Peng, C. Y.; Ayala, P. Y.; Chen, W.;

Wong, M. W.; Andres, J. L.; Replogle, E. S.; Gomperts, R.; Martin, R. L.; Fox, D. J.; Binkley, J. S.; Defrees, D. J.; Baker, J.; Stewart, J. P.; Head-Gordon, M.; Gonzalez, C.; Pople, J. A. Gaussian, Inc., Pittsburgh PA, 1995.
(27) Houk, K. N.; Condroski, K. R.; Pryor, W. A., *J. Am. Chem. Soc.* **1996**, *118*, 13002.
(28) Dryuk, V. G. *Tetrahedron* **1976**, *32*, 2855.
(29) Murray, R. W. *Chem. Rev.* **1989**, *89*, 1187.
(30) Goldstein, E.; Beno, B.; Houk, K. N. *J. Am. Chem. Soc.* **1996**, *118*, 6036.
(31) Macromodel V5.0: Mohamadi, F.; Richards, N. G. J.; Guida, W. C.; Liskamp. R.; Lipton, M.; Caufield, C.; Chang, G.; Hendrickson, T.; Still W. C. *J. Comput. Chem.* **1990**, *11*, 440. MM2* is the MacroModel implementation of MM2.
(32) Houk, K. N.; DeMello, N.; Condroski, K.; Fennen, J.; Kasuga, T. Electronic Conference on Heterocyclic Chemistry (www.choic.ac.uk/ectoc/echet96/echet_ma.html), 1996.
(33) Katsuki, T. *J. Mol. Catalysis A: Chemical* **1996**, *113*, 87-107.
(34) Palucki, M.; Finney, N. S.; Pospisil, P. J.; Güler, M. L.; Ishida, T.; Jacobsen, E. N. *J. Am. Chem. Soc.* **1998**, *120*, 948-954, and references therein.
(35) Shaik, Sason, Filatov, M.; Schröder, D.; Schwarz, H. *Chem. Eur. J.* **1998**, *4*, 193-199.
(36) Schäfer, A.; Horn, H.; Ahlrichs, R. *J. Phys. Chem.* **1992**, *97*, 2571.
(37) Schäfer, A.; Huber, C.; Ahlrichs, R. *J. Phys. Chem.* **1994**, *100*, 5829.
(38) Strassner, T.; Houk, K. N. In preparation.
(39) Irie, R.; Noda, K.; Ito, Y.; Matsumoto, N.; Katsuki, T. *Asymmetry* **1991**, *2*, 481.

Chapter 4

Transition States in Catalysis and Biochemistry

Margareta R. A. Blomberg and Per E. M. Siegbahn

**Department of Physics, University of Stockholm, Box 6730,
S-113 85 Stockholm, Sweden**

Three different examples are given from recent applications
where transition states for catalytic processes are obtained. The
size of the model systems range from 15 to 40 atoms and
transition metals are present. The first example is a model
study of homogeneous alkane activation. In the reaction between nickel cations and n-butane, it is shown that an unusual
type of multi-center transition states determine the fractions
of different elimination products. The second example is concerned with methane hydroxylation in methane monooxygenases
(MMO) present in microorganisms termed methanotrophs. The
H-atom abstraction from methane is studied using a dinuclear
iron model complex. The final example is the substrate mechanism of ribonucleotide reductase (RNR), transforming RNA
nucleotides into DNA nucleotides.

In the present survey of recent work concerned with studies of chemical reactions, three different examples will be described. In all of these the determination of transition states is the key problem. The examples serve to show
possibilities and limitations of present approaches. For many years the accuracy of the methods used was the main limitation, but this is no longer true
for the present type of problems rel 12 ating to reaction mechanisms. In the
present work the DFT (Density Functional Theory) method termed B3LYP is
used (1). For benchmark tests of this method comprising 55 common first and
second row molecules performed using slightly larger basis sets than used here
(2), an average absolute deviation compared to experiments of 2.2 kcal/mol was
obtained for the atomization energies, of 0.013 Å for the bond distances and of
0.62 degrees for the bond angles. The present accuracy should be almost as

high as in this benchmark test and should be enough to discriminate between different reaction mechanisms. Instead, the main limitation is now the size of the system that can be treated. This means that one has to be careful in selecting proper models for the real situation. Since the most realistic approach is to start out by performing a calculation on an isolated gas phase model, the actual environment also has to be considered before the final answer is reached. The exception to this rule is, of course, when gas phase reactions of reasonably small systems are studied.

In the first example given here, the reaction between nickel cations and butane, the simplest case of a direct comparison to gas phase experiments is discussed. This example is chosen to point out the possibility to make detailed evaluations of the accuracy of calculated potential surfaces by direct comparisons to experiments. It also shows the complexity of even seemingly simple reactions where the models chosen are known to be the best possible. In contrast, for the other two examples taken from biochemistry, an accurate modeling of the true situation is one of the major problems. This is particularly true for the first example of methane activation by methane monooxygenase (*3*), where even the metal complex directly involved in the reactions is too big to be a useful first model. The second problem, concerned with the RNR substrate reaction (*4*), is simpler in this respect and all residues directly involved could eventually be included in the model. In general for biochemical reactions, also the effects of the surrounding protein environment should be taken into consideration. This was done in the RNR substrate case but, as will be shown, for properly chosen models these effects can be kept very small.

Computational Details

Methods and basis sets. All calculations were made using the GAUSSIAN-94 program (*5*). The calculations were performed in several steps. First, an optimization of the geometry was performed using the B3LYP method (*1*). Double zeta basis sets were used in this step (the LANL2DZ set of the Gaussian-94 program). In the second step the energy was evaluated for the optimized geometry using very large basis sets including diffuse functions and with two polarization functions on each atom (the 6-311+G(2d,2p) basis sets). The final energy evaluation was also performed at the B3LYP level. All energies discussed below include zero-point vibrational effects. These were calculated using all electron basis sets of essentially double zeta quality. For the Ni^+ + butane reaction, the zero-point vibrational effects were calculated at the B3LYP level, while for MMO and RNR Hartree-Fock results, scaled by 0.9, were used. All transition states are characterized by having only one imaginary frequency.

In the RNR calculations dielectric effects from the surrounding protein were calculated at the optimized structures, using the Self-Consistent Reaction Field (SCRF) method. In the present study the self-consistent isodensity polarized continuum model (SCI-PCM) was used. In this method the solute cavity is determined from a surface of constant charge density around the solute molecule. The dielectric constant of the protein is the main empirical parameter of the model and it was chosen to be equal to 4 in line with previous suggestions for proteins. This choice has recently been found to give very good agreement with experiment for two different electron transfer processes in the bacterial photosynthetic reaction center (6). In the present case where neutral models are chosen throughout, the dielectric effects are found to be extremely small.

Localisation of transition states. A major part of the studies discussed here is spent on the determination of relevant transition states. To some extent the procedure to find the transition states requires chemical intuition and each problem has its own solution, but there are some technical points which are quite general and could be useful to mention. In some of the present systems the final models used are quite large, particularly for the biochemical problems. One useful general experience here is that the local structure of a transition state for a particular reaction is usually quite independent of the size of the system. Therefore, to obtain a good starting point the model should first be made as small as possible. The second useful technical point is that freezing procedures should be extensively used to approach the transition state. Following some assumed reaction path, the positions of some key atoms should be fixed at different intervals and all other coordinates optimized until a barrier region is located. At this stage, a Hessian matrix should be calculated to be used for the final localisation of the transition state. Since Hessian matrices are expensive to obtain, it is useful to try to use as simple methods and as small basis sets as possible. Even Hessians obtained at the Hartree-Fock level using minimal basis sets can be quite adequate. Once the transition state has been determined for the smallest possible model, the molecular model should be extended by adding all necessary parts of the system that might affect the energy. Even though these extensions may be necessary for the final accurate energetic results, they normally do not modify the structure and character of the transition states very much. Environmental effects from e.g. a surrounding protein can simply be added on afterwards since it is very unusual that they will affect the structures significantly, at least when the present type of neutral models are used. In this way, transition states can today be obtained for systems with up to 50 atoms using the methods described above. Examples of such cases will be given in the subsections below.

Results

Ni$^+$ and Butane. An important long-term goal is to understand the fundamental mechanisms of homogeneous catalysis involving transition metal complexes. An attractive route in this direction is to study the reactions of the simplest model systems, the naked metal atoms, in the gas phase. This approach has been used by both experimentalists and theoreticians, creating a useful meeting point for the different techniques, and in particular makes it possible to obtain a thorough evaluation of the accuracy of calculated potential energy surfaces. Below we summarize the preliminary results from a recent study of the reaction between naked nickel cations and n-butane (7). The purpose of this study is to combine energies, geometries, and vibrational frequencies from quantum chemical calculations on all key reaction intermediates and transition states to build a detailed statistical rate model for the $Ni^+ + n\text{-}C_4H_4$ reaction. Such a model would provide absolute time scales for complex fragmentation and time-dependent branching fractions that are directly comparable to experimental results (7). In the present chapter we report some of the main features of the calculated potential surfaces. The results from molecular beam experiments on this reaction can be summarized such that at 0.2 kcal/mol collision energy, and after 6 μs, 63 % C_2H_6 elimination, 1 % CH_4 elimination, 26 % H_2 elimination, and 11 % $NiC_4H_{10}^+$ complexes are observed (8).

Nickel cations and n-butane form an electrostatic complex, $NiC_4H_{10}^+$, calculated to be bound by 32 kcal/mol relative to ground state reactants (Ni^+, ^2D plus $n\text{-}C_4H_4$), see Fig. 1. The elimination reaction paths from this electrostatic complex can start with either C-C activation or C-H activation. Earlier experimental work on alkane activation by transition metal cations has been interpreted in terms of this initial C-C or C-H insertion as the rate-limiting step, and the subsequent step has been expected to be a β-hydrogen or β-alkyl migration, followed by a three-center, reductive elimination of H_2 or alkane. In contrast, the present calculations show that both C-H and C-C activation have low barriers, well below (12-15 kcal/mol) the separated reactants. Furthermore, the highest potential energy along the paths to H_2 or alkane activation occurs at intriguing *multi-center transition states* (7). These involve concerted motion of several atoms along a segment of the reaction path connecting each insertion intermediate to the corresponding exit-channel ion-induced-dipole complex. Similar results have previously been obtained for the reaction between n-propane and different transition metal cations (9,10).

In ref. 7 it is shown that, both H_2 and C_2H_6 elimination from n-butane, can occur via either an initial C-C, or an initial C-H insertion. However, the subsequent multi-center transition states are found to be substantially lower for the reaction paths that start with C-C bond activation, and therefore we restrict the present discussion only to those reactions. The calculated potential surfaces for the reaction paths initiated by insertion of Ni^+ into a C-C bond of n-butane are shown in Fig. 1. From this figure it can be seen that the main differences between the reactions appear at the multi-center transition states that follow the C-C insertions, and as will be discussed below, the ordering of the multi-center transition states agree with the experimentally observed yields of the different elimination products.

From the central C-C insertion product, $Ni(C_2H_5)_2^+$, two different multi-center transition states are possible, MCTS1 and MCTS2. In MCTS1, which gives the lowest barrier, 11 kcal/mol below the separated reactants, a hydrogen atom moves directly from one ethyl group to the other, finally leading to the $Ni(C_2H_4)(C_2H_6)^+$ electrostatic complex. From this complex ethane can be eliminated, see Fig. 1, and in the molecular beam experiment this is the most abundant elimination product, 63 % yield. In MCTS2, shown in Fig. 2, one hydrogen atom from each ethyl group leaves and directly forms H_2, finally leading

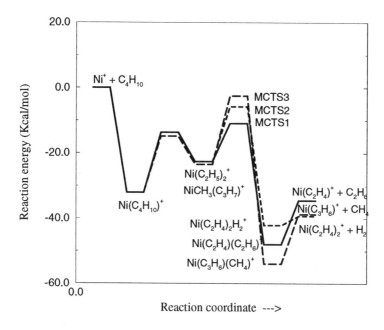

Figure 1: Calculated potential energy surfaces for C-C insertion of Ni$^+$ into n-C$_4$H$_{10}$.

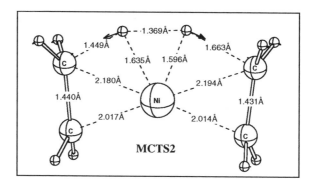

Figure 2: Structure of one of the optimized multi-center transition states for Ni$^+$ + n-C$_4$H$_{10}$ discussed in the text. The arrows indicate the imaginary vibrational mode.

to the Ni(C$_2$H$_4$)$_2$H$_2$$^+$ electrostatic complex. This reaction path has the second lowest barrier, 6 kcal/mol below the separated reactants, and subsequently H$_2$

can be eliminated, see Fig. 1, yielding the second most abundant elimination product, 26 % yield in the molecular beam experiment.

From the terminal C-C insertion product, $Ni(CH_3)(C_3H_7)^+$, one possible multi-center transition state, MCTS3, has been investigated. In this transition state a hydrogen atom moves directly from the ethyl group to the methyl group. This reaction path, thus yielding the $Ni(C_3H_6)(CH_4)^+$ electrostatic complex, gives the third lowest barrier, 2 kcal/mol below the separated reactants and it can finally lead to methane elimination, see Fig. 1. The experimental yield of the methane elimination product is only 1 %.

In conclusion, the calculations are in good agreement with the experimental observations, giving a qualitatively correct distribution of the possible elimination products. However, the quantitative assessment of the calculated potential surface, in particular the height of the rate determining barriers, will have to await the more detailed comparisons between experiment and theory based on the statistical rate modeling in progress (7). Furthermore, the calculations give a new interpretation of the experimental results, in terms of an unusual type of multi-center transition states as the rate-limiting step.

Methane Monooxygenase (MMO). Methane monooxygenases are a group of enzymes which convert methane to methanol via a monooxygenase pathway in which the dioxygen molecule is activated via reduction :

$$CH_4 + O_2 + NADH + H^+ \longrightarrow CH_3OH + H_2O + NAD^+ \qquad (1)$$

The X-ray structure of the 251 kD hydroxylase from *M. capsulatus* (11), has a dinuclear iron center bridged by a hydroxide, a glutamate, and an acetate from the buffer, where waters are expected to be bound in vivo. The terminal ligands are mostly oxygen derived but there is also a histidine ligand on each iron center. Experiments have indicated that the active species in the reaction with methane, commonly denoted compound **Q**, probably has bridging μ-oxo groups and with both irons having oxidation state IV (12,13). The complex is diamagnetic. To model this compound, a complex with only oxygen derived ligands was adopted for simplicity. Ferromagnetic coupling is furthermore used to simplify the calculations.

The first part of the study of methane activation by MMO is to optimize the structure of compound **Q** without methane, see Fig. 3. The geometry obtained after a B3LYP optimization was quite surprising. First, both iron centers are only five coordinated. In fact, models with 6-coordinated starting structures for the optimization, quickly rearrange to adopt 5-coordination. The second, and perhaps even more surprising feature, is that the two oxo groups are asymmetrically placed between the Fe atoms so that the two Fe-O (oxo) bond lengths for each metal are substantially unequal. The short Fe-O distances are found to be 1.74 and 1.77 Å and the long distances 2.00 and 2.05 Å. This is very different from what is known from similar structures, most notably Mn(III) dimeric complexes. After this work was finished a striking confirmation of this structure appeared (14). New EXAFS data on intermediate **Q** of MMO was interpreted in terms of an Fe(IV)-$(\mu$-O$)_2$-Fe(IV) diamond core structure

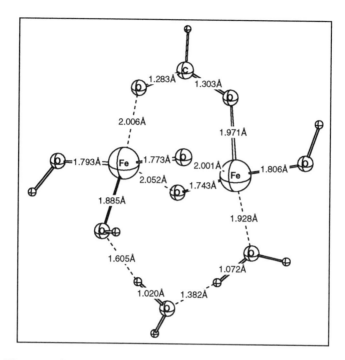

Figure 3: Optimized structure of the ^{9}A model of compound **Q**.

with asymmetric bridging oxo groups, essentially identical with the intermediate proposed here. Fe-O (oxo) distances of 1.77 Å and 2.05 Å were reported. Even though ferromagnetic coupling and simple water and hydroxyl ligands are used, a structure is thus obtained from the calculations which in the most critical region for the chemistry is almost identical to the true structure. The structure of compound **Q** can be regarded as being derived from an octahedron, but showing strong Jahn Teller (JT) distortions. The JT axis, along which the M-L bonds are elongated, is not the axis normal to the M-(μ-O)$_2$-M plane, as in the known Mn(III) cases, but is the axis defined by the long M-oxo bond and the vacant site trans to it; this vacancy can be interpreted as an extreme case of JT distortion leading to the departure of the H_2O ligand, assisted by the relatively strong second sphere hydrogen bonding that it can engage in when no longer bound to iron. The four short M-L bonds are in the plane normal to the JT axis. Of these four short bonds, one is formed to a bridging oxo.

Since the simple model of the reactant complex of MMO is so well in line with experimental information, the results for the methane activation reaction should also be interesting. The resulting transition state is shown in Fig.4. It should first be noted that the starting guess for the optimization was actually a four-center transition state with a rather short Fe-C distance, but at convergence

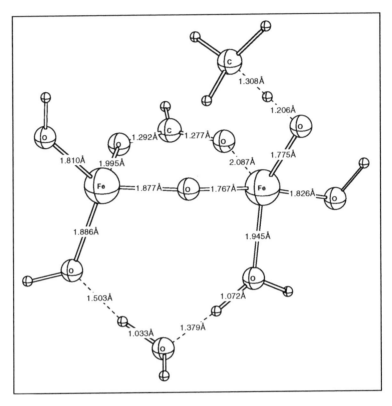

Figure 4: Optimized structure of the ⁹A transition state of the methane reaction.

the character of the transition state is very clearly one of a pure abstraction.

With the identification of the transition state structure for the methane reaction, explanations for the main experimental results can be suggested. First, the large KIE (Kinetic Isotope Effect) for the reaction ($12,13$), is clearly consistent with the type of transition state shown in Fig.4. There is a large zero-point effect for the reaction barrier and for the same reason, there is also a large KIE. The calculated KIE for CD_4 at 298.15 °K is 8.0, which is somewhat larger than the experimental value, indicating that the real transition state could be a little bit tighter with more bonding towards the metal. Quite recently, new experiments by Nesheim and Lipscomb (15) have led to much larger KIE values for the methane reaction than the previous ones. For CD_2H_2 a value of 9 was found in close agreement with the calculated value. However, for $CH_4:CD_4$ an even larger value of 19 was measured. Large effects of tunneling were suggested to rationalize these large KIE values. Tunneling effects are not accounted for in the present simple estimate of the KIE value.

The second main experimental result, 65 % retention of configuration with chiral ethane, is in apparent contradiction with the formation of a nearly free methyl radical in the transition state of Fig.4. A reasonable explanation could be that after the transition state is passed, an Fe-CH_3 bond is very rapidly formed. The calculations actually confirm that the most stable product of the methane activation indeed has an Fe-CH_3 bond, although a quite weak one with a bond strength of 9 kcal/mol. The final structure on the methane reaction pathway is the methanol product which is quite stable by 40.3 kcal/mol with respect to the starting reactants. The rather simple model used for the present calculations is thus able to explain the main experimental results. At present, calculations are in progress to study this reaction further with more realistic models including all the actual residues of the first coordination sphere. The results obtained to date essentially confirm the picture described above.

RNR Substrate Reaction. DNA differs chemically from RNA in two major respects. First, its nucleotides contain 2'-deoxyribose residues (DNA nucleotides) rather than ribose residues (RNA nucleotides). Secondly, DNA contains the base thymine whereas RNA contains uracil. The enzymes that catalyze the set of reactions required for the first of these transformations, see scheme 1, are named Ribonucleotide Reductases (RNR). The present knowledge of RNR has been summarized in recent reviews (*16,17*). The *E. coli* RNR is an $\alpha_2\beta_2$ tetramer that can dissociate into two catalytically inactive homodimers, R1 and R2. The X-ray structures of both R1 (*18,19*) and R2 (*20*) have been determined. The reactions leading to substrate conversion can be described in the following way. An Fe(II)-dimer in R2 is first oxidized, probably to a bis-μ-oxo Fe(IV)-dimer, which then is reduced in two steps probably by abstracting two hydrogen atoms from surrounding residues to become a resting Fe(III)-dimer complex. In the reduction process a Tyr122 radical is produced, which can be stored for long periods waiting for the substrate to arrive 30 Å away in R1. When the substrate arrives, the radical character will move along a hydrogen bonded chain from Tyr122 to Cys439 at the substrate site in R1. The substrate reactions are then catalyzed by the Cys439 radical in a process where two other cysteines and probably also a glutamate participates. The leading experimental model for the substrate reactions has been given by Stubbe (*21*).

Scheme 1

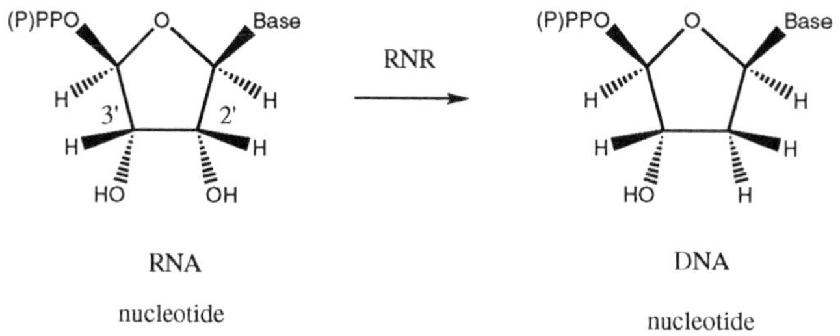

RNA nucleotide   DNA nucleotide

58

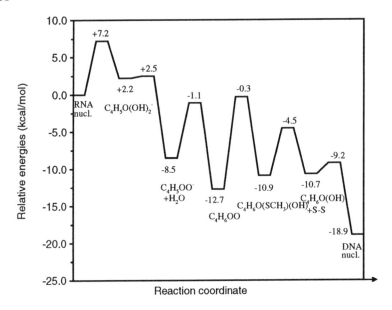

Figure 5: Energy-diagram for the transformation from the RNA to the DNA nucleotide.

The calculations show that the apparently simple transformation in scheme 1 goes over several steps. The computed energy diagram is shown in Fig.5. The rate limiting step is the fourth one, where a rather stable keto intermediate is attacked. After considerable experimentation a transition state was found, shown in Fig.6. There are some important aspects of this transition state that should be noted here. First, the attack on the keto group is a complex concerted action where Cys225 and Glu441 are mainly involved but where also the involvement of Asn437 is significant. This type of cyclic transition state is found in many other standard transformations common in enzyme reactions. In a simplified picture, the reaction can be described as a proton transfer from Cys225 via a water molecule and Glu441 to the keto oxygen, and a simultaneous attack of the cysteine anion on the carbon of the keto group. It is extremely important to note that in spite of this description in terms of ions, the model chosen is neutral. The choice of neutral models for this type of reaction is quite different from what is normally done, for example using molecular mechanics models, and the present modeling is therefore controversial at the present stage. However, the calculations definitely show that this is a strongly concerted process and any attempt to divide this process into separate attacks using ionic models does not work. There are several important implications of this finding. The main one is that since the model is neutral the effects of the surrounding protein should be very small. Using a dielectric continuum model with a cavity shaped

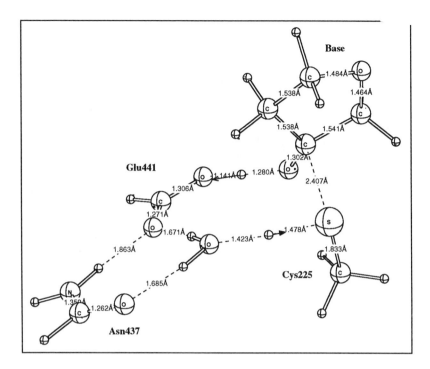

Figure 6: Optimized transition state structure for cysteine attack on ribose C3′ in the RNR substrate reaction.

after the molecule and a dielectric constant equal to four, the effects of the surrounding protein were found to be truly negligible. The effect on the barrier for the reaction shown in Fig.6 is only 0.9 kcal/mol and the effect on the endothermicity is even smaller with 0.4 kcal/mol. In many other enzyme reactions studied recently the same observation is made that neutral gas phase models give a very good description of the reaction. The consistency of the model is in these reactions demonstrated by the small effects found by simple models of the surrounding and, of course, the reasonable agreement with experimental facts available.

Conclusions

It has been shown that if appropriate procedures are used rather complicated transition states for systems containing up to 50 atoms can be obtained. It is furthermore shown that the accuracy of the calculated energy profiles should be good enough to discriminate between different possible reaction mechanisms in both homogeneous catalysis and biochemistry.

60

Literature Cited

1. Becke, A. D. *Phys. Rev.* **1988**, *A38*, 3098; Becke, A. D. *J. Chem. Phys.* **1993**, *98*, 1372; Becke, A. D. *J. Chem. Phys.* **1993**, *98*, 5648.
2. Bauschlicher, C.W.,Jr; Ricca, A.; Partridge, H.; Langhoff, S.R., in *Recent Advances in Density Functional Methods, Part II*, Ed. D.P. Chong, p.165 (World Scientific Publishing Company, Singapore, **1997**).
3. Siegbahn, P.E.M.; Crabtree, R.H. *J. Am. Chem. Soc.* **1997**, *119*, 3103.
4. Siegbahn, P.E.M., submitted.
5. Frisch, M. J.; Trucks, G. W.; Schlegel, H. B.; Gill, P. M. W.; Johnson, B. G.; Robb, M. A.; Cheeseman, J. R.; Keith, T.; Petersson, G. A.; Montgomery, J. A.; Raghavachari, K.; Al-Laham, M. A.; Zakrzewski, V. G.; Ortiz, J. V.; Foresman, J. B.; Cioslowski, J.; Stefanov, B. B.; Nanayakkara, A.; Challacombe, M.; Peng, C. Y.; Ayala, P. Y.; Chen, W.; Wong, M. W.; Andres, J. L.; Replogle, E. S.; Gomperts, R.; Martin, R. L.; Fox, D. J.; Binkley, J. S.; Defrees, D. J.; Baker, J.; Stewart, J. P.; Head-Gordon, M.; Gonzalez, C.; Pople, J. A. *Gaussian 94 Revision B.2*; Gaussian Inc.: Pittsburgh, PA, **1995**.
6. Blomberg, M.R.A.; Siegbahn, P.E.M.; Babcock, G.T., submitted.
7. Yi,S.S.; Blomberg, M.R.A.; Weisshaar, J.C., to be published.
8. Noll, R.J.; Weisshaar, J.C. *J. Am. Chem. Soc.* **1994**, *116*, 10288.
9. Yi, S.S.; Blomberg, M.R.A.; Siegbahn, P.E.M.; J.C. Weisshaar *J. Phys. Chem. A* **1998**, *102*, 395.
10. Holthausen, M.C.; Fiedler, A.; Schwarz, H.; Koch, W. *J. Phys. Chem.* **1996**, *100*, 6236, Holthausen, M.C.; Koch, W. *J. Am. Chem. Soc.* **1996**, *118*, 9932, Holthausen, M.C.; Koch, W. *Helv. Chim. Acta* **1996**, *79*, 1939.
11. Rosenwieg, A.C.; Frederick, C.A.; Lippard, S.J.; Nordlund, P. *Nature* **1993**, *366*, 537.
12. Feig, A.L.; Lippard, S.J. *Chem. Rev.* **1994**, *94*, 759.
13. Wallar, B.J.; Lipscomb, J.D. *Chem. Rev.* **1996**, *96*, 2625.
14. Shu, L.; Nesheim, J.C.; Kauffmann, K.; Munck, E.; Lipscomb, J.D.; Que, Jr., L. *Science* **1997**, *275*, 515.
15. Nesheim, J.C.; Lipscomb, J.D. *Biochemistry* **1996**, *35*, 10240.
16. Sjöberg, B.M. *Structure* **1994**, *2*, 793, Sjöberg, B.-M. *Structure and Bonding* **1997**, *88*, 139.
17. Gräslund, A.; Sahlin M. *Annu. Rev. Biophys. Biomol. Struct.* **1996**, *25*, 259.
18. Uhlin, U.; Eklund, H. *Nature* **1994**, *370*, 533.
19. Eriksson, M.; Uhlin, U.; Ramaswamy, S.; Ekberg, M.; Regnström, K.; Sjöberg, B.-M.; Eklund, H. *Structure* **1997**, *5*, 1077.
20. Nordlund, P.; Sjöberg, B.-M.; Eklund, H. *Nature* **1990**, *345*, 593.
21. Stubbe, J. *Biol. Chem.* **1990**, *265*, 5330; Mao, S.S.; Holler, T.P.; Yu, G.X.; Bollinger, J.M.; Booker, S.; Johnston, M.I.; Stubbe, J. *Biochemistry* **1992**, *31*, 9733.

Chapter 5

Enzymes, Abzymes, Chemzymes—Theozymes?

Claudia Müller, Li-Hsing Wang, and Hendrik Zipse

Institute of Organic Chemistry, TU Berlin, Str. d. 17. Juni 135, D-10623 Berlin, Germany

The reaction of amines with esters has been investigated theoretically as well as experimentally. The uncatalyzed reaction proceeds through a direct displacement pathway for esters with good leaving groups. The reaction barrier is mainly caused by unfavorable proton transfer geometries. The reaction can be accelerated substantially through bifunctional catalysts such as 2-pyridones to such an extent, that proton transfer is not rate limiting anymore. Formation of a tetrahedral zwitterionic structure, which becomes rate limiting in the pyridone-catalyzed case, can be facilitated through the presence of a specifically oriented second pyridone molecule.

1. Introduction.

In recent years, various new approaches for the development of catalysts that rival enzymes in terms of their reactivity and selectivity have been devised. The concept of catalytically active antibodies (1,2) ("antibody enzymes", "abzymes") takes advantage of the ability of the immune system to produce antibodies which specifically bind a selected antigen. Provided a hapten can be found that resembles the transition state of a reaction, one must expect that antibodies raised against this hapten are catalytically active for the respective reaction. Following a more traditional approach combining mechanistic insight and empirical optimization, the development of man-made catalysts ("chemzymes") (3) has probably been the most successful route to develop catalysts for many organic transformations. The development of catalysts can also be based on theoretical studies of the influence of various functional groups on the mechanism of a given reaction. This "theozyme" approach (4) has been pioneered by Houk et al. and applied to reactions as diverse as the Diels-Alder reaction, the intramolecular ring opening of epoxides, and the ring opening of isoxazoles. In all these examples, the theoretical studies provide a mechanistic basis for the rationalization of the catalytic activities of catalytic antibodies in these reactions. We are now trying here to use the theozyme approach to develop catalysts for the reaction of amines with esters, in which a peptide bond is formed. In order to keep the experimental work as close to the theoretical studies as possible all experiments discussed in the following have been conducted in a medium of low polarity (chlorobenzene). The reaction of primary amines with active esters such as *para*-nitrophenyl actetate has been studied in apolar media already in quite some detail before.

$$H_2N-R \; + \; \text{ester} \; \xrightarrow[\text{chlorobenzene}]{\text{catalyst}} \; \text{amide} \; + \; \text{phenol} \qquad (1)$$

Polyethers as well as various basic compounds have been found to accelerate the rate of the reaction (5,6). In the following we will first take a look at theoretical results from model studies for the uncatalyzed reaction. As a second step, catalysts for the ester aminolysis will be derived from the theoretical results following various strategies.

2. Theoretical study of the uncatalyzed reaction

The smallest possible model system for the reaction of esters with amines consists of ammonia (**1**) and formic acid (**2**). This system has been studied at several different levels of theory in the gas phase (7,8). In all cases the reaction has been found to proceed through two distincly different reaction pathways, which can best be described as "addition/elimination" and "direct substitution" (Figure 1). The addition/elimination pathway involves formation of an uncharged tetrahedral intermediate **6** through addition of ammonia to the formic acid C-O double bond. Elimination of water (**3**) through transition state **7** yields formamide (**4**). The same products are formed in a single kinetic step from the reactants through transition state **8** for the direct displacement pathway. Somewhat surprisingly, the reaction barriers are comparable for both pathways. In order to verify that both pathways are viable options in more realistic model systems, the reaction between phenyl acetate (**10**) and methyl amine (**9**) has been studied at the same level of theory (Figure 2).

In this larger model system the barriers for the addition/elimination and the direct displacement pathways are reduced by 6.9 and by 15.6 kcal/mol, respectively, relative to the parent system shown in Figure 1. This leaves the direct substitution process as the preferred mode of action in this system. The structure as well as the charge distribution of transition state **12** was consequently used in the development of an AMBER force field representation of this transition state that can be used to investigate the interaction of **12** with potential catalysts. The complex between **12** and tetraglyme depicted in Figure 3 was obtained in this way (8).

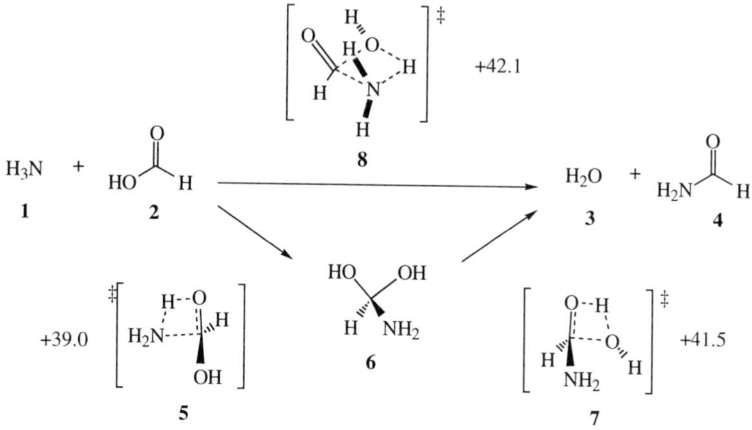

Figure 1. Addition/elimination and direct substitution pathways in the reaction of ammonia with formic acid. Energy differences between reactants and transition states are given in kcal/mol at the MP2/6-31G(d,p)//HF/6-31G(d,p) + ΔZPE level of theory.

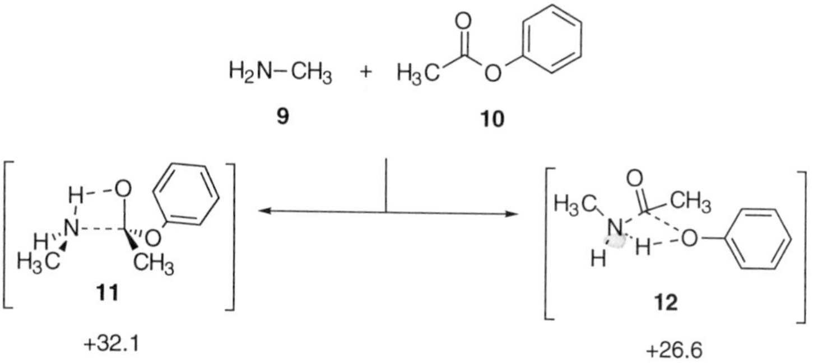

Figure 2. Transition states for the addition/elimination and direct substitution pathways in the reaction of methyl amine (**9**) with phenyl acetate (**10**). Energy differences between reactants and transition states are given in kcal/mol at the MP2/6-31G(d,p)//HF/6-31G(d,p) + ΔZPE level of theory.

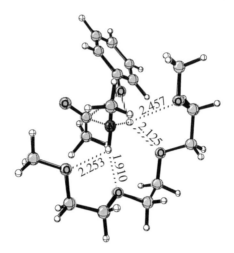

Figure 3. Complex between tetraglyme and transition state **12** as optimized with the AMBER force field.

Using this force field approach, the interaction energy between various catalysts and the transition state can be estimated. The difference between transition state and ground state complexation energy should form a sufficient basis for the evaluation of the catalytic potential of various catalysts. The catalytic potential of a number of polyethers and polyalcohols has been evaluated using this force field approach. The estimated barrier lowering has then been compared to the experimentally determined catalytic rate constant for reaction of butyl amine with *para*-nitrophenyl acetate (Figure 4). Unfortunately, no good correlation between theoretically predicted and experimentally determined catalytic activity could be found. The limited predictive accuracy of this approach apparent from Figure 4 might be due to several factors. First of all, the differences in catalytic activity found experimentally are rather small and the intrinsic inaccuracy of the force field used might simply not allow for a prediction of relative barriers to within 0.5 kcal/mol. Inspection of the ground state complexes formed by polyalcohols also points to the fact that reactions involving catalysts with hydroxyl groups might proceed through a different pathway, eliminating **12** as a good representation of the rate limiting transition state.

This suggests that an accurate prediction of the catalytic activity of a wide range of structurally diverse catalysts is only possible under the condition that the transition structure is reoptimized for each new catalyst (8).

2. Pyridone Catalysis

A second approach for the *de novo* design of catalysts based on the results of quantum chemical calculations involves an analysis of the barrier formation of the uncatalyzed reaction. For the reaction of methyl amine with *para*-nitrophenyl acetate shown in Figure 5, this turns out to be remarkably easy. While formation of the C-N bond and cleavage of the C-O bond starts quite early along the reaction coordinate, no proton transfer can be observed before the transition state. It is only

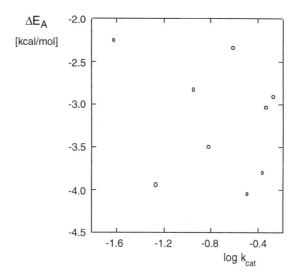

Figure 4. Correlation between the theoretically predicted and experimentally found catalytic efficiency of various polyethers and polyalcohols in the reaction of butyl amine with *para*-nitrophenyl acetate in chlorobenzene at room temperature.

after this point that N to O proton transfer commences, indicating the difficulty of this part of the overall process.

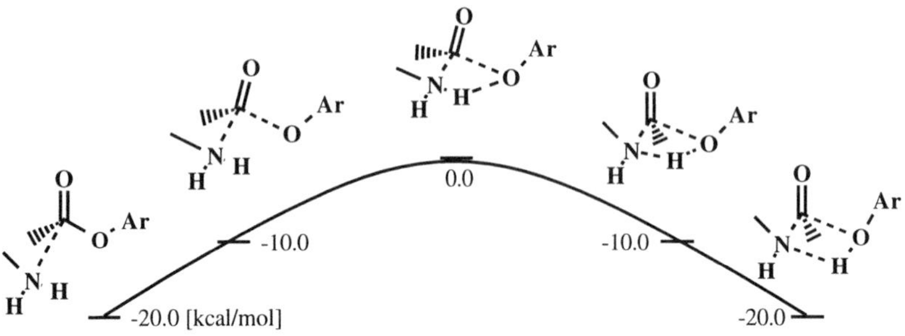

R(C-N) [Å]	2.43	1.80	1.57	1.45	1.37
R(C-O) [Å]	1.41	1.51	1.98	2.30	2.35
R(N-H) [Å]	1.00	1.01	1.03	1.32	2.10

Figure 5. Selected structural data along the direct substitution pathway for the reaction of methyl amine with *para*-nitrophenyl acetate as calculated at the HF/3-21G level of theory.

Inspection of the transition state structure involved in this reaction reveals a rather acute proton transfer angle of 115°, far from the optimum value of 180°. This suggests that inclusion of any structural motive that leads to a more favorable proton transfer angle will lower the reaction barrier and thus function as a catalyst. This goal can be achieved in several ways as shown schematically in Figure 6. Inclusion of a structure X-H, in which X may be any electronegative atom, will lead to a six-membered ring transition structure, in which proton transfer occurs through a more favorable geometry. The ideal situation of collinear proton transfer can, however, only be achieved with a bridging unit C(=X)YH as shown in Figure 6. In this structure the center X acts as a proton donor, while the center Y delivers a proton to the ester leaving group. One additional requirement for these bifunctional catalysts is that the two possible tautomers should not be too different energetically and that the catalyst be neither strongly acidic nor basic in order to avoid unwanted side reactions. One compound that fits all these requirements is 2-pyridone (**13**) that can exist as either tautomer **13a** or **13b**, with **13a** slightly preferred in the gas phase (9,10). The catalytic potential of **13** was evaluated theoretically in the reaction of methyl acetate with methyl amine as model substrates.

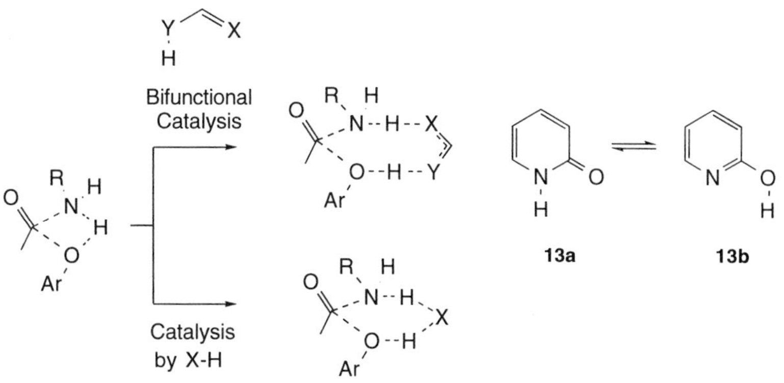

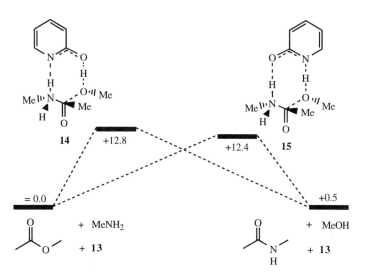

Figure 6. Two different strategies for the improvement of the proton transfer step in the reaction of amines with esters.

Figure 7. Most favorable reaction pathways in the pyridone-catalyzed reaction of methyl acetate with methyl amine as calculated at the Becke3LYP/6-31G(d)//HF/3-21G + ΔZPE level of theory. Energies are given in kcal/mol.

Several different reaction pathways have been investigated (9), and the most favorable pathway shown in Figure 7 corresponds to a fully concerted mechanism, in which the products are formed in a single kinetic step starting from the reactant complex. In this case there is little difference between the catalytic ability of tautomer **13a**, which leads to transition state **15**, and tautomer **13b**, which leads to transition state **14**. Despite the fact that the reaction is concerted in both cases, the double proton transfer involved in these reactions is decidedly asynchronous. While proton transfer from pyridone to the methoxide leaving group is halfway complete in **14** as well as **15**, the second proton transfer from methyl amine back to the pyridone molecule has not yet begun. This indicates that the energetically most demanding step in the pyridone catalyzed reaction of methyl amine with methyl acetate corresponds to proton transfer from pyridones to the methoxy leaving group. Somewhat surprisingly then, pyridones function as acids in this reaction.

Table 1. Theoretically predicted and experimentally found catalytic activity of pyridones in the reaction of n-butyl amine with *para*-nitrophenyl acetate in chlorobenzene at room temperature.

	13[a]	16[a]	17[a]	18[b]
ΔE_A^{calc} [kcal/mol]	=0.0	-1.24	-0.6	-2.4 (R=H)
$k_{cat}^{22°}$ [$M^{-2}s^{-1}$]	5.0 ±13 %	15.1 ±7 %	170 ±32 %	1.4 ±5 % (R=C_6H_{13})
K [M^{-1}]	71 (±30%)	-	-	-
Solubility [mM]	246	3.0	0.3	> 4.0 (R=C_6H_{13})

[a] Exerimental results from Ref. 9. [b] Experimental results from Ref. 11

Based on this observation catalysts with improved activity can be generated by simply enhancing the acidity through introduction of electron withdrawing substituents into the pyridone ring. Reinvestigation of the reaction shown in Figure 7 using several cyano-substituted pyridones supports this hypothesis. The relative barriers given in Table 1 correspond to the reaction pathway through the pyridone N-H-tautomers as in transition structure **15**. Relative activation barriers have been calculated at the Becke3LYP/6-31G(d)//HF/3-21G + ΔZPE level of theory and the barrier for the unsubstituted pyridone **13** has been taken as the point of reference. Introduction of a cyano group at the C4-position of the pyridone ring system as in **17** is predicted to lead to a small barrier lowering of 0.6 kcal/mol, while introduction of a cyano group in the 3-position as in **18** leads to a much larger effect. Introduction of an additional methyl group in the 6-position as in **16** exerts steric effects, which reduces the catalytic efficiency somewhat relative to that for pyridone **18**. In order to test these theoretical predictions the reaction of butyl amine with *para*-nitrophenyl acetate has again been investigated experimentally in chlorobenzene and the pyridones shown in Table 1 (as well as some other pyridones not shown here) have been used as catalysts.

How do the theoretical predictions compare to the experimental results? Introduction of a cyano group does indeed lead to improved catalytic activities, as is most evident for pyridone **17**. The catalytic rate constant for this catalyst is more than 30 times larger than that for the parent pyridone **13**. Unfortunately, the high catalytic potential of pyridone **18** did not materialize in the experiments, indicating a lack of correlation between the theoretically predicted and the experimentally found catalytic efficiencies for substituted pyridones in this study. Moreover, kinetic measurements are complicated through two properties of pyridones that have not been apparent in the modeling studies. The first complication arises from the formation of dimers that bind substantial amounts of the catalyst in a catalytically unreactive dimeric form. That the equilibrium constant for dimer formation could only be determined in some of the cases is related to the low solubility of some of the substituted pyridones in chlorobenzene, which represents the second problem encountered here. This problem is most severe for catalyst **17** that can hardly be dissolved in chlorobenzene at all.

3. Catalysis by Bispyridones

The lack of correlation between the theoretically predicted and the experimentally determined catalytic activity described in Table 1 suggests that the reaction between methyl amine and methyl acetate might not be representative enough for the experiments performed with much more reactive esters. The catalytic potential of pyridone **13** was therefore also studied theoretically for the reaction of methyl amine with *para*-nitrophenyl acetate. Replacement of the methoxy by the *para*-nitrophenoxy leaving group leads to a number of significant changes on the potential energy surface for the pyridone-catalyzed aminolysis reaction. While the reaction depicted in Figure 7 for methyl acetate is complete after one single kinetic step, there are three transition states now on the way from reactant complex **19** to the product complex **25** at the HF/3-21G level of theory (Figure 8). It is only the last of these transition states **24** that corresponds to the proton transfer step found to be of sole importance in the smaller model system.

At all theoretical levels considered the rate limiting step now corresponds to initial attack of the amine nitrogen atom at the carbonyl carbon atom. Whether the transition structure for this step is reactant like (as predicted at the HF/3-21G level of theory) or corresponds more to the zwitterionic structure **21** (at the Becke3LYP/6-31G(d)//HF/3-21G level) cannot be determined unambiguously.

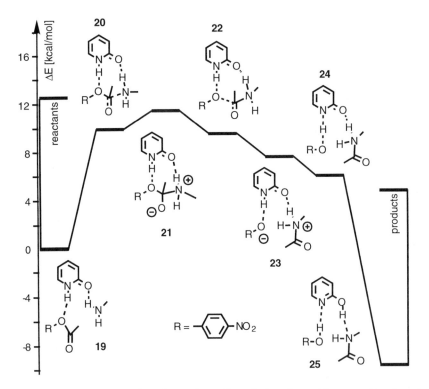

Figure 8. Pyridone-catalyzed reaction of methyl amine with *para*-nitrophenyl acetate at the Becke3LYP/6-31G(d)//HF/3-21G + ΔZPE level of theory.

Irrespective of this latter detail pyridone catalysis turns out to be so efficient in this system that proton transfer does not represent the rate limiting step anymore. That some of the cyano-substituted pyridones turn out to be more powerful catalysts than 2-pyridone **13** itself must therefore be considered purely accidental! To further improve pyridone catalysis of ester aminolysis, it is the energy of the zwitterionic structure **24** that must be lowered. To this end we have considered an extended model system containing two pyridone moieties. The potential energy surface for the aminolysis reaction catalyzed by this "theozyme" constructed from two pyridone units is shown in Figure 9.

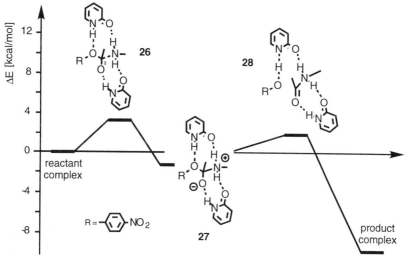

Figure 9. Potential energy surface for the reaction of methyl amine with *para*-nitrophenyl acetate catalyzed by two pyridones at the Becke3LYP/6-31G(d)// HF/3-21G + ΔZPE level of theory.

Stabilization of the tetrahedral zwitterionic structure is most efficient if the second pyridone molecule forms two hydrogen bridges to the carbonyl oxygen and the amine hydrogen atoms of the substrate as shown in Figure 9. Oriented in this fashion the second pyridone molecule lowers the reaction barrier for the initial amine attack on the carbonyl carbon atom by more than 5 kcal/mol. The rate limiting transition state **26** corresponds to an early transition state and the tetrahedral intermediate **27** is a real minimum at all levels of theory considered here. The only other transition state **28** on the way to the product complex corresponds to the hydrogen transfer step. But despite the fact that introduction of the second pyridone molecule was specifically aimed at reducing the energy of the zwitterionic intermediate, the initial transition state for formation of this intermediate remains rate limiting. In order to verify the high catalytic potential of bispyridones predicted by the computational studies described in Figure 9 synthesis of molecular structures that contain two pyridone moieties at the distance and orientation as in structure **26** and that are soluble in apolar media are underway at the moment in this laboratory.

4. Conclusions

Two pathways are predicted to exist for the uncatalyzed reaction between amines and esters by various *ab initio* methods. One of these pathways involves intermediate addition to the C-O double bond followed by a subsequent elimination step. The second pathway corresponds to a direct substitution mechanism, in which the amide products are formed in a single kinetic step from the reactants. The direct substitution mechanism is preferred in those cases, in which the ester carries a good leaving group. The reaction barrier along this reaction pathway is caused by the unfavorable proton transfer geometry. Development of catalysts through utilization of the direct substitution transition state as a "guest" that is bound by a "host" catalyst, turns out to be only moderately successful, probably due to the variable nature of the transition state structure. Efficient catalysts can be designed by specifically improving the proton transfer step involved in the direct substitution pathway. Pyridones turn out to be powerful catalysts due to bifunctional catalysis of the proton transfer step. Theoretical studies involving methyl acetate as the ester component predict the hydrogen transfer step to be still rate limiting and electron deficient pyridones to be better proton donors and thus better catalysts. Experimental verification of this prediction does indeed show some enhancement of the catalytic efficiency. In a more detailed theoretical study using *para*-nitrophenyl acetate as the ester, however, it became obvious that this experimental finding is purely accidental. For reactive esters with good leaving groups pyridone catalysis is efficient enough to render the proton transfer step not rate limiting anymore. Instead, initial attack of the amine nitrogen atom at the carbonyl carbon represents the energetically most demanding step now. The barrier for the initial addition step can be lowered substantially through inclusion of a second pyridone oriented such that multiple hydrogen bonds are formed between the two pyridone molecules and the substrates. On a more general note it must be realized that experimental verification of theoretically predicted catalytic activities will also have to address such down-to-earth issues as solubility, chemical stability, and the state of aggregation of the catalysts.

Acknowledgments. This work was made possible through a grant from the Deutsche Forschungsgemeinschaft. H. Z. gratefully acknowledges additional support by the Fonds der chemischen Industrie and by Prof. H. Schwarz.

References

1. Lerner, R. A.; Benkovic, S. J.; Schultz, P. G. *Science* **1991,** *252,* 659.
2. Schultz, P. G.; Lerner, R. A. *Acc Chem.Res.* **1993,** *26*, 391.
3. Corey, E. J. *Pure Appl. Chem.* **1990,** *62*, 1209.

4. (a) Na, J.; Houk, K. N.; Hilvert, D. *J. Am. Chem. Soc.* **1996**, *118*, 6462. (b) Na, J.; Houk, K. N. *J. Am. Chem. Soc.* **1996**, *118*, 9204. (c) Heine, A.; Stura, E. A.; Yli-Kauhaluoma, J. T.; Gao, C.; Deng, Q.; Beno, B. R.; Houk, K. N.; Janda, K. D.; Wilson, I. A. *Science* **1998**, *279*, 1934.

5. (a) Gandour, R. D.; Walker, D. A.; Nayak, A.; Newkome, G. R. *J. Am. Chem. Soc.* **1978**, *100*, 3608. (b) Hogan, J. C.; Gandour, R. D. *J. Org. Chem.* **1991**, *56*, 3608. (c) Hogan, J. C.; Gandour, R. D. *J. Org. Chem.* **1992**, *57*, 55.

6. (a) Su, C.; Watson, J. W. *J. Am. Chem. Soc.* **1974**, *96*, 1854. (b) Openshaw, H. T.; Whittacker, N. *J. Chem. Soc. C* **1969**, 89. (c) Rony, P. R. *J. Am. Chem. Soc.* **1969**, *91*, 6090.

7. Oie, T.; Loew, G. H.; Burt, S. K.; Binkley, J. S.; MacElroy, R. D. *J. Am. Chem. Soc.* **1982**, *104*, 6169.

8. Zipse, H.; Wang, L.-h.; Houk. K. *Liebigs Ann.* **1996**, 1511.

9. Zipse, H.; Wang, L.-h. *Liebigs Ann.* **1996**, 1501.

10. (a) Beak, P. *Acc.Chem. Res.* **1977**, *10*, 186. (b) For a review of theoretical work on pyridones see: Cramer, C. J.; Truhlar, D. G. *Rev. Comp. Chem.*. **1995**, *6*, 1.

11. Müller, C. *Diplomarbeit,* TU Berlin, **1997**.

Chapter 6

Solvent as Catalyst: Computational Studies of Organic Reactions in Solution

Dongchul Lim, Corky Jenson, Matthew P. Repasky, and William L. Jorgensen

Department of Chemistry, Yale University, New Haven, CT 06520–8107

Quantum and statistical mechanics are being used to study the origin of solvent effects on the rates of organic reactions. Typically, a reaction path is obtained from ab initio molecular orbital calculations in the gas phase. The effects of solvation are then determined for this pathway with free-energy perturbation calculations using Monte Carlo simulations in several solvents. The latter calculations feature a classical description of the intermolecular interactions with partial charges for the reacting system determined from the quantum mechanical wavefunctions. Application to a variety of reactions including substitutions, additions, and pericyclic reactions has revealed insightful details on variations in solvation along the reaction paths. Such information is valuable for the rational design of catalysts.

Reaction rates are subject to substantial medium effects, which can have great practical relevance for organic synthesis as well as enzyme catalysis (*1*). If a transition state is more polar than the reactants, rate increases can be expected with increasing solvent polarity. Ionizations, such as the dissociation of a carboxylic acid or an S_N1 reaction of an alkyl halide, are extreme examples in which two ions can be created, but only in adequately polar surroundings. Intermediate cases involving charge delocalization, such as in an S_N2 reaction of a halide ion with an alkyl halide, can still exhibit rate ranges of 10^{20} upon transfer from the gas phase to polar solvents. Even pericyclic reactions, which were once thought to be relatively immune to solvent effects, upon closer scrutiny and extension of the media to water are found to show rate ranges up to 10^4 (*2,3*). The origin of such variations must contain valuable messages about catalyst design.

In the context of transition state theory (4), the rate constant is expressed by eq. 1 where κ, the transmission coefficient, is near 1 for typical organic reactions, the exponential term represents the relative probabilities of reactants

$$k_f = \kappa(k_B T / h)\exp(-\Delta G^* / k_B T) \qquad (1)$$

and transition state, and ΔG^* is the free energy of activation, which consists of an intrinsic, gas-phase component and the change in free energy of solvation in proceeding from reactants to transition state (eq. 2). Though dynamical effects in

$$\Delta G^* = \Delta G^*_{gas} + \Delta G^*_{sol} \qquad (2)$$

solution can lead to variations in κ, the dominant source of a change in rate between two solvents comes from the change in ΔG^*_{sol}. The free energy of solvation along a reaction path can be computed in several ways. A key decision is whether to treat the solvent as a continuum or to represent the solvent molecules explicitly. The continuum treatments can be quantum mechanical as in self-consistent reaction field (SCRF) calculations or classical as with generalized Born models (5,6). These approaches reflect the electrostatic interactions with the structureless solvent and the energetics of cavity formation for the solutes. They do not provide any information on specific interactions with solvent molecules, which may affect the energetics and be critical to understanding the origin of the rate variations, particularly in hydrogen-bonding solvents.

With explicit solvent molecules, molecular dynamics (MD) or Monte Carlo (MC) simulations are employed and can readily provide appropriate Boltzmann averaging at specified temperatures and pressures (7). A classical description of the solute-solvent and solvent-solvent interactions is normally used and allows for facile changes of solvent. Alternatively, mixed quantum (QM) and classical (MM) mechanics simulations can be performed with the solutes, and perhaps a few solvent molecules, treated by a QM method in the electric field of the MM solvent molecules (8,9). The difficulty with QM/MM calculations coupled with MD or MC simulations is that the QM level has been largely restricted to semiempirical methods, e.g., AM1 or PM3, since the QM calculation has to be performed at least every time the solute moves. Consequently, our approach has been to (a) obtain a gas-phase reaction path and atomic charges for the solutes from ab initio QM calculations, and then (b) perform MC simulations to obtain the change in free energy of solvation along the reaction path.

In the MC calculations, the reacting system is placed in a periodic cell containing 500-1500 solvent molecules. The intermolecular interaction energies, ΔE_{ab}, are represented classically by Coulombic and Lennard-Jones terms between the atoms i in molecule a and the atoms j in molecule b (eq. 3), where the q_i, σ_i, and ε_i are the partial charges and Lennard-Jones parameters. The principal concerns here are that the same reaction path is used in all media and that polarization of the solutes by the solvent is not included. The former issue is

$$\Delta E_{ab} = \sum_i \sum_j \{ q_i q_j e^2 / r_{ij} + 4 \varepsilon_{ij} [(\sigma_{ij}/r_{ij})^{12} - (\sigma_{ij}/r_{ij})^6] \} \qquad (3)$$

diminished by avoiding reactions that feature creation or annihilation of charges, i.e., ionizations or neutralizations, or by obtaining the reaction path from SCRF calculations (10). The latter concern can be moderated by using effective partial charges that compensate for the polarization.

Free Energy Calculations

Direct and Umbrella Sampling. The most straightforward procedure for computing the free energy change along the reaction coordinate ζ is to obtain the population distribution function $g(\zeta)$ by direct sampling. That is, ζ can be allowed to vary during the MD or MC simulation. The frequencies of occurrence of different values of ζ can be accumulated in $g(\zeta)$, which is related to the relative free energy or "potential of mean force" (pmf), $G(\zeta)$, by eq. 4 where c is a constant. The approach is only effective if the range of ζ is not too large and if

$$G(\zeta) = -k_B T \ln g(\zeta) + c \qquad (4)$$

there are not substantial energy barriers, e.g., >5 $k_B T$, along ζ. This is the case for some conformational equilibria (11); however, it is not expected to be applicable for most reactions on both counts. The utility of the approach was much enhanced through the development of umbrella sampling by Valleau and co-workers (12). An artificial biasing function $u(\zeta)$ can be added to the potential energy to extend the range of sampling of ζ, e.g., by flattening the problematic barriers. The true distribution can subsequently be recovered from the biased one, $g'(\zeta)$, via eq. 5,

$$g(\zeta) = c g'(\zeta) \exp[u(\zeta) / k_B T] \qquad (5)$$

where c is a normalization constant. If an adequate range of ζ is still not covered, then a series of simulations can be run with different biasing functions that provide sampling in overlapping regions of ζ. The individual distribution functions can then be spliced together to give the overall $g(\zeta)$ and potential of mean force. Valleau originally called this procedure "multistage sampling", though it is also known as "importance sampling". This approach was used in the earliest solution-phase studies for the S_N2 reaction of $Cl^- + CH_3Cl$ and then for the addition reaction of $OH^- + H_2C=O$ (13-16).

Free Energy Perturbation Theory. The principal alternative arises from Zwanzig's perturbation theory (17). Eq. 6 gives the statistical mechanical expression for the free energy change between systems A and B in terms of their partition functions Z. It can be expanded and rearranged to yield eq. 7, which is the key relationship in free energy perturbation (FEP) calculations. The average

$$\Delta G(A \to B) = -k_B T \ln(Z_B / Z_A) \qquad (6)$$

$$\Delta G(A \to B) = -k_B T \ln < \exp[-(E_B - E_A) / k_B T] >_A \qquad (7)$$

involves the total potential energy difference, $E_B - E_A$, selected by sampling configurations of system A. Although the expression is exact, convergence is slow if A and B are not similar. Consequently, a series of simulations may be run in which A is gradually converted to B and the incremental free energy changes are accumulated. A major advantage is that biasing functions are not needed and it is easier to be confident in the convergence of the calculations.

For reactions, A and B may be the reactants and a transition structure (TS) and they could be interconverted along the minimum energy reaction path (MERP) obtained from the gas-phase QM or SCRF calculations. Other pathways leading to the transition structure can also be considered, though judicious choice will speed convergence of the results. A now standard procedure consists of generating a movie for the gas-phase MERP using reaction-path following procedures (18) and by obtaining atomic charges for each structure by fitting to the ab initio electrostatic potential surface (19). The charges are used for the solute-solvent Coulombic interactions and the remaining non-bonded interactions come from standard Lennard-Jones potentials. The MC program then only has to perturb between the sequence of structures, A_i, that are in the movie. In practice, two free energy increments are normally obtained from one simulation by perturbing from A_i to both A_{i+1} and to A_{i-1}, which is termed "double-wide sampling" (20). Applications of the FEP approach have elucidated the origin of solvent effects on Diels-Alder reactions (19,21), the Claisen rearrangements of allyl vinyl ether (22,23) and chorismate (24), ring openings of cyclopropanones (25), the Mislow-Evans rearrangement of allyl p-tolyl sulfoxide (26), the [2 + 2] cycloaddition of 1,1-dicyanoethylene and methyl vinyl ether (10), prototypical cation-olefin additions and carbenium ion rearrangements (27,28), the epoxidation of olefins by dimethyldioxirane (29), and the 1,3-dipolar cycloaddition of methyl azide and ethene (30).

Results

Several factors that control solvent effects on reaction rates have been reflected in or emerged from the computational studies.

Changes in Polarity. A reaction for which the transition structure (TS) is more polar than the reactants is expected to experience rate enhancements with increasing solvent polarity. Of course, this is the case when ion pairs or zwitterions are created from neutral precursors, as in polar [2 + 2] cycloadditions, which often exhibit rate ratios of ca. 10^4 in going from CCl_4 to acetonitrile (10). However, it is also true when there is a smaller polarity change such as just an increase in dipole moment. Thus, Bell's classical expression (eq. 8) for the free energy of solvation of a dipole with radius R in a continuum with dielectric constant ε embodies a quadratic dependence on the dipole moment, μ (31). Such effects are easiest to identify for reactions in a series of non-polar and dipolar

$$\Delta G_{sol} = -(\mu^2 / 3R^3)\left[\frac{\varepsilon - 1}{2\varepsilon + 1}\right] \qquad (8)$$

78

aprotic solvents; with protic solvents effects due to changes in hydrogen bonding and polarity are often intermingled.

For example, the stereomutation of cyclopropanones is expected to proceed through oxyallyl intermediates (*32,33*). CASSCF/6-31G* calculations for the parent reaction found that the transition structure with μ = 3.52 D is nearly

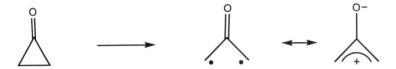

identical to oxyallyl with μ = 3.47 D, which is best represented as the diradical (*25*). However, as expected from the zwitterionic resonance structure, there is a polarity increase from cyclopropanone with μ = 2.79 D. Consistently, MC/FEP calculations for the ring opening revealed more exoergic solvation of oxyallyl than cyclopropanone by ca. 0.6, 1.3, and 2.1 kcal/mol in tetrahydrofuran (THF, ε = 7.6), methylene chloride (ε = 8.9), and acetonitrile (ε = 35.9) relative to propane (ε = 1.7). The corresponding rate increases of a factor of 3-35 at 25 °C are in good accord with experimental data for alkyl substituted cases (*32,33*). The consistency of the computed and observed results strongly supports the CASSCF description of oxyallyl as the diradical.

Another reaction that is accelerated by increasing solvent polarity is the epoxidation of alkenes by dimethyldioxirane (*34*). In this case, the B3LYP/6-31G* transition structure (Figure 1) for the reaction with *cis*-2-butene reflects an increase in dipole moment to 4.9 D from 2.9 D for dimethyldioxirane (DMD). The lengthening of the OO bond from 1.51 Å in DMD to 1.89 Å in the TS is

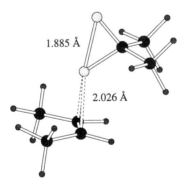

1.885 Å

2.026 Å

Figure 1. Transition structure for epoxidation of *cis*-2-butene by DMD.

accompanied by the oxygens proximal and distal to the alkene becoming more negative by 0.33 and 0.52 e (*29*). For the relatively modest solvent change from methyl acetate (ε = 7.1) to methylene chloride (ε = 8.9), the MC/FEP calculations predicted a rate acceleration by a factor of 6, which is similar to the experimental rate ratio of 3 for the reaction of cyclohexene with DMD in these solvents as 1:1

v:v mixtures with acetone (*35*). Substantially larger effects are observed in protic solvents.

Naturally, there are also cases for which increasing solvent polarity decreases reaction rates. Examples include S_N2 reactions of anions with alkyl halides (*15*), Wittig reactions of phosphonium ylides, and the Mislow-Evans rearrangement of allylic sulfoxides to sulfenates. In these cases, the reactants include a charge-localized ion or for the ylides and sulfoxides, they are highly

polar with a significant contribution from a zwitterionic resonance structure. For the Mislow-Evans rearrangement of allyl *p*-tolyl sulfoxide, RHF/6-31G* calculations found a reduction in dipole moment from 4.3 D for the reactant to 2.7 D for the TS and to 2.2 D for the sulfenate product (*26*). The log of the rate constant is observed to decrease linearly with the solvent polarity index E_T; the deceleration is 14-fold in going from methylcyclohexane to acetonitrile (*36*).

Hydrophobic Effects. The burial of non-polar surface area in aqueous solution is energetically favorable and has wide-ranging influence from protein folding to rates of organic reactions (*3*). The relationship between the change in solvent-accessible surface area (SASA) and free energy of hydration is linear with a proportionality constant of ca. 0.01 cal/mol-Å^2 for reduction in exposed hydrocarbon surface area (*37,38*). Thus, rate accelerations are expected for reactions in water which reduce hydrocarbon SASA. This can be readily achieved by the formation of C-C bonds and led to the discovery of the acceleration of Diels-Alder reactions in water by Rideout and Breslow (*39*).

The change in SASA for formation of a C-C bond can be estimated from models with a 1.4 Å probe radius for a water molecule. For example with geometries from optimizations with the OPLS-AA force field, for 2 $CH_3\cdot \rightarrow$ CH_3CH_3, 2 $H_2C=CH_2 \rightarrow$ cyclobutane, and $H_2C=CH_2$ + *s-trans*-1,3-butadiene → cyclohexene, one obtains ΔSASAs of -112, -130, and -148 Å^2. Thus, since cycloadditions have very product-like transition structures (*40*), the burial of ca. 140 Å^2 of non-polar surface area leads to a 1.4 kcal/mol hydrophobic lowering of ΔG*. This estimate is completely consistent with the experimental changes in free energies of hydration for the reactions of ethene with *s-trans*-1,3-butadiene and isoprene to yield the cyclohexenes, -1.5 and -1.3 kcal/mol (*41*). Hydrophobic enhancements beyond this level require concomitant burial of additional surface area, presumably involving substituents. For the reactions of cyclopentadiene with methyl vinyl ketone (MVK) and acrylonitrile, Rideout and Breslow obtained rate accelerations in water over isooctane of factors of 741 and 31, respectively, which translate to ΔΔG*s of -3.8 and -2.1 kcal/mol. Though a hydrophobic effect of ca. -1.5 kcal/mol accounts for much of these results, the data indicate an additional effect is operative, which may be associated with the functional groups in the dienophiles.

For a purely hydrophobic case, MC/FEP calculations were performed for the dimerization of cyclopentadiene (21). From the reactants to the TS with 2.2-Å

lengths for the forming CC bonds, the change in free energy of hydration was -1.8 ± 0.3 kcal/mol, and there was negligible difference in continuing to the product. The computed ΔSASA for reactants to TS is -163 Å^2, which is only 10% more negative than the value above for the parent reaction. Thus, all of the computational results are consistent with the hydrophobic component of the acceleration for typical Diels-Alder reactions in water being near 1.5 kcal/mol or a factor of 10-20 at 25 °C. Experimentally, kinetic measurements for cyclopentadiene dimerization in water are difficult owing to problems with solubility and volatility. van der Wel et al. circumvented this problem by studying the Diels-Alder reaction of acridizinium bromide with cyclopentadiene; the rate in water is 3-12 times faster than the rates in a variety of polar aprotic solvents (42), which is in-line with the above estimates.

Variations in Hydrogen Bonding. Variations in numbers and strengths of hydrogen bonds have consistently emerged from the MC simulations as a key to understanding rate changes in polar, protic solvents. For the S_N2 reaction of Cl⁻ + CH_3Cl, it was found that hydration converts the gas-phase double-well pmf to an almost unimodal profile and much enhances the activation barrier through weakening of 6-7 hydrogen bonds in progressing to the transition structure (13,14). Hydrogen-bond weakening accompanying charge delocalization is also responsible for the barrier for the addition reaction of $H_2C=O$ + OH⁻ in water (16).

In later work, MC/FEP calculations revealed that the parent Claisen rearrangement is accelerated in water by the change from one hydrogen bond for the ether oxygen of the reactant to two as it becomes partially enolic in the transition structures (22,23). This led to the prediction that catalysts for Claisen rearrangements should "incorporate two or more hydrogen-bond-donating groups positioned to interact with the oxygen in the transition state" (22). The prediction was borne out by subsequent crystal structures for chorismate mutase with a bound transition-state analog (43). The uncatalyzed rearrangement of chorismate to prephenate is also notable for its striking 100-fold acceleration upon transfer from methanol to water. MC/FEP calculations reproduced the effect (24). Its origin was traced to an enhanced population of the diaxial conformer of chorismate in water, which arises largely from a unique water molecule acting as a double hydrogen-bond donor to the C4 hydroxyl group and the side-chain

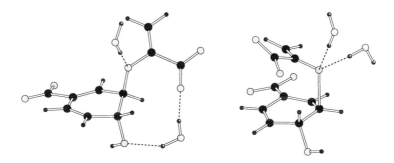

chorismate   prephenate

carboxylate (Figure 2). The need for explicit representation of the solvent molecules is readily apparent in these studies.

Figure 2. Left - the diaxial conformer of chorismate with the single hydrogen bond to the ether oxygen and the bridging water molecule. Right - the chorismate TS with two hydrogen-bonded water molecules on the ether oxygen. Structures from MC simulations in periodic cells containing 1647 water molecules.

Returning to cycloadditions, the 741-fold (3.8 kcal/mol) aqueous acceleration for the Diels-Alder reaction of cyclopentadiene and MVK was attributed to arise in about equal measure from hydrophobic effects and from strengthening of the two hydrogen bonds to the carbonyl oxygen in progressing to the more polarized transition structure (*19,21*). Gas-phase RHF/6-31G* calculations found a 2.0 kcal/mol strengthening of the optimal C=O⋯H_2O hydrogen bond in going from the MVK-water complex to the TS-water complex (*44*). The difference is reduced to 1.5 kcal/mol with acrylonitrile as the dienophile. Thus, the lessened rate acceleration in the latter case follows from the diminished polarization and the fact that nitriles form one hydrogen bond in water as compared to two for ketones. Similarly, the larger rate enhancements observed by Engberts and co-workers with naphthoquinones as dienophiles follow from their still greater hydrogen-bond accepting ability (*2*). On the other hand, studies of solvent effects on 1,3-dipolar cycloadditions of azides and norbornene or ethene indicate that enhanced hydrogen-bonding is not a factor and the limited accelerations in water arise from hydrophobic effects (*30,45*). The problem is that

82

the 1,3-dipoles with zwitterionic resonance structures are similar in polarity to the transition structures. In order to obtain rate accelerations in protic solvents, one again needs to exploit hydrogen bonding to polarized activating groups in the TS. The most likely candidates are electron-deficient dipolarophiles reacting with electron-rich 1,3-dipoles.

Alkene epoxidation by DMD provides a final illustration of a reaction that is promoted by protic solvents. For instance, a 13-fold acceleration is observed for the DMD epoxidation of *p*-methoxystyrene in going form pure acetone to aqueous acetone with a 0.64 mol fraction of water (*46*). For the pure solvents, the MC/FEP calculations for the reaction of DMD with *cis*-2-butene yielded predicted rate ratios of 81:6:1 for methanol, methylene chloride, and methyl acetate (*29*). There is again striking enhancement of hydrogen bonding in going from DMD to the TS; RHF/6-31G* results predict a 5 kcal/mol strengthening of the optimal hydrogen bond with a water molecule. The MC simulations also revealed that in methanol at 25 °C, the TS participates in two hydrogen bonds with strengths greater than 2.5 kcal/mol (Figure 3), while DMD only has an average of 0.5 interaction in this range. The build-up of negative charge on the DMD oxygens in the TS, which was mentioned above, is clearly responsible for these findings. The results also indicate that placement of a hydrogen-bond donating group in the substrate to interact with both dioxirane oxygens in the TS would yield high reactivity and a basis for alkene facial selectivity.

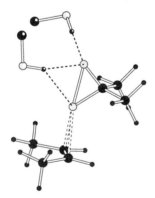

Figure 3. TS for epoxidation of *cis*-2-butene by DMD with the hydrogen-bonded methanol molecules, as extracted from a MC configuration in a periodic box containing 395 solvent molecules.

Other Work. Related MD and MC calculations have been carried out for a variety of organic reactions by several groups; some recent examples include the following. Supercritical water was considered as a medium for the $Cl^- + CH_3Cl$ reaction by Balbuena et al.; at 587 K and a water density of only 0.405 g cm^{-3} the reaction still has a similar pmf as for 298 K and 1 g cm^{-3} (*47*). This S_N2 reaction was also the test case for the earliest combined QM(AM1)/MM calculation with explicit representation of the solvent and MD sampling (*48*). QM/MM methodology has been applied as well to the S_N1 reaction of t-butyl chloride in water by Hartsough and Merz; this case is particularly challenging owing to the evolution of the electrostatic interactions as the solvolysis proceeds (*49*). Previously, Peng and Merz performed standard MD/FEP calculations for the reaction of $OH^- + CO_2 \rightarrow HCO_3^-$ in water as a preamble to further studies of carbonic anhydrase (*50*). Gao and Furlani have used QM/MM calculations to reinvestigate Diels-Alder reactions and the Claisen rearrangement of allyl vinyl ether (*51,52*). A simplified QM(AM1)/MM method was also applied to the latter process (*53*); all studies have found that the effects of hydration are sensitive to the geometry of the TS with the overly tight TS from the AM1 calculations leading to underestimation of the rate enhancement. The rearrangement of chorismate was studied by Davidson et al. with MC/FEP calculations in water (*54*). Relative to the diequatorial form, the diaxial conformer and TS were computed to be better hydrated by about 7 and 20 kcal/mol, as in the prior work (*24*). Gao and Xia went on to study the Menshutkin reaction of $NH_3 + CH_3Cl$ and the Kemp decarboxylation of 3-carboxybenzisoxazole (*55-57*). The latter reaction, which is particularly sensitive to solvent variation, was also treated with MC/FEP calculations by Zipse et al (*58*). Both studies found the expected hydrogen-bond weakening in progressing to the charge-delocalized transition structure, which slows the reaction in protic solvents. In other work, Zipse has examined solvent effects on several radical reactions (*59-61*). The $S_{RN}2$ reaction of Cl^- + CH_2CH_2Cl was found to have a lower activation energy, but similar hydration effects, as the S_N2 reaction of ethyl chloride. An alternative $S_{RN}2'$ pathway was investigated and found to be competitive in the gas-phase, but disfavored by solvation.

From this overview, it is apparent that there has been great progress in the quantitative theoretical treatment of organic reactions in solution. The same methodology is also being applied in a rapidly expanding opus of simulations of enzymatic reactions (*62,63*). These studies are providing atomic details on the participation of solvent molecules and protein side chains in chemical transformations. Such information is essential in advancing the understanding and control of solvent effects on reaction rates and catalysis.

Acknowledgments. Gratitude is expressed to the National Science Foundation for support and to the co-workers on the described projects, who are listed in the references.

Literature Cited

(1) Reichardt, C. *Solvents and Solvent Effects in Organic Chemistry*; VCH: Weinheim, 1990.

(2) Engberts, J. B. F. N. *Pure & Appl. Chem.* **1995**, *67*, 823.

(3) Li, C.-J.; Chan, T.-K. *Organic Reactions in Aqueous Media*; Wiley: New York, 1997.

(4) Truhlar, D. G.; Garrett, B. C.; Klippenstein, *J. Phys. Chem.* **1996**, *100*, 12771.

84

(5) Cramer, C. J.; Truhlar, D. G. *Rev. Comp. Chem.* **1995**, *6*, 1.
(6) Orozco, M.; Alhambra, C.; Barril, X.; López, J. M.; Busquets, M. A.; Luque, F. J. *J. Mol. Model.* **1996**, *2*, 1.
(7) Jorgensen, W. L. *Acc. Chem. Res.* **1989**, *22*, 184.
(8) Åqvist, J.; Warshel, A. *Chem. Rev.* **1993**, *93*, 2523.
(9) Cramer, C.J.; Truhlar, D. G. In *Solvent Effects and Chemical Reactivity*; Tapia, O., Bertran, J., Eds.; Kluwer: Dordrecht, 1995.
(10) Lim, D.; Jorgensen, W. L. *J. Phys. Chem.* **1996**, *100*, 17490.
(11) Jorgensen, W. L. *J. Phys. Chem.* **1983**, *87*, 5304.
(12) Valleau, J. P.; Torrie, G. M. *A Guide to Monte Carlo for Statistical Mechanics,* Berne, B. J., Ed.; Statistical Mechanics, Part A; Plenum: New York, 1977; 169-194.
(13) Chandrasekhar, J.; Smith, S. F.; Jorgensen, W. L. *J. Am. Chem. Soc.* **1984**, *106*, 3049.
(14) Chandrasekhar, J.; Smith, S. F.; Jorgensen, W. L. *J. Am. Chem. Soc.* **1985**, *107*, 154.
(15) Chandrasekhar, J.; Jorgensen, W. L. *J. Am. Chem. Soc.* **1985**, *107*, 2974.
(16) Madura, J. D.; Jorgensen, W. L. *J. Am. Chem. Soc.* **1986**, *108*, 2517.
(17) Zwanzig, R. W. *J. Chem. Phys.* **1954**, *22*, 1420.
(18) Gonzalez, C.; Schlegel, H. B. *J. Phys. Chem.* **1990**, *94*, 5523.
(19) Blake, J. F.; Jorgensen, W. L. *J. Am. Chem. Soc.* **1991**, *113*, 7430.
(20) Jorgensen, W. L.; Ravimohan, C. *J. Chem. Phys.* **1985**, *83*, 3050.
(21) Jorgensen, W. L.; Blake, J. F.; Lim, D.; Severance, D. L. *J. Chem. Soc. Faraday Trans.* **1994**, *90*, 1727.
(22) Severance, D. L.; Jorgensen, W. L. *J. Am. Chem. Soc.* **1992**, *114*, 10966.
(23) Severance, D. L.; Jorgensen, W. L. In *Structure and Reactivity in Aqueous Solution*; Cramer, C. J., Truhlar, D. G., Eds.; ACS Symposium Series 568; American Chemical Society: Washington, DC, 1994; pp 243.
(24) Carlson, H. A.; Jorgensen, W. L. *J. Am. Chem. Soc.* **1996**, *118*, 8475.
(25) Lim, D.; Hrovat, D. A.; Borden, W. T.; Jorgensen, W. L. *J. Am. Chem. Soc.* **1994**, *116*, 3494.
(26) Jones-Hertzog, D. K.; Jorgensen, W. L. *J. Am. Chem. Soc.* **1995**, *117*, 9077.
(27) Schreiner, P. R.; Severance, D. L.; Jorgensen, W. L.; Schleyer, P. v. R.; Schaefer, H. F., III *J. Am. Chem. Soc.* **1995**, *117*, 2663.
(28) Jenson, C.; Jorgensen, W. L. *J. Am. Chem. Soc.* **1997**, *119*, 10846.
(29) Jenson, C.; Liu, J.; Houk, K. N.; Jorgensen, W. L., *J. Am. Chem. Soc.* **1997**, *119*, 12982.
(30) Repasky, M. P.; Jorgensen, W. L. *J. Chem. Soc., Faraday Disc.* **1998**, *110*, 000.
(31) Bell, R. P. *Trans. Faraday Soc.* **1931**, *27*, 797.
(32) Sclove, D. B.; Pazos, J. F.; Camp, R. L.; Greene, F. D. *J. Am. Chem. Soc.* **1970**, *92*, 7488.
(33) Cordes, M. H.; Berson, J. A. *J. Am. Chem. Soc.* **1992**, *114*, 11010.
(34) Murray, R. W. *Chem. Rev.* **1989**, *89*, 1187.
(35) Murray, R. W.; Gu, D. *J. Chem. Soc., Perkin Trans 2* **1993**, 2203.
(36) Bickart, P.; Carson, F. W.; Jacobus, J.; Miller, E. G.; Mislow, K. *J. Am. Chem. Soc.* **1968**, *90*, 4869.
(37) Eisenberg, D.; McLachlan, A. D. *Nature* **1986**, *319*, 199.

(38) McDonald, N. A.; Carlson, H. A.; Jorgensen, W. L. *J. Phys. Org. Chem.* **1997**, *10*, 563.

(39) Rideout, D. C.; Breslow, R. *J. Am. Chem. Soc.* **1980**, *102*, 7816.

(40) Houk, K. N.; Li, Y., Evanseck, J. D. *Ang. Chem., Int. Ed. Engl.* **1992**, *31*, 682.

(41) Hine, J.; Mookerjee, P. K. *J. Org. Chem.* **1975**, *40*, 292.

(42) van der Wel, G. K.; Wijene, J. W.; Engberts, J. B. F. N. *J. Org. Chem.* **1996**, *61*, 9001.

(43) For a review, see: Ganem, B. *Angew. Chem., Int. Ed. Engl.* **1996**, *35*, 936-945.

(44) Blake, J. F.; Lim, D.; Jorgensen, W. L. *J. Org. Chem.* **1994**, *59*, 803.

(45) Wijnen, J. W.; Steiner, R. A.; Engberts, J. B. F. N. *Tetrahedron Lett.* **1995**, *36*, 5389.

(46) Baumstark, A. L.; McCloskey, C. J. *Tetrahedron Lett.* **1987**, *28*, 3311.

(47) Balbuena, P. B.; Johnston, K. P.; Rossky, P. J. *J. Am. Chem. Soc.* **1994**, *116*, 2689.

(48) Bash, P. A.; Field, M. J.; Karplus, M. *J. Am. Chem. Soc.* **1987**, *109*, 8092.

(49) Hartsough, D. S.; Merz, K. M., Jr. *J. Phys. Chem.* **1995**, *99*, 384.

(50) Peng, Z.; Merz, K. M., Jr. *J. Am. Chem. Soc.* **1993**, *115*, 9640.

(51) Furlani, T. R.; Gao, J. *J. Org. Chem.* **1996**, *61*, 5492.

(52) Gao, J. *J. Am. Chem. Soc.* **1994**, *116*, 1563.

(53) Kaminski, G. A.; Jorgensen, W. L. *J. Phys. Chem. B* **1998**, *102*, 1787.

(54) Davidson, M. M.; Guest, J. M.; Craw, J. S.; Hillier, I. H.; Vincent, M. A. *J. Chem. Soc., Perkin Trans. 2* **1997**, 1395.

(55) Gao, J. *J. Am. Chem. Soc.* **1991**, *113*, 7796.

(56) Gao, J.; Xia, X. *J. Am. Chem. Soc.* **1993**, *115*, 9667.

(57) Gao, J. *J. Am. Chem. Soc.* **1995**, *117*, 8600.

(58) Zipse, H.; Apaydin, G.; Houk, K. N. *J. Am. Chem. Soc.* **1995**, *117*, 8608.

(59) Zipse, H. *Angew. Chem., Int. Ed. Engl.* **1994**, *33*, 1985.

(60) Zipse, H. *J. Am. Chem. Soc.* **1994**, *116*, 10773.

(61) Zipse, H. *J. Am. Chem. Soc.* **1995**, *117*, 11798.

(62) Åqvist, J.; Warshel, A. *Chem. Rev.* **1993**, *93*, 2523.

(63) Kollman, P. A. *Acc. Chem. Res.* **1996**, *29*, 461.

ORGANOMETALLIC CATALYSTS

Chapter 7

Molecular Reaction Modeling from Ab-Initio Molecular Dynamics

Peter E. Blöchl[1], Hans Martin Senn[2], and Antonio Togni[2]

[1]IBM Research Division, Zurich Research Laboratory, CH-8803 Rüschlikon, Switzerland
[2]Swiss Federal Institute of Technology ETH, Laboratory of Inorganic Chemistry, CH-8092 Zürich, Switzerland

A tutorial-like introduction to the methodology of transition state search within the framework of AIMD is given. We describe how to locate transition states, how to obtain dynamical reaction paths, and we discuss free energy integration. As an illustrating example we present results for the catalytic hydroamination of alkenes using $\{NiCl(PH_3)_2\}^+$ complexes.

Ab-initio molecular dynamics (AIMD), invented 1985 by Car and Parrinello (1), has established itself as a useful tool for the study of catalytic processes. The methodology used in connection with AIMD takes advantage of a number of concepts which originated in solid state physics and adapts others from chemistry. While being based on state-of-the-art density functional methodology (2, 3), AIMD opens a wide range of options that have been accessible in the past only in combination with simpler semi-empirical or molecular mechanics methods. It thus expands the spectrum of theoretical chemistry into new directions.

The calculations presented here are based on the projector-augmented wave (PAW) method (4), which has been developed to make the full wave functions available within the AIMD approach, avoiding the usual pseudopotential approximations (5, 6). Quantities depending on the charge density near the nucleus, such as electric field gradients, are therefore directly accessible with high accuracy (Petrilli, H. M.; Blöchl, P. E.; Blaha, P.; Schwarz, K. *Phys. Rev. B* **1998**, in press).

We describe in a tutorial-like way aspects of the methodology related to transition states as used within the framework of AIMD. We discuss how to locate transition states, how to analyze the dynamics of the reaction event, and how to perform free energy integrations. These methods have been used earlier in connection with molecular mechanics, semiempirical, and AIMD simulations, and they belong now, in variants, to the toolset of state-of-the-art AIMD simulations performed by ourselves and others.

Methodology

All our calculations are based on density functional theory (2, 3). In particular, the results presented here are obtained using the gradient corrections due to Becke (7) and Perdew (8).

We use the general framework of AIMD (1), termed the "Car–Parrinello method" after its inventors. The approach used here is based on the fictitious Lagrangian methodology (1, 9), in which the electron wave functions are propagated in time via a differential equation. If properly done, the wave functions follow the nuclei adiabatically on, or in practice very close to, the Born–Oppenheimer surface, that is with the electron wave functions in the instantaneous ground state. Both atoms and wave functions follow dynamical equations of motion:

$$M_i \ddot{R}_i = -\frac{\partial E}{\partial R_i} \tag{1}$$

$$m_\Psi \big| \ddot{\Psi}_n \big\rangle = -H \big| \Psi_n \big\rangle + \sum_m \big| \Psi_m \big\rangle \Lambda_{m,n}$$

Here, E is the Kohn–Sham total energy functional, which depends on the atomic positions R_i and on the one-particle electron wave functions $|\Psi_n\rangle$; M_i are the nuclear masses, and m_Ψ is the fictitious mass of the wave functions. H is the one-particle Hamiltonian acting on the wave functions, and $\Lambda_{m,n}$ are the Lagrange parameters for the constraint that all wave functions are orthonormal. The analogy between the two differential equations becomes evident, when we express the Hamiltonian via a partial derivative $H|\Psi_n\rangle = \partial E / \partial\langle\Psi_n|$ of the Kohn–Sham total energy functional.

The advantage of this approach lies in the absence of self-consistency cycles as the atoms move. Other variants of AIMD (10) use self-consistency cycles, while still exploiting the information of the previous steps. Those approaches allow larger time steps for the atom dynamics, but at the cost of self-consistency cycles.

The AIMD approach can be implemented in various electronic structure methods. Most commonly, pseudopotentials are used (5, 6). The electronic structure problem in our work, on the other hand, has been solved using the PAW method (4). It is an all-electron method in the sense that the full, valence and core, wave functions and densities are constructed to evaluate the total energy and forces.

However, we use the frozen core approximation (11), that is, we import the core densities of an isolated atom into the molecular calculation. This turns out to be a minor approximation, if the exact core densities are used, because then the errors in the total energy are only of second order in the deviation of the true and the atomic core density. Note, however, that the frozen core states cannot be identified on a one-to-one basis with one-particle states of the molecule. If one-particle states are needed, the consistent approach is to diagonalize the Hamiltonian, whose matrix elements are obtained from core and valence states resulting from the frozen core calculation. Such a transformation mixes individual states, but leaves the total energy, charge density, and forces untouched.

The main characteristics of the PAW method are the choice of the basis functions and the procedure to calculate total energy and forces. We use basis functions composed of plane waves, which are augmented by atomic "orbitals" near the nucleus in order to incorporate the proper nodal structure of the wave functions. A wave function takes the form

$$\big| \Psi_n \big\rangle = \big| \tilde{\Psi}_n \big\rangle + \sum_{R,l,m,q} \big[\big| \phi_{R,l,m,q} \big\rangle - \big| \tilde{\phi}_{R,l,m,q} \big\rangle \big] \big\langle \tilde{p}_{R,l,m,q} \big| \tilde{\Psi}_n \big\rangle \tag{2}$$

where $\left| \tilde{\Psi}_n \right\rangle$ are the so-called pseudo wave functions, which are the variational parameters, that is they contain the expansion coefficients $c_{G,n}$ as $\tilde{\Psi}_n(\mathbf{r}) = \sum_G e^{iGr} c_{G,n}$. The true wave function is obtained by adding and subtracting one-center expansions in terms of partial waves $\left| \phi_{R,l,m,q} \right\rangle$ and $\left| \tilde{\phi}_{R,l,m,q} \right\rangle$, which are atom-centered radial functions multiplied by spherical harmonics. The index R denotes a nuclear site, l, m are angular momentum indices, and q is an additional counter. In the following, we will collect the indices R, l, m, q into one integer counter i or j. The all-electron partial waves $\left| \phi_i \right\rangle$ are obtained from an atomic calculation and contain the entire nodal structure of the real wave functions. The pseudo partial waves $\left| \tilde{\phi}_i \right\rangle$ are constructed to be pairwise identical to the all-electron partial waves beyond a certain radius – about the size of the covalent radius – and smooth inside. The coefficients of the two expansions are obtained as the scalar product of the pseudo wave function with so-called projector functions $\left| \tilde{p}_i \right\rangle$, which obey the requirement $\left\langle \tilde{p}_i \middle| \tilde{\phi}_j \right\rangle = \delta_{i,j}$.

The total energy functional in the PAW method is expressed in such a way that integrations are done partly in a Fourier representation and partly on radial grids in a spherical harmonics expansion (and partly analytically). All integrals obtained on a Fourier grid are smooth, and the combined error of the one-center expansions is rapidly convergent. For a more detailed description see ref. (4).

A plane wave calculation always introduces periodic images as a result of discretizing the Fourier space. This feature is of course desired in solid state calculations, because infinitely extended systems can be described, but it is problematic, if isolated molecules with a large dipole moment or even charged molecules are to be treated. Several remedies for this problem are in use (12-15). In our work, we correct for the interaction between periodic images (15) by creating a model density composed of spherical, atom-centered Gaussians in such a way that the model density closely reproduces the long-range electrostatic potential of the true density. Once the model density is obtained, the electrostatic interaction between periodic images can be subtracted using standard Ewald summation techniques. We thus obtain total energies and forces from truly isolated molecules.

Constraints

The method used to locate transition states, described in the next section, uses constraint forces to move the system "up-hill" towards a transition state and to stabilize the system at the transition state. Therefore, we summarize here the main features of the methodology (16).

A constraint is defined by an equation $X(R_1, \ldots, R_N) = c$, which defines a hypersurface in coordinate space. The aim is to obtain equations of motion such that the system, given that it fulfills the constraint equation at $t = 0$, will also obey it in future. In order to keep the motion on the hypersurface defined by the constraint equation, additional forces must act on the system to avoid deviations. These forces shall perturb the system as little as possible, and therefore are chosen perpendicular to this hypersurface, i.e. as $(\lambda \nabla_{R_1} X, \ldots, \lambda \nabla_{R_N} X)$, where λ is the Lagrange parameter, which is adjusted such that the atomic motion remains on the hypersurface. This force of constraint does not change the energy of the system, because it acts strictly perpendicular to the atomic velocities.

The above sketches the basic idea of constraints. The remainder of this section will be devoted to the description of the actual implementation.

The equations of motion for the atoms are, generalized to n independent constraints,

$$M_i \ddot{R}_i = -\frac{\partial E}{\partial R_i} + \sum_{j=1}^{n} \frac{\partial X_j}{\partial R_i} \lambda_j \tag{3}$$

When we discretize the equations of motion using the Verlet algorithm (17), which replaces the differentials by the respective discrete differential quotient, we obtain

$$R_i(t + \Delta) = 2 R_i(t) - R_i(t - \Delta) - \frac{\partial E}{\partial R_i} \frac{\Delta^2}{M_i} + \sum_{j=1}^{n} \frac{\partial X_j}{\partial R_i} \frac{\Delta^2}{M_i} \lambda_j \tag{4}$$

where Δ is the time step. If not specified otherwise, the derivatives are taken at $R_i(t)$. The Lagrange parameters are determined from the requirement that the positions fulfill the constraint equations for $t + \Delta$, i.e.

$$X_j(R_1(t + \Delta),\ldots,R_N(t + \Delta)) = c_j \tag{5}$$

In general, this is still a nonlinear equation which needs to be solved iteratively. This is done in the following way: We first introduce $\overline{R}_i$ as

$$\overline{R}_i = 2 R_i(t) - R_i(t - \Delta) - \frac{\partial E}{\partial R_i} \frac{\Delta^2}{M_i} \tag{6}$$

so that the atomic positions for the next time step are

$$R_i(t + \Delta) = \overline{R}_i + \sum_j \frac{\partial X_j}{\partial R_i} \frac{\Delta^2}{M_i} \lambda_j \tag{7}$$

We start with an ansatz $\lambda_j = 0$ for the Lagrange parameters and linearize the constraint equation (5). The resulting system of equations

$$\sum_k \left(\sum_i \frac{\partial X_j}{\partial R_i} \frac{\Delta^2}{M_i} \frac{\partial X_k}{\partial R_i} \right) \delta \lambda_k = c_j - X_j(R_1(t + \Delta),\ldots,R_N(t + \Delta)) \tag{8}$$

can be solved for an improved estimate of the Lagrange parameters $\lambda_j + \delta \lambda_j$, which is re-inserted into equation (7). This process is repeated until self-consistency.

For non-linear constraints, we simplify the procedure by noting that the Verlet algorithm has itself only limited accuracy. As it introduces errors of order t^4 in the positions, we use in practice a Taylor expansion of the constraint equations up to second order in $|R(t + \Delta) - R(t)|$.

Locating Transition States

Determining transition states is the key to understanding reaction mechanisms. The free energy of the transition state largely determines the reaction rates. Furthermore, as noted by Pauling, an enzyme's action results from "binding transition states", thus reducing the energy of activation, and the same is true for most catalytic systems. This statement touches not only the energetics, but it points to the fact that the atomic and electronic structure of a transition state provides us with valuable insight on how to find different, potentially more effective, catalysts.

Considerable effort has been spent in finding ways to locate transition states effectively. For related works, we refer the reader to the lucid review by Schlegel (18) and more recent papers (19-21), including references cited therein. Here, we present an approach that is particularly suitable within the AIMD methodology (22).

Within the AIMD approach, moving atoms is relatively efficient, because force calculations are not time-consuming in a plane-wave-based method and, more importantly, because there is no self-consistency cycle in every time step. On the other hand, the atomic motion must be sufficiently continuous to avoid deviations of the wave functions from the Born–Oppenheimer surface, which would make expensive self-consistency cycles necessary.

These features are exploited in a natural way to find transition states. The approach corresponds to a dynamical simulation with external forces that drive the system across the barrier. The external forces are obtained from a constraint, which closely resembles the reaction coordinate. As an example, the constraint chosen for a simple bond-breaking or bond-forming reaction between two atoms A and B would have the form $|R_A - R_B| = c$. The value of the reaction coordinate c is continuously varied with time so that the system proceeds from the initial configuration to the transition state.

While the system moves from the initial configuration across the barrier, all atoms are free to relax under the given constraint. If the motion were sufficiently slow, the exact transition state would be obtained as the maximum of the total energy as a function of the reaction coordinate. In practice, however, the ascent to the transition state is too fast for keeping all degrees of freedom relaxed. Therefore, we fix the value of the constraint at the maximum of the total energy and relax the system fully under this condition. The constraint force will now have a finite value, whose sign indicates on which side of the barrier we have stopped. We then continue the approach to the transition state from this point onward. This iterative process has converged when the force acting on the constraint is sufficiently small. Finally, the position of the transition state is refined by interpolating energy and constraint force from the last few relaxation steps.

As we did not evaluate the vibrational frequencies, how can we be sure that we have indeed obtained a transition state of first order? This is demonstrated in the following. We need to show that all forces at the transition state thus obtained vanish and, secondly, that there is exactly one imaginary frequency.

Obviously, the forces vanish: The system is fully relaxed under the constraint condition, that is all forces perpendicular to the reaction coordinate vanish. As the energy is maximized along the reaction coordinate, also the force acting in parallel to the reaction coordinate vanishes.

In order to show that there is exactly one imaginary frequency, we expand the total energy E up to second order in the deviation from the transition state at $R_{i,T}$ and we expand the constraint X up to first order.

$$E(R_1,\ldots,R_N) = E_T + \tfrac{1}{2}\sum_{i,j}(R_i - R_{i,T})d_{i,j}(R_j - R_{j,T}) + O\big((R_j - R_{j,T})^3\big) \qquad (9)$$

$$X(R_1,\ldots,R_N) = \sum_{i,j}b_i(R_i - R_{i,T}) + O\big((R_j - R_{j,T})^2\big)$$

where E_T is the energy of the transition state. (The reason for going only to first order in the constraint lies in the fact that the constraint forces are multiplied by the Lagrange parameters, which depend themselves to first order on the deviation from the transition state.)

The equations of motion near the transition state are

$$M_i \ddot{R}_i = -\sum_j d_{i,j}(R_j - R_{j,T}) + b_i \lambda \qquad (10)$$

where $d_{ij} = \partial^2 E / (\partial R_i\, \partial R_j)$ is the Hessian matrix, and $b_i \lambda = (\partial X / \partial R_i)\lambda$ is the force of constraint.

A variable transformation simplifies these equations of motion to

$$\ddot{S}_i = -\sum_j D_{i,j} S_j + B_i \lambda \tag{11}$$

where $S_i = \sqrt{M_i}(R_i - R_{i,\mathrm{T}})$, $D_{i,j} = d_{i,j} / \sqrt{M_i M_j}$ is the dynamical matrix, and $B_i = b_i / \sqrt{M_i}$ is the transformed constraint vector. The eigenvalues of the dynamical matrix are the squared frequencies, and the eigenvectors divided by the square-root of the masses are the vibrational eigenmodes.

As the energy of the transition state is a local maximum along the reaction coordinate, the dynamical matrix cannot be positive definite, so it must contain at least one negative eigenvalue or, in other words, at least one imaginary frequency.

Let us now show that there is at most one imaginary frequency. If there were more than one imaginary frequency, we would find at least two eigenvectors U_i and V_i of the dynamical matrix with negative eigenvalues, i.e. imaginary frequencies, of which at least one had a projection onto B. Let us begin with this assumption and show that such a system is unstable and does not result in a solution of the transition state search. Out of two vectors U and V we can construct at least one vector $W_i = U_i \sum_j V_j B_j - V_i \sum_j U_j B_j$ that is orthogonal to B. This vector lies within the constraint hypersurface. Evaluating the energy along this direction, that is for $S_i = \alpha W_i$, where α is an arbitrary small parameter, we find that it is negatively curved, hence unstable. When the system is relaxed along all unconstrained distortions, the system will therefore move away from such a state.

A converged result can thus only be obtained for the true transition state with exactly one negative eigenvalue, i.e. one imaginary frequency.

The constraints used to find reaction barriers need to fulfill the requirement that they do not have any projection onto a translation or rotation of the molecule as a whole. If this were the case, any value of the constraint could be reached in the ground state by just translating or rotating the molecule. One remedy is to define the constraints in terms of relative coordinates, such as bond lengths, bond angles, etc. Another solution is to impose additional constraints that prohibit any translation or rotation of the molecule.

Dynamical Reaction Path and Intrinsic Reaction Coordinate

Once the transition state has been located, one may be interested in the reaction path. It is widely known that reactions at finite temperature pass the barrier in a variety of ways. In an attempt to single out one particular path, the so-called "intrinsic reaction coordinate" (IRC) has been defined (*23, 24*); it is obtained from $M_i \ddot{R}_i = -\partial E / \partial R_i$. Recent developments are reviewed in ref. (*25*).

There are other possible definitions of a reaction pathway. We discuss here one, which may be termed "dynamical reaction path" (DRP). This path is the zero-temperature limit of a method suggested by Keck (*26*), and was later termed "dynamical reaction coordinate" by Stewart et al. (*27*). The rationale behind the definition of the DRP is as follows: Let us consider an arbitrarily low temperature. In this case the reaction is a very rare event. However, if we monitor only those trajectories that cross the barrier, we find that they become increasingly similar to each other as the temperature is lowered. In the limit, this path will pass with probability one through the exact saddle point with zero velocity.

The DRP is obtained in practice by first determining the saddle point of the total energy surface. Then, two trajectories, obeying Newton's equations of motion, are created, starting from the saddle point with arbitrarily small velocity. One of the trajectories is reversed in time, and the two paths are joined at the saddle point. The DRP corresponds to a system at $T = 0$, with negligible coupling to the heat bath. The IRC approach, on the other hand, corresponds to a system at $T = 0$ with strong coupling to the heat bath, so that the excess kinetic energy, produced while descending from the barrier, is instantaneously dissipated. Both methods can therefore be considered as two limiting cases of a $T = 0$ approach. The zero-temperature limit is reasonably justified, if the reaction barrier is large compared to thermal energies. This approximation serves to give a unique definition of the reaction path.

The DRP has the advantage that it would represent the true dynamics at zero temperature, if the heat bath were explicitly included in the calculation. The DRP has relevance also at higher temperatures, in the sense that it is the most probable path across the barrier. It has, however, the disadvantage that it does not directly approach the ground state, but maintains the reaction energy indefinitely. The reaction energy is dissipated only infinitely slowly due to the infinitesimal coupling to the heat bath. (Correctly formulated, the DRP corresponds to a microcanonical [E = const.] ensemble.)

It is conceivable to extend this approach to an intermediate coupling, which would correspond to applying a constant friction that reflected the heat transfer from the molecule to the environment and that were proportional to the atomic masses and a scaling factor estimated from the heat equation.

Finite-Temperature Simulations

Simulations at finite temperatures offer additional ways to study reactions. Besides making true finite temperature quantities such as entropies of reaction accessible, they are a tool to explore phase space in an unbiased and more effective way than it is possible at zero temperature.

In order to perform a simulation at finite temperature, a mechanism such as a thermostat is required, establishing a canonical ensemble with the corresponding average kinetic energy and its fluctuations. The state-of-the-art method used is the so-called Nosé–Hoover thermostat (28, 29). It consists of one additional dynamical variable, which creates an exact canonical ensemble by a feedback mechanism. In addition to the thermostat coupled to the atoms, a second thermostat is coupled to the electron wave functions, which prevents deviations from the Born–Oppenheimer surface from accumulating during a long-time simulation (30).

The most straightforward and most unbiased approach is a finite-temperature simulation of the system, monitoring the frequency with which the reaction takes place. This approach is best when the barriers are low or the temperatures sufficiently large so that a significant number of reaction events can be observed in a simulation of typically 5–20 ps, needed to obtain a reasonably small statistical error bar. As an example for a direct simulation, we will highlight a study to explore the adsorption properties of water and methanol at acid sites in zeolites by Nusterer et al. (31, 32). The cluster approximation used in previous studies has been avoided and the configuration space explored using finite temperature simulations. We thus not only could determine the proper ground state structures, but the analysis of the dynamical behavior allowed the comparison of IR spectra. By comparing our results with experiment, we could establish that these polar molecules are not protonated by the acid sites of the zeolite at low coverage, but only at coverages of more than one molecule per acid site.

On the other hand, when the barriers are large, techniques other than direct simulation are more appropriate. The zero-temperature transition state search can be extended to finite temperatures. Here, the system is constrained to one value of the reaction coordinate c, defined in equation (5), but at finite temperature. Then, the reaction coordinate is slowly moved across the barrier. The free energy as a function of the reaction coordinate $F(c)$ is obtained by integrating the force of constraint over the reaction coordinate ($16, 33, 34$), which is equivalent to integrating the product of Lagrange multipliers and the velocity of the reaction coordinate over the time of simulation.

$$F(c) = F(c_0) + \int_{c_0}^{c} dc \frac{dF}{dc} = F(c_0) + \int_{t(c_0)}^{t(c)} dt\, \lambda(t) \frac{dc}{dt} \qquad (12)$$

Note that to be exact, the so-called Blue-Moon correction (35) needs to be added. As this correction is usually small and vanishes identically for linear and bond lengths constraints (35), it is presently neglected in our simulations.

The conceptual advantage of the free energy integration is that the entropy is explicitly included in the reaction barrier. Another important advantage is that phase space is explored more effectively than by using the zero-temperature transition state search. Hence, in many cases unexpected reaction mechanism become evident that would have been missed in the more controlled zero-temperature regime; for an illustrative example, see ref. (36). This approach is therefore most appropriate when the reaction mechanism is not immediately evident. Further applications of the slow-growth thermodynamic integration method are shown in another chapter of this book (Woo, T. K.; Margl, P. M.; Deng, L.; Cavallo, L.; Ziegler, T., this book). Recently, the free energy integration has been generalized to solvated systems (34) and to QM-MM coupling approaches applied to organometallic systems with large ligands (37).

Hydroamination of Alkenes

The hydroamination of alkenes is a very important and useful transformation in the synthesis of fine chemicals. However, so far no efficient homogeneous catalytic system has been found for this reaction ($38, 39$); for a recent experimental contribution, see ref. (40). Therefore, we started to investigate novel hydroamination catalysts using electronic structure techniques. As an almost unlimited number of potential catalysts is conceivable, we have limited our search to complexes containing d^8 transition metals from groups 9 and 10. Here we show preliminary results of this study.

We have investigated catalysts containing the complex moiety $\{MCl(PH_3)_3\}^{n+}$, where the metal can be a group 9 element [Co(I), Rh(I), Ir(I)] with $n = 0$ or a group 10 element [Ni(II), Pd(II), Pt(II)] with $n = 1$.

Two reaction mechanisms may be envisaged.

• The first one (Figure 1) involves formation of an alkene complex **1**, external nucleophilic attack onto the coordinated alkene by an amine, thus forming the ammonioalkyl complex **2**, with subsequent transfer of a proton from the ammonium group to C-α (i.e. the carbon atom directly bonded to the metal), which results in protonolytic cleavage of the metal–carbon bond and elimination of the reaction product.

• The second mechanism involves coordination of the amine to the metal and oxidative addition of one N–H bond, yielding an amido-hydrido species. After coordination of the alkene, either the hydride is inserted into the metal–alkene bond – forming an amido-alkyl complex – or the amido moiety is inserted, yielding an aminoalkyl complex. C–N or C–H reductive elimination, respectively, liberates the product and closes the catalytic cycle.

We find that external nucleophilic attack is favorable for group 10 metal centers, whereas N–H activation is favored for group 9 metal centers. This can be rationalized as follows: N–H activation increases the oxidation state by two, hence this process is favored for a metal center in a lower oxidation state, such as Co(I), Rh(I), Ir(I), as compared to Ni(II), Pd(II), or Pt(II) centers. On the other hand let us consider the nucleophilic attack on the alkene. In this case, an charge is transferred from the lone pair of the amine to the carbon atom coordinated to the metal. This charge transfer is favored for a positive metal center as compared to a neutral one, and thus this mechanism is more favorable for Ni(II), Pd(II), and Pt(II) centers than for Co(I), Rh(I), Ir(I).

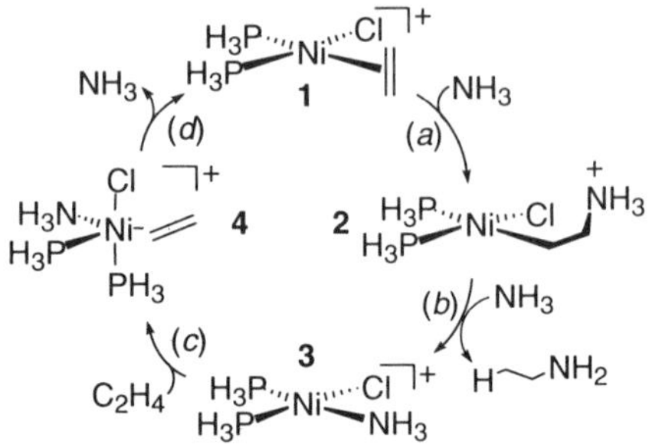

Figure 1. Catalytic cycle for hydroamination via external nucleophilic attack on the alkene.

In the following we will describe our results for the first mechanism involving external attack on the alkene with NH_3 and C_2H_4 as substrates and M = Ni(II).

The first step (a) involves the attack of the ammonia onto the coordinated alkene. This step is driven by the interaction of the lone pair of ammonia with the empty π and π^* orbitals of the complex stemming from antibonding combinations of the π and π^* orbitals of the alkene with metal d orbitals of corresponding symmetry. Despite the fact that the LUMO is the π orbital and the π^* orbital lies above it, the more relevant frontier orbital is the π^* orbital, as deduced from the reaction pathway. The reason is that the π orbital is more delocalized than the π^* orbital, which has a larger weight on the alkene. During the reaction, the alkene slips along its axis, and a σ bond is formed between the metal and the second carbon atom not being attacked by the amine.

The rate-delimiting step of the reaction is the protonolytic cleavage (b). The activation energy is calculated as 113–132 kJ mol^{-1}, depending on the model used. This barrier is expected to be dependent on solvent effects, as the proton transfer may be mediated by an amphiphilic solvent. In order to obtain a rough estimate of the solvent effects, we performed the calculation with none, one, and two additional molecules of ammonia.

• Without any additional ammonia, the transition state locates the proton approximately midways between the nitrogen atom and C-α. Once the proton has been transferred, the product is not eliminated, but rather remains loosely bound to the metal via an agostic-like interaction (H–Ni 1.76 Å, C–H 1.15 Å).

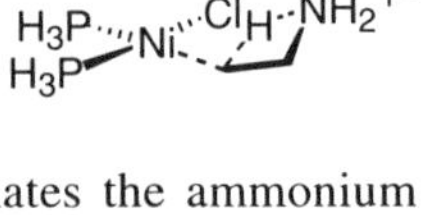

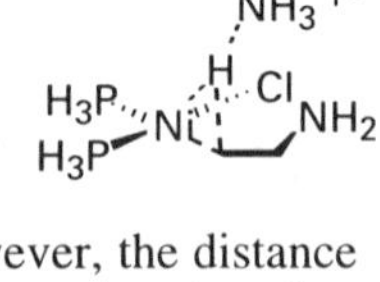

• Offering one additional ammonia molecule, it deprotonates the ammonium group, and transfers a different proton almost simultaneously to C-α. At the transition state, the proton attracted to the carbon atom experiences additional agostic stabilization from the metal. The formed aminoalkane remains still bonded via the proton to the otherwise free coordination site of the metal. Now, the ammonia is electrostatically attracted to the axial position of the complex. However, the distance between nitrogen and nickel is larger than a covalent bond. The attraction therefore stems mostly from the interaction between the positive charge of the complex with the dipole of the ammonia. Finally, the product is expelled while the ammonia shifts from the axial to an equatorial position, re-establishing a square-planar coordination sphere.

• In a third calculation we offered a second ammonia, which is loosely bound in the axial position. The reaction proceeds similarly, but now the second ammonia in the axial position occupies the free coordination site.

Somewhat surprisingly, the calculated barriers differ by less than 20 kJ mol^{-1}. This finding is attributed to similar solvent stabilization of the initial and the transition states.

In the last steps (c) and (d), the ammine ligand is substituted by an alkene, thus completing the catalytic cycle. We found that the dissociative mechanism, namely elimination of the ammonia before addition of the alkene, is unfavorable compared to the associative mechanism, where first the penta-coordinate complex **4** is formed before the ammonia is eliminated. This is due to the intermediate three-coordinate complex being highly unstable. In the associative mechanism an alkene adds to the axial position, and the ammine ligand makes way, such that the new main axis of the distorted trigonal bipyramid thus formed passes through chlorine and a phosphine.

Finally, the ammine ligand is eliminated and the initial alkene complex is recovered.

Dynamical Reaction Path of the Rate-Determining Step. We analyzed the rate-determining step, namely the protonolytic cleavage (b), in more detail. We have chosen here the mechanisms with one additional ammonia molecule, which we deemed most realistic. Figure 2 shows the energy profile of the dynamical reaction path of this step with three snapshot structures.

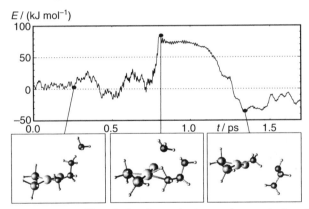

Figure 2. Energy profile of the dynamical reaction path for the rate-delimiting reaction step (b).

Initially, the ammonia molecule is attracted to the ammonium group and forms a transient hydrogen bond. The freed potential energy can however not be dissipated rapidly enough so that this hydrogen bond breaks and reforms again twice. Then the ammonia approaches the ammonium group again, forms a hydrogen bond, and after only one single N–H stretch period, the proton is definitely transferred to the ammonia molecule.

The formed ammonium ion slightly rotates, presenting a second proton to the C-α. In a very sudden and forceful process this proton is transferred to C-α. At the same time, the proton is attracted by the metal atom, forming the transition state, where it is shared about equally (H–Ni 1.69 Å, H–C 1.34 Å) between the metal and the C-α carbon atom, and stays at a somewhat longer distance (H–N 1.86 Å) to the now deprotonated ammonium ion. The proton rests in this position for about 0.3 ps. Meanwhile, the ammonia molecule, which retreated after the proton transfer, slowly drifts towards the axial position of the complex. Here it forms a bond, and in another violent motion sweeps from the axial to the equatorial position, displacing the product which is now eliminated. As the product drifts away from the complex, an intermittent hydrogen bond between a proton of the ammine ligand and the nitrogen of the product is formed.

Conclusion

The AIMD approach opens new doors to quantum chemistry. It provides an efficient tool to study reaction mechanisms using quasi-statistical and dynamical approaches. It provides a means to study reactive processes in a variety of ways, thus providing flexibility to adapt the approach to the problem at hand. It allows to study isolated and solvated molecules, surfaces, liquids, microporous materials and solids.

References

1. Car, R.; Parrinello, M. *Phys. Rev. Lett.* **1985**, *55*, 2471.
2. Hohenberg, P.; Kohn, W. *Phys. Rev. B* **1964**, *136*, 864.
3. Kohn, W.; Sham, L. J. *Phys. Rev. A* **1965**, *140*, 1133.
4. Blöchl, P. E. *Phys. Rev. B* **1994**, *50*, 17953.
5. Hamann, D. R.; Schlüter, M.; Chiang, C. *Phys. Rev. Lett.* **1979**, *43*, 1494.
6. Vanderbilt, D. *Phys. Rev. B.* **1990**, *41*, 7892.
7. Becke, A. D. *Phys. Rev. A* **1988**, *38*, 3098.
8. Perdew, J. P. *Phys. Rev. B* **1986**, *33*, 8822.
9. Pastore, G.; Smargiassi, E.; Buda, F. *Phys. Rev. A* **1991**, *44*, 6334.
10. Payne, M. C.; Teter, M. P.; Allan, D. C.; Arias, T. A.; Joannopoulos, J. D. *Rev. Mod. Phys.* **1992**, *64*, 1045.
11. Harris, J.; Jones, R. O.; Müller, J. E. *J. Chem. Phys.* **1981**, *75*, 3904.
12. Hockney, R. W. *Computer Simulation Using Particles*; McGraw-Hill: New York, 1981, pp 211–214.
13. Barnett, N. B.; Landman, U. *Phys. Rev. B* **1993**, *48*, 2081.
14. Makov, G.; Payne, M. C. *Phys. Rev. B* **1995**, *51*, 4014.
15. Blöchl, P. E. *J. Chem. Phys.* **1995**, *103*, 7422.
16. Ciccotti, G.; Ferrario, M.; Hynes, J. T.; Kapral, R. *Chem. Phys.* **1989**, *129*, 241.
17. Verlet, L. *Phys. Rev.* **1967**, *1959*, 98.
18. Schlegel, H. B. *Adv. Chem. Phys.* **1987**, *67*, 249.
19. Ionova, I. V.; Carter, E. A. *J. Chem. Phys.* **1995**, *103*, 5437.
20. Ulitsky, A.; Shalloway, D. *J. Chem. Phys.* **1997**, *106*, 10099.
21. Ayala, P. Y.; Schlegel, H. B. *J. Chem. Phys.* **1997**, *107*, 375.
22. Margl, P.; Schwarz, K.; Blöchl, P. E. *J. Am. Chem. Soc.* **1994**, *116*, 11177.
23. Fukui, K. *J. Phys. Chem.* **1970**, *74*, 4161.
24. Fukui, K. *Acc. Chem. Res.* **1981**, *14*, 363.
25. Deng, L.; Ziegler, T. *Int. J. Quantum Chem.* **1994**, *52*, 731.
26. Keck, J. *Discuss. Faraday Soc.* **1962**, *33*, 173.
27. Stewart, J. J. P. *J. Comp. Chem.* **1987**, *8*, 1117.
28. Nosé, S. *J. Chem. Phys.* **1984**, *84*, 511.
29. Hoover, W. G. *Phys. Rev. A* **1985**, *31*, 1695.
30. Blöchl, P. E., Parrinello, M. *Phys. Rev. B* **1992**, *45*, 9413.

31. Nusterer, E.; Blöchl, P. E.; Schwarz, K. *Angew. Chem. Intl. Ed. Engl.* **1996**, *35*, 175.
32. Nusterer, E.; Blöchl, P. E.; Schwarz, K. *Chem. Phys. Lett.* **1996**, *253*, 448.
33. Margl, P. M.; Ziegler, T.; Blöchl, P. E. *J. Am. Chem. Soc.* **1996**, *118*, 5412.
34. Curioni, A.; Sprik, M.; Andreoni, W.; Schiffer, H.; Hutter, J.; Parrinello, M. *J. Am. Chem. Soc* **1997**, *119*, 7218.
35. Carter, E. A.; Ciccotti, G.; Hynes, J. T.; Kapral, R. *Chem. Phys. Lett.* **1989**, *156*, 472.
36. Margl, P. M.; Woo, T. K.; Blöchl, P. E.; Ziegler, T. *J. Am. Chem. Soc.* **1998**, *120*, 2174.
37. Woo, T. K.; Margl, P. M.; Blöchl, P. E.; Ziegler, T. *J. Phys. Chem.* **1997**, *101*, 7877.
38. Taube, R. In *Applied Homogeneous Catalysis with Organometallic Compounds*; Cornils, B., Herrmann, W. A., Eds.; VCH: Weinheim, 1996; Vol. 1, pp 507–520.
39. Roundhill, D. M. *Catal. Today* **1997**, *37*, 155.
40. Dorta, R.; Egli, P.; Zürcher, F.; Togni, A. *J. Am. Chem. Soc.* **1997**, *119*, 10857.

Chapter 8

Transition States for Proton Transfer Reactions in Late Transition Metal Chemistry and Catalysis: σ-Bond Metathesis Pathways

A. Dedieu, F. Hutschka, and A. Milet

Laboratoire de Chimie Quantique, Université Louis Pasteur and CNRS, UMR 7551, 4 rue Blaise Pascal, F 67000 Strasbourg, France

Transition states for intra- and intermolecular proton transfer reactions involving late transition metal complexes are analyzed. The reactions under study are key steps of metal catalyzed or metal mediated multistep processes. This includes a sigma bond metathesis step between H_2 and the rhodium formate intermediate in the hydrogenation of CO_2 to formic acid, homogeneously catalyzed by rhodium phosphine complexes; its activation by an external amine; the heterolytic recombination of H_2 from a Pd-H···H-N dihydrogen bond; The factors accounting for the feasibility of these reactions are delineated.

Proton transfer reactions are probably the most ubiquitous reactions of chemistry. They take place in organic, inorganic, organometallic and bioinorganic chemistry, either in the stoichiometric or in the catalytic regime. The heterolytic cleavage of H_2 by transition metal complexes or its reverse reaction, the heterolytic recombination of H_2 between a metal hydride and a protonic hydrogen atom, both belong to this class of reactions. They involve in many instances a four center transition state characteristic of a σ-bond metathesis. Such processes were initially discovered for early transition metal complexes or electron poor early f-block element complexes (1-9). Subsequent theoretical studies showed that in this case empty d metal orbitals were crucial for allowing these [2_s+2_s] processes (10-17) which should be thermally forbidden according to the Woodward-Hoffmann rules (18,19). But there has been recently increasing experimental evidence that σ-bond metathesis could also take place with late - electron rich - transition metal complexes (20-35), i.e. systems with very few, if any, low lying empty d metal orbitals. Simultaneously, a few theoretical studies - including ours - started to put forward a σ-bond metathesis mechanism in H-H or X-H bond forming and breaking reactions (36-44). Somewhat analogous to these reactions, but not necessarily involving a four center transition state, are H_2

elimination reactions or H/D exchange reactions that take place in complexes displaying a M-H⋯H-X "dihydrogen bond" *(45-51)*. Theoretical studies have adressed the structural features and the strength of dihydrogen bonds but there has been no determination of a transition state for these reactions *(52-57)*.

We have been involved ourselves - through various collaborations with experimental groups - in the study of σ-bond metathesis reactions or proton transfer reactions that might take place in processes catalyzed or mediated by complexes of late transition metals *(38,40,41,58-61)*. The purpose of the present article is to review our most recent advances. In order to delineate some of the factors that account for these processes we will focus on the geometry and on the electronic structure of the transition states. Our theoretical protocol relies on geometries optimized either at the SCF or at the MP2 level, the energetics being obtained either at the MP2 or at the QCISD(T) level. More details can be found in the original publications.

The [2+2] σ-bond Metathesis Reaction in Late Transition Complexes.

Rhodium Complexes. We were first faced with the possibility of a σ-bond metathesis reaction in the course of a theoretical study of the catalytic hydrogenation of CO_2 to formic acid. This hydrogenation reaction, catalyzed by various Rh or Ru complexes, equation (1), has attracted much interest in the recent years *(62-71)*. For

$$\text{Rh or Ru catalysts}$$
$$H_2 + CO_2 \rightleftharpoons HCO_2H \quad (1)$$

rhodium several mechanisms have been proposed, depending on the actual catalyst. They all involve at a given stage of the process a Rh(III) intermediate with a hydride and a formate ligand. Reductive elimination of HCO_2H from this intermediate produces a Rh(I) species which then incorporates H_2 and CO_2, completing the catalytic cycle. Thus the main characteristic of these mechanisms is the involvement of the two oxidation states I and III for rhodium. For the catalytic system where the catalyst is a bisphosphine rhodium hydrido complex L_2RhH (L_2 = [$Ph_2P(CH_2)_nPPh_2$], n=3,4) Leitner *et al.* proposed the path A of Scheme 1 *(63,65)*. It comprises, after the insertion of CO_2 into the Rh-H bond of L_2RhH, a sequence of

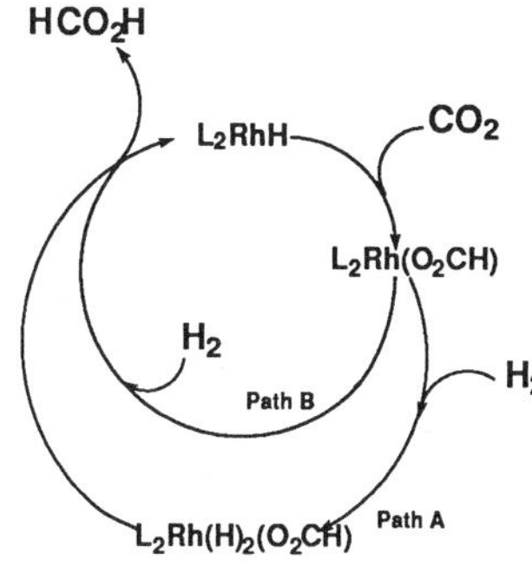

Scheme 1

(L₂=diphosphine)

oxidative addition and reductive elimination steps for the H_2 activation process. Our theoretical analysis of the catalytic process showed that another pathway involving a

102

σ-bond metathesis reaction of H_2 with the rhodium formate intermediate, see path B of Scheme 1, could be competitive with the pathway A (*38,40*). In particular the energy barrier for the [2+2] addition of H_2 to the Rh-O bond of the three coordinated d^8 [Rh(PH$_3$)$_2$(O$_2$CH)] formate intermediate, reaction (2), is smaller than the barrier

$$[Rh(PH_3)_2(O_2CH)] + H_2 \longrightarrow [Rh(H)(PH_3)_2...(HCO_2H)] \qquad (2)$$

$$[Rh(H)_2(PH_3)_2(O_2CH)] \longrightarrow [Rh(H)(PH_3)_2...(HCO_2H)] \qquad (3)$$

for the reductive elimination of HCO$_2$H from [Rh(H)$_2$(PH$_3$)$_2$(O$_2$CH)], reaction (3): 14.8 instead of 24.7 kcal mol^{-1}, QCISD(T)//MP2 values. The corresponding MP2//MP2 values are 11.2 and 23.6 kcal mol^{-1}. The reasonable agreement between the QCISD(T)//MP2 and MP2//MP2 values indicates that the MP2//MP2 level of calculation is appropriate for assessing this type of chemistry in the rhodium systems. The Figure 1 shows the computed energy profile for reaction (2), together with the structure of the corresponding transition state **TS(2)**, and the eigenvector associated with the imaginary frequency of 1136cm^{-1}. **TS(2)** is characterized by an appreciable elongation of the H-H bond (from 0.798 Å in the dihydrogen precursor

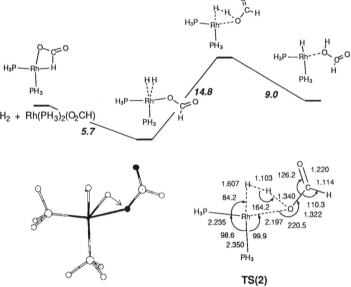

Figure 1. Energy profile (at the QCISD(T)//MP2 level) for the σ-bond metathesis reaction (2) and geometry (at the MP2 level) of the transition state **TS(2)**, together with the eigenvector corresponding to the imaginary frequency. Energies are in kcal mol^{-1}, bond lengths in Å and angles in deg.

complex to 1.103Å) and a relatively short O..H distance, 1.34 Å. Most noticeably the H...H...O unit is close to linearity: the H-H-O angle amounts to 164.2°. We may thus anticipate that the process should bear some similarity with a linear proton transfer between free H_2 and one oxygen lone pair *(72)*.

An analysis of the wavefunction of **TS(2)** revealed two factors that come into play for allowing the σ-bond metathesis process. One is related to the nature of the molecular orbitals that are directly involved in the process, see the interaction diagram of Figure 2, the other one is the respective charges of the reacting entities. As in early transition metal chemistry one finds in the d^8 three-coordinate T-shaped [Rh(PH$_3$)$_2$(O$_2$CH)] fragment a relatively low lying empty orbital, mostly of metal

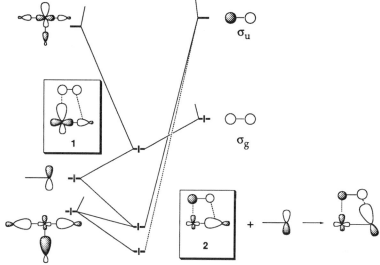

Figure 2. Schematic interaction diagram between the orbitals of [(PH$_3$)$_2$Rh(O$_2$CH)] and the orbitals of H_2 for the transition state **TS(2)** of the σ bond metathesis reaction of H_2 with [(PH$_3$)$_2$Rh(O$_2$CH)].

d_σ character. This orbital is a hybridized $d_{x^2-y^2}$ orbital interacting in an antibonding way with one of the lone pairs of the formate ligand. It has a phase relationship such that it can interact positively with the σ_g orbital of the incoming H_2, see **1**. Its bonding counterpart, which is essentially this formate lone pair, can also interact positively with the empty σ_u orbital of H_2, see **2**. The interactions **1** and **2** provide the basis for the allowance of the process. They are relatively weak, especially **2**, due to some energy mismatch. But this detrimental feature is overcome by the involvement in the interaction pattern of the second lone pair on the formate ligand, see Figure 2. The mixing of this second lone pair allows a continuous rehybridization of the interacting orbitals throughout the reaction that maximises the interaction between the oxygen atom and the transferred hydrogen atom.

Thus a situation symmetric to the one described for σ-bond metathesis reactions of H-H or C-H bonds in early transition metal complexes prevails. There, the mutual mixing of suitable s, p, and d *empty orbitals on the metal* maintains the optimal overlap with the occupied σ_g orbital, thereby triggering in the heterolytic splitting process the dissociation of the incipient *hydride* and its binding to the metal (*12*). In the late transition metal complex [Rh(H$_2$)(PH$_3$)$_2$(O$_2$CH)] the mixing of the additional *occupied lone pair on the ligand* maintains the optimal overlap with the empty σ_u orbital. It triggers therefore the dissociation of the incipient *proton* and its association with the ligand. In that sense the lone pair plays the same role as an external base, a feature that was also recognized experimentally with Rh(III) or Ir(III) complexes bearing sulfur ligands (*26,28,35*).

The second factor that one should not overlook is the charge distribution in the dihydrogen adduct of the formate intermediate, [Rh(H$_2$)(PH$_3$)$_2$(O$_2$CH)]. Dihydrogen complexes are known to be generally acidic (*5-7,23*), especially those with unstretched H-H bonds (complexes having relatively stretched H-H bonds give less definite results (*73-78*)). In the [Rh(H$_2$)(PH$_3$)$_2$(O$_2$CH)] system the H-H bond is hardly elongated (*40*) and the hydrogen atom adjacent to the formate ligand is positively charged (+0.16e). The oxygen atom of the formate ligand bears a large negative charge (-0.64e). This large separation of the charges favors the transfer of the proton.

Is this σ-bond metathesis process of H$_2$ with the Rh-O bond of [Rh(PH$_3$)$_2$(O$_2$CH)] peculiar to this system? A d^8 three-coordinate T-shaped fragment is isolobal to a d^6 five coordinate square pyramidal fragment, and thus a σ-bond metathesis process between the M-L bond of a square pyramidal d^6 ML$_5$ system and H$_2$ should be also possible. Experimentally, a σ-bond metathesis process has been corroborated very recently for Rh(III) complexes with sulfur ligands (*35*). In that context it is worth noting that a Rh^{III} system, *viz.* [H$_2$Rh(PMe$_2$Ph)$_3$(solvent)]$^+$ had been found some years ago to catalyze the transformation of H$_2$ + CO$_2$ into HCO$_2$H (*64*). The catalytic cycle shown on the left part of the Figure 3 was proposed (*64*). On the basis of the above mentioned isolobal analogy, we can think to

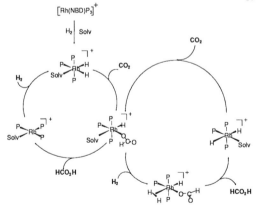

Figure 3. a) left part: Catalytic cycle experimentally proposed (*64*) for the [H$_2$Rh(PMe$_2$Ph)$_3$(solvent)]$^+$ catalyzed transformation of H$_2$ + CO$_2$ into HCO$_2$H . b) right part: Alternative pathway proposed on the basis of the calculations (*38,60*). (NBD=norbornadiene, P=PMe$_2$Ph, Solv=solvent)

a competitive pathway (see the right part of Figure 3) that would involve the σ-bond metathesis reaction of H_2 with the Rh-O bond of the five coordinate d^6 cationic $[Rh(H)(H_2)(PMe_2Ph)_3(O_2CH)]^+$ intermediate (after substitution of the solvent by the dihydrogen). We investigated at the MP2//MP2 level the model reaction (4).

$$[Rh(H)(H_2)(PH_3)_3(O_2CH)]^+ \longrightarrow [Rh(H)_2(PH_3)_3\cdots(HCO_2H)]^+ \qquad (4)$$

The corresponding geometries and energies, shown on Figure 4, do indeed point to the analogy between reactions (3) and (4), see for instance the geometry of the transition state TS(4). The associated energy barrier is somewhat greater, however, 24.4 kcal mol^{-1}. The relatively large increase in the barrier may be tentatively ascribed to the greater *trans* effect of the hydride as compared to the phosphine. This *trans* effect

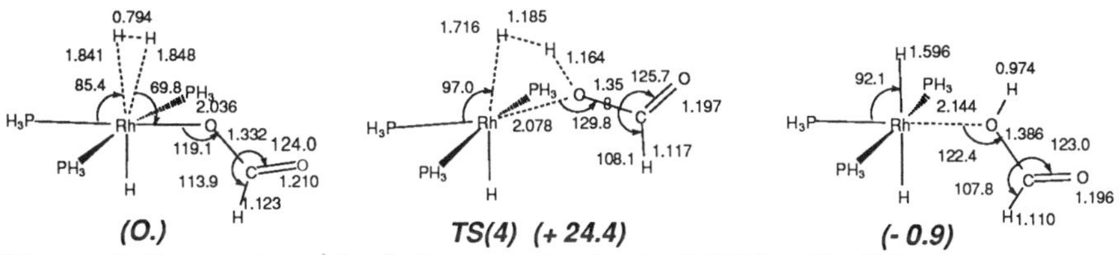

(O.)   TS(4) (+ 24.4)   (- 0.9)

Figure 4. Geometries and relative energies (at the MP2 level) of the structures relevant to the σ-bond metathesis reaction of H_2 with $[Rh(H)(PH_3)_3(O_2CH)]^+$. Energies are in kcal mol^{-1}, bond lengths in Å and angles in deg.

leads to an elongation of the bond between Rh and H_2 which in turn results in a longer H...O distance in the reactant (2.23Å instead of 2.15Å). We did not find any other factors that could explain this feature and that would be also consistent with our other investigations (*vide infra*) such as the charge of the overall complex, the respective partial charges of the interacting atoms or the respective orbital energies.

The possibility for a σ-bond metathesis process had been adressed previously by Versluis and Ziegler (*36*) in their theoretical study of the H_2-induced acetaldehyde elimination from $[Co(CO)_3\{C(O)CH_3\}]$, the last step in the hydroformylation process catalyzed by $HCo(CO)_4$. On the basis of LDF calculations they characterized an intermediate lying 19.9 kcal mol^{-1} higher in energy than the parent η^2-H_2 complex. Interestingly its geometry is similar to the one of TS(2). A transition state, located very near to this intermediate was also obtained from an approximate reaction profile. Orbital interactions similar to 1 and 2 were put forth to account for the feasibility of this process which does not seem to require an additional lone pair. Calculations at a better level (*e.g.* including a more precise transition state determination and a gradient corrected DFT calculation of the energy) should be performed before a more complete discussion of the lone pair involvement in this case can be done. A σ-bond metathesis pathway between the Rh-H bond of a bisphosphine chloro rhodium(I) complex and the B-H bond of a boron hydride was also proposed by Morokuma *et al.* (*37*) as a competitive mechanism in the catalyzed olefin hydroboration reaction. It is somewhat different, since the Lewis acid properties of the boron hydride are likely to come into play through some interactions between the p vacant orbital on boron and the doubly occupied d_{z^2} orbital on Rh. The reaction may therefore be seen as an electrophilic addition of $B(OR)_2^+$ on coordinated ethylene, concomitantly with a hydride transfer to Rh. One could also consider it as a $[2_s +2_s +2_s]$ reaction between the B-H bond, the C=C π bond and the d_π orbital on rhodium. Such a reaction is clearly symmetry allowed.

Palladium Complexes. The ability of H_2 to undergo a σ-bond metathesis reaction with a late transition metal complex should not be restricted to these d^8 and d^6 rhodium complexes. Any d^8 T-shaped ML_3 or d^6 square pyramidal ML_5 system that has one ligand with such an additional lone pair not engaged in the bond to the metal should also be a good candidate for these [2+2] reactions. This would be the case for *e.g.* metal hydroxo or alkoxy complexes. Furthermore one might expect that the reverse process, *i.e.* the σ-bond metathesis reaction of water or of an alcohol with a metal hydride or a metal alkyl bond, Scheme 2, should be feasible. In order to

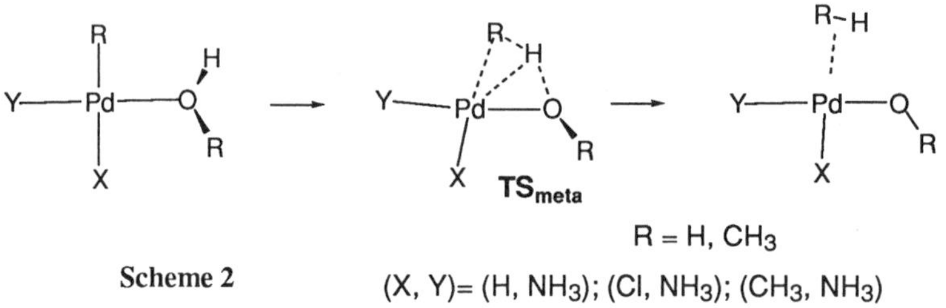

Scheme 2

R = H, CH₃

(X, Y)= (H, NH₃); (Cl, NH₃); (CH₃, NH₃)

delineate more precisely the factors that account for the [2+2] mechanism for such systems, and in connection with our studies that were aimed at finding the mechanism of the reaction of water with tris(pyrazolyl)borate Pd(II) complexes, *vide infra*, we carried out a comprehensive study on a variety of such reactions for model Pd(II) and Pd(IV) complexes. The detail of the calculations and the characteristics of the transition states can be found in the original publication (*41*). One may simply note here that as in **TS(2)**, all the transition states are characterized by a quite large R...H...O angle, varying from 149° to 155°. The non interacting R group is found out of the plane of the reaction, as shown on Scheme 2. We found (*41*) that this is due a preferential interaction of the p lone pair of ROH (instead of the σ lone pair) with the empty $d_{x^2-y^2}$ orbital. This p lone pair is a better electron donor than the σ lone pair. Having the R group out of the reaction plane allows its interaction with the $d_{x^2-y^2}$ orbital.

We also found that for *a given oxidation state of the metal* the energy barrier for the [2+2] addition of the O-H bond of ROH on Pd-R depends greatly -as expected- on the nature of the Pd-R bond: When the Pd-R bond has a greater ionic character, the energy barrier is relatively low, in the range 10 to 20 kcal mol⁻¹(*41*). In contrast, when the Pd-R bond is quite covalent (this happens for instance when there is a strong electron withdrawing ligand such as chlorine, or when R lies *cis* to a hydride or another alkyl *(61)*), then a much higher barrier is found (between 32 and 45 kcal mol^{-1}). However this rough correlation is subject to some levelling effect due to the large negative charge of the oxygen atom *(41)*.

We may mention here that reactions involving Pt complexes instead of Pd complexes were found to have consistently higher barriers (by about 4 kcal mol^{-1}). Also, as for rhodium, higher oxidation state (either Pd(IV) or Pt(IV)) led to higher energy barriers *(41, 61)*. However the increase (3 to 4 kcal mol^{-1}) is much less important than in the rhodium case. This may be due to the fact that in the Pd as in Pt systems that we investigated no change was made in the ligands *trans* to the reacting entities (the +IV oxidation state was simply obtained by adding two hydrides above and below the Pd(R)(X)(Y)(ROH) plane).

The Assistance of the Heterolytic Cleavage of H$_2$ by an External Base.

We have mentioned in the previous section that the lone pair of the formate ligand was triggering the dissociation of the incipient H$^+$ and could be therefore considered as playing the same role as an external base. It is known experimentally that the heterolytic splitting of H$_2$ can be accelerated by amines, the role of the amine being likely to abstract a proton *(20,79-85)*. Interestingly too, experimental investigations of the catalytic process of Scheme 1 had indicated that the hydrogenation is enhanced by the presence of an amine in the reaction cell *(65)*. Thus it might be anticipated that in reaction (2) an amine could mediate the proton transfer between the coordinated H$_2$ and the adjacent formate ligand. Indeed MP2//MP2 calculations that we carried out *(58,60)*, using NH$_3$ as a model for the amine, showed how such an additional Lewis base can act as a relay, by first abstracting proton from the coordinated H$_2$ to form an ion pair from which the proton is then released to the formate ligand, equations (5) and (6).

$$[Rh(H_2)(PH_3)_2(O_2CH)\cdots(NH_3)] \longrightarrow [Rh(H)(PH_3)_2(O_2CH)^-\cdots(NH_4)^+] \quad (5)$$

$$[Rh(H)(PH_3)_2(O_2CH)^-\cdots(NH_4)^+] \longrightarrow [Rh(H)(PH_3)_2\cdots(HCO_2H)\cdots(NH_3)] \quad (6)$$

The Figure 5 summarizes our findings: The overall picture is slightly complicated by an extra rearrangement of the formate ligand in the ion-pair intermediate [Rh(H)(PH$_3$)$_2$(O$_2$CH)$^-$...NH$_4$$^+$] (this rearrangement is due to our choice of NH$_3$ as a model of a tertiary amine. Its corresponding transition state **TSrear** is shown in the original publication *(58)*), But the most salient result is that the energy barriers associated with the two transition states **TS(5)** and **TS(6)** are quite low, much lower than the barrier for the σ-bond metathesis process in the absence of an external amine. Also of interest is the fact that the overall process is exothermic by 13.6 kcal mol^{-1} (the value corrected from the basis set superposition error is 12.7 kcal mol^{-1}). At the MP2//MP2 level the non assisted reaction in endothermic by 1 kcal mol^{-1} and has a barrier of 11 kcal mol⁻¹. Thus, both the kinetics and the thermodynamics of the heterolytic splitting of H_2 are favored in the presence of the an external base.

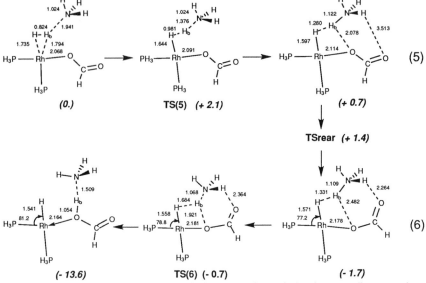

Figure 5. MP2 geometries and relative energies of the intermediates and transition states along the reaction path of the σ-bond metathesis in the presence of the amine. Energies are in kcal mol⁻¹, bond distances in Å and angles in deg.

The Elimination of H_2 from a Dihydrogen Bond.

A more careful look at the microscopic reverse of reaction (5) leads immediately to the conclusion that it can be viewed as the elimination of H_2 from a system having a Rh-H⋯H-N "dihydrogen bond". This term coined by Crabtree (*55*) refers to an intra- or intermolecular interaction M-H⋯H-X between a conventional hydrogen bond donor as the weak acid component and a metal hydride bond as the weak base component. The computed geometry for **TS(5)** therefore corresponds to the geometry of the transition state for H_2 elimination from a dihydrogen bond. The corresponding energy barrier is very low, 1.4 kcal mol⁻¹. This indicates that such a process can be very easy, provided that some criteria are fullfilled. In a combined experimental and theoretical study, Crabtree, Eisenstein *et al.* showed (*52*) that the strength of the Ir-H⋯H-N interaction in the [IrH₃(PPh₃)₂(2-aminopyridine)] and [IrH₂(X)(PPh₃)₂(2-aminopyridine)], (X=F, Cl, Br, I) systems can be as large as 5.8 kcal mol⁻¹, and that it is depends on the polarizability of the Ir-H bond, which in turn is greatly affected by the nature of the ligand X *trans* to it. The calculations were restricted to ground state properties only and did not assess the geometric and electronic characteristics of the *transition state* for the H_2 elimination.

We have obtained - to our knowledge for the first time - such characteristics in the course of our study of the reduction of water to H_2 mediated by bis-alkyl tris(pyrazolyl)borate Pd(II) complexes, see Scheme 3. This reaction which was discovered by Canty *et al.*(86,87) is quite interesting in view of the fundamental importance of water reduction by inorganic species. However nothing was known about its mechanism .

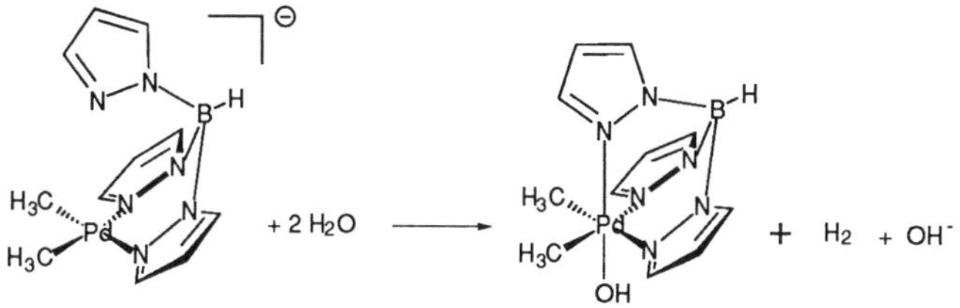

Scheme 3

We modelled the tris(pyrazolyl)borate ligand $[(pz)_3BH]^-$ by the tris(hydrazonyl)borate ligand, $[(H_2C=N-NH)_3BH]^-$. The geometries were optimized at the SCF level and the energies computed at the MP2 level, see reference (*59*) for more details. The overall mechanism, as deduced from the calculations (*59*), can be summarized by equations (7) and (8)

$$[Pd(Me_2)\{(pz)_3BH\}]^- + H_2O \longrightarrow Pd(Me_2)(H)(OH)\{(pz)_3BH\}^- \qquad (7)$$

$$[Pd(Me_2)(H)(OH)\{(pz)_3BH\}]^- + H_2O \rightarrow Pd(Me_2)(OH)\{(pz)_3BH\} + H_2 + OH^- \quad (8)$$

Note here that the oxidative addition of H_2O that takes place in reaction (7) may be either *cis* or *trans* (*i.e.* leading to $[Pd(Me_2)(H)(OH)\{(pz)_3BH\}]^-$ isomers with H and OH either *cis* or *trans* to each other). The theoretical study showed that the *trans* stereochemical course is the operative one in the overall process (*59*). But for the sake of comparison we will analyze here both possibilities for reaction (8). Reaction (8) starts with a protonation of the non coordinated nitrogen atom and hence the two isomers shown on the top of Figure 6 have to be considered. From these, H_2 elimination can proceed through a recombination of the proton and of the hydride.

It is clear from Figure 6 that the energy barrier for this elimination is highly dependent on the charge of the hydride, which in turn depends on the nature of the Pd-H bond: in the *trans* isomer the three pure σ-donor ligands, *i.e.* the two methyl groups and the hydride are in a *fac* disposition. This leads to a quite covalent nature of the corresponding bonds with the metal (*61*). In contrast, in the *cis* isomer the three pure σ-donor ligand are in a *mer* disposition and the corresponding bonds to the metal have a ionic character (*61*). In line with these features the energy barrier associated with **TS(8a)** is much higher than the energy barrier associated with **TS(8b)**. One might argue that in the isomer leading to **TS(8a)**, the positive charge of the hydrogen atom bound to Pd (+0.01e) should preclude any heterolytic recombination. Moreover the H...H distance is too long for bonding, 2.95Å. But the rotation around the B-N bond is easy, 5.6 kcal mol^{-1}(*59*). It brings the proton at 1.61Å from the hydride. As a result the hydride is polarized: it acquires a charge of -0.08e. It can thus undergo the coupling with the proton bound to nitrogen. The energy difference between this

110

structure and **TS(8a)**, 18.8 kcal mol^{-1}, remains nevertheless greater than the energy barrier associated with **TS(8b)**. We have already mentioned that the polarizability of the metal-hydrogen bond had been recognized as one of characteristic features for establishing a M-H...H-X dihydrogen bond *(52)*. *The present results show that this polarizablity is also important for the H$_2$ elimination process itself.*

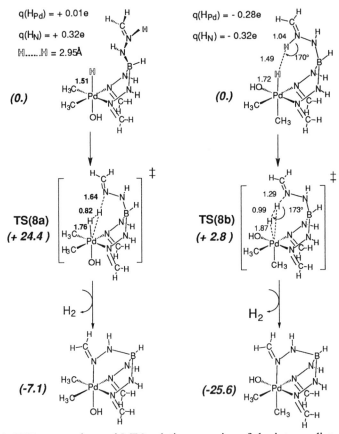

Figure 6. SCF geometries and MP2 relative energies of the intermediates and transition states for the H$_2$ elimination from the Pd-H···H-N dihydrogen bond in [Pd(Me$_2$)(H)(OH){(pz)$_2$(pzH)BH}]. Energies are in kcal mol^{-1}, bond distances in Å and angles in deg.

Conclusion

The results reported in this account show that σ-bond metathesis reactions between H$_2$ and late transition metal complexes should be considered as possible pathways, especially when involving ligands with additional lone pairs (σ-bond metathesis reactions involving *e.g.* alkyl ligands *(42,88)* lie out of the scope of the rationalization provided). The geometrical and the electronic structure of the corresponding transition states point to the proton transfer nature of these metathesis reactions. The reverse reaction, H_2 elimination from a M-H bond and a X-H bond (with either X bound to the metal or in the vicinity of the metal) is strongly dependent on the polarizability of the metal hydrogen bond. It would be now of great interest to check theoretically the feasibility for similar reactions in bioinorganic systems.

Literature Cited

(1)   Watson, P. L. *J. Am. Chem.Soc.* **1983**, *105*, 6491.
(2)   P. L. Watson, P. L.; Parshall, G. W. *Acc. Chem.Res.* **1985**, *17*, 51.
(3)   Crabtree, R. H.; Hamilton, D. G. *Adv. Organomet. Chem.* **1988**, *28*, 299.
(4)   Jordan, R. F. *Adv. Organomet. Chem.* **1991**, *32*, 325.
(5)   Jessop, P. G.; Morris, R. H. *Coord. Chem.Rev.* **1992**, *121*, 155.
(6)   Crabtree, R. H. *Angew. Chem. Int. Ed. Engl.* **1993**, *32*, 789.
(7)   Heinekey, D. M.; Oldham, Jr., W. J. *Chem. Rev.* **1993**, *93*, 913.
(8)   Arndtsen, B. A.; Bergman, R. G.; Mobley, T. A.; Peterson, T. H. *Acc. Chem. Res.* **1995**, *28*, 154.
(9)   Crabtree, R. H. *Chem. Rev.* **1995**, *95*, 987.
(10) Steigerwald, M. L; Goddard III, W. A. *J. Am. Chem. Soc.* **1984**, *106*, 308.
(11) Upton, T. H. *J. Am. Chem. Soc.* **1984**, *106*, 1561.
(12) Folga, E.; Woo, T.; Ziegler, T. In *Theoretical Aspects of Homogeneous Catalysis;* van Leeuwen, P. W. N. M.; Morokuma, K.; van Lenthe, J. H., Eds.; Kluwer, Dordrecht 1995, p. 115 and references therein.
(13) Folga, E.; Ziegler, T.; Fan, L. *New. J. Chem.* **1991**, *15*, 741.
(14) Ziegler, T.; Folga, E.; Berces, A. *J. Am. Chem. Soc.* **1993**, *115*, 636.
(15) Cundari, T. R. *Int. J. Quantum Chem. Proc. Sanibel Symp.* **1992**, *26,*793.
(16) Cundari, T. R. *Chem. Phys.*. **1993**, *178*, 235.
(17) Cundari, T. R. *Organometallics* **1994**, *13*, 2987.
(18) Woodward, R. B.; Hoffmann, R. *The Conservation of Orbital Symmetry*, Academic Press, New York, **1970**.
(19) Fukui, H. *Acc. Chem. Res.* **1971**, *4*, 57.
(20) Crabtree, R. H.; Lavin, M.; Bonneviot, L. *J. Am. Chem. Soc.* **1986**, *108*, 4032.
(21) Joshi, A. M., James, B. R. *Organometallics*, **1990**, *9*, 199.
(22) Albeniz, A. C.; Heinekey D. M.; Crabtree, R. H. *Inorg. Chem.* **1991**, *30*, 3632.
(23) Morris, R. H. *Inorg. Chem.* **1992**, *32*, 1471.
(24) Albeniz, A. C.; Schulte G.; Crabtree, R. H. *Organometallics* **1992**, *11*, 242.
(25) Bianchini, C.; Meli, A.; Peruzzini, M.; Frediani, P.; Bohanna, C.; Esteruelas M. A.; Oro, L. A. *Organometallics* **1992**, *11*, 138;
(26) Sellmann, D.; Käppler J.; Moll, M. *J. Am. Chem. Soc.* **1993**, *115*, 1830.
(27) Burger, P.; Bergman, R. G.; *J. Am. Chem. Soc.* **1993**, *115*, 10462 and references therein
(28) Jessop, P.G.; Morris, R. H. *Inorg. Chem.* **1993**, *32*, 2236
(29) Hartwig, J. F.; Bhandari S.; Rablen, P. R. *J. Am. Chem. Soc.* **1994**, *116*, 1839.
(30) Lee, J. C.; Peris, E.; Rheingold A. L.; Crabtree, R. H. *J. Am. Chem. Soc.* **1994**, *116*, 11014.
(31) Kunai, A.; Sakurai, T.; Toyoda, E.; Ishikawa M.; Yamamoto, Y. *Organometallics* **1994**, *13*, 3233.

112

(32) Schlaf M.; Morris, R. H. *J. Chem. Soc. Chem. Commun.* **1995**, 625.
(33) Arndtsen, B. A.; Bergman, R. G.*Science*, **1995**, *270*, 1970.
(34) Schlaf M.; Lough, A. J.; Morris, R. H. *Organometallics* **1996**, *15*, 4423.
(35) Sellmann, D.; Rackelmann, G. H.; Heinemann, F. W. *Chem. Eur. J.* **1997**, *3*, 2071 and references therein.
(36) Versluis, L.; Ziegler, T. *J. Am. Chem. Soc.* **1990**, *9*, 2985.
(37) Musaev, D. G.; Mebel, A. M.; Morokuma, K. *J. Am. Chem. Soc.* **1994**, *116*, 10693.
(38) Hutschka, F.; Dedieu, A.; Leitner, W. *Angew. Int. Ed. Engl.* **1995**, *34*, 1742.
(39) Siegbahn, P. E. M.; Crabtree, R. H. *J. Am. Chem. Soc.* **1996** *118*, 4442.
(40) Hutschka, F.; Dedieu, A.; Eichberger, M.; Fornika, R.; Leitner, W. *J. Am. Chem. Soc.* **1997**, *119*, 4432.
(41)
(41) Milet, A.; Dedieu, A.; Kapteijn, G.; van Koten, G. *Inorg. Chem.* **1997**, *36*, 3223.
(42) Su, M.-D.; Chu, S.-Y. *J. Am. Chem. Soc.* **1997**, *119*, 5373.
(43) Musaev, D. G.; Svensson, M.; Morokuma, K.; Strömberg, S.; Zetterberg, K. Siegbahn, P. E. M. *Organometallics* **1997**, *16*, 1933.
(44) Musaev, D. G.; Froese, R. D. J.; Svensson, M.; Morokuma, K. *J. Am. Chem. Soc.* **1997**, *119*, 367.
(45) Lough, A. J.; Park, S.; Ramachandran R.; Morris, R. H. *J. Am. Chem. Soc.* **1994**, *116*, 8356.
(46) Lee, Jr., J. C.; Peris, E; Rheingold, A. L.; Crabtree, R. H. *J. Am. Chem. Soc.* **1994**, *116*, 11014.
(47) Xu, W.; Lough A. J.; Morris, R. H. *Inorg. Chem.* **1996**, *35*, 1549.
(48) Schlaf, M.; Lough A. J.; Morris, R. H. *Organometallics* **1996**, *15*, 4423.
(49) Park, S.; Lough, A. J.; Morris, R. H. *Inorg. Chem.* **1996**, *35*, 3001.
(50) Shubina, E. S.; Belkova, N. V.; Krylov, A. N.; Vorontsov, E. V.; Epstein, L. M.; Gusev, D. G.; Niedermann, M.; Berke, H. *J. Am. Chem. Soc.* **1996**, *118*, 1105.
(51) Ayllón, J. A.; Gervaux, C.; Sabo-Etienne, S.Chaudret, B. *Organometallics* **1997**, *16*, 2000.
(52) Peris, E.; Lee, Jr., J. C.; Rambo, J.; Eisenstein, O.; Crabtree, R. H. *J. Am. Chem. Soc.* **1995**, *117*, 3485.
(53) Liu, Q.; Hoffmann, R. *J. Am. Chem. Soc.* **1995**, *117*, 10108.
(54) Wessel, J.; Lee, J. C.; Peris, E.; Yap, G. P. A.; Fortin, J. B.; Ricci, J. S.; Sini, G.; Albinati, A.; Koetzle, T. F.; Eisenstein, O.; Rheingold, A. L.; Crabtree, R. H. *Angew. Chem. Int. Ed. Engl.* **1995**, *34*, 2507.
(55) Crabtree, R. H.; Siegbahn, P. E. M.; Eisenstein, O.; Rheingold, A. L.; Koetzle, T. F. *Acc. Chem. Res.* **1996**, *29*, 348.
(56) Belkova, N. V.; Shubina, E. S. ; Ionidis, A. V.; Epstein, L. M. ; Jacobsen, H.; Messmer, A.; Berke, H. *Inorg. Chem.* **1997**, *36*, 1522.
(57) Bosque, R.; Maseras, F.; Eisenstein, O.; Patel, B. P.; Yao, W.; Crabtree, R. H. *Inorg. Chem.* **1997**, *36*, 5505.
(58) Hutschka, F.; Dedieu, A. *J. Chem. Soc. Dalton Trans.* **1997**, 1899.
(59) Milet, A.; Dedieu, A.; Canty, A. J. *Organometallics* **1997**, *16*, 5331.
(60) Hutschka, F., Ph.D. Thesis, Université Louis Pasteur, Strasbourg, France, **1997**.

(61) Milet, A., Ph.D. Thesis, Université Louis Pasteur, Strasbourg, France, **1997**.
(62) Jessop, P. G.; Ikariya, T.; Noyori, R. *Chem. Rev.* **1995**, *95*, 259
(63) Leitner, W. *Angew. Chem. Int. Ed. Engl.* **1995**, *34*, 2207.
(64) Tsai, J. C.; Nicholas, K. M. *J. Am. Chem. Soc.* **1992**, *114*, 5117.
(65) Leitner, W.; Dinjus, E.; Gaßner, F. *J. Organomet. Chem.* **1994**, *475*, 257
(66) Lau, C.-P.; Chen, Y.-Z. *J. Mol. Catal. A: Chem.* **1995**, *101*, 33.
(67) Fornika, R.; Görls, H.; Seemann, B.; Leitner, W. *J. Chem. Soc. Chem. Commun.* **1995**, 1479.
(68) Jessop, P. G.; Hsiao, Y.; Ikariya, T.; Noyori, R. *J. Am. Chem. Soc.* **1996**, *118*, 344.
(69) Zhang, J. Z.; L;, Z.; Wang, H.; Wang, C. Y. *J. Mol. Cat. A: Chem.* **1996**, *112*, 9.
(70) Graf, E.; Leitner, W. *Chem. Ber.* **1996**, *129*, 91.
(71) Gassner, F.; Dinjus, E.; Görls, H.; Leitner, W. *Organometallics* **1996**, *15*, 2078.
(72) Scheiner, S.; *Acc. Chem. Res.* **1994**, *27*, 402.
(73) Chin, B.; Lough, A., Morris, R. H., Schweitzer, C. T.; D'Agostino, C. *Inorg. Chem.* **1994**, *33*, 6278.
(74) Craw, J. S.; Bacsay, G. B.; Hush, N. S. *J. Am. Chem. Soc.* **1994**, *116*, 5937.
(75) Bytheway, I.; Bacsay, G. B.; Hush, N. S. *J. Phys. Chem.* **1994**, *100*, 6023.
(76) Maseras, F.; Lledos, A.; Costas, M.; Poblet, J.P. *Organometallics* **1996**, *15*, 2947.
(77) Schlaf M.; Lough, A. J.; Maltby, P. A.; Morris, R. H. *Organometallics* **1996**, *15*, 2270.
(78) Maltby, P. A.; Schlaf, M.; Steinbeck, M.; Lough, A. J.; Morris, R. H.; Klooster, W. T.; Koetzle, T. F.; Srivastava, R. C. *J. Am. Chem. Soc.* **1996**, *118*, 5396.
(79) Brothers, P. J. In *Prog. Inorg. Chem. Vol. 28*, Lippard, S. J., Ed.; Wiley: New York **1981**, p. 1.
(80) Crabtree R. H. In *The Organometallic Chemistry of the Transition Metals (2nd Edition)* ; Wiley: New York, **1994**, pp. 220-221.
(81) Zimmer, M.; Schulte, G.; Luo, X.-L.; Crabtree, R. H. *Angew. Chem. Int. Ed. Engl.* **1991**, *30*, 193.
(82) Chinn, M. S.; Heinekey, D. M. *J. Am. Chem. Soc.* **1987**, *109*, 5865.
(84) Bianchini, C.; Marchi, A.; Marvelli, L.; Peruzzini, M.; Romerosa, A.; Rossi, R.; Vacca, A. *Organometallics* **1995**, *14*, 3203.
(85) Chan, W.-C.; Lau, C.-P.; Chen, Y.-Z., Fang, Y.-Q.; Ng, Si--M.; Jia, G. *Organometallics* **1997**, *16*, 34.
(86) Canty, A. J.; Fritsche, S. D.; Jin, H.; Skelton, B. W.; White, A. H. *J. Organomet. Chem.* **1995**, *490*, C18.
(87) Canty, A. J.; Jin, H.; Roberts, A. S.; Skelton, B. W. *Organometallics*, **1996**, *15*, 5713.
(88) Strout, D. L.; Zaric, S.; Niu, S. Hall, M. B. *J. Am. Chem. Soc.* **1996**, *118*, 6068.

Chapter 9

Reaction Mechanisms of Transition Metal Catalyzed Processes

Christian Boehme, Stefan Dapprich, Dirk Deubel, Ulrich Pidun, Martin Stahl, Ralf Stegmann, and Gernot Frenking

Fachbereich Chemie, Philipps-Universität Marburg, Hans-Meerwein-Strasse, D-35032 Marburg, Germany

This chapter gives a summary of four different projects in which the reaction mechanisms of transition metal homogeneously catalyzed reactions have been studied with quantum chemical methods. The geometries of the reacting agents, intermediates and transition states are optimized at the DFT level of theory (B3LYP) using relativistic effective core potentials for the metals and valence basis set of DZ+P quality.

1. Introduction

Transition metal compounds belong to the most powerful and important catalysts in synthetic and industrial chemistry. Although enormous efforts have been made to find out experimentally the mechanisms of transition-metal catalyzed processes, most reactions are still little understood. Theoretical work in this field has blossomed in the last 10 years, because DFT methods (1) and the use of effective core potentials (2-5) provided reliable and economic means to calculate accurately the structures and properties of heavy-atom molecules (6,7). Quantum chemical calculations are also used now to gain insight into the mechanisms of homogeneous catalysis (8). In this work we report about theoretical studies of three reactions which are catalyzed by transition metal compounds.

2. The Rearrangement of Acetylen to Vinylidene in the Coordination Sphere of a Transition Metal

Vinylidene is a shallow minimum on the C_2H_2 singlet potential energy surface. The barriere for rearrangement to acetylene, which is ~45 kcal/mol lower in energy than vinylidene, is < 1 kcal/mol (9). The reaction proceeds via migration of one C-H bond towards the empty p(π) orbital at C_α of ($^1A_1$) CCH$_2$ (Figure 1). The vinylidene-acetylene rearrangement has a much higher barrier if this orbital is partially occupied, as e.g. in the $^3B_1$ state or in the anion ($^2B_2$) $C_2H_2^-$ (Figure 1). Moreover, the transition state for the 1,2-hydrogen migration is nonplanar in the latter cases, while it is planar for the reaction of the neutral singlet state (10). This is important for the reaction of vinylidene complexes $L_nM=CCH_2$, because the metal->vinylidene π-backdonation leads to a partially filled p(π) orbital at C_α.

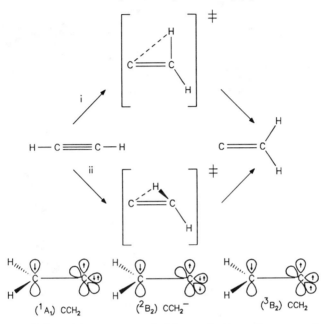

$(^1A_1)$ CCH$_2$ $(^2B_2)$ CCH$_2^-$ $(^3B_2)$ CCH$_2$

Figure 1. Schematic representation of the planar and nonplanar reaction path for the acetylene-vinylidene rearrangement (top). Relevant orbitals of vinylidene (bottom).

We calculated the reaction mechanism for the rearrangement of the acetylene complex $F_4W(HCCH)$ (**1**) to the vinylidene isomer $F_4(CCH_2)$ (**2**) at CCSD(T)/II//B3LYP/II (11,12). Two different pathways were found. One pathway is a two-step reaction via the alkynylhydrido complex $F_4WH(CCH)$ (**3**), which has high-lying transition states for the hydrogen migrations **1** -> **3** (**TS2**) and **3** -> **2** (**TS3**) (Figure 2). The second pathway is the direct 1,2-hydrogen migration **1** -> **2** via transition state **TS1**. Figure 2 shows that **TS2** has a nonplanar C_2H_2 moiety! The barrier for the one-step process via **TS2** (84.8 kcal/mol) is slightly lower than

116

the transition state **TS3** (85.5 kcal/mol). However, both reactions pathways have very high activation barriers which means that the <u>intra</u>molecular rearrangement **1** -> **2** is energetically very unfavorable and should rather proceed <u>inter</u>molecularly. This is

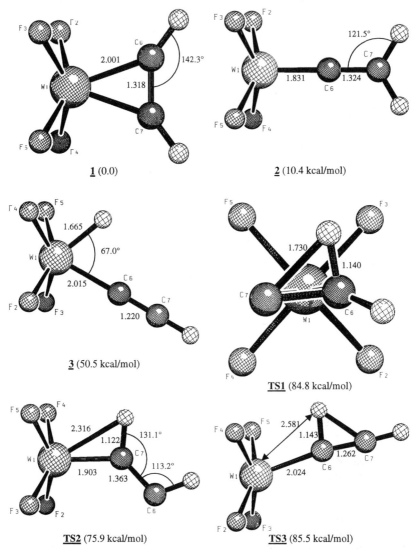

Figure 2. Optimized (B3LYP/II) energy minima **1 - 3** and transition states **TS1 - TS3** on the $F_4W(C_2H_2)$ potential energy surface

in agreement with recent work of Morokuma et al. (13) about the rearrangement of $Cl(PH_3)_2Rh(HCCH)$ to the vinylidene isomer (14).

3. The Mechanism of the McMurry Reaction

The reductive coupling of carbonyl compounds in the presence of low-valent titanium compounds is an important reaction in organic chemistry, which is used to synthesize a variety of olefines (15).

$$2\ R_1R_2C=O\ +\ [Ti]\quad \dashrightarrow\quad R_1R_2C=CR_1R_2\ +\ [Ti,O]$$

The active Ti agent is not known, but it is believed that the reaction proceeds heterogeneously via radical intermediates (15). It has recently been proposed that under specific reaction conditions, a nucleophilic pathway may be operative (16). The crucial steps of the postulated mechanism (Figure 3) are the formation of the pinacolate 7 and compound 6, which was suggested as the precursor for the final product, i.e. the olefin. Because no experimental evidence for intermediates 4, 5 and 7 could be given, the suggestions were highly speculative. We carried out quantum mechanical calculations in order to shed some light on the postulated mechanism (17).

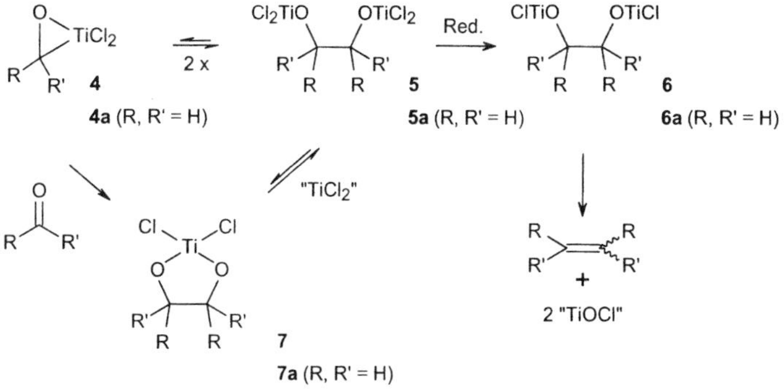

Figure 3. Postulated nucleophilic pathway of the McMurry reaction

Figure 4 shows the calculated reaction profile at B3LYP/II (12) for the formation of the model pinacolate 7a. Theory predicts that the initially formed Ti(II) compound binds the first carbonyl molecule in an side-on fashion yielding 4a, which then adds a second carbonyl compound side-on in complex 8a. The latter compound rearranges in an exothermic reaction with a low barrier (8.0 kcal/mol) to the pinacolate 7a.

118

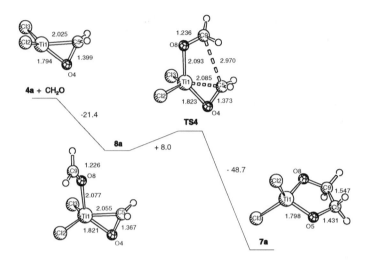

Figure 4. Calculated reaction profile (B3LYP/II) of the formation of the model pinacolate **7a**.

The more speculative reaction step concerns the formation of **5**. Figure 5 shows that dimerization of model compound **4a** yields first the complex **9a**, which then rearranges via **TS5** to the cyclic dimetallapinacolate **5a**, which has a rather long Ti—Ti bond. In spite of an extensive search we could not locate a diradical species on the potential energy surface as originally postulated for **5** (Figure 3). There is no experimental evidence for the formation of open-shell species under these reaction conditions (15). Compound **6** and not **5** has been suggested to be the immediate precursor for the formation of the olefin (Figure 3), because one more equivalent reduction agent is needed for the overall reaction.

Figure 6 shows that the fragmentation of the model compound **6a** yielding ethylene is exothermic, while the analogous fragmentation reaction of **5a** is clearly endothermic. This is in agreement with the experimental results. The optimization of **6a** leads also to a cyclic structure and not to a diradical species as originally suggested for **6** (Figure 3). The calculated results show that the proposed nucleophilic reaction pathway (16) is energetically feasible. The optimized geometries of the intermediates may help to identify them experimentally.

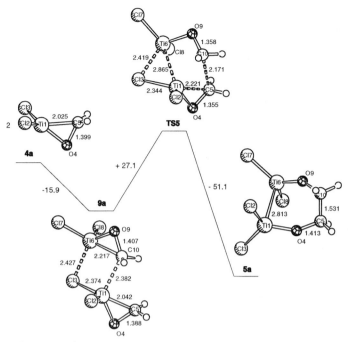

Figure 5. Reaction profile (B3LYP/II) for the formation of **5a**.

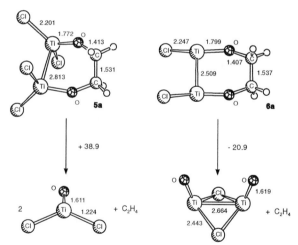

Figure 6. Energies (kcal/mol) for ethylene formation from **5a** and **6a**.

120

4. Substituent Effects on the Mechanism of OsO₄ Addition to Ethylene

Theoretical calculations have clearly shown that the addition of OsO_4 to olefines proceeds as a [3+2] cycloaddition, and not stepwise via a [2+2] addition with an osmaoxetane intermediate as originally suggested (18). This means that the substituent effect on the reaction rate might already be predicted from the frontier orbitals of the reactands. To this end we calculated at B3LYP/II (12) the barriers for the addition of substituted ethylene RHCCH₂ for R = F, CH₃, CHO, and NH₂. Figure 7 shows the optimized transition state structures. Table 1 gives the calculated activation barriers and reaction energies for the [3+2] addition reactions.

Table 1. Calculated (B3LYP/II) activation barriers $\Delta E^{\neq}$ and reaction energies ΔE_R [kcal/mol] for the addition of OsO_4 to olefins C_2H_3R. Energy levels [eV] of the HOMO and LUMO of C_2H_3R relative to ethylene.

R	$\Delta E^{\neq}$	ΔE_R	ϵHOMO	ϵLUMO
H	5.0	-32.3	0	0
F	7.0	-36.8	+0.08	-0.01
CHO	8.7	-24.2	-0.76	+0.85
CH₃	4.9	-32.5	+0.48	+0.31
NH₂	no barrier	-31.6	+2.18	+1.30

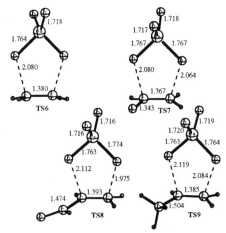

Figure 7. Optimized transition states (B3LYP/II) of OsO_4 addition to olefins.

The theoretically predicted activation barrier for the addition of OsO_4 to the parent olefine C_2H_4 (5.0 kcal/mol) becomes higher for R = F (7.0 kcal/mol) and for R = CHO (8.7 kcal/mol). This means that electron withdrawing substituents raise the activation barrier. The OsO_4 addition to propene is a little lower (4.9 kcal/mol) than for C_2H_4, while the addition of aminoethylene to OsO_4 proceeds without an activation energy. It follows that electron releasing substituents lower the activation barrier. There is no correlation between the activation barriers and the reaction energies.

Table 2 shows the results of the CDA partitioning scheme for the [3+2] transition states (19). For the parent system, the ethylene->OsO_4 donation is stronger than the ethylene<-OsO_4 backdonation by a factor of 1.32. Examination of the individual orbital contribution to the donation and backdonation shows that the dominant terms arise as expected from the HOMO(ethylene)-> LUMO(OsO_4) donation and LUMO(ethylene)<-HOMO(OsO_4) backdonation. It follows that the energy levels of the frontier orbitals of the olefin and particular ϵ_{HOMO} should already indicate if the barrier for the [3+2] addition increases or decreases.

Table 2. CDA results for the transition states of the [3+2] addition of OsO_4 to olefins C_2H_3R

R	donation (d) C_2H_3R->OsO_4	backdonation (b) C_2H_3R<-OsO_4	d/b	repulsion r C_2H_3R<->OsO_4
H	0.179	0.136	1.32	-0.290
F	0.192	0.161	1.19	-0.369
CHO	0.193	0.184	1.05	-0.406
CH_3	0.196	0.153	1.28	-0.348

Table 1 gives also the relative eigenvalues of ϵ_{HOMO} and ϵ_{LUMO} of the olefines with respect to ethylene (22) and the calculated activation barriers. Acrolein is the only olefin which has a lower lying C-C π MO than ethylene, while the C-C π^* orbital is raised in energy (20). Accordingly, the activation barrier is higher than for C_2H_4. The CDA results give nearly equal weight to the donation and backdonation (Table 2). The HOMO of ethyleneamine is much higher than for ethylene, which explains that there is no barrier for the addition reaction for R = NH_2. The HOMO and the LUMO of propene are slightly higher than in ethylene, which is in agreement with the small lowering of the energy barrier. The HOMO of fluoroethylene is a little bit higher in energy than in C_2H_4. This seems to be in conflict with the increase of the barrier for the [3+2] addition. Note that the π-

orbitals of the substituted olefins do not have equal weight at the two carbon atoms, and that they extend to the substituents X with a node between the CC-X π bond, which diminuishes the attractive interaction between C_α and the approaching oxygen atom. This may compensate for small changes in the eigenvalues of the frontier orbitals. In general, however, it can be said that the frontier orbitals of the substituted ethylene, and particularly the eigenvalue of the HOMO, may be used to estimate if the addition of OsO_4 becomes faster or slower relative to ethylene.

5. The Addition of Metal Oxides to Olefines

While the addition of OsO_4 to ethylene does not proceed via a metallaoxetane as intermediate (18), it is still possible that other metal oxides may add olefines in a stepwise mechanism via a [2+2] addition as a first step. In particular, the results of kinetic studies of the the the addition of Cp^*ReO_3 to various olefines have been interpreted as evidence for a stepwise mechanism, although a direct proof for the formation of a metallaoxetane intermediate could not be given (21). We calculated therefore the relevant transition states and energy minima of the C_2H_4 addition to ReO_4^-, $ClReO_3$ and $CpReO_3$ and compared them with the OsO_4 addition. The most important structures are shown in Figure 8. The calculated energies are given in Table 3.

Table 3. Calculated relative energies (B3LYP/II) of the stationary points for the [3+2] and [2+2] addition of metal oxides MO to ethylene [kcal/mol].

MO	MO+C_2H_4	TS(3+2)	TS(2+2)	oxetane	TS(Rearr)	Product
OsO_4	0.0	5.0	44.0	5.0	36.2	-32.3
ReO_4^-	0.0	37.7	47.2	21.6	97.0	19.4
$ClReO_3$	0.0	24.0	31.8	5.6	54.6	6.0
$CpReO_3$	0.0	14.1	25.4	-2.2	49.3	-21.0

The addition of ReO_4^- to ethylene is energetically much less favorable than the osmylation reaction. The formation of the metalladioxolane is endothermic with 19.4 kcal/mol, and the barrier for the [3+2] addition is 37.7 kcal/mol. This may be compared with the analogous reaction of isoelectronic OsO_4, which proceeds with a low barrier of 5.0 kcal/mol. The OsO_4 addition is exothermic with -32.3 kcal/mol. Note that the activation

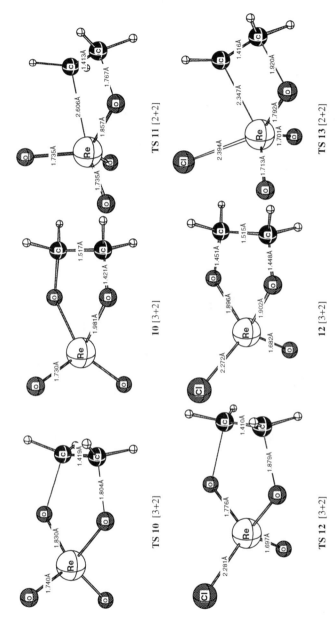

Figure 8 (a). Optimized (B3LYP/II) transition states of the [3+2] and [2+2] addition and the [3+2] reaction product for the addition of ReO₄⁻ (top) and ClReO₃ (bottom) to ethylene.

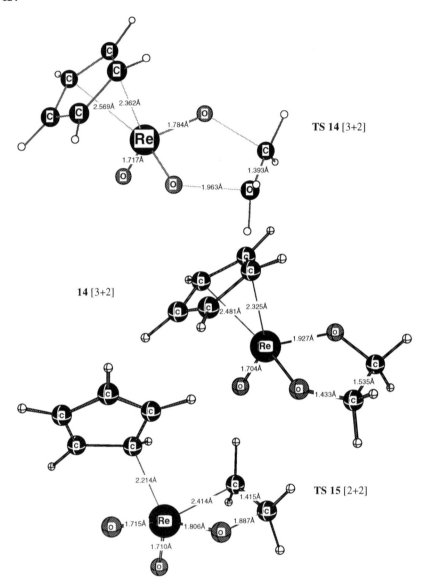

Figure 8 (b). Optimized (B3LYP/II) transition states of the [3+2] and [2+2] addition and the [3+2] reaction product for the addition of CpReO₃ to ethylene.

barrier for the [2+2] addition of ReO_4^- is only slightly higher (47.2 kcal/mol) than for OsO_4 (44.0 kcal/mol). The barrier for the rearrangement of the rheniumoxetane to the rheniumdioxolane (97.0 kcal/mol) increases substantially, however.

The calculations predict that $ClReO_3$ has a higher barrier for the [3+2] addition (24.0 kcal/mol) and a lower activation energy for the [2+2] addition (31.8 kcal/mol) than OsO_4. The transition state for the rearrangement of the metallaoxetane to the diolate remains very high (54.6 kcal/mol), which is even higher than the corresponding transition state for the osmylation reaction (Table 3). The transition states for the addition of $CpReO_3$ to ethylene are lower in energy than those of the $ClReO_3$ addition. However, the barrier for the [3+2] addition (14.1 kcal/mol) remains significantly lower than the activation energy for the [2+2] addition (25.4 kcal/mol). Thus, theory predicts that the addition of $CpReO_3$ should also proceed as a concerted [3+2] cycloaddition.

Acknowledgment

We thank the Deutsche Forschungsgemeinschaft (SFB 260-D19 and Graduiertenkolleg Metallorganische Chemie) and the Fonds der Chemischen Industrie for financial support. Excellent service was provided by the computer centers HRZ Marburg, HLRZ Darmstadt, HLRS Stuttgart and HLRZ Jülich.

Literature Cited

(1) Ziegler, T. Chem. Rev. **1991**, _91_, 651.

(2) (a) Dolg, M.; Wedig, U.; Stoll, H; Preuss, H. J. Chem. Phys. **1987**, _86_, 866. (b) Andrae, D.; Häußermann, U.; Dolg, M.; Stoll, H; Preuss, H. Theor. Chim. Acta **1990**, _77_, 123. (c) Bergner, A.; Dolg, M.; Küchle, W.; Stoll, H; Preuss, H. Mol. Phys. **1993**, _80_, 1431; (d) Dolg, M.; Stoll, H.; Savin, A.; Preuss, H. Theor. Chim. Acta **1989**, _75_, 173. (e) Dolg, M.; Stoll, H; Preuss, H. J. Chem. Phys. **1989**, _90_, 1730.

(3) (a) P. J. Hay and W. R. Wadt, J. Chem. Phys. _82_, 270 (1985). (b) W.R. Wadt and P.J. Hay, J. Chem. Phys. _82_, 284 (1985). (c) P.J. Hay and W.R. Wadt, J. Chem. Phys. _82_, 299 (1985).

(4) (a) Pacios, L. F.; Christiansen, P. A. J. Chem. Phys. **1985**, _82_, 2664. (b) Hurley, M.M.; Pacios, L.F.; Christiansen, P.A.; Ross B.A.; Ermler, W.C. J. Chem. Phys. **1986**, _84_, 6840. (c) LaJohn, L.A.; Christiansen, P.A.; Ross, R.B.; Atashroo T.; Ermler, W.C. J. Chem. Phys. **1987**, _87_,

126

2812. (d) Ross, R.B.; Powers, J.M.; Atashroo, T.; Ermler, W.C.; LaJohn, L.A.; Christiansen, P.A. J. Chem. Phys. 93, 6654 (1990).

(5) Stevens, W.J.; Krauss, M.; Basch H.; Jasien, P.G. Can. J. Chem. 1992, 70, 612.

(6) Frenking, G.; Antes, I.; Böhme, M.; Dapprich, S.; Ehlers, A.W.; Jonas, V.; Neuhaus, A.; Otto, M.; Stegmann, R.; Veldkamp, A.; Vyboishchikov, S.F. in Reviews in Computational Chemistry, Vol. 8, K.B. Lipkowitz and D.B. Boyd (Eds), VCH, New York, 1996, p. 63.

(7) Cundari, T.R.; Benson, M.T.; Lutz, M.L.; Sommerer, S.O. in Reviews in Computational Chemistry, Vol. 8, K.B. Lipkowitz and D.B. Boyd (Eds), VCH, New York, 1996, p. 145.

(8) Theoretical Aspects of Homogeneous Catalysis, P.W.N.M. van Leuwen, K. Morokuma, J.H. van Lenthe (Eds), Kluwer Academic Publishers, Dordrecht, 1995.

(9) (a) Gallo, M.M.; Hamilton, T.P.; Schaefer III, H.F.; J. Am. Chem. Soc. 1990, 112, 8714. (b) Petersson, G.A.; Tensfeldt, T.G.; Montgomery, J.A., Jr. J. Am. Chem. Soc. 1992, 114, 6133. (c) Jensen, J.H.; Morokuma, K.; Gordon, M.S.; J. Chem. Phys. 1994, 100, 1981. (d) Chen, Y.; Jonas, D.M.; Kinsey, J.L.; Field, R.W.; J. Chem. Phys. 1989, 91, 3976. (e) Ervin, K.M.; Ho, J.; Lineberger, W.C. J. Chem. Phys. 1989, 91, 5974. (f) Ervin, K.M.; Gronert, S.; Barlow, S.E.; Gilles, M.K.; Harrison, A.G.; Bierbaum, V.M.; DePuy, C.H.; Lineberger, W.C.; Ellison, G.B. J. Am. Chem. Soc. 1990, 112, 5750.

(10) Frenking, G. Chem. Phys. Lett. 1983, 100, 484.

(11) Stegmann, R.; Frenking, G. Organometallics, in print.

(12) Basis set II has a small-core relativistic ECP at W with a DZP-quality valence basis set and 6-31G(d) at the other atoms (6). B3LYP is the three-parameter fit of the nonlocal functionals introduced by Becke: Becke, A.D. J. Chem. Phys. 1993, 98, 5648. The calculations have been carried out using Gaussian 94: Frisch, M.J.; Trucks, G.W.; Schlegel, H.B.; Gill, P.M.W.; Johnson, B.G.; Robb, M.A.; Cheeseman, J.R.; Keith, T.A.; Petersson, G.A.; Montgomery, J.A.; Raghavachari, K.; Al-Laham, M.A.; Zakrzewski, V.G.; Ortiz, J.V.; Foresman, J.B.; Cioslowski, J.; Stefanov, B.B.; Nanayakkara, A.; Challacombe, M.; Peng, C.Y.; Ayala, P.Y.; Chen, W.; Wong, M.W.; Andres, J.L.; Replogle, E.S.; Gomberts, R.; Martin, R.L.; Fox, D.J.; Binkley, J.S.; Defrees, D.J.; Baker, I.; Stewart, J.J.P.; Head-Gordon, M.; Gonzalez, C.; Pople, J.A. Gaussian Inc., Pittsburgh, PA 1995

(13) Wakatsuki, Y.; Koga, N.; Werner, H.; Morokuma, K. <u>J. Am. Chem. Soc.</u> **1997**, <u>119</u>, 360.

(14) A nonplanar transition state for the direct 1,2-hydrogen migration for the system $Cl(PH_3)_2Rh(C_2H_2)$ was not reported in ref. 13, but it has later been found: Stegmann, R.; Frenking. G. to be published.

(15) McMurry, J.E. <u>Chem. Rev.</u> **1989**, <u>89</u>, 1513.

(16) (a) Fürstner, A.; Jumban, D.N. <u>Tetrahedron</u> **1992**, <u>48</u>, 5991. (b) Fürstner, A.; Hupperts, A.; Ptock, A.; Janssen, E. <u>J. Org. Chem.</u> **1994**, <u>59</u>, 5215. (c) Bogdanovic, B.; Bolte, A. <u>J. Organomet.</u> **1995**, <u>502</u>, 109.

(17) Stahl, M.; Pidun, U.; Frenking, G. <u>Angew. Chem.</u> **1997**, <u>109</u>, 2308; <u>Angew. Chem., Int. Ed. Engl.</u> **1997**, <u>36</u>, 2234.

(18) (a) Pidun, U.; Boehme, C.; Frenking, G. <u>Angew. Chem.</u> **1996**, <u>108</u>, 3008; <u>Angew. Chem., Int. Ed. Engl.</u> **1996**, <u>35</u>, 2817. (b) Dapprich, S.; Ujaque, G.; Maseras, F.; Lledós, A.; Musaev, D.G.; Morokuma, K. <u>J. Am. Chem. Soc.</u> **1996**, <u>118</u>, 11660. (c) Torrent, M.; Deng, L.; Duran, M.; Sola, M.; Ziegler, T. <u>Organometallics</u> **1997**, <u>16</u>, 13. (d) Del Monte, A.J.; Haller, J.; Houk, K.N.; Sharpless, K.B.; Singleton, D.A.; Straßner, T.; Thomas, A.A. <u>J. Am. Chem. Soc.</u> **1997**, <u>119</u>, 9907.

(19) Dapprich, S.; Frenking, G. <u>J. Phys. Chem.</u> **1995**, <u>99</u>, 9352.

(20) The C-C π-bonding orbital is the second highest occupied MO of acrolein, and the C-C antibonding π^* MO is the second lowest unoccupied orbital. The HOMO of acrolein is an oxygen lone-pair orbital, and the LUMO is a C-O π^* antibonding MO, which are not relevant for the [3+2] cycloaddition.

(21) (a) Gable, K.P.; Phan, T. N. <u>J. Am. Chem. Soc.</u> **1994**, <u>116</u>, 833. (b) Gable, K.P.; Juliette, J.J.J <u>J. Am. Chem. Soc.</u> **1995**, <u>117</u>, 955. (c) Gable, K.P.; Juliette, J.J. <u>J. Am. Chem. Soc.</u> **1996**, <u>118</u>, 2625.

(22) The absolute values for the frontier orbital energies are: $\epsilon_{HOMO}(C_2H_4) = -7.25$ eV, $\epsilon_{LUMO}(C_2H_4) = +0.51$ eV, $\epsilon_{HOMO}(OsO_4) = -10.06$ eV, $\epsilon_{LUMO}(C_2H_4) = -2.47$ eV.

Chapter 10

Catalysis of the Hydrosilation and Bis-Silylation Reactions

Brett M. Bode, Farhang Raaii, and Mark S. Gordon

Department of Chemistry, Iowa State University, Ames, IA 50011

Ab initio electronic structure calculations using RHF, MP2, and CCSD(T) levels of theory have been used to investigate a reaction path for the hydrosilation reaction catalyzed by divalent titanium (modeled by TiH_2 and $TiCl_2$). Optimized structures and energies are presented. All levels of theory predict a barrierless reaction path compared to a barrier of at least 55 kcal/mol at the MP2 level for the analogous uncatalyzed reactions. The use of correlated methods (MP2 or CCSD(T)) is required to obtain accurate structures and energies.

The hydrosilation reaction is a general method for adding an Si-H bond across a C-C double bond. This method encompasses a wide variety of substituted alkenes, dienes, and alkynes leading to many different organosilicon products. Thus the method is very useful; indeed it is the second most important method of producing organosilanes on a large scale (*1*). The general hydrosilation reaction may be written as:

$$R_3Si-H + R'_nA{=}BR''_n \xrightarrow{\text{catalyst}} R'_nA{-}BR''_n \text{ with } R_3Si \text{ and } H$$

One of the simplest examples known experimentally is the addition of trichlorosilane to ethylene, which will occur rapidly at room temperature and give nearly 100% yields with a variety of homogeneous transition metal based catalysts (*2*).

Several analogous uncatalyzed reactions ($HSiCl_3$, SiH_4 + ethylene, SiH_4 + propene) were studied previously (*3*); all were found to have large ($\geq$54 kcal/mol) barriers. Thus, the catalyst is crucial in making the process economically viable. Industrially one active catalyst is believed to be a divalent Cp_2Ti species (Cp = C_5H_5) (*1*).

This paper will consider first the simplest prototypical example of a catalyzed hydrosilation reaction, in which A and B are carbon; R, R' and R" are hydrogen and the catalyst is TiH_2. This choice of reactants and catalyst allows mapping the entire reaction path at a high level of theory. Particularly, the choice of TiH_2 as the catalyst allows the use of high-level all-electron *ab initio* wavefunctions which would not be possible if the more complex catalysts such as $TiCl_2$ or $TiCp_2$ were used. The significant differences between the all hydrogen calculations and calculations involving Cl substituents will be discussed.

Computational Methods

The minimum energy reaction path connecting reactants to products was determined using all electron *ab initio* wavefunctions. The reaction paths were fully optimized using Møller-Plesset second order perturbation theory (MP2). Single-point energies were computed at the MP2 optimized stationary points using coupled cluster singles and doubles plus perturbative triples (CCSD(T)). The basis sets used were a triple-ζ quality valence for the all hydrogen reaction (4) and a double-ζ quality valence using SBKJC effective core potentials for the reactions involving chlorine (5).

The GAMESS (6) program was used for all of the RHF calculations and most of the MP2 optimizations. The Gaussian 92 suite of programs (7) was used for the remainder of the MP2 calculations and the CCSD(T) calculations.

Results and Discussion

Figure 1 shows the energy profile of the proposed catalyzed reaction for the unsubstituted system at the RHF, MP2, and CCSD(T) levels of theory. Figure 2 shows the MP2 energy profiles for each of the reaction studied. The zero of energy on the curve for each level of theory is the sum of the reactant energies at that level of theory (structures **a**, **b**, and **c** in Figure 3). The MP2 structures at each stationary point are given in Figure 3. Only structures for the unsubstituted reaction are shown, unless there is an important change upon Cl substitution. MP2 and CCSD(T) ZPE corrected energies are listed relative to the zero of energy in Table I.

Unsubstituted Reaction. It is important to note that all points on the energy plot in Figure 1 lie below the energy of the reactants, in contrast to the large barrier in the uncatalyzed reaction. Note also that there are large differences between the SCF and MP2 energy profiles, while the differences between MP2 and CCSD(T) are much smaller. So, electron correlation is essential for a correct description of this reaction surface, and MP2 is qualitatively correct. This provides some confidence in the MP2 results for the reactions containing Cl (see below).

There are two possibilities for the first step of the reaction, both of which are barrierless processes. The first, and more exothermic, is to add the TiH_2 catalyst across the ethylene double bond to form the three membered ring compound shown in Figure 3d. This process is downhill in energy by 61.9 (53.4) kcal/mol at the ZPE corrected MP2 (CCSD(T)) level of theory. Note that, based on the large exothermicity and the large (0.016Å) increase in the CC bond length, structure **d** is a three-membered ring, not a π complex. Silane will then add to form the complex depicted in Figure 3e. This second barrierless addition is downhill by 6.5 (6.0) kcal/mol.

The electronic structure of TiH_2 was considered in detail previously (8). Like CH_2, the ground state is a triplet, and the lowest singlet state is 21 kcal/mol higher in energy. Since TiH_2 has an electronic structure similar to singlet CH_2 or SiH_2, $(s^2d^2$ vs. $s^2p^2)$ a reasonable expectation is that an alternative mechanism would start with an insertion of TiH_2 into an Si-H bond of silane. Indeed, this occurs with no barrier to produce structure **d'**, a Ti-Si analog of ethane. This step is downhill by 31.9 (27.8)

130

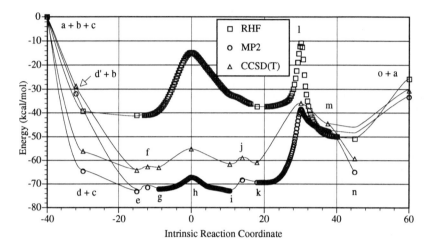

Figure 1: Energies for the Unsubstituted Reaction at RHF, MP2, and CCSD(T) Levels of Theory. (Reproduced with Permission from ref. 4. Copyright 1998 American Chemical Society)

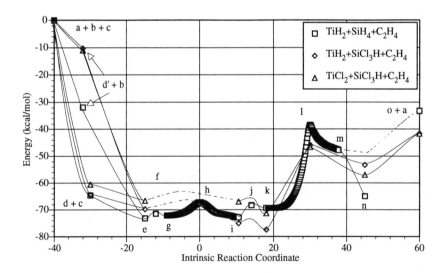

Figure 2: Energies for all reactions at the MP2 Level of Theory.

Table I: MP2 and CCSD(T) relative energies (kcal/mol) with MP2 ZPE correction

Geometry point	$TiH_2+SiH_4+C_2H_4$ MP2 + MP2 ZPE	$TiH_2+SiH_4+C_2H_4$ CCSD(T) + MP2 ZPE	$TiH_2+SiCl_3H+C_2H_4$ MP2 + MP2 ZPE	$TiCl_2+SiCl_3H+C_2H_4$ MP2 + MP2 ZPE
a+b+c (reactants)	0	0	0	0
d'+b	-31.1	-27.8	-8.3	-10.6
d+c	-61.9	-53.4	-61.8	-59.6
e	-68.4	-59.4	-66.2	-64.9
f	-66.6	-57.8		
g	-66.6	-57.4		
h	-61.3	-49.2		
i	-67.2	-55.8	-70.0	-64.1
j	-64.0	-54.6		
k	-65.4	-56.8	-74.3	-70.5
l	-33.5	-31.1	-42.3	-44.5
m	-39.0	-37.0		
n	-59.2	-53.3	-46.9	-51.8
o+a (products)	-28.0	-25.4	-37.2	-37.2

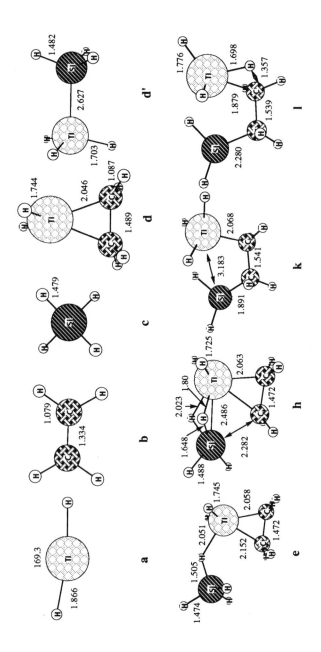

133

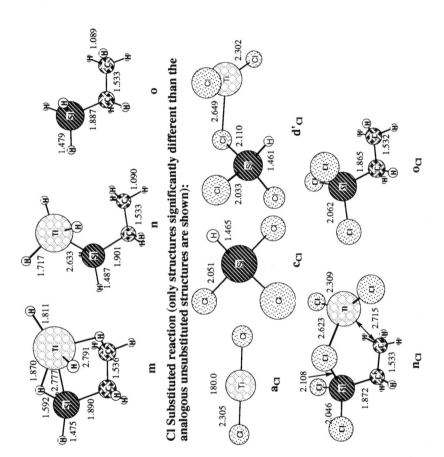

Cl Substituted reaction (only structures significantly different than the analogous unsubstituted structures are shown):

Figure 3: Significant MP2 Structures Along the Minimum Energy Reaction Path.

kcal/mol. When ethylene is added to this compound, it rearranges with no barrier to the same structure as in Figure 3e.

So, whether the catalyst adds to ethylene or silane initially, the net result after the two reactants and the catalyst have been added together is the formation of compound e, with no intervening barrier. The overall exothermicity to this point is 68.4 (59.4) kcal/mol. This very large drop in energy drives the entire reaction path down in energy. In fact, the reaction path is forced down enough that all subsequent points are below the reactants in energy.

Recall that the final desired product is ethylsilane. So, starting from compound 3e, the silyl group needs to migrate to the nearest (α) carbon, and a hydrogen needs to be transferred to the adjacent (β) carbon, with the ultimate removal of the TiH$_2$ catalyst. Therefore, the next step in the reaction is to transfer a H from the complexed silane to the Ti and to attach the Si to the α C.

The first part of this step involves rotation of the silane such that there are 2 bridging hydrogens between the Si and Ti; that is e->g via f in Figures. 1 and 3. At the MP2 level there is a small barrier (at structure f) of 1.7 kcal/mol to this process, but after the ZPE corrections are added the barrier disappears. The reaction then proceeds through transition state h with a barrier of 5.3 (8.2) kcal/mol, leading to the four-membered ring shown in Figure 3i. This ring is 5.9 (6.6) kcal/mol below the TS h. The four-membered ring can be opened up by breaking the Si-Ti bond to give the compound k. The TS for this step is shown in Figure 3j. It has a barrier height of 3.2 (1.2) kcal/mol. Compound k is 1.4 (2.1) kcal/mol below the TS j.

The final step in the process is to regenerate the catalyst by transfer of a hydrogen from Ti to C and elimination of TiH$_2$. The transition state for this process is shown in Figure 3l; the associated barrier height is 31.9 (25.7) kcal/mol. This TS is still 33.5 (31.1) kcal/mol lower in energy than the initial reactants. The IRC from this TS leads to the structure shown in Figure 3m which is 5.2 (5.9) kcal/mol below the TS. The structure shown in Figure 3m is not a stationary point, but it illustrates that the reaction path goes through a structure in which the TiH$_2$ is complexed to two hydrogens. Optimization from this point leads to the insertion of TiH$_2$ into an Si-H bond, as shown in Figure 3n. The insertion product is 25.7 (22.2) kcal/mol below the TS in energy. However, the TiH$_2$ in Figure 3m is not tightly bound to the ethylsilane as evidenced by both the relatively long Ti-Si and Ti-H bond distances and the fact that it is a modest 11.0 (11.6) kcal/mol in energy below separated products. Thus, we do not expect a transition state for the process of simply abstracting the TiH$_2$ to form separated products. This can occur readily due to the excess energy available to the system, since the separated products are 28.0 (25.4) kcal/mol below the reactants in energy.

Once TiH$_2$ is removed, the process is complete with ethylsilane as the product. The overall process is exothermic by 28.0 (25.4) kcal/mol at the ZPE corrected MP2 (CCSD(T)) level of theory. This compares with the value of 29.1 kcal/mol computed by Day and Gordon at the MP2/6-311G(d,p) level of theory and a value of 27.4 kcal/mol computed by McDouall et. al. (9) at the MP4/6-31G(d)//HF/3-21G level of theory. There does not seem to be a good experimental ΔH_0 for this reaction, but we can estimate the value to be 28.9 kcal/mol based on the experimental heats of formation for ethylene and silane (10), and the best previous theoretical heat of formation for ethylsilane

The driving force for the entire reaction comes in the first two steps with the formation of the compound shown in Figure 3e which is 68.4 (59.4) kcal/mol below the reactants in energy and is the global minimum on the reaction surface. The first two steps in the reaction illustrate the reasons this structure is so stable. In the first step, the electron deficient TiH_2 adds to the ethylene across the π bond in much the same manner as the addition of CH_2 to ethylene to form cyclopropane. The second step is much less exothermic and is driven mostly by the electrostatic attraction between the positively charged titanium (+0.83) and the negative hydrogen (-0.12) on the silicon.

Effect of Cl substitution. Now consider the effect of substituting Cl atoms for hydrogens in either the catalyst or the reactants. The relevant structures are shown in Figure 3 and the potential energy curves in Figure 2. Since, as noted above, MP2 and CCSD(T) give similar results for the unsubstituted reaction, only MP2 results are reported for the reactions containing Cl. The overall features of the Cl-substituted reactions are quite similar to those for the unsubstituted case. In particular, the highest point in all cases corresponds to the separated reactants + catalyst. The proximity of the catalyst once again drives the energy strongly downhill in order to form stable intermediate complexes, and this stabilization is sufficient keep all subsequent transition states well below the reactants in energy. In view of the similarities among the reactions considered, we will only briefly discuss the essential differences:

(1) Recall that there are two alternative initiating reactions, both of which are considerably downhill without barriers. One of these corresponds to insertion of TiH2 into an Si-H bond of silane to form structure **d'**. However, when the substrate is $SiCl_3H$, this insertion (by either TiH_2 or $TiCl_2$) is not a barrierless process and thus does not occur. Instead, as illustrated in Figure 3, the catalyst forms a complex (**d'**) in which the Ti is interacting with one of the chlorines on Si. These complexes, which are downhill by 8.3 (10.6) kcal/mol for TiH_2 ($TiCl_2$) will also rearrange readily to structure **e** when ethylene is added in the next step.

(2) In the unsubstituted reaction, Figure 2 illustrates the tendency of the catalyst to reinsert into the departing product. When $SiCl_3H$ is used instead of SiH_4, the catalyst would have to insert into a Si-Cl bond instead of a Si-H bond. A barrier exists for this insertion, so instead a complex (**n**) is formed between the departing product and the catalyst. This complex is only 9.7 (14.6) kcal/mol below the final products when the catalyst is TiH_2 ($TiCl_2$), so it is not difficult to remove the catalyst from the ethyltrichlorosilane product.

(3) The reaction that forms ethyltrichlorosilane is exothermic by 37.2 kcal/mol. This is the same as the exothermicity predicted earlier using the MP2/6-311G(d,p) level of theory.

Bis-Silylation. The addition of a Si-Si bond across a C=C bond,

$$X_3Si\!\!-\!\!SiX'_3 + R_2C\!\!=\!\!CR'_2 \longrightarrow \begin{array}{cc} X_3Si & SiR'_3 \\ | & | \\ (R_2)C\!\!-\!\!\!-\!\!C(R'_3) \end{array}$$

is referred to as bis-silylation. Although the Si-Si single bond is fairly weak, this addition in the absence of a catalyst requires a high energy barrier, about 50 kcal/mol at the MP4/6-311G(d,p) level of theory (11). The first investigation of the catalyzed bis-silylation reaction used TiH_2 as a prototype catalyst to be consistent with our previous studies of hydrosilation (12); however, the most common catalysts for this reaction are phosphino complexes of Pt and Pd (13-18). Several potential reaction paths were explored, with the catalyst interacting first with ethylene in some cases and first with disilane in others. All structures were optimized using a triple ζ quality basis set and Møller-Plesset second-order perturbation (MP2) theory. The results of these calculations suggest that TiH_2 is such an effective catalyst for hydrosilation that

136

an Si-H bond of disilane persistently adds across the C=C bond of ethylene. No transition state or route to bis-silylation (addition of Si_2H_6 to ethylene) has been found in the presence of TiH_2. This may change for a fully substituted disilane, since then there would be no Si-H bond into which the divalent Ti could insert.

Conclusions

The results presented here clearly show that divalent titanium is an effective catalyst for the hydrosilation reaction. The simple model system using all hydrogen substituents seems to get all of the important features of the reaction pathway qualitatively correct. The effect of the substituents on the catalyst seems to be minor. More significant is the effect of substituents on the silicon with the chlorine substituted system being less prone to Ti insertion into an Si-X bond. The overall catalyzed reaction has no net barrier, because of the very stable cyclic $TiX_2CH_2CH_2$ intermediate. However, the energy profile of the multistep process (Figure 1) does offer the possibility of finding some of the intermediate structures if the process was carried out at low temperature.

Acknowledgments

The work described in this chapter was supported by grants from Iowa State University in the form of a Department of Education GAANN fellowship awarded to BMB and in the form of a grant to purchase computers used in this project, the National Science Foundation (CHE-9633480) and the Air Force Office for Scientific Research (F49-620-95-1-0073). The computations were performed in part on computers provided by Iowa State University and through a grant of computer time at the San Diego Supercomputer Center.

References

1. Barton, T.J.; Boudjouck, P "Organosilicon chemistry - a brief overview", in Advances in Silicon-Based Polymer Science, Advances in Chemistry Series No. 224, Ziegler J.; Fearon, F. W. G. (eds) American Chemical Society, Washington, DC. 1990, 3-46.
2. Speier, J. L. Adv. Organomet. Chem. 1979, 17, 407-447.
3. Day, P.N.; Gordon, M.S. Theor. Chim. Acta. 1995, 91, 83-90.
4. Bode, B.M.; Gordon, M.S.; Day, P.N. J. Am. Chem. Soc. 1998, 120, 1552-1555.
5. The details of the calculations involving non-hydrogen substituents will be presented in an upcoming paper in J. Am. Chem. Soc.
6. Schmidt, M. W.; Baldridge, K. K.; Boatz, J. A.; Elbert, S. T.; Gordon, M. S.; Jensen, J. H.; Koseki, S.; Matsunaga, N.; Nguyen, K. A.; Su, S.; Windus, T. L.; Dupuis, M.; Montgomery, J. A. Jr. The general atomic and molecular electronic structure system. J. Comp. Chem. 1993, 14, 1347-1363.
7. Gaussian 92 rev C, Frisch, M. J.; Trucks, G. W.; Head-Gordon, M.; Gill PM. W.; Wong, M. W.; Foresman, J. B.; Johnson, B. G.; Schlegel, H. B.; Robb, M. A.; Replogie, E. S.; Gomperts, R.; Andres, J. L.; Raghavachari, K.; Binkley, J. S.; Gonzalez, C.; Martin, R. L.; Fox, D. L.; Defrees, D. J.; Baker, J.; Stewart JJ. P.; Pople, J. A. Gaussian Inc., Pittsburg PA, 1992.
8. Kudo, T.; Gordon, M.S. J. Chem. Phys. 1995, 102, 6806-6811.

9. McDouall, J. J. W.; Schlegel, H. B.; Francisco, J. S. *J. Am. Chem. Soc.* **1989**, *111*, 4622-4627.
10. Wagman, D. D.; Evans, W. H.; Parker, V. B.; Schumm, R. H.; Halow, I.; Bailey, S. M.; Churney, K. L.; Nuttall, R. L. *J. Phys. Chem. Ref. Data Suppl.* **1982**, *11*, 2.
11. Raaii, F.; Gordon, M. S.*J. Phys. Chem.*, in press.
12. Raaii, F.; Gordon, M. S.unpublished results.
13. Kobayashi, T.; Hayashi, T.; Yamashita, H.; Tanaka, M. *Chem. Lett.* **1989**, 467.
14. Hayashi, T.; Kobayashi, T.; Kawamoto, A.M.; Yamashita, H.; Tanaka, M. *Organometallics*, **1990**, *90*, 33.
15. Okinoshima, H.; Yamamoto, K.; Kumada, M. *J. Organomet. Chem.* **1975**, *86*, C27.
16. Ito, Y.; Suginome, M.; Murakami, M. *J. Org. Chem.* **1991**, *56*, 1948.
17. Yamashita, H.; Catellani, M.; Tanaka, M. *Chem. Lett.* **1991**, 241.
18. Yamashita, H.; Tanaka, M. *Chem. Lett.* **1992**, 1547.

Chapter 11

Theoretical Studies of Inorganic and Organometallic Reaction Mechanisms 13: Methane, Ethylene, and Acetylene Activation at a Cationic Iridium Center

Shuqiang Niu, Douglas L. Strout[1], Snežana Zarić[2], Craig A. Bayse, and Michael B. Hall

Department of Chemistry, Texas A&M University, College Station, TX 77843-3255

The oxidative-addition/reductive-elimination (OA/RE) reactions of methane, ethylene and acetylene with the $CpIr(PH_3)(CH_3)^+$ complex are investigated by ab initio methods and density functional theory (DFT). The calculated results shows that the OA reaction from $CpIr(PH_3)(CH_3)(agostic-alkane)^+$ to $CpIr(PH_3)(CH_3)(H)(alkyl)^+$ is endothermic by 4.4 and 0.8 kcal/mol with a low barrier of 11.5 and 10.0 kcal/mol at the DFT-B3LYP and coupled cluster with singles and doubles (CCSD) levels of theory, respectively. The RE reaction from $CpIr(PH_3)(CH_3)(H)(alkyl)^+$ to a β-agostic complex, $CpIr(PH_3)(alkyl)^+$, is exothermic with a low barrier of 7.1 and 9.2 kcal/mol. A strong stabilizing interaction between either ethylene or acetylene and $CpIr(PH_3)(CH_3)^+$ leads to a high activation barrier (24-36 kcal/mol) for the OA processes of either one. Compared to ethylene, the OA/RE reaction of acetylene with $CpIr(PH_3)(CH_3)^+$ complex is more favorable. Thus, the dimerization of terminal alkynes catalyzed by cationic iridium complexes is plausible.

Transition metal-catalyzed reactions can be regarded as consisting of several elementary reactions such as oxidative addition, reductive elimination, migratory insertion, β-hydride elimination, σ-bond metathesis, and nucleophilic addition (1). Numerous experimental and theoretical studies have been undertaken in order to understand these pivotal fundamental steps (1,2). Because of enormous progress in computational chemistry, today one can also determine accurate geometries and relative energies of the species involved in these reactions from first principles. This is especially important in those cases where experimental results are difficult to obtain (1,3).

[1] Current address: Environmental Molecular Sciences Laboratory, Pacific Northwest National Laboratory, Richland, WA 99352.
[2] Permanent address: Faculty of Chemistry, University of Belgrade, Studentski trg 16, 11001 Beograd, Yugoslavia.

Both early and late transition metals undergo many of the elementary reactions mentioned above, but the mechanism often differs in important details. In some instances, the differences are so large that the reactions are only superficially similar. For example, consider carbon—hydrogen bond activation, both early and late transition metals can break C—H bonds, but the mechanism is very different. Early transition metals, often d^0 system such as Sc(III), activate C—H bond through a σ-bond metathesis (4) as shown in equation 1.

$$[M]\!-\!CH_3 + RH \longrightarrow \left[\begin{array}{c} R\cdots\cdots H \\ \vdots \quad\quad \vdots \\ [M]\text{-}\text{-}\text{-}CH_3 \end{array}\right]^{\ddagger} \longrightarrow [M]\!-\!R \quad + \quad CH_4 \quad (1)$$

By contrast, late transition metals, such as Ir(I) or Rh(I), accomplish C—H bond activation through an oxidative-addition (5) process as shown in equation 2. Thus, the early metals more often display catalytic hydrogen exchange while the late metals, especially the heavier ones, tend toward stoichiometric reactions.

$$[M]\!-\!CH_3 + RH \longrightarrow \left[R\!-\!\begin{array}{c} H \\ | \\ [M] \end{array}\!-\!CH_3 \right] \longrightarrow [M]\!-\!R \quad + \quad CH_4 \quad (2)$$

Bergman and co-workers recently reported an Ir(III) system which catalyzes both hydrogen exchange in methane and alkanes in the solution-phase at room temperature and generates olefin complexes (6). It shows a striking similarity in hydrogen exchange to the reaction of early metals. Because the behavior of this Ir(III) system, $Cp^*Ir(PR_3)(CH_3)^+$ [$Cp^* = \eta^5\text{-}C_5(CH_3)_5$], is quite different from the now more common Ir(I) systems, Bergman and co-workers suggested two possible reaction mechanisms: (i) the reaction resembles an early metal reaction and proceeds through a σ-bond metathesis, as in equation 1, or (ii) the reaction resembles a late metal and proceeds through an oxidative-addition/reductive-elimination (OA/RE) pathway, as in equation 2. The former is unexpected for a late metal, while the latter involves Ir(III) going to Ir(V), a fairly high oxidation state for an Ir system without electronegative ligands. Recently, several experimental and theoretical studies have been performed to elucidate the details of this reaction mechanism (7).

In this work, the hydrogen exchange of methane (**1**), ethylene (**2**), and acetylene (**3**) with the Ir(III)-methyl complex are investigated with density functional theory (DFT) and ab initio molecular orbital theory.

Computational Details

The geometry optimizations in this work have been performed using DFT (8) methods, specifically the Becke three parameter hybrid exchange functional (8b-d) and the Lee-Yang-Parr correlation functional (B3LYP) (8e). The transition states were optimized using a quasi-Newton method and characterized by determining the number of imaginary frequencies (9). For a direct estimation of electron correlation effects and accurate reaction energetics, coupled cluster with singles and doubles (CCSD) (10) calculations were carried out on the DFT-optimized geometries. To simplify calculations we replaced the phosphine group and the Cp^* ring in the actual molecules by a PH_3 group and a Cp ring (Cp = $\eta^5\text{-}C_5H_5$) (11). Differences in the zero-point energy (ZPE) and thermal corrections to the reaction energy tend to be small, so they are not included in the energy calculations. For example, we determined ZPE and thermal corrections for the transformations $H_2Ti(CHCH_2)(H) \rightarrow H_2Ti(\eta^2\text{-}CH_2CH_2)$ and $H_2Ti(\eta^2\text{-}CH_2CH_2)(H)^+ \rightarrow H_2TiCH_2CH_3^+$. The ZPE corrections were 2.6 and 2.5 kcal/mol and the thermal corrections were 2.0 and 1.4 kcal/mol, respectively (12).

Iridium was described by a modified version of the Hay and Wadt basis set with effective core potentials (ECP) (13a). The modifications to the double-ζ basis set were made by Couty and Hall (13b) and give a better representation of the 6p space. The result is a [3s3p2d] contracted basis set for iridium, where the 5s and 5p basis functions are left totally contracted but the 6s, 6p, and 5d are split (41), (41), and (21), respectively. The carbons and hydrogens are described using the Dunning-Hay double-ζ basis functions (14a). The Hay and Wadt ECP double-ζ basis set is used for phosphorus (14b). At the CCSD//B3LYP level, the association energies of methane, ethylene, and acetylene with CpIr(PH$_3$)(CH$_3$)$^+$ complex have been corrected for the basis-set superposition error (BSSE) (14c).

All ab initio and DFT calculations were performed with GAMESS-UK (15) and GAUSSIAN94 programs (16), at the Cornell Theory Center on IBM ES6000 and Scalable Powerparallel (SP2), at Cray Research, Inc. on the Cray C90, at the Supercomputer Center of Texas A&M University and the Department of Chemistry on Silicon Graphics Power Challenge servers, and on Silicon Graphic Power Indigo II IMPACT 10000 workstations in our laboratory and at the Institute of Scientific Computation (ISC) of Texas A&M University.

Results and Discussion

A. Methane C—H Bond Activation. The experimental work of Bergman and co-workers shows that alkanes can be activated by Cp*Ir(PMe$_3$)(CH$_3$)$^+$ at room temperature to generate olefin complexes (6). The reaction initially produces a new alkyl Ir complex through either oxidative-addition/reductive-elimination (OA/RE) or σ-bond metathesis and then produces an Ir-olefin complex through β-H transfer as illustrated in Scheme 1. Here, the methane C—H bond activation process by the two mechanisms will be examined.

Scheme 1

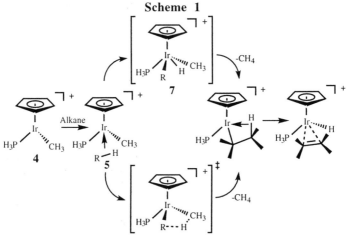

The B3LYP optimized geometries for the reactant (**4**), agostic intermediate (**5**), transition state (**6**), and oxidative-addition intermediate (**7**) along the OA/RE pathway are shown in Figure 1.

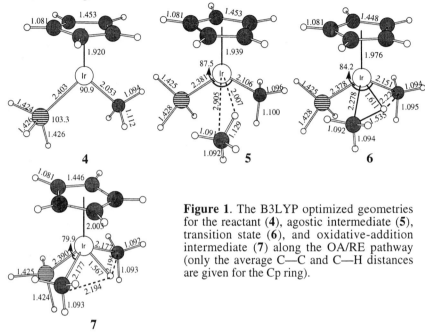

Figure 1. The B3LYP optimized geometries for the reactant (**4**), agostic intermediate (**5**), transition state (**6**), and oxidative-addition intermediate (**7**) along the OA/RE pathway (only the average C—C and C—H distances are given for the Cp ring).

The complete pathway for methyl exchange at the iridium center is symmetric with respect to the oxidative-addition intermediate (**7**). In intermediate **5**, a C—H bond of methane is close to the metal center with an Ir—H distance of 2.007 Å, and the inside C—H distance is longer by 0.034 Å than that of a free methane molecule. Such structural features point to the presence of an Ir—H—C agostic interaction in **5** (17). In transition state **6**, there is significant Ir—C and Ir—H bond making. In intermediate **7**, the C—H bond is completely broken, while the Ir—H bond and a new Ir—CH$_3$ bond are fully formed. It is noteworthy that the Cp ring of **7** is clearly slipped with respect to reactant **4** as shown in Scheme 2 (18a). Clearly, with the oxidative-addition going from **1** + **4** to **7**, the Ir—Cp and Ir—CH$_3$ distances of **5**, **6**, and **7** are increasing and the Cp ring slips more at each step.

Scheme 2

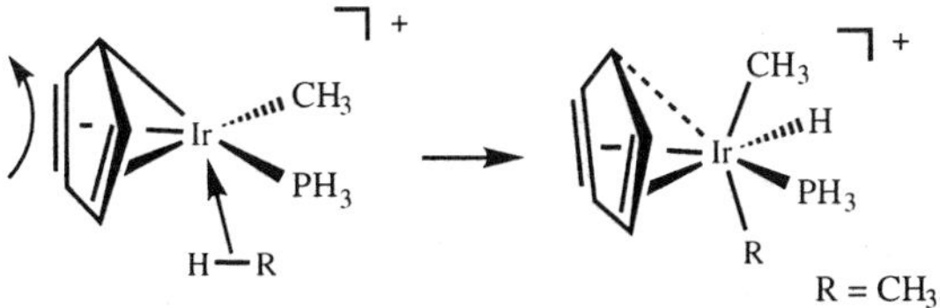

The energy profile at the B3LYP level is presented in Figure 2 and the relative energies at the B3LYP and CCSD//B3LYP levels are summarized in Table I (Reaction 1). Compared to the reactants (**1** and **4**), the agostic structure **5** is more stable by 1.0

and 6.4 kcal/mol at the B3LYP and CCSD//B3LYP levels. However, after corrections for the basis-set superposition error (BSSE) at the CCSD//B3LYP level, **5** is only 0.9 kcal/mol more stable than **1** and **4**. Calculations of the alkane association energy in better basis sets would likely yield a value close to 4 kcal/mol for methane and higher values for larger alkanes (18b). The data for the C—H activation of methane (**1**) shows that the barrier to bond activation, with respect to the agostic structure, is similar for both B3LYP and CCSD//B3LYP, 11.5 and 10.0 kcal/mol, respectively. The oxidative-addition intermediate (**7**) is slightly less stable, by 4.4 (B3LYP) and 0.8 kcal/mol (CCSD//B3LYP), than the agostic one (**5**). The steric effect between the Cp ring and larger phosphine ligand used in the experimental work will lead to an increase in the energy difference between **7** and **5** (19a), which is consistent with the fact that intermediate **7** has not been observed experimentally (6). On the other hand, a decrease of the steric effect between the phosphine and alkyl ligands may lead to an increase in stability of an Ir(V) complex. According to our results, the CpIr(PH$_3$)(H)$_2$(CH$_3$)$^+$ complex could be trapped at low temperature (19b). The reductive-elimination from **7** through the OA/RE TS, **6**, to the agostic complex, **5**, is exothermic with a low barrier of 7.1 and 9.2 kcal/mol for B3LYP and CCSD//B3LYP, respectively.

Table I. Relative Energies (ΔE) of the Reactions (1)-(3) by B3LYP and CCSD//B3LYP (kcal/mol).

Structure		B3LYP	CCSD//B3LYP
		ΔE	ΔE
The C—H Activation Reaction of Methane: Reaction 1			
CpIr(PH$_3$)(CH$_3$)$^+$, CH$_4$	**1+4**	0.00	0.00
CpIr(PH$_3$)(CH$_3$)(CH$_4$)$^+$	**5**	-1.03	-6.43 (-0.90)[a]
CpIr(PH$_3$)(CH$_3$)(H)(CH$_3$)$^+$	**6 (TS$_{5-7}$)**	10.51	3.54
CpIr(PH$_3$)(CH$_3$)(H)(CH$_3$)$^+$	**7**	3.38	-5.62
The C—H Activation Reaction of Ethylene: Reaction 2			
CpIr(PH$_3$)(CH$_3$)$^+$, C$_2$H$_4$	**2+4**	0.00	0.00
CpIr(PH$_3$)(CH$_3$)(C$_2$H$_4$)$^+$	**8**	-32.04	-43.08 (-28.68)[a]
CpIr(PH$_3$)(CH$_3$)(H)(C$_2$H$_3$)$^+$	**9 (TS$_{8-10}$)**	1.84	-8.57
CpIr(PH$_3$)(CH$_3$)(H)(C$_2$H$_3$)$^+$	**10**	-1.88	-13.54
CpIr(PH$_3$)(CH$_3$)(H)(C$_2$H$_3$)$^+$	**11 (TS$_{10-12}$)**	4.89	-4.45
CpIr(PH$_3$)(CH$_4$)(C$_2$H$_3$)$^+$	**12**	-10.14	-12.04
The C—H Activation Reaction of Acetylene: Reaction 3			
CpIr(PH$_3$)(CH$_3$)$^+$, C$_2$H$_2$	**3+4**	0.00	0.00
CpIr(PH$_3$)(CH$_3$)(C$_2$H$_2$)$^+$	**13**	-27.97	-32.63 (-20.05)[a]
CpIr(PH$_3$)(CH$_3$)(H)(C$_2$H)$^+$	**14 (TS$_{13-15}$)**	-4.31	-0.01
CpIr(PH$_3$)(CH$_3$)(H)(C$_2$H)$^+$	**15**	-9.46	-17.06
CpIr(PH$_3$)(CH$_3$)(H)(C$_2$H)$^+$	**16 (TS$_{15-17}$)**	-3.76	-9.36
CpIr(PH$_3$)(CH$_4$)(C$_2$H)$^+$	**17**	-22.35	-18.74

a. The BSSE correction is included.

Despite a careful search for the pathway of a σ-bond metathesis mechanism along a reaction coordinate (RC) involving both Ir—H and C—H distances, we could only find a monotonic increase in energy. This result which implies that the OA/RE pathway is the only plausible mechanism for the hydrogen transfer (7a,19a). Thus, with confidence we conclude that the σ-bond metathesis pathway does not exist for the hydrogen transfer process by the CpIr(PH$_3$)(CH$_3$)$^+$ complex.

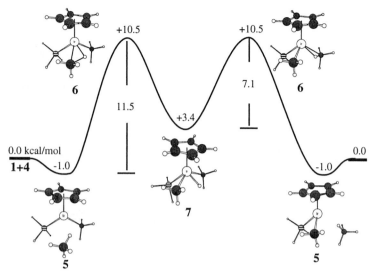

Figure 2. The B3LYP energy profiles along the OA/RE and β-H transfer pathways from reactants, **1** and **4**, to **7**.

B. Ethylene C—H Bond Activation. The possible reactions of an iridium-alkene complex may result in new complexes and products as illustrated in Scheme 3.

Scheme 3

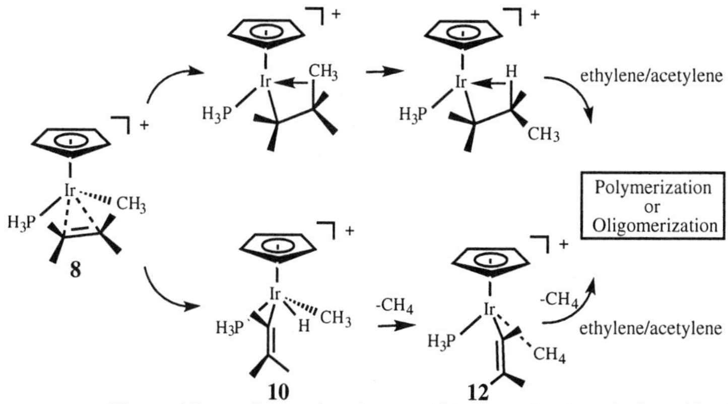

In this part, we will consider and examine the possible steps in the ethylene C—H bond activation process.

The B3LYP optimized geometries for the π-complex (**8**), oxidative-addition TS (**9**), oxidative-addition intermediate (**10**), reductive-elimination TS (**11**), and reductive-elimination product (**12**) along the OA/RE pathway are shown in Figure 3. The Ir-ethylene π-complex, **8**, displays very similar structural features to complex **7** in the longer Ir—Cp and Ir—CH₃ bonds, which are characteristic of an Ir(V) complex as mentioned above (see Scheme 2). The substantial difference between the π-complex (**8**) and that of an early transition metal arises because of the particularly strong back-bonding between ethylene and Ir. Compared to a free ethylene molecule, the C—C

144

double bond of **8** is longer by 0.080 Å, and the distances between Ir and the carbons of the ethylene are close to that of normal Ir—C bonds.

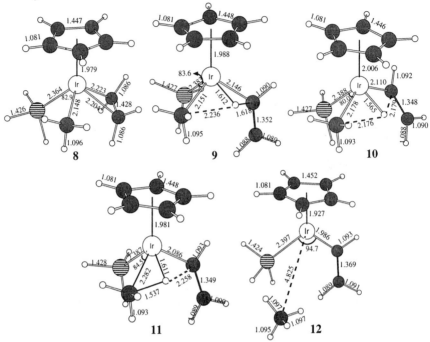

Figure 3. The B3LYP optimized geometries for the π-intermediate (**8**), oxidative-addition TS (**9**), oxidative-addition intermediate (**10**), reductive-elimination TS (**11**), and reductive-elimination product (**12**) along the OA/RE pathway (only the average C—C and C—H distances are given for the Cp ring).

As ethylene rotates to put a C—H bond close to the metal center, the C—H bond begins to break and the Ir—H bond begins to form on reaching the transition state (**9**). Compared to the TS, **6**, the TS of ethylene C—H bond activation, **9**, is slightly later, where the C—H distance is longer by 0.083 Å than that of **6**. The oxidative-addition intermediate, **10**, is similar structurally to the alkane oxidative-addition intermediate, **7**. As the methyl ligand migrates and the hydrogen inserts into the Ir—CH₃ bond, an iridium-vinyl complex, **12**, is formed, which differs from the methane iridium-alkyl complex, **5**, by having a much weaker agostic interaction between the methane and metal center.

Scheme 4

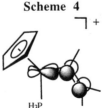

H₃P

Compared to **10**, the Ir—vinyl bond in **12** is shorter by 0.124 Å and the C—C double bond is longer by 0.021 Å. These differences are indicative of a strong π-donating interaction between the vinyl ligand and the metal center of **12** as illustrated in Scheme 4. This interaction is favored over a β-agostic interaction between metal center and vinyl ligand and over the agostic interaction between metal center and methane. Thus, the long Ir—methane distance in **12** is due to this electronic effect rather than to a steric effect.

The energy profiles at the B3LYP and CCSD//B3LYP levels are presented in Figure 4 and Table I (Reaction 2). The first step in the OA/RE reaction is the π-complexation of ethylene to the iridium center. The π-complex is very stable; the ethylene association energy is -32.4 kcal/mol at the B3LYP level and -28.7 kcal/mol at the CCSD//B3LYP level (with the BSSE correction), respectively. This stabilization leads to a high barrier of 34 kcal/mol in the oxidative-addition step and a larger endothermicity of about 21 kcal/mol for the OA/RE product, **12**. The high barriers and endothermicities make this reaction much more difficult than the OA/RE of methane, which is consistent with the fact that experimental observation of ethylene C—H bond activation and catalytic alkane dehydrogenation were not reported (6).

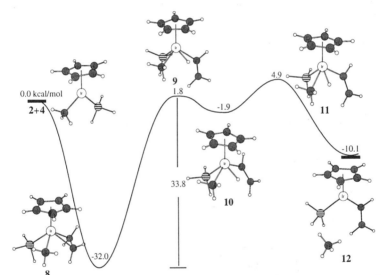

Figure 4. The B3LYP energy profiles along the OA/RE pathways and from reactants, **2** and **4**, to **12**.

C. Acetylene C—H Bond Activation. Recent experimental work shows that OA/RE and insertion reactions play important roles in the dimerization of terminal alkynes by late-transition metal catalysts (20). The cationic iridium-acetylene complex is similar to the iridium-ethylene complex in chemical behavior. As shown in Scheme 5, the acetylene π-complex (**13**) can undergo either oxidative-addition to give an iridium-acetylide complex or insert into an Ir—alkyl bond to give an iridium-vinyl complex. Thus, reactions of acetylene with the $CpIr(PH_3)(CH_3)^+$ complex may result in several new products. In this part, we will consider and examine the possible steps in the acetylene C—H bond activation process.

146

Scheme 5

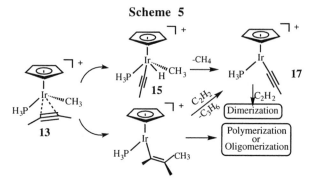

The B3LYP optimized geometries for the π-intermediate (**13**), oxidative-addition TS (**14**), oxidative-addition intermediate (**15**), reductive-elimination TS (**16**), and reductive-elimination product (**17**) along the OA/RE pathway are shown in Figure 5. The acetylene π-complex, **13**, is similar to the ethylene π-complex, **8**, both show longer Cp—Ir and Ir—CH$_3$ distances than those of iridium-alkane complex. The C—C triple bond is longer by 0.049 Å than the C—C triple bond in free acetylene. These structural features show that there is a strong back-donating interaction between acetylene and the metal center. However, this interaction should be weaker than that of the iridium-ethylene π-complex as illustrated by smaller changes in C—C and Cp—Ir distances (*vide infra*).

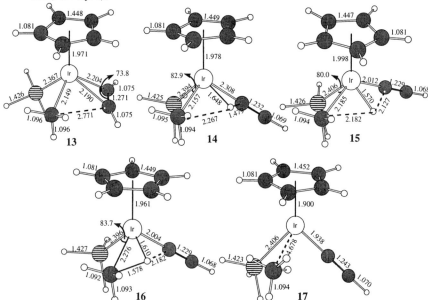

Figure 5. The B3LYP optimized geometries for the π-intermediate (**13**), oxidative-addition TS (**14**), oxidative-addition intermediate (**15**), reductive-elimination TS (**16**), and reductive-elimination product (**17**) along the OA/RE pathway (only the average C—C and C—H distances are given for the Cp ring).

As the acetylene rotates to put a C—H bond close to the metal center, the transition state, **14**, forms in which the C—H bond is partly broken, while the Ir—acetylide and Ir—H bond are almost formed. Compared to the transition states for methane and ethylene C—H activation, **6** and **9**, the transition state for acetylene C—H bond activation, **14**, is slightly earlier with the shortest C—H and longest Ir—H distance. In the reductive-elimination TS, **16**, the methyl—H distance is longer by 0.04 Å than that of **6** and **11**. After the hydrogen inserts into the Ir—methyl bond, a "non-agostic" iridium methane acetylide complex, **17**, is formed, where the Ir—acetylide bond and the C—C distance are shorter than those of the oxidative-addition intermediate (**15**). Clearly, there is a strong π-donating interaction between the acetylide ligand and the metal center of **17** as there was in the Ir-vinyl complex, **12** (see Scheme 4).

The energy profiles at the B3LYP and CCSD//B3LYP levels are presented in Figure 6 and Table I (Reaction 3).

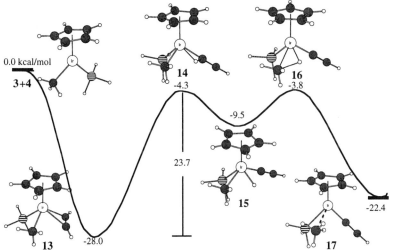

Figure 6. The B3LYP energy profiles along the OA/RE pathway from reactants, **3** and **4**, to **17**.

The acetylene OA/RE mechanism shares with its ethylene counterpart the feature of strong π-complexation between the hydrocarbon and the iridium center. The association energy of acetylene to the iridium complex, **4**, -28.0 (B3LYP) and -20.1 (CCSD//B3LYP with the BSSE correction), is smaller by 4.0-8.6 kcal/mol than that of ethylene. Compared to ethylene, the acetylene reaction from the iridium π-complex, **13**, to the iridium acetylide complex, **17**, is less endothermic with a smaller barrier for the oxidative-addition step. Thus, the OA/RE reaction of acetylene with the iridium complex, although not facile, does take place more easily than that of the ethylene. In fact, there are several experimental reports about C—H bond activation of acetylene with late-transition metal complexes (20).

Conclusions

The alkane oxidative-addition from $CpIr(PH_3)(CH_3)(alkane)^+$ to $CpIr(PH_3)(CH_3)$-$(H)(alkyl)^+$ is endothermic by about 0.8-4.4 kcal/mol with a low barrier of 10.0-11.5 kcal/mol, and reductive-elimination from $CpIr(PH_3)(CH_3)(H)(alkyl)^+$ to a β-agostic

148

structure iridium-alkyl complex, $CpIr(PH_3)(alkyl)^+ + CH_4$, is exothermic with a low barrier of 7.1-9.2 kcal/mol. Thus, the alkane C—H bond activation is a low temperature reaction process.

Because of the strong association of ethylene and acetylene with $CpIr(PH_3)(CH_3)^+$, the oxidative-addition/reductive-elimination (OA/RE) processes of ethylene and acetylene with $CpIr(PH_3)(CH_3)^+$ are high energy processes, with barriers of 24-36 kcal/mol. Compared to ethylene, the OA/RE reaction of acetylene with $CpIr(PH_3)$-$(CH_3)^+$ complex is more favorable. On the other hand, the strong stabilizing interaction of ethylene with $CpIr(PH_3)(CH_3)^+$ may lead to the following chemical equilibrium: the alkane dehydrogenation catalyzed by $CpIr(PH_3)(CH_3)^+$ generates $CpIr(PH_3)(H)(\eta^2-C_2H_4)^+$, lying toward the iridium-olefin product.

Previous works have mentioned that the third-row transition-metal complexes undergo oxidative-addition, $M^I + A—B \rightarrow M^{III}(A)(B)$, more easily than their second-row transition-metal congeners since late third-row transition metals have either $d^n s^1$ ground states or $d^n s^1$ low-lying excited states, while late second-row transition metals have d^{n+1} ground states with high-lying $d^n s^1$ excited states (12,21). This preference emphasizes the importance of forming sd hybrid for the two new covalent bonds in the product. On the other hand, since the redox ability of transition-metals is reflected by their ionization potential (IP), the energy gap between the metal and ligand is directly proportional to the IP of the transition-metal. Thus, the oxidative-addition reaction, $M^{III} + A—B \rightarrow M^V(A)(B)$, for M = Ir proceeds more easily than that for M = Rh because Ir(III) has a smaller ionization energy than Rh(III) (7b,22). Following the ionization potential of late transition-metals, one can predict that the ease of oxidative-addition is: Os(II) > Ru(II) ≥ Pt(II) > Pd(II).

Acknowledgment. We thank the Robert A. Welch Foundation (Grant A–648) and the National Science Foundation (grants 94–23271 and 95-28196) for financial support. This research was conducted in part with use of the Cornell Theory Center, a resource for the Center for Theory and Simulation in Science and Engineering at Cornell University, which is funded in part by the National Science Foundation, New York State, and IBM Corporation.

References

(1) (a) Elschenbroich, Ch.; Salzer, A. *Organometallics*, VCH Publishers, New York, **1989**.
(b) Crabtree, R. H. *The Organometallic Chemistry of the Transition Metals*, John Wiley & Sons, New York, **1988**.
(b) Parshall, G. W. *Homogeneous Catalysis*, Wiley: New York, **1980**
(2) (a) Koga, K.; Morokuma, K. *Chem. Rev.* **1991**, *91*, 823.
(b) Musaev, D. G.; Morokuma, K. *Advances in Chemical Physics, Volume XCV*, Prigogine, I.; Edited by Rice, S. A. John Wiley & Sons, New York, **1996**, 61.
(c) Siegbahn, P. E. M.; Blomberg, M. R. A. *Theoretical Aspects of Homogeneous Catalysts, Applications of Ab Initio Molecular Orbital Theory*, Edited by Leeuwen, van P. W. N. M.; Lenthe, van J. H.; Morokuma, K. Kluwer Academic Publishers, Hingham, MA, **1995**.
(3) *Reviews in Computational Chemistry, Vol. 1-7*, Edited by Lipkowitz, K. B.; Boyd, D. B., New York, N.Y. : VCH, **1990-1996**
(4) (a) Watson, P. L. *J. Am. Chem. Soc.* **1983**, *105*, 6491.
(b) Watson, P. L.; Parshall, G. W. *Acc. Chem. Res.* **1985**, *18*, 51.
(5) (a) Janowicz, A. H.; Bergman, R. G. *J. Am. Chem. Soc.* **1982**, *104*, 352.

149

(b) Hoyano, J. K.; Graham, W. A. G. *J. Am. Chem. Soc.* **1982**, *104*, 3723.
(c) Crabtree, R. H.; Mellea, M. F.; Mihelsie, J. M.; Quick, J. M. *J. Am. Chem. Soc.* **1982**, *104*, 107.
(d) Jones, W. D.; Feher, F. J. *J. Am. Chem. Soc.* **1982**, *104*, 4240.
(6) (a) Arndtsen, B. A.; Bergman, R. G. *Science* **1995**, *270*, 1970.
(b) Burger, P.; Bergman, R. G. *J. Am. Chem. Soc.* **1993**, 115, 10462.
(c) Lohrenz, J. C. W.; Hacobsen, H. *Angew. Chem. Int. Ed. Engl.* **1996**, *35*, 1305.
(7) (a) Strout, D.; Zaric, S.; Niu, S.-Q., Hall, M. B. *J. Am. Chem. Soc.* **1996**, *118*, 6068.
(b) Su, M.-D.; Chu, S.-Y. *J. Am. Chem. Soc.* **1997**, *119*, 5373.
(c) Hinderling, C.; Plattner, D. A.; Chen, P. *Angew. Chem. Int. Ed. Engl.* **1997**, *36*, 243.
(d) Hinderling, C.; Feichtings, D.; Plattner, D. A.; Chen, P. *J. Am. Chem. Soc.* **1997**, *119*, 10793.
(8) (a) Parr, R. G.; Yang, W.; *Density-functional theory of atoms and molecules* , Oxford Univ. Press: Oxford, **1989**.
(b) Becke, A. D. *Phys. Rev.* **1988**, A38, 3098.
(c) Becke, A. D. *J. Chem. Phys.* **1993**, 98, 1372.
(d) Becke, A. D. *J. Chem. Phys.* **1993**, 98, 5648.
(e) Lee, C.; Yang, W.; Parr, R. G. *Phys. Rev.* **1988**, B37, 785.
(9) Schlegel, H. B. *Theor. Chim. Acta.* **1984**, 66, 33
(10) (a) Bartlett, R. J. *Ann. Rev. Phys. Chem.* **1981**, 32, 359.
(b) Scuseria, G. E.; Schaefer, III H. F. *J. Chem. Phys.* **1989**, 90, 3700.
(11) (a) Song, J.; Hall, M. B. *J. Am. Chem. Soc.* **1993**, 115, 327.
(b) Sulfab, Y.; Basolo, F.; Rheingold, A. L. *Organometallics* **1989**, 8, 2139.
(12) Jiménez-Cataño, R.; Niu, S.-Q.; Hall, M. B. *Organometallics* **1997**, 16, 1962.
(13) (a) Hay, P. J.; Wadt, W. R. *J. Chem. Phys.* **1985**, *82*, 299.
(b) Couty, M.; Hall, M. B. *J. Comput. Chem.* **1996**, *17*, 1359.
(14) (a) Dunning, T. H., Jr.; Hay, P. J. in *Modern Theoretical Chemistry* , ed. H.F. Schaefer, III. Plenum: New York, 1976.
(b) Wadt, W. R.; Hay, P. J. *J. Chem. Phys.* **1985**, *82*, 284.
(c) The BSSE correction for the reaction systems were performed using model systems: $(H)Ir(PH_3)(H)(L)^+$ (L = CH_4, C_2H_2, and C_2H_4).
(15) Guest, M. F.; Kendrick, J.; van Lenthe, J. H.; Schoeffel, K.; Sherwood, P. GAMESS-UK; Daresbury Labratory: Warrington, WA4 4AD, U. K., **1994**.
(16) Frisch, M. J.; Trucks, G. W.; Schlegel, H. B.; Gill, P. M. W.; Johnson, B. G.; Robb, M. A.; Cheeseman, J. R.; Keith, T. A.; Petersson, G. A.; Montgomery, J. A.; Raghavachari, K.; Al-Laham, M. A.; Zakrzewski, V. G.; Ortiz, J. V.; Foresman, J. B.; Cioslowski, J.; Stefanov, B. B.; Nanayakkara, A.; Challacombe, M.; Peng, C. Y.; Ayala, P. Y.; Chen, W.; Wong, M. W.; Andres, J. L.; Replogle, E. S.; Gomperts, R.; Martin, R. L.; Fox, D. J.; Binkley, J. S.; Defrees, D. J.; Baker, J.; Stewart, J. P.; Head-Gordon, M.; Gonzalez, C.; Pople, J. A. Gaussian 94 (Revision D.2); Gaussian, Inc., Pittsburgh PA, **1995**.
(17) (a) Bailey, N. A.; Jenkins, J. M.; Mason, R.; Shaw, B. L. *J. Chem. Soc., Chem. Commun.* **1965**, 237.
(b) LaPlaca, S. J.; Ibers, J. A. *Inorg. Chem.* **1965**, *4*, 778.
(c) Cotton, F. A.; Day, V. W. *J. Chem. Soc., Chem. Commun.* **1974**, 415.
(d) Dwoodi, Z.; Green, M. L. H.; Mtetwa, V. S. B.; Prout, K.; Schultz, A. J.; Williams, J. M.; Koetzle, T. F. *J. Chem. Soc., Dalton Trans.* **1986**, *802*, 1629.
(e) Dwoodi, Z.; Green, M. L. H.; Mtetwa, V. S. B.; Prout, K. *J. Chem. Soc., Chem. Commun.* **1982**, *802*, 1410.
(f) Brookhart, M.; Green, M. L. H.; Wong, L.-L. *Inorg. Chem.* **1988**, *30*, 1.

(18) (a) The optimized geometry of $CpIr(PH_3)(CH_3)^{3+}$ shows a similar structural feature to $CpIr(PH_3)(H)(CH_3)(R)^+$ [R = CH_3 (7), C_2H_3 (14), and C_2H (26)] with a slipped Cp ring and coordinating fashion. Niu, S.-Q.; Hall, M. B. unpublished work.
(b) Zarić, S.; Hall, M. B. *J. Phy.Chem.* **1991**, *101*, 4646.
(19) (a) Niu, S.-Q.; Hall, M. B. *J. Am. Chem. Soc.* submitted.
(b) Niu, S.-Q.; M. B. unpublished work.
(c) Roubi, M. *C&E News* **1997**, *75(49)*, 25.
(d) Xu, W.-W.; Rosini, G. P.; Gupta, M.; Jensen,, C. M.; Kaska, W. C.; Krogh-Jespersen, K.; Goldman, A. S. *Chem. Commun.* **1997**, 2273.
(20) (a) Crackness, R. B.; Orpen, A. G.; Spencer, J. L. *J. Chem. Soc., Chem. Comm.* **1984**, 326.
(b) Brookhart, M.; Schmidt, G. F.; Lincoln, D.; Rivers, D. S. *Transition Meral Catalyzed Polymerizations*, Quirk, R., Ed., Cambridge Univ. Press, **1988**,
(c) Schmidt, G. F.; Brookhart, M. *J. Am. Chem. Soc.* **1985**, *107*, 1443.
(d) Trost, B. M.; Sorum, M. T.; Chan, C.; Harms, A. E.; Rühter, G. *J. Am. Chem. Soc.* **1997**, *119*, 698.
(e) Slugovc, C.; Mereiter, K.; Zobetz, E.; Schmid, R.; Kirchner, K. *Organometallics* **1996**, 15, 5275.
(21) (a) Low, J. J.; Goddard, W. A., III, *J. Am. Chem. Soc.* **1986**, *108*, 6115.
(b) Low, J. J.; Goddard, W. A., III, *J. Am. Chem. Soc.* **1984**, *106*, 8321.
(c) Low, J. J.; Goddard, W. A., III, *J. Am. Chem. Soc.* **1984**, *106*, 6928.
(22) (a) Shriver, D. F.; Atkins, P. W.; Langford, C. H. *Inorganic Chemistry*, W. H. Freeman and Company, New York, **1990**.
(b) The IP values of transition-metal for high valence states can be obtained from the experimertal IP values for lower valence states by extrapolation.

Chapter 12

Transition States for Oxidative Addition to Three-Coordinate Ir(I): H–H, C–H, C–C, and C–F Bond Activation Processes

Karsten Krogh-Jespersen and Alan S. Goldman

Department of Chemistry, Rutgers, The State University of New Jersey, New Brunswick, NJ 08903

We have examined elementary oxidative addition reactions involving the $Ir(PH_3)_2X$ (X = Cl, H, Li, BH_2, NH_2, F, Ph) system and small substrates such as dihydrogen (H_2), methane (CH_4) and fluoromethanes (CH_3F, CHF_3, CF_4), and ethane (C_2H_6). Electronic structure calculations employed the B3LYP hybrid density functional, an effective core potential on the metal atom, and basis sets of valence double-zeta or better quality. H-H and, in some cases, C-H bond activation occurs with no apparent or only modest activation energy barriers. The cleavage of C-F bonds, in contrast, requires surmounting a sometimes substantial activation energy barrier as does C-C cleavage, even though these addition reactions are also thermodynamically favorable. A substituent effect study shows that π-donation from the ancillary ligand X favors the addition reaction, whereas σ-donation from X is unfavorable.

A catalytic cycle may involve many elementary reaction steps, but catalytic efficiency will in general be determined by the free energy of a single transition state relative to the catalyst resting state (*1*). While de novo catalyst design remains an unfeasible goal, applications of electronic structure methodology offer an increasingly powerful approach toward understanding the factors that determine transition state energies and can potentially provide valuable guidance in efforts to improve known catalysts or develop catalysts based on known systems.

Among the most important current goals in catalysis are the development of systems for the selective conversions of several simple yet unreactive substrates. Perhaps most noteworthy of these are alkanes; efficient methods for alkane functionalizations, such as oxidation or dehydrogenation, have tremendous value from industrial and environmental perspectives (*2*). A problem with interesting similarities to alkane functionalization is that of selective C-F bond conversion (*3*). Fluorocarbons have increasingly recognized importance in pharmaceutical and industrial applications, but development of methods for their derivatization may prove even more challenging than the analogous alkane problem.

Complexes containing the moiety ML_2X (M = Rh, Ir; L = tertiary phosphine; X = a formally anionic ligand) are included among the most important and widely used of catalysts, promoting a diverse array of organic transformations. For example, $Rh(PPh_3)_3Cl$ (Wilkinson's catalyst) is perhaps the best known catalyst for the addition of H_2 (as well as H-Si or H-B bonds) across the double bond of olefins (4,5). Related reactions catalyzed by $Rh(PR_3)_2Cl$-containing complexes, including the photo- and transfer-dehydrogenation of alkanes, have been reported (6,7,8). Very recently, $(PCP)IrH_2$ [$PCP = \eta^3$-1,3-$C_6H_3(PBu^t_2)_2$] was reported to catalyze the simple thermochemical dehydrogenation of alkanes to give alkene and dihydrogen (9). All of these catalyses undoubtedly involve formally oxidative addition reactions to give five-coordinate, 16e M(III) complexes. A number of computational studies describing oxidative addition of H-H, C-H, and Si-H bonds to $Rh(PR_3)_2X$ have been reported (10). In this contribution, we will focus on elementary oxidative addition reactions involving the coordinatively unsaturated $Ir(PH_3)_2X$ fragment and small substrates such as dihydrogen (H_2), methane (CH_4) and fluoromethanes (CH_3F, CHF_3, CF_4), and ethane (C_2H_6). Prior computational studies by Eisenstein et al. have focused on the electronic properties determining the structures of five-coordinate d^6 $IrL_2ZZ'X$ complexes (11); methane C-H activation by $Ir(PH_3)_2X$ (X = Cl, H) has been studied by Cundari (12).

Computational Methods

We made use of electronic structure methods implemented in the GAUSSIAN 94 series of computer programs (13). The large majority of the calculations employed the B3LYP hybrid density functional (14), an effective core potential (ECP) on the metal atom, and basis sets of valence double-zeta or better quality. The Hay-Wadt relativistic small-core ECP and corresponding basis set (split valence double-zeta) was used for Ir (LANL2DZ model) (15a). We used Dunning/Huzinaga all-electron, full double-zeta plus polarization function basis sets for the second and third row elements (15b). Hydrogen atoms formally existing as hydrides in the product complexes were described by the 311G(p) basis set (15c); all other hydrogen atoms (methyl and phosphine groups, etc.) carried a 21G basis set (15d). Proper basis set substitutions were made in the reactants for the presence of "hydrides". A few calculations were carried out using second-order Moller-Plesset theory (MP2) (16) or the coupled-cluster, single and double excitation method (with triple excitations treated non-iteratively), CCSD(T) (17), and the basis sets outlined above.

Geometries of reactants, transition states and products were fully optimized (18) under appropriate symmetry constraints (typically C_{2v} or C_S). The total internal energies at the stationary points were used to determine reaction or activation energies (ΔE, $\Delta E^{\ddagger}$). In most cases, the stationary points were further characterized by normal mode analysis (numerical differentiation of analytical gradients). The (unscaled) vibrational frequencies formed the basis for the calculation of vibrational zero-point energy corrections and, together with thermodynamic corrections for finite temperature and changes in molecularity upon dissociation or complexation, provided the data needed to convert from internal energies to reaction or activation enthalpies (ΔH, $\Delta H^{\ddagger}$; T = 298 K) and free energies (19).

Reactant and Product Bond Strengths

The reactions of interest here involve the cleavage of H-H, C-H, C-C and C-F bonds in small molecules. In Table 1, we show relevant experimental and computed bond dissociation energies.

Table 1. Comparison of computed bond dissociation energies (ΔE) and enthalpies (ΔH) with experimentally determined bond dissociation enthalpies ($\Delta H(exp)$). All values in kcal/mol.

Bond	Molec.	ΔE(B3LYP)	ΔH(B3LYP)	ΔH(exp)[a]	ΔH(MP2)[b]	ΔH(CCSD(T))[c]
H-H	H_2	109.1	105.1	104.2	96.8	101.9
C-H	CH_4	112.2	104.9	105.1	100.1	101.2
C-H	CHF_3	111.7	105.5	106	103.7	105.2
C-C	C_2H_6	102.3	95.5	88	96.1	93.2
C-F	CH_3F	113.0	109.1	108	108.5	102.1
C-F	CF_4	128.8	127.5	129	127.8	121.6

[a] Reference (20)

[b] $\Delta H(MP2) = \Delta E(MP2) + [\Delta H(B3LYP) - \Delta E(B3LYP)]$

[c] $\Delta H(CCSD(T)) = \Delta E(CCSD(T)) + [\Delta H(B3LYP) - \Delta E(B3LYP)]$

The dissociation energies computed using B3LYP compare very favorably, within 1 kcal/mol or better, with experimental energies for the H-H bond and the C-H bond in CH_4 or CHF_3. The C-F bond strength in CH_3F is modeled accurately as is the considerable larger C-F bond strength observed in CF_4. The C-C bond strength in C_2H_6 is, however, overestimated in the B3LYP calculations by almost 8 kcal/mol. Included in Table 1 are the dissociation enthalpies computed with the MP2 and CCSD(T) methods. MP2 theory underestimates the H-H and C-H bond strengths but performs well on the C-F bond strengths. The CCSD(T) method is generally considered to be highly accurate, and it does perform better than MP2 for the H-H and C-H bond strengths, but the method underestimates the strength of the C-F bonds. Of the three methods tested, CCSD(T) returns the value for the C-C bond strength closest to the experimental value. The average absolute error for the six bond strengths of interest here is 2 kcal/mol for B3LYP but twice that for both MP2 and CCSD(T). It should be noted that atomization energies were included in the data set used by Becke to determine the three parameters in the B3LYP hybrid functional (14a).

It is more difficult to assess the performance of the computational method on product (M-H, M-CH_3 or M-CF_3, M-F) bond strengths, since the availability of accurate experimental determinations of metal-ligand bond energies is generally quite limited. A number of supportive studies may be cited, however (21). For example, based on data gathered from a large number of calculations on organometallic complexes, Ziegler has offered the opinion that the current generation of density functional methods (such as B3LYP), which include nonlocal exchange and correlation energy corrections, may afford metal-ligand bond energies of almost experimental accuracy (~5 kcal/mol) (21a). Extensive comparisons by Siegbahn et al. (21c, d) of B3LYP to high level ab initio methods on a variety of transition metal species, including oxidative addition reactions, were largely favorable, in particular for Ir-containing systems. We have recently completed calorimetric studies of the thermodynamics associated with addition of H_2, CO, N_2, and C-H bonds to $M(P^iPr_3)_2Cl$ (M = Ir, Rh) complexes (22). Our B3LYP electronic structure calculations on these systems (P^iPr_3 replaced by PMe_3) produced reaction energies in excellent agreement with the experimental results.

Oxidative Addition Reactions to Three-Coordinate Ir(I)

Addition of Small Substrates to Ir(PH3)2Cl. Computed reaction and activation energies for a number of small substrates adding to *trans*-IrL2Cl, L = PH3, are summarized in Table 2.

Table 2. Calculated Addition and Activation Energies (B3LYP; kcal/mol)

Metal Complex	Addendum	ΔE	$\Delta E^{\ddagger}$
Ir(PH3)2Cl	H-H	-54.1	---
Ir(PH3)2Cl	H-CH3	-27.2	---
Ir(PH3)2Cl	H-CF3	-42.7	---
Ir(PH3)2Cl	H3C-CH3	-12.9	31.6
Ir(PH3)2Cl	F-CH3	-37.6	14.1
Ir(PH3)2Cl	F-CHF2	-32.1	21.1
Ir(PH3)2Cl	F-CF3	-30.8	31.9

H2 adding to Ir(PH3)2Cl proceeds with a large reaction energy (-54 kcal/mol) and leads to a trigonal-bipyramidal (TBP) product, **1**, featuring a highly acute H-Ir-H angle (YCl shape) (*11,23*). The addition takes place without the prior formation of an adduct or the appearance of any activation energy barrier.

Cl-Ir-H = 146.1
H-Ir-H = 67.8

2.446
Cl—— Ir
1.562 H

1

Cl-Ir-H = 159.8
Cl-Ir-C = 171.2
H-Ir-C = 29.0

2.391 2.395 CH3
Cl—— Ir
1.719 H /1.218

2a

Cl-Ir-H = 158.3
Cl-Ir-C = 169.0
H-Ir-C = 32.7

2.321 2.322 CH3
Cl—— Ir
1.658 H 1.288

2b

Cl-Ir-H = 141.6
Cl-Ir-C = 144.8
H -Ir-C = 73.6

2.460 2.104 CH3
Cl—— Ir
1.557 H

2c

A highly stabilized adduct (**2a**) forms between CH4 and Ir(PH3)2Cl, bound by about 12 kcal/mol with respect to the reactants. The geometry of the adduct shows an elongated C-H bond length of 1.22 Å (1.09 Å in CH4) and short Ir-H (1.72 Å) and Ir-C (2.40 Å) distances. A transition state structure (**2b**; imaginary frequency of 279*i* cm^{-1}) was located near the geometry of the adduct with a very small activation energy relative to the adduct complex ($\Delta E^{\ddagger} = 0.25$ kcal/mol). In **2b**, the methane fragment is very close to the Ir(PH3)2Cl fragment, the C-H bond (1.29 Å) has lengthened further relative to the adduct **2a**, and the Ir-CH$_3$ (2.32 Å) and Ir-H distances (1.66 Å) are very close to their values in the product (2.10 Å and 1.56 Å, **2c**). The eigenvector for the imaginary frequency in **2b** has C-H lengthening as its dominant component with a significant admixture of relative translation between the two molecular fragments (*12*). On the enthalpy surface, however, **2b** is 1.04 kcal/mol below **2a** and the free energy difference is 0.75 kcal/mol in favor of **2b**. It would appear that there are many, highly stabilized "adduct-type" geometries available for this system (*10c,12*) and considering the relative energetics presented above, we do not believe that in solution at ambient temperature there could be any significant activation energy barrier for the addition of CH$_4$ to Ir(PH$_3$)$_2$Cl. This is in accord with previous calculations on this reaction by Cundari using Hartree-Fock based geometries and MP2 energies (*12*). Bergman has observed experimentally that C-H oxidative addition to a related low valent metal system (Cp*Rh(CO)) exhibits a very small activation energy (~5-6 kcal/mol, relative to a precursor complex) (*24*), a result supported by several computational studies (*21d,25*).

The C-H bond strength in CHF$_3$ is almost identical to that in CH$_4$ (Table 1) but the exothermicity of the addition reaction to Ir(PH$_3$)$_2$Cl for the former species is greater (-42.7 kcal/mol vs. -27.2 kcal/mol). Neither an adduct nor a transition state could be located for this process. The product (**3**) is TBP of Y$_{Cl}$ shape but there is larger asymmetry around Ir in **3** than in **2c** as the CF$_3$ group is distorted toward an apical position in a square-pyramidal (SP) arrangement (*26*). The Ir-CF$_3$ bond length in **3** is shorter than the Ir-CH$_3$ bond length in **2c** by 0.08 Å, and the difference in exothermicity of the two C-H additions permits us to estimate that the Ir-CF$_3$ bond dissociation energy is approximately 15 kcal/mol higher than that of Ir-CH$_3$.

Structure **3**:
2.024 CF$_3$
2.439
Cl — Ir
1.552 H
Cl-Ir-H = 150.9
Cl-Ir-C = 131.7
H-Ir-C = 77.4

Structure **4a**:
2.592 CH$_3$ 1.757
2.383
Cl — Ir ------ F
2.132
Cl-Ir-F = 182.9
Cl-Ir-C = 134.8
F-Ir-C = 42.3

Structure **4b**:
CH$_3$
2.067
2.424
Cl — Ir
2.008 F
Cl-Ir-F = 173.5
Cl-Ir-C = 96.1
F-Ir-C = 90.4

Structure **5**:
CF$_3$
1.995
2.409
Cl — Ir
2.005 F
Cl-Ir-F = 166.6
Cl-Ir-C = 106.3
F-Ir-C = 87.1

Can Ir(PH$_3$)$_2$Cl efficiently cleave terminal sigma bonds other than C-H? We had particular hopes and interest in C-F bonds. The exothermicity for F-CH$_3$ addition is moderate (-37.6 kcal/mol) as is the computed barrier of 14.1 kcal/mol. The C-F bond is significantly elongated in the transition state (1.39 Å in F-CH$_3$, 1.76 Å in TS **4a**; imaginary frequency = 574i cm^{-1}) but interestingly, as noted above for C-H activation, the distance between the approaching terminal atom (F in this case) and Ir is very close in the TS (2.13 Å) to what it will become in the product **4b** (2.01 Å). The F-Ir-Cl angle is close to 180° in both **4a** and **4b** (SP, T$_{CH3}$). For CHF$_3$, addition of a C-F bond to Ir(PH$_3$)$_2$Cl shows a reaction energy of -32.1 kcal/mol and an activation energy of 21.1 kcal/mol; i.e., both values are about 6 kcal/mol less favorable than the corresponding values for F-CH$_3$. However, experimentally it would of course be impossible to prevent the seemingly barrier-free (and thermodynamically more favorable) C-H addition of either H-CH$_2$F or H-CF$_3$ to

Ir(PH$_3$)$_2$Cl; thus, selective C-F bond breaking by this complex does not appear feasible. Furthermore, upon full substitution of F for H the reaction energy increases only slightly to -30.8 kcal/mol but the activation energy barrier increases sharply in magnitude to 31.9 kcal/mol; the greater value of $\Delta E^{\ddagger}$ is possibly attributable to the increasing C-F bond strength in tetrafluoromethane. The transition state (imaginary frequency = 372i cm^{-1}) for the CF$_4$ + Ir(PH$_3$)$_2$Cl addition reaction looks very similar to **4a** but lies earlier along the reaction coordinate (e.g. Ir-F = 2.17 Å, Ir-C = 2.64 Å and C---F = 1.623 Å) in accord with the increased barrier (27); the product, **5**, is square-pyramidal with CF$_3$ at the apex (SP, T$_{CF_3}$). We emphasize that the thermodynamic parameters for addition of fluoromethanes to Ir(PH$_3$)$_2$Cl are quite favorable - but the kinetics parameters are not. Independently, similar conclusions have been drawn by Eisenstein, Perutz, and Caulton with respect to C-F addition to Os(0) (3e).

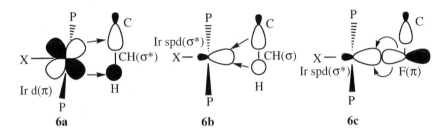

In the process of H-H or C-H addition the Ir center acts as both a Lewis base and a Lewis acid. Electron donation by the metal may occur from the filled d(π)-orbital lying in the equatorial plane into the σ^* bond orbital of the substrate (**6a**). The substrate may donate into the empty Ir spd(σ^*)-orbital which is oriented along the Ir-X axis (**6b**). The ease or difficulty of oxidative addition to the Ir(PH$_3$)$_2$Cl complex may be traced to the directionality of the orbitals forming the bond which is broken (28) and, to some extent, also to the inherent strength of this bond. The bond orbitals in H$_2$ are composed of spherical s–type orbitals and the absence of directionality enables these orbitals to continue to maintain intramolecular H-H bonding while reaching out to the metal fragment along the reaction coordinate and developing Ir-H bonding. The large Ir-H bond strength in the product (~75-80 kcal/mol) provides a strong thermodynamic driving force, and there is no barrier for H$_2$ addition to Ir(PH$_3$)$_2$Cl. In H-CH$_3$ activation, the H atom still uses a spherical s-orbital for bonding and the dominant interaction in the adduct **2a** and in the "transition state" **2b** is the substantially developed formation of the strong Ir-H bond. Concomitantly, the C atom with its sp^3-type highly directional hybrids attempts to maintain bonding with the departing H and simultaneously reorient to interact with the metal, a difficult task. The reaction energy for CH$_4$ is diminished relative to that of H$_2$, because Ir-CH$_3$ bonds are significantly weaker than Ir-H bonds (29). In the case of H$_3$C-CH$_3$ adding to Ir(PH$_3$)$_2$Cl, the thermodynamic driving force is relatively low (ΔE = -12.9 kcal/mol) and the computed activation energy barrier becomes large ($\Delta E^{\ddagger}$ = 31.6 kcal/mol). The pronounced directionality of the C(sp^3) hybrids at both termini of the C-C bond demands significant structural distortions in the transition state in preparation for Ir-CH$_3$ bond formation. For example, the C-C distance in the transition state is 1.82 Å (1.54 Å in the reactant) and the Ir-C distance is 2.32 Å (2.10 Å in the TBP product (Y$_{Cl}$)).

The C-F bond activation process incorporates features from both the C-H and C-C cleavage reactions. The F terminus is very close to the Ir atom in the transition state (viz. **4a**), almost taking up the position it will occupy in the product (**4b**). Direct donation of lone-pair electron density from F into the vacant Ir $spd(\sigma^*)$-orbital appears to be the dominant interaction in the transition state (**6c**). This interaction stabilizes the transition state for C-F cleavage, however, an activation energy barrier does arise because of the directionality of the C-F σ-bond hybrids. The barrier for C-C cleavage is of the same magnitude as that for $F-CF_3$ cleavage and much higher than that for $F-CH_3$ cleavage, even though the latter bonds are stronger than the C-C bond by 40 and 20 kcal/mol, respectively (Table 1). Comparison of reaction energies for $H-CH_3$ and $F-CH_3$, or $H-CF_3$ and $F-CF_3$, provided in Tables 1 and 2 leads to the conclusion that the Ir-F bond is more than 10 kcal/mol stronger than the Ir-H bond.

Oxidative Addition to $Ir(PH_3)_2X$: X = H, F, and Ph. Experimentally, substitution of Cl by an aryl group appears to dramatically affect the catalytic activity of ML_2X (M = Rh, Ir) fragments (9). In order to investigate the electronic effects of ligand X in more detail, we computed the energetic parameters for a number of addition reactions where Cl was substituted by H, F, or Ph. Relevant data are shown in Table 3.

Table 3. Calculated Addition and Activation Energies (B3LYP; kcal/mol)

Metal Complex	Addendum	ΔE	$\Delta E^{\ddagger}$
$Ir(PH_3)_2H$	H-H	-21.6^a	---
$Ir(PH_3)_2H$	$H-CH_3$	-2.9	11.8^b
$Ir(PH_3)_2H$	$H-CF_3$	-12.8	8.0
$Ir(PH_3)_2H$	$F-CH_3$	-28.8	23.5
$Ir(PH_3)_2H$	$F-CF_3$	-28.5	39.1
$Ir(PH_3)_2F$	H-H	-47.0	---
$Ir(PH_3)_2F$	$F-CH_3$	-34.8	14.5
$Ir(PH_3)_2F$	$F-CF_3$	-29.8	31.6
$Ir(PH_3)_2Ph$	H-H	-23.0	---
$Ir(PH_3)_2Ph$	$H-CH_3$	$+4.4$	14.6^b
$Ir(PH_3)_2Ph$	$F-CH_3$	-21.6	24.7

^{a} Singlet, Y-shape. ^{b} Relative to adduct.

We notice that the reaction exothermicities decrease significantly for H-H or C-H addition (~25-30 kcal/mol) upon the substitution of the strong, purely σ-donating H^- ligand for the weakly π- and σ-donating Cl^-. A significantly stabilized $(H_2):Ir(PH_3)_2H$ adduct may be located on the internal energy surface with a highly elongated H-H bond distance of 0.98 Å and Ir-H distances of 1.64 Å, but a transition state is structurally nearby (H-H = 1.11 Å, Ir-H = 1.64 Å) and only 0.08 kcal/mol higher in energy. On the enthalpy and free energy surfaces the "transition state" appears to be energetically below the adduct and does not appear to present any real barrier to the addition of H_2 to $Ir(PH_3)_2H$.

A weakly bound adduct ($\Delta E = -3.6$ kcal/mol) was formed between $Ir(PH_3)_2H$ and CH_4, **7a**. In this adduct, the C-H bond is only weakly stretched to 1.11 Å (compared to 1.10 Å for the other C-H bonds) and the Ir-H and Ir-C distances are well over 2 Å. In the transition state, **7b**, the C-H distance is significantly longer (1.58 Å) and the Ir-C and, especially, Ir-H distances are much shorter at 2.25 Å and 1.60 Å, respectively. The energy of the transition state (**7b**) is 11.1 kcal/mol above that of the adduct (**7a**). Whereas the transition state structure appears to lead toward a $SP(T_H)$

structure, large angular movements take place and the preferred product structure (**7c**) is Y-shaped (Y_{CH_3}) with a highly acute H-Ir-H angle (61.4°), Ir-C = 2.16 Å and Ir-H = 1.59 Å. For both $Ir(PH_3)_2Cl(CH_3)H$ and $Ir(PH_3)_2H(CH_3)H$ the preferred B3LYP structure of the product is TBP whereas Hartree-Fock theory appears to favor square-pyramidal structures with a hydride at the apex (*this work, 12*). Adduct **7a** is in fact lower in energy than the product by 0.7 kcal/mol; the enthalpy and free energy <u>favor</u> the product by 1.3 and 0.5 kcal/mol, respectively. In any case, we do not attribute any significant stability to the alkyl-dihydride complex. Compared to $CH_4 + Ir(PH_3)_2H$, the addition of H-CF$_3$ to $Ir(PH_3)_2H$ displays approximately the same barrier (8.0 kcal/mol) as addition of H-CH$_3$. However, the reaction is significantly exothermic (-12.8 kcal/mol); accordingly, the transition state is "earlier" than **7b** --- the Ir-H (1.646 Å) and Ir-C (2.386 Å) distances are slightly longer and the H-C bond is less activated (1.365 Å). We did not look for adduct formation with this reactant system.

$$\text{7a}$$

H1-Ir-H2 = 159.9
H1-Ir-C = 178.2
H2-Ir-C = 21.9

$$\text{7b}$$

H1-Ir-H2 = 148.8
H1-Ir-C = 166.8
H2-Ir-C = 44.5

$$\text{7c}$$

H1-Ir-H2 = 61.4
H1-Ir-C = 149.6
H2-Ir-C = 149.0

Additions of F-CH$_3$ and F-CF$_3$ bonds to $Ir(PH_3)_2H$ display similar reaction energies, values which are not significantly less than those of the analogous $Ir(PH_3)_2Cl$ reactions. The products are TBP (Y_F) with C-Ir-H angles of 70°-75°; thus one of the added moieties now occupies the unique (π-donating) equatorial position. The transition state structures are very similar to what we obtained when X = Cl (i.e. **4a**-like) though the activation energy barriers have increased by almost 10 kcal/mol. The increases in activation barriers (X = H vs. Cl) indicate that increased σ-donation from X is disfavorable for oxidative addition to IrL_2X. It increases the electron density in the Ir spd(σ^*)-orbital in the reactant fragment, thus rendering this orbital a poorer acceptor and diminishing the stabilizing interaction **6b**.

What effects would a weaker σ- but moderately strong π-donor, such as F$^-$, exert on these addition reactions (*30*)? There is, as expected, no barrier for addition of H$_2$ to $Ir(PH_3)_2F$ but the exothermicity is slightly reduced relative to X = Cl (-54 kcal/mol vs. -47 kcal/mol). The reaction and activation energies for F-CH$_3$ and F-CF$_3$ addition when X = F (Table 3) are almost superimposable on those obtained when X= Cl (Table 2).

A convincing isolation of π-effects can be made with the model complex $Ir(PH_3)_2(NH_2)$ (*11,31*). If we arrange the NH$_2$ plane perpendicular to the P-Ir-P axis (**8a**), the addition of H$_2$ (in the equatorial plane containing NH$_2$) has a reaction energy $\Delta E = -37.0$ kcal/mol to give **9a**. However, following a 90° rotation of the NH$_2$ group (**8b**), the exothermicity for H$_2$ addition in the equatorial plane is increased to $\Delta E = -52.3$ kcal/mol. The **8a-8b** energy difference is 2.9 kcal/mol in favor of **8a**, whereas **9b** is 18.1 kcal/mol more stable than **9a**. The **8b** -> **9b** transformation has the N lone-pair oriented for maximum interaction with the Ir $d(\pi)$-orbital responsible for donation into the σ^* orbital of the H_2 molecule (**6a**). This interaction is effectively turned off in the **8a** -> **9a** transformation. Eisenstein previously made a similar observation when comparing the relative stabilities of TBP and SP isomers (*11*).

Furthermore, when calculations are conducted with X = BH_2, potentially a π-acceptor, it is found that H_2 addition is much less favorable than to any of the derivatives discussed so far. $Ir(PH_3)_2(BH_2)$ with BH_2 vertically in the P-C-P plane (cf. **8b**) produces only a weakly bound dihydrogen adduct ($\Delta E \sim$ -2 kcal/mol) and no dihydride. A dihydride ($\Delta E \sim$ -7 kcal/mol) may form when BH_2 is in the equatorial plane (perpendicular to the P-C-P plane, cf. **8a**), which shows that, rather than acting as a π-acceptor, it is preferable to have the BH_2 group acting as a π-donor through hyperconjugation via the B-H bond orbitals. Independent of π-interactions, the low exothermicity of H_2 addition to either boryl isomer suggests that strongly σ-donating ligands X disfavor addition. Consistent with this view, the reaction energy when X = Li is about -6 kcal/mol for the dihydride product.

Major problems in the practical applications of the tri-coordinate Ir-halogen complexes include phosphine degradation and dimer formation through bridging halogens. Both these problems are apparently resolved through the use of the rigid tridentate 1,5-bis(dialkylphosphinomethyl)phenyl ligand (PCP). Metal catalysts based on the PCP ligand retain long-term stability at elevated temperatures. Here we use a simple phenyl group (Ph = -C_6H_5) to model the aryl group in the PCP ligand and keep the Ph plane constrained to eclipse the Ir-P bonds (**10**), although this is not inherently the favored orientation for the Ph ligand.

For the $Ir(PH_3)_2Ph + H_2$ reaction system, there appears to be no barrier to product formation and no sign of intermediate adduct formation. The product is a

TBP dihydride complex (Ir-H = 1.560 Å, Ir-C = 2.137 Å, H-Ir-H = 61.7°). The computed reaction energy of -23.0 kcal/mol is similar to that of $Ir(PH_3)_2H + H_2$.

The $Ir(PH_3)_2Ph + H-CH_3$ addition shows a weakly stabilized adduct, about 0.90 kcal/mol lower in energy than the reactants, in which the Ir-C (3.00 Å) and Ir-H (2.27 Å) distances are long and the C-H bond is only weakly affected (1.11 Å). The reaction appears to be slightly uphill in internal energy (ΔE = 4.4 kcal/mol) with a relatively low barrier of 14.6 kcal/mol. Based on our experiences using PMe_3 instead of PH_3, we expect the reaction energy to increase by approximately 10 kcal/mol, if alkylated phosphines were used in the calculations (22). The product geometry is perhaps best characterized as SP with H at the apex but it shows a quite narrow H-Ir-$C(CH_3)$ angle (11).

Finally, the $Ir(PH_3)_2Ph + F-CH_3$ reaction is moderately exoergic (ΔE = -21.6 kcal/mol) but shows, as expected, a considerable barrier for C-F bond cleavage ($\Delta E^{\ddagger}$ = 25.1 kcal/mol). Using X = Ph as a model ligand in these addition reactions leads to activation energies very similar to those obtained for X = H, indicating that in the fixed orientation the Ph ligand behaves as a moderately strong σ-donor with no particular π-attributes.

Concluding Remarks

Exothermicity of H_2 addition to $Ir(PH_3)_2X$ increases as follows for various X: $BH_{2(vert)} \sim Li < BH_{2(eq)} \ll H \sim Ph_{(vert)} < NH_{2(eq)} < F < NH_{2(vert)} \sim Cl$. The favorable effect of π-donation by X is highlighted most succinctly by the effect of NH_2 group rotation; π-donation may also be used to rationalize the very favorable addition to $Ir(PH_3)_2F$ and, to some extent, $Ir(PH_3)_2Cl$. However, the very large range in exothermicities spanned from X = Li to X=Ph cannot be explained solely in terms of π-donation, particularly since phenyl should be the better (although weak) π-acceptor. Our calculations also indicate that increased σ-donation by X disfavors addition of H_2 to $Ir(PH_3)_2X$. However, axial substitution of PH_3 by PMe_3 was previously found to favor H_2 (and CH_4) addition to $Ir(PH_3)_2Cl$ and $Ir(PH_3)_2H$ (22). Thus, it appears that increased σ-donating ability of ancillary ligands may be favorable for ligands that are not trans to the site of addition.

Calculations on C-H addition reactions, although done on a smaller set of X ligands, indicate that C-H and H-H additions are influenced similarly by variations in the nature of X. Inspection of data in Table 3 for C-F and C-H cleavage shows that the kinetics of these reactions, i.e. the activation energy barriers, are affected similarly by substituent variations: π-donors (F, Cl) decrease the barrier, σ-donors (X = H, Ph) increase the barrier. The thermodynamics of C-F addition, however, are much less sensitive to these factors. Probably, this is because the fluorocarbon acts as a σ-donor in the transition state (6c) but, in the product, the fluoride ligand is a π-donor. Thus, while C-H and C-F addition to $Ir(PH_3)_2Cl$ give comparable reaction energies, the addition of C-F bonds to $Ir(PH_3)_2H$ or $Ir(PH_3)_2Ph$ is thermodynamically much more favorable than C-H addition.

We showed recently that addition of H_2 to neutral Vaska complexes $Ir(PH_3)_2(CO)X$ was kinetically and thermodynamically disfavored by π-donating ability of X, although σ-donating ability seemed to have a slightly favorable effect (31). These results are exactly opposite to the substituent effects found for H_2 and C-H addition to $Ir(PH_3)_2X$ in the present study. Clearly, the commonly held belief, that "oxidative" addition is intrinsically favored by electron-rich ligands is a gross oversimplification at best (32), and is probably best disregarded in the case of addenda such as H_2 and C-H which form relatively covalent bonds with the metal

(*10c,28*). In other words, at least for these substrates, substituent effects in oxidative addition reactions are dominated by the individual orbital interactions specific to each system, rather than by any transfer of charge to or from the addenda (*33*).

Acknowledgments. We gratefully acknowledge the Division of Chemical Sciences, Office of Basic Energy Sciences, Office of Energy Research, U.S. Department of Energy for financial support and the National Center for Supercomputing Applications, University of Illinois-Champaign, for a grant of computer time on the SGI Power Challenge Array.

Literature Cited
(1) Parshall, G. W. *Homogeneous Catalysis*, Wiley: New York, 1980. *Theoretical Aspects of Homogeneous Catalysis, Applications of Ab Initio Molecular Orbital Theory*, Eds. van Leeuwen, P. W. N. M.; van Lenthe, J. H.; Morokuma, K.; Kluwer: Dordrecht, The Netherlands, 1994. Koga, N.; Morokuma, K. *Chem. Rev.* **1991**, *91*, 823-842.
(2) (a) *Activation and Functionalization of Alkanes*; Hill, C., Ed.; John Wiley & Sons: New York, 1989. (b) Davies, J. A.; Watson, P. L.; Liebman, J. F.; Greenberg, A. *Selective Hydrocarbon Activation*; VCH Publishers, Inc.: New York, 1990. (c) Arndtsen, B. A.; Bergman, R. G.; Mobley, T. A.; Peterson, T. H. *Acc. Chem. Res.* **1995**, *28*, 154-162.
(3) (a) Burdeniuc, J.; Jedlicka, B.; Crabtree, R. H. *Chem. Ber./Recueil* **1997**, *130*, 145-154. (b) Whittlesey, M. K.; Perutz, R. N.; Greener, B.; Moore, M. H. *Chem. Commun.* **1997**, 187-188. (c) Selmeczy, A. D.; Jones, W. D.; Partridge, M. G.; Perutz, M. N. *Organometallics*, **1994**, *13*, 522-532. (d) Aizenberg, M.; Milstein, D. *Science* **1994**, *265*, 359-360. (e) Bosque, R.; Fantacci, S.; Maseras, F.; Clot, E.; Eisenstein, O.; Perutz, R. N.; Renkema, K. B.; Caulton, K. G., in preparation.
(4) Collman. J. P.; Hegedus, L. S.; Norton, J. R.; Finke, R. G. *Principles and Applications of Organotransition Metal Chemistry*; University Science Books: Mill Valley, Ca 1987; pp 523-576.
(5) Daniel, C.; Koga, N.; Han, J.; Fu, X. Y.; Morokuma, K. *J. Am. Chem. Soc.* **1988**, *110*, 3773.
(6) Nomura, K.; Saito, Y. *Chem. Commun.* **1988**, 161. Sakakura, T.; Sodeyama, T.; Tokunaga, M.; Tanaka, M. *Chem. Lett.* **1988**, 263-264.
(7) Maguire, J. A.; Boese, W. T.; Goldman, A. S. *J. Am. Chem. Soc.* **1989**, *111*, 7088-7093. Maguire, J. A.; Goldman, A. S. *J. Am. Chem. Soc.* **1991**, *113*, 6706-6708. Maguire, J. A.; Petrillo, A.; Goldman, A. S. *J. Am. Chem. Soc.* **1992**, *114*, 9492-9498.
(8) Koga, N.; Morokuma, K. *ACS Symposium Series* **1989**, *394*, 77.
(9) Xu, W.; Rosini, G. P.; Gupta, M.; Jensen, C. M.; Kaska, W. C.; Krogh-Jespersen, K.; Goldman, A. S. *Chem. Commun.* **1997**, 2273-2274.
(10) (a) Hutschka, F.; Dedieu, A.; Eichberger, M.; Fornika, R.; Leitner, W. *J. Am. Chem. Soc.* **1997**, *119*, 4432-4443. (b) Margl, P.; Ziegler, T.; Bloechl, P. E. *J. Am. Chem. Soc.* **1995**, *117*, 12625-12634. (c) Koga, N.; Morokuma, K. *J. Am. Chem. Soc.* **1993**, *115*, 6883-6892; *J. Phys. Chem.* **1990**, 94, 5454-5462. (d) Blomberg, M. R. A.; Siegbahn, P. E. M.; Svensson, M. *J. Am. Chem. Soc.* **1992**, *114*, 6095-6102.
(11) Riehl, J.-F.; Jean, Y.; Eisenstein, O.; Pelissier, M. *Organometallics* **1992**, *11*, 729-737. Rachidi, I. E.-I.; Eisenstein, O.; Jean, Y. *New J. Chem.* **1990**, *14*, 671.
(12) Cundari, T. R. *J. Am. Chem. Soc.* **1994**, *116*, 340-347.
(13) Frisch, M. J.; Trucks, G. W.; Schlegel, H. B.; Gill, P. M. W.; Johnson, B. G.; Robb, M. A.; Cheeseman, J. R.; Keith, T.; Petersson, G. A.; Montgomery, J. A.; Raghavachari, K.; Al-Laham, M. A.; Zakrzewski, V. G.; Ortiz, J. V.; Foresman, J. B.; Cioslowski, J.; Stefanov, B. B.; Nanayakkara, A.; Challacombe, M.; Peng, C. Y.; Ayala, P. Y.; Chen, W.; Wong, M. W.; Andres, J. L.; Replogle, E. S.; Gomperts, R.; Martin, R. L.; Fox, D. J.; Binkley, J. S.; Defrees, D. J.; Baker, J.; Stewart, J. J. P.; Head-Gordon, M.; Gonzalez, C.; Pople, J. A. GAUSSIAN 94, Revision D.2; Gaussian, Inc.: Pittsburgh PA, 1995.

162

(14)   (a) Becke, A. D. *J. Chem. Phys.* **1993**, *98*, 5648-5652. (b) Lee, C.; Yang, W.; Parr, R. G. *Phys. Rev. B* **1988**, *37*, 785.
(15)   (a) Hay, P. J.; Wadt, W. R. *J. Chem. Phys.* **1985**, *82*, 270. (b) Dunning, T. H.; Hay, P. J. in *Modern Theoretical Chemistry*; Ed. H. F. Schaefer III; Plenum: New York, 1976; pp. 1-28. (c) Krishnan, R.; Binkley, J. S.; Seeger, R.; Pople, J. A. *J. Chem. Phys.* **1980**, *72*, 650. (d) Binkley, J. S.; Pople, J. A.; Hehre, W. J. *J. Am. Chem. Soc.* **1980**, *102*, 939.
(16)   Head-Gordon, M.; Pople, J. A.; Frisch, M. J. *Chem. Phys. Lett.* **1988**, *153*, 503.
(17)   Pople, J. A.; Head-Gordon, M.; Raghavachari, K. *J. Chem. Phys.*, **1987**, *87*, 5968. Purvis, G. D.; Bartlett, R. J. *J. Chem. Phys.* **1982**, *76*, 1910.
(18)   Schlegel, H. B. in *Modern Electronic Structure Theory*, Ed. Yarkony, D. R., World Scientific Publishing: Singapore, 1994.
(19)   Hehre, W. J.; Radom. L.; Pople, J. A.; Schleyer, P. v. R. *Ab Initio Molecular Orbital Theory*, Wiley-Interscience: New York, 1986. McQuarrie, D. A. *Statistical Thermodynamics*, Harper and Row: New York, 1973.
(20)   *CRC Handbook of Chemistry and Physics*, CRC Press, Inc.: Boca Raton, 1984.
(21)   (a) Ziegler, T. *Can. J. Chem.* **1995**, *73*, 743-761. (b) Szilagyi, R. K.; Frenking, G. *Organometallics* **1997**, *16*, 4807-4815. (c) Eriksson, L. A.; Pettersson, L. G. M.; Siegbahn, P. E. M.; Wahlgren, U. *J.Chem. Phys.* **1995**, *102*, 872-878. Siegbahn, P. E. M., in *Advances in Chemical Physics, Vol. XCIII*; Eds. Prigogine, I.; Rice, S.A.; John Wiley and Sons, Inc., 1996, pp. 333-387. (d) Siegbahn, P. E. M. *J. Am. Chem. Soc.* **1996**, *118*, 1487-1496. (e) Ricca, A.; Bauschlicher, C. W. Jr. *Theor. Chim. Acta* **1995**, *92*, 123-131; *J. Phys. Chem.* **1994**, *98*, 12899.
(22)   Rosini, G. P.; Liu, F.; Krogh-Jespersen, K.; Goldman, A. S.; Li, C.; Nolan, S. P. *J. Am. Chem. Soc.*, in press.
(23)   The "spectator" phosphines, which are in all cases perpendicular to the Ir-X axis and *trans* to each other with Ir-P bond lengths in the range 2.31 - 2.36 Å, will not be shown in illustrations.
(24)   Wasserman, E. P.; Moore, C. B.; Bergman, R. G. *Science* **1992**, *255*, 315. Schultz, R. H.; Bengali, A. A.; Tauber, M. J.; Weiler, B. H.; Wasserman, E. P.; Kyle, K. R.; Moore, C. B.; Bergman, R. G. *J. Am. Chem. Soc.* **1994**, *116*, 7369-7377. Bengali, A. A.; Schultz, R. H.; Moore, C. B.; Bergman, R. G. *J. Am. Chem. Soc.* **1994**, *116*, 9585-9589.
(25)   Song, J.; Hall, M. B. *Organometallics*, **1993**, *12*, 3118-3126. Couty, M.; Bayse, C. A.; Jimenez-Catano, R.; Hall, M. B. *J. Phys. Chem.* **1996**, *100*, 13976-13978. Musaev, D. G.; Morokuma, K. *J. Am. Chem. Soc.* **1995**, *117*, 799-805. Ziegler, T.; Tschinke, V.; Fan. L.; Becke, A. D. *J. Am. Chem. Soc.* **1989**, *111*, 9177.
(26)   Rossi, A. R.; Hoffmann, R. *Inorg. Chem.* **1975**, *14*, 365.
(27)   Hammond, G. S. *J. Am. Chem. Soc.* **1955**, *77*, 334.
(28)   Low, J. J.; Goddard, W. A. III *Organometallics* **1986**, *5*, 609-622. Saillard, J.-Y.; Hoffmann, R. *J. Am. Chem. Soc.* **1984**, *106*, 2006-2026.
(29)   Ziegler, T.; Tschinke, V. ; Becke, A. *J. Am. Chem. Soc.* **1987**, *109*, 1351-1358.
(30)   Cooper, A. C.; Folting, K.; Huffman, J. C.; Caulton, K. G. *Organometallics* **1997**, *16*, 505-507.
(31)   Abu-Hasanayn, F.; Goldman, A.S.; Krogh-Jespersen, K. *J. Phys. Chem.* **1993**, *97*, 5890-5896; *Inorg. Chem.* **1994**, *33*, 5122-5130.
(32)   Crabtree, R. H.; Quirk, J. M. *J. Organomet. Chem.* **1980**, *199*, 99-106.
(33)   A reviewer has suggested that the computed differences in substituent effects in oxidative addition to $Ir(PH_3)_2X$ and $Ir(PH_3)_2COX$ species may be related to differences in Lewis acid/base characteristics with the former species acting as Lewis acids (empty acceptor orbital, **6b**) and the latter species acting more as Lewis bases (filled d_{z^2} orbital). We thank the reviewer for this and other insightful comments.

Chapter 13

Molecular Mechanics as a Predictive Tool in Asymmetric Catalysis

Per-Ola Norrby

Department of Medicinal Chemistry, Royal Danish School of Pharmacy, Universitetsparken 2, DK-2100 Copenhagen, Denmark

Several methods for applying force field calculations in predictions of selectivities in asymmetric catalysis are discussed. Starting from an initial QM study of the reaction mechanism, force fields can be rapidly developed for various points on the reaction path using a recently developed parameter refinement method. Ground state calculations for high energy intermediates may be used either directly, in biased form, augmented by LFER models, or in combination to yield transition state structures at intersection points. TS force fields can be produced by appropriate modification of a set of QM normal modes. Recent results from application of these methods to the Pd-assisted allylation and the HWE reaction are presented.

Asymmetric synthesis can be defined as being the art and science of synthesizing chiral compounds in optically active form, starting with achiral precursors. Few areas of organic chemistry have attracted so much interest in recent years, one of the major reasons being the fact that the biological activity of chiral compounds often resides mainly or exclusively in only one of the mirror-image forms; the other enantiomer may be devoid of activity or even have totally different and undesirable properties. There is thus a requirement, in both academia and industry, for reliable and general methods of stereoselective synthesis capable of producing chiral compounds in optically pure (or highly enriched) form. This can be accomplished by many methods, including resolution of product mixtures, synthesis starting from the chiral pool, or stereoselective synthesis employing enantiopure reagents or auxiliaries. However, in terms of chiral material consumed in the synthesis, the most economical method is asymmetric catalysis. A few decades ago, Nature was the sole designer of asymmetric catalysts (enzymes), but lately man-made catalysts have become available. For specific substrates, enzymes still shows the highest selectivities, rates, and turnovers available, but designed catalysts can now compete favorably, particularly in scope.

Many catalytic systems are designed around a transition metal, which is the actual reactive site. The reactivity is tuned by variation of the electronic properties of the ligands, and by choice of suitable solvent and other reaction conditions. Tuning the selectivity of the catalyst is commonly accomplished by adjusting the geometry of the ligand, in order to optimize the interactions with the substrate. Rational ligand design thus requires a model showing the expected interactions between ligand and substrate around the reaction center. Preferably, the model should describe the selectivity determining transition state(s) in the reaction, but a qualitative picture might be obtained from reaction intermediates. In the latter case, the model can sometimes be based on experimentally determined structures, but many high energy intermediates cannot be observed, and transition state structures are of course only available from theoretical methods.

Catalyst design is still to a very large extent based on the chemical intuition of the synthetic chemist, often augmented by some type of computational analysis. Computational tools in modeling of reactions range from two-dimensional tools used to elucidate electronic and steric features of the reaction center (e.g., linear free energy relationships based on substituent effects (1)), up through high level quantum mechanical (QM) determinations of the entire reaction path.

The modeling of a reaction can be divided into two parts. The first is to gain a thorough understanding of all bond forming and bond breaking events along the reaction path, and in particular which steps are rate limiting and selectivity determining. This initial mechanistic elucidation is heavily dependent on experimental information, but high level QM methods are becoming more and more important in describing the potential energy surfaces (PES) of reactions. At this stage, it is often sufficient to use small model systems, incorporating only the immediate environment around the reaction center. The second part consists of rationalizing reaction selectivities dependent on non-bonded interactions between remote parts of the reaction complex, frequently necessitating inclusion of a hundred or more atoms in the model. The problem is exaggerated by the flexibility of large systems, requiring evaluation of several thousand conformations. For such cases, the problem is clearly beyond any accurate QM method available today. Force field methods are fast enough, but are hampered by the lack of accurately determined parameters. Here is described a method for rapid determination of molecular mechanics (MM) parameters from available data about the reaction path, in particular from a preceding high-level QM investigation of small model systems. It is demonstrated that the resulting force fields are of sufficient accuracy to allow synthetically useful predictions of reaction selectivities to be made.

Modeling Reaction Paths

A catalytic cycle can be divided into several elementary steps. These generally include formation of a catalyst-substrate complex, transformation into a product complex (sometimes assisted by external reagents or substrates), dissociation of the product, and possibly regeneration of the active catalyst (Figure 1).

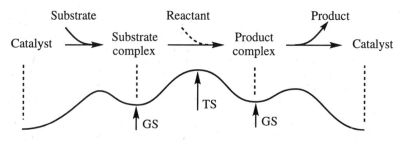

Figure 1. General energy profile for catalytic reactions.

From the viewpoint of catalyst efficiency and turnover, the most important feature of the catalytic cycle is the rate-limiting step. However, when designing a catalyst for a specific selectivity, it is important to realise that the selectivity determining step(s) is not always be the same as the rate limiting step. Reaction selectivity may be determined already in the substrate coordination step, but if the catalyst-substrate complex is formed reversibly or can isomerize, the transformation into catalyst-product complex can become controlling. If two transition states are close in energy, it may be necessary to model both in order to rationalize observed selectivities.

The selectivity in the reaction is completely decided by relative energies of diastereomers of the selectivity determining transition state(s). Ideally, the model should therefore describe the transition states accurately. However, transition states are generally much harder to determine than ground states. It can sometimes be better to base the model on approximate extrapolations from a well-determined ground state than on direct comparisons of badly determined transition states. Modeling approaches based on both ground and transition states will be discussed below. Irrespective of which method is used, it will generally be necessary to determine accurate force field parameters for one or more stationary points along the reaction path.

Parameterization. For the vast majority of metal complexes, no accurate force field parameters are available. After the appropriate point on the reaction path has been located, an MM force field must be defined and the corresponding parameters refined. The parameterization procedure depends on what structural data are available. Intermediates can sometimes be isolated and their structure determined, but in most cases, the parameterization is heavily dependent on QM data, both for structures, force constants, and electrostatics.

The work described herein uses a recently developed method for refinement of MM parameters. (Norrby, P.-O.; Liljefors, T. *J. Comput. Chem.*, in press.) By defining the merit function for a force field as a weighted sum of squared deviations of the MM results from a defined set of reference data points, the parameter values may be optimized by minimizing the merit function using least squares techniques. Hagler et al. recently showed how the consistent force field (CFF) could be optimized using QM energies and energy derivatives calculated for fixed geometries. (2) The current methodology is slower than the methods of Hagler et al., as it relies upon numerical instead of analytical derivatives, but is also more general, as it can be utilized for a wide variety of force fields and with any type of data that can be calculated from the force field. The types of data that has been used to date include bond lengths, angles, dihedral angles, and inverse atom-atom distances from reference structures and conformational energies (QM and experimental), as well as QM-calculated energy derivatives and charges. In particular the QM-calculated energy second derivatives (Hessians) are essential in the parameterization to ensure a proper description of the PES immediately surrounding the stationary points of interest.

The final results are not only dependent on the parameterization, but also on the force field used as a basis for the parameter refinement. Choice of the underlying force field requires several considerations. Firstly, the force field should be good enough to describe the remote parts of the system (for which parameters are not refined) with high accuracy. To be useful in selectivity predictions, conformational energies should be reproduced within ca 2 kJ/mol, a level that is reached by several, but by no means all current force fields. (3) Secondly, the functional form of the force field should be complex enough to reproduce the reference data, in particular the QM Hessians used to define the area of the PES immediately surrounding the point of interest. The second requirement limits the choice to force fields that have been developed to reproduce simultaneously molecular structures and vibrational spectra. Moreover, for metal-catalyzed systems, the force field must also be able to handle complex coordinations. A few force fields are available that fulfil the listed

166

criteria. All models described herein have been developed by modification of the MM3* force field in the MacroModel package. (*4*)

Transition State Parameterization. For ground state force fields, the goal of the parameterization is simply to reproduce all reference data points as closely as possible. For transition states, the picture is not as clearly defined. Molecular mechanics force fields in general are not well suited to describe bond breaking and other transition state phenomena. Even if a true transition state force field could be obtained, most program packages in common use today lack good tools for finding transition states. The traditional solution has been to define the transition state structure as a minimum in the force field. This will allow the standard energy minimization routines to find the correct structures, and possibly to reproduce the correct relative energies between diastereomeric transition states. The use of structures, energies, and charges as reference data in the parameterization is straightforward, as all of these should still be reproduced as closely as possible. However, the Hessians pose a special problem. Each mass-weighted Hessian corresponds to one unique set of eigenvectors and eigenvalues, with exactly one negative eigenvalue for transition states. Reproducing the Hessian exactly will, by necessity, result in the point being a transition state also in the force field. The solution is to modify the reference Hessian before parameterization. The relationship between eigenvectors (**X**), eigenvalues (diagonal matrix **W**), and the mass-weighted Hessian (**A**) is shown in equation 1. The eigensystem is obtained by diagonalization of the mass-weighted Hessian. The negative eigenvalue which defines the TS can now be replaced with a large positive value, generating an eigensystem corresponding to a minimum. A new, modified Hessian can be produced by a simple matrix multiplication (Figure 2). Using the modified Hessian in the parameterization will result in a force field where the original transition state structure is perceived as a minimum. Moreover, small geometrical distortions perpendicular to the reaction coordinate will result in the same energy changes as expected from the true transition state PES.

$$A = X^T \cdot W \cdot X \qquad (1)$$

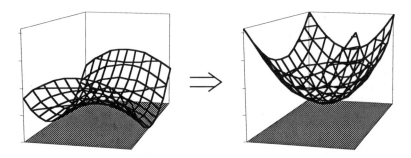

Figure 2. 2-Dimensional representation of the transformation of a transition state PES into a ground state PES to be used in standard force field modeling.

Ground State Modeling

All modeling of ground states in selectivity predictions is based on the assumption that the intermediate modelled is structurally similar to the selectivity determining transition state. The correspondence is expected to be close when the intermediate is high in energy (Hammond postulate). The simplest possible approach is to calculate the relative energies of the intermediates by an appropriate force field, to achieve a ranking of possible diastereomers. If the intermediate and TS are similar enough in structure, qualitative predictions can be made from such a model. When the force field for the intermediate is developed, it is possible to bias the results by including experimental selectivities in the reference data (Figure 3). This can drastically improve quantitative predictions, and can aid in understanding which interactions underlie the observed selectivities, (5-6) but it is not an entirely safe procedure. Biasing is guaranteed to improve the agreement with the data points included as reference, but possibly for the wrong reasons. Results from such a study should therefore be carefully validated.

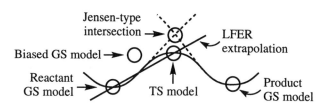

Figure 3. Pictorial representation of various types of selectivity modeling and their position on the reaction path.

Force Field-based QSAR Modeling. An extension of the above methodology is to develop a model for transition state properties in the form of a linear free energy relationship (LFER) based on calculated properties of the intermediate (7) and possibly on well-established empirical parameters (1) (Figure 3). The activation energy for a given reaction path can then be calculated by a simple Boltzmann summation over all contributing isomers. It is of outmost importance to choose the descriptors for the LFER model carefully. Using too many variables will, as always, lead to a serious risk of overfitting. Only descriptors that correlate strongly with the change in energy on going to the transition state should be included. For example, if a bond is breaking in the transition state, the length of that bond in the intermediate will be expected to correlate with the activation energy (with a negative slope), as an already strained bond will require less energy to break.

The LFER model is developed by standard QSAR methods, including statistical tests to remove redundant descriptors, and cross-validation of the final model. The results are dependent on how well the underlying force field can describe the intermediates. However, it can be expected that systematic errors in the underlying force field model will be to some extent compensated in the LFER model. Using again the length of a breaking bond as an example descriptor, the final selectivity model is not too sensitive to errors in the reference bond length parameter, as the effect of adding an LFER equation is that only relative bond lengths in diastereomeric complexes will influence the final results. Also, a slightly erroneous force constant for the bond will be automatically compensated by a change in the slope for the descriptor in the LFER model. A corollary is that the LFER model must always be used in conjunction with the method for which it was developed. A change in the underlying method for determining structures of the intermediates must be accompanied by a redetermination of the LFER model, as the systematic errors are bound to change.

Interpolation to Transition States. All transition states in a catalytic cycle are flanked by two intermediates. It is conceivable that if the PES around each intermediate can be described accurately at a sufficient distance from the ground state, a transition state can be located where the two surfaces intersect (Figure 3). Jensen has recently developed a method for locating transition states by minimizing along the intersection of two potential energy surfaces. (8) This very interesting method has the advantage that an approximation of the transition state structure can be calculated from ground state force fields. In ligand design, the structure of the transition state is very useful when optimizing substituents for specific steric interactions. The exact position of the transition state along the reaction coordinate is very dependent on the ability of the force fields to model large distortions from the ground states, and also on the relative energy of the intermediates, a quantity that is not defined in force field methods and may not be accurately known for the reaction. As force fields generally do not include a term for absolute bond energies, the relative energy of the intermediates must be adjusted by a constant. If transition state reference data is available (in the form of a QM-calculated structure or experimental activation energies), the energy difference between intermediates may be treated as a single parameter, and used to optimize the correspondence with the available data.

Palladium-assisted Allylation. There has been much recent interest in finding efficient ligands for asymmetric palladium-assisted allylation. The reaction is known to proceed through an (η^3-allyl)palladium complex (Figure 4). The basic mechanism involves replacement of the leaving group by palladium with inversion, followed by attack of external nucleophile, also with inversion, leading to an overall retention of configuration. This simple picture is confused by rapid equilibrations in the intermediate, and the possibility of two regioisomeric nucleophilic attacks, allowing formation of several products. Depending on substrate, nucleophile, and reaction conditions, either formation or reaction of the intermediate (η^3-allyl) complex can be rate limiting. (9) The former is mainly the case in desymmetrization reactions. In this work, the focus will be on reactions where nucleophilic attack must be selectivity determining. It can be noted that the two types of transition states are very similar, only differing in the nature of the nucleophile/leaving group.

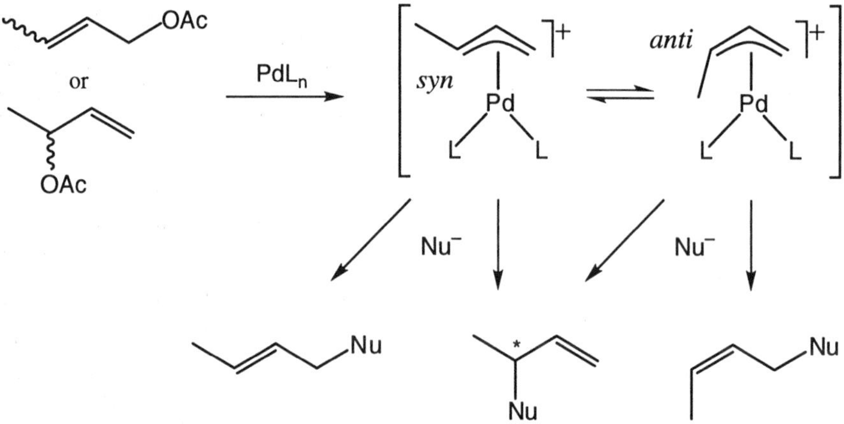

Figure 4. The palladium-assisted allylation.

Looking at the transition state of nucleophilic attack, it can be expected to show similarities to both the (η^3-allyl)Pd complex and the primary product, a Pd(0)-olefin complex. The latter has seldom been detected, (10) and accurate structural determinations of such complexes are scarce. On the other hand, (η^3-allyl)palladium complexes are stable, and many X-ray crystallographic structures are available. A few years ago, molecular mechanics parameters were developed within the MM2 force field for (η^3-allyl)palladium complexes with nitrogen ligands. (11) The force field allowed an estimation of the relative population of *syn* and *anti* forms of the intermediate, (12) and through this a correlation with the product pattern in allylation reactions. (13) Later, it was shown that combination of the force field structures with empirical knowledge about reactivities of the two allyl termini allowed semi-quantitative predictions of the product pattern also in cases where the regioselectivity in nucleophilic attack is expected to be the dominant selectivity-determining factor. (*14*) This is the case in the reaction that has become the standard test for new ligands: addition of malonates to the 1,3-diphenyl allyl moiety (Figure 5).

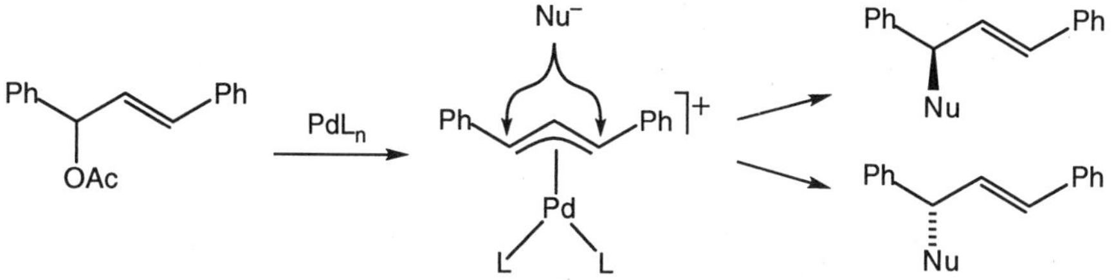

Figure 5. Palladium-assisted allylation proceeding through a symmetrically 1,3-disubstituted intermediate.

In the reaction depicted in Figure 5, only one principal intermediate is formed (in the presence of a non-symmetric bidentate ligand, the intermediate shows two rapidly equilibrating rotameric forms). The enantiomeric products arise from attacks on the same intermediate, and thus arguments based on relative populations of intermediates cannot be used. To calculate product ratios, it is here necessary to quantify transition state energies for the diastereomeric attacks. This has recently been done utilizing an LFER model. (*7*) Several descriptors were evaluated, but only a few showed a statistically significant correlation with the observed selectivities. Descriptors in the final model included steric interactions with the approaching nucleophile (evaluated by a probe technique), strain in the breaking Pd-C bond, and rotation of the allyl towards a product-like conformation. The final model showed good cross-validated internal predictivity ($Q^2 = 0.86$) and moreover did not yield good cross-validation when tested with 10 sets of randomized data, indicating that the effects identified are real. (*15*) The model has also shown good performance with electronically different ligands outside the original training set (Malone, Y.; Guiry, P. J.; Norrby, P.-O., manuscript in preparation). Very recently, it has been possible to locate a transition state for the reaction using high level QM methods in combination with solvation models. The structure of this TS correlates well with expectations from the LFER modeling (Hagelin, H.; Åkermark, B.; Norrby, P.-O., submitted).

Transition State Modeling

So far, no transition state force fields for catalytic systems have been finalized with the methods described above. Several are in progress, but the bottle-neck in the case of metal-catalyzed systems is the necessary initial QM study, with the determination of accurate normal mode analyses. However, from a modeling standpoint, there is no difference between catalytic and stoichiometric reactions, since modeling is done on a molecular level. The technique will therefore be exemplified with a recent transition state force field developed for a stoichiometric reaction utilizing a chiral auxiliary to effect the selectivity.

Asymmetric Horner-Wadsworth-Emmons Reactions. The Wittig reaction and its variants are powerful methods for formation of carbon-carbon double bonds. In the Horner-Wadsworth-Emmons (HWE) reaction, the nucleophile is a carbanion stabilized by a phosphonate and an additional electron withdrawing group (frequently a carboxylic ester). The mechanism of the HWE reaction, depicted in Figure 6, has recently been studied by high level QM methods. (*16*) Two transition states were shown to influence the reaction: addition of the phosphonate to the aldehyde, and ring closure of the intermediate oxyanion to an oxaphosphetane, which in turn eliminates rapidly to the final alkene products. The products do not contain any new chirality, but the transition states contain two additional stereocentra which will influence the path of the reaction. The asymmetric version of the reaction has interesting applications in the areas of kinetic resolution and desymmetrization. (*17*) In the former case, the reaction is unusual in that both enantiomers of racemic aldehydes react, but frequently yield opposite double bond isomers. The HWE reaction thus avoids one of the common problems in kinetic resolution, reduced selectivity at high conversion due to accumulation of the slower reacting isomer.

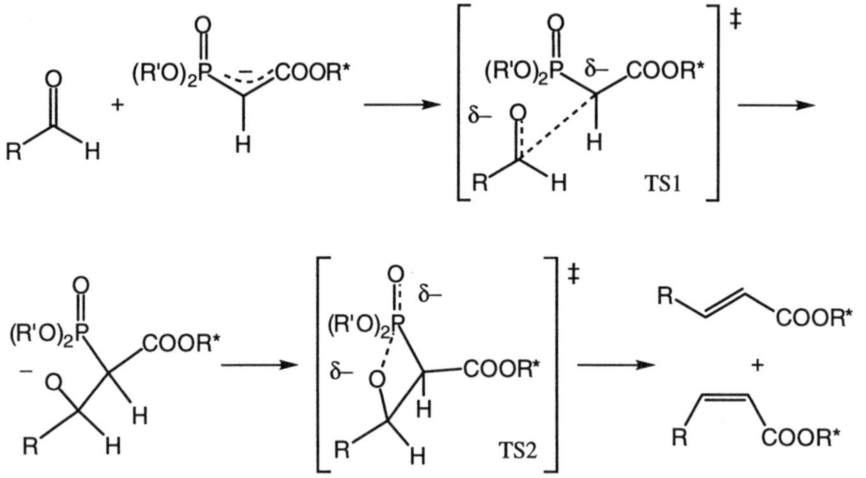

Figure 6. The mechanism of the HWE reaction.

Force fields were developed for both transition states in the HWE reaction, starting from several QM transition states for reactions of formaldehyde or acetaldehyde with trimethyl phosphonoacetate anion. From the QM study, it could be seen that the energy difference between the two transition states was low and sensitive to the reaction conditions. It was therefore treated as a variable to be optimized by fitting to experimental data.

Starting from enantiopure phosphonate reagents (R*=chiral auxiliary) and either *meso*-dialdehydes or racemates of chiral aldehydes, the reaction can follow eight diastereomeric paths. For each path, the overall rate depends on all available conformations of both transition states, as shown in equation 2 (E_Δ is the adjustable energy difference between TS1 and TS2, whereas indices "i" and "j" are used to identify conformers of one diastereomer of TS1 and TS2, respectively).

$$r = \left[\left(\sum e^{-E_i/RT} \right)^{-1} + \left(\sum e^{(E_\Delta - E_j)/RT} \right)^{-1} \right]^{-1} \tag{2}$$

By evaluating the relative rates for all paths, it is possible to predict three selectivities in each reaction: the (*E*)/(*Z*) ratio and, for each double bond isomer, the diastereomeric ratio. Naturally, all three selectivities must be rationalized using the same relative TS energy E_Δ. The model has recently been tested against experimental results for a small set of phosphonate reagents reacting with aldehydes having a stereogenic α-carbon (a total of 14 data points). In 12 cases could the major isomer be identified, and in no case did the maximum error exceed 3 kJ/mol (Norrby, P.-O.; Brandt, P; Rein, T., submitted). The variable E_Δ parameter showed physically reasonable values, in that the sterically more encumbered systems showed higher influence from the tighter TS2. A somewhat surprising result was that in no case could one single transition state be considered completely selectivity determining. By analysing the transition state structures, it was found that the function of the chiral auxiliary is mainly to block one face of the phosphonate anion, resulting in high TS1 energies for four of the eight diastereomeric paths. The main effect in the tight TS2 was a strong coupling between the configuration at the aldehyde α-carbon and the newly formed stereocenter at the former carbonyl carbon, again blocking four paths selectively, two of which were also blocked by TS1. It can be seen from symmetry considerations that the two remaining reaction paths must lead from opposite substrate enantiomers to opposite double bond isomers, in good agreement with experimental observations.

Concluding Remarks

Here has been outlined several methods for achieving rapid modeling of selectivity in catalytic reactions, with accuracies good enough to be of use to synthetic chemists. The most general, but also the hardest to develop, is the type of model where the true transition state is treated as a force field minimum. Well-determined force fields of this type will be applicable to any type of reaction. For reactions that are strongly dependent upon the environment (solvent or enzyme active sites), it may be sufficient to add the environment for the final force field, but it would probably be advantageous if the environment could be included in the entire process. Solvation can be achieved using continuum models, both in the initial QM and in the force field modeling. Including an enzyme as the environment could best be realized by performing the initial mechanistic study using a hybrid QM/MM method.

All ground state methods are dependent on the transition state being reasonably similar structurally. For reactions where the intermediate is well defined and can only lead to one unique TS, a biased ground state model may give very good results, but requires extensive validation. LFER models are more general, and can be applied also to cases where each intermediate can lead to more than one TS. It is also less sensitive to systematic errors in the underlying force field. Jensen-type interpolations show one distinct advantage over other methods based on ground state force fields: a structure of the TS is produced, allowing identification of specific interactions which may be modified in ligand design.

Two important goals in the modeling of the palladium-assisted allylation reaction have been realized very recently: the QM location of the reaction transition state (*vide supra*), and a ground state force field for the primary reaction product, the Pd-olefin complex (Hagelin, H.; Svensson, M.; Åkermark, B.; Norrby, P.-O., manuscript). These will in turn allow true transition state modeling, and location of Jensen-type intersections, respectively. For the first time, it will thus be possible to compare several modes of transition state modeling for the same type of organometallic reaction.

As with all force field based methods, it must be noted that even though changes in the steric environment can be handled well, changes in the electronic structure may require redetermination of many parameters. Thus, the force fields are useful in designing the geometry of ligands, but cannot safely predict the effects of changing electronic properties of ligands. Each such change will at least require a revalidation, and possibly an extensive reparameterization.

Timing Considerations. To be useful to synthetic chemists, predictions in asymmetric synthesis must reach an accuracy around 2-4 kJ/mol, corresponding approximately to the difference between a catalyst with low selectivity yielding around 0-30% ee, and a highly selective catalyst yielding enantioselectivities in the range 90-99%. To be useful in ligand design, the time required to test a new ligand should be substantially less than what is needed to synthesize it and test it. The initial phase of each of these projects involves a thorough QM study of the reaction mechanism, with a project time most commonly measured in months or years.

172

However, in the case of the HWE reaction, with the final QM results in hand the entire force field development required about a month, using a standard workstation. At this point, with the final force field available, a new reaction can be tested in a few days using only a single CPU. This includes generating and minimizing at least 100.000 trial conformations. It should be noted here that initial attempts to use this technique failed, due solely to incomplete conformational searches where only a few thousand isomers were generated for each reaction. The systems considered here contained around 70-100 atoms. Thus, single conformations could plausibly be evaluated in QM calculations of good accuracy, but the necessity to include several thousands of conformations to achieve good results puts these predictions beyond any QM-based modeling technique available today.

Acknowledgment. Collaborations with B. Åkermark and his research group on the Pd-assisted allylations, with T. Rein and P. Brandt on the HWE reaction, and T. Liljefors on force field development have made possible the work described herein. Financial aid from The Danish Medical Research Council and The Danish Technical Research Council is gratefully acknowledged.

Literature Cited

1. Hansch, C; Leo, A.; Taft, R. W. *Chem. Rev.* **1991**, *91*, 165.
2. Maple, J. R.; Hwang, M.-J.; Stockfisch, T. P.; Dinur, U.; Waldman, M.; Ewig, C. S.; Hagler, A. T. ; *J. Comput. Chem.* **1994**, *15*, 162.
3. Gundertofte, K.; Liljefors, T.; Norrby, P.-O.; Pettersson, I. *J. Comput. Chem.* **1996**, *17*, 429.
4. MacroModel V6.0; Mohamadi, F.; Richards, N. G. J.; Guida, W. C.; Liskamp, R.; Lipton, M.; Caulfield, C.; Chang, G.; Hendrickson, T.; Still, W. C. *J. Comput. Chem.* **1990**, *11*, 440.
5. Norrby, P.-O.; Kolb, H. C.; Sharpless, K. B. *J. Am. Chem. Soc.* **1994**, *116*, 8470.
6. Norrby, P.-O.; Linde, C.; Åkermark, B. *J. Am. Chem. Soc.* **1995**, *117*, 11035.
7. Oslob, J. D.; Åkermark, B.; Helquist, P.; Norrby, P.-O. *Organometallics* **1997**, *16*, 3015.
8. Jensen, F. *J. Comput. Chem.*, **1994**, *15*, 1199.
9. Trost, B. M.; van Vranken, D. L. *Chem. Rev.*, **1996**, *96*, 395.
10. Steinhagen, H.; Reggelin, M.; Helmchen, G. *Angew. Chem., Int. Ed. Eng.* **1997**, *36*, 2108.
11. Norrby, P.-O.; Åkermark, B.; Hæffner, F.; Hansson, S.; Blomberg, M. *J. Am. Chem. Soc.* **1993**, *115*, 4859.
12. Sjögren, M.; Hansson, S.; Norrby, P.-O.; Åkermark, B.; Cucciolito, M. E.; Vitagliano, A. *Organometallics* **1992**, *11*, 3954.
13. Sjögren, M. P. T.; Hansson, S.; Åkermark, B.; Vitagliano, A. *Organometallics* **1994**, *13*, 1963.
14. Peña-Cabrera, E.; Norrby, P.-O.; Sjögren, M.; Vitagliano, A.; deFelice, V.; Oslob, J.; Ishii, S.; Åkermark, B.; Helquist, P. *J. Am. Chem. Soc.*, **1996**, *118*, 4299.
15. Oslob, J. D.; Åkermark, B.; Helquist, P.; Norrby, P.-O. *Organometallics,* **1997**, *16*, 3015.
16. Brandt, P.; Norrby, P.-O.; Martin, I.; Rein, T. *J. Org. Chem.* **1998**, *63*, 1280.
17. Rein, T.; Reiser, O. *Acta Chem. Scand.* **1996**, *50*, 369.

Chapter 14

Combined QM/MM and Ab Initio Molecular Dynamics Modeling of Homogeneous Catalysis

T. K. Woo[1], P. M. Margl[1], L. Deng[1], L. Cavallo[2], and T. Ziegler[1]

[1]Department of Chemistry, University of Calgary, 2500 University Drive N.W., Calgary, Alberta T2N 1N4, Canada
[2]Dipartimento di Chimica, Università "Federico II" di Napoli, Via Mezzocannone 4, I-80134 Napoli, Italy

Abstract: With the methodologies presented here, the combined QM/MM method and the *ab initio* molecular dynamics method, we are moving towards more realistic computational models. The QM/MM method allows for the simulation of large systems at the *ab initio* level without completely neglecting the groups and substituents not within the active site. As demonstrated by our QM/MM calculations of the Brookhart Ni(II) based polymerization catalyst, these outer groups can often play a crucial role in the chemistry. Since the stereoselectivity in many catalytic systems is controlled by steric interactions there is great potential for the combined QM/MM method to be utilized effectively in such areas. The *ab initio* molecular dynamics method also shows great potential for becoming a standard computational chemistry tool particularly for exploring processes which have a high degree of configurational variability. We have applied the methodology to several transition metal based homogenous catalytic systems, clearly demonstrating the usefulness of the method. With electrostatic coupling, the combed QM/MM molecular dynamics method is a very promising tool for including solvent effects. Indeed with the combination of these techniques we are developing more sophisticated models of catalytic systems which can take into account large ligands, finite temperature effects and potentially solvent effects.

1. INTRODUCTION

Homogenous catalytic systems have often been used to model more complicated heterogeneous systems. However, even seemingly simple homogenous systems pose many challenges for computational quantum chemists. Often times a first principle's quantum mechanical calculation involves a stripped down model that only vaguely resembles the true system. If large ligand systems are involved they are most often neglected in high level calculations with the hope that they do not substantially influence the nature of the reaction mechanisms. Unfortunately, the surrounding ligand system, solvent or matrix can often play a critical mechanistic role. One reasonable approach to constructing a more sophisticated computational model which approximates these effects is the combined quantum mechanics and molecular mechanics (QM/MM) method[1-11], which has recently received significant attention in addressing these issues[4] with contributions from Singh and Kollman[1], Maseras and Morokuma[3,] Gao and coworkers[4,6,7],Warshel and Levitt,[8] Field et al.[2], Théry et al.[9] Caldwell[10] et al. and Bakowies and Thiel[5,11]. In this hybrid method part of the molecule, such as the active site, is treated quantum mechanically while the remainder of the system is treated with a molecular mechanics force field. This allows extremely large systems that are out of the reach of pure QM calculations to be studied in an efficient and detailed manner. In addition to neglecting solvent or ligand effects, many conventional electronic structure studies involve the exploration of the potential energy surface at the zero temperature limit. Therefore, finite temperature(non-zero) effects are neglected. *Ab initio* molecular dynamics methodologies allow for the exploration of potential energy surfaces of chemical reactions at finite temperatures. In this article we attempt to review our experiences with the application of the QM/MM and AIMD methodologies to transition metal homogenous catalysis.

2. APPLICATION OF THE QM/MM METHOD TO TRANSITION METAL BASED HOMOGENOUS CATALYSIS

Most applications of the QM/MM methodology (where the boundary cuts covalent bonds) have been concentrated in the area of protein modeling.[12-16] To date applications to transition metal based catalysis have been limited. Morokuma *et al* has demonstrated the methodology to the organometallic reaction of $Pt(PR_3)_2 + H_2$.[17] We have recently applied the QM/MM methodology to examine the role of the bulky substituents in the Brookhart Ni-diimine olefin polymerization catalyst.[18]

Ni(II) diimine based single site homogeneous catalysts of the type $(ArN=C(R)=C(R)=NAr)Ni-R'^+$ have emerged as promising alternatives to both traditional Ziegler-Natta and metallocene catalysts for olefin polymerization.[19-21]. Brookhart and coworkers have shown that these catalysts are able to efficiently convert ethylene into high molecular weight polymers with a controlled level of polymer branching. In this polymerization system the bulky aryl groups play a crucial role, since without the bulky substituents the catalyst acts only as a dimerization catalysts due to the favorability of the hydrogen-transfer chain termination process. From the structure of the catalyst, it is evident that the bulky aryl substituents partially block the axial coordination sites of the Ni center. It is likely this steric feature which impedes the termination to the insertion process thereby promoting the intrinsically poor polymerization catalyst into a commercially viable one.

With the purpose of examining, in detail, the role of the bulky substituents in the Brookhart polymerization catalyst we have performed both pure QM calculations of the system without the bulky ligands and combined QM/MM calculations on the "real" system. In the QM/MM model, the bulky R=Me and Ar=2,6-C_6H_3(i-Pr)$_2$ groups are treated by the AMBER95 molecular mechanics potential whereas the Ni-diimine core including the growing chain and monomer are treated by a density functional potential. Figure 1 shows the full catalyst system examined where the carbon atoms with asterisks represent the link atoms at the QM/MM boundary which are replaced by dummy hydrogen atoms in the QM model system.

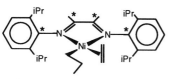

Figure 1. Model for the active complex in the Brookhart Ni-diimine olefin polymerization system. The growing chain is modeled with a propyl group while the monomer is taken as an ethene. Atoms highlighted with an asterisk represent link atoms that are replaced by dummy hydrogen atoms in the model QM system.

Using the prescription of Maseras and Morokuma[3] we have combined the AMBER95[22] molecular mechanics force field with the Amsterdam Density Functional (ADF)[23,24] program system. The QM system was calculated at the non-local density functional level with Becke's[25] 1988 exchange and Perdew's[26,27] 1986 correlation functionals. Full computational details are provided elsewhere.[18]

Three processes are believed to dominate the polymerization chemistry of the catalyst system, namely, chain propagation, chain termination and chain branching, as shown in Figure 2. The propagation commences from an olefin π-complex which has been determined experimentally to be the catalytic resting state. Insertion of the olefin into the M-C$_\alpha$ bond forms a metal alkyl cation. Uptake of the monomer returns the system to the resting state. Chain termination occurs via monomer assisted β-elimination, either in a fully concerted fashion as illustrated in Figure 2b, or in a multistep associative mechanism as implicated by Johnson et al.[20] The unique short chain branching is proposed to occur via an alkyl chain isomerization process as sketched in Figure 2c. In this proposed process, β-hydride elimination first yields a putative hydride olefin π-complex. Rotation of the π-coordinated monomer followed by rotation yields a secondary carbon unit and therefore a branching point. It is important to note that the branching process commences from the Ni-alkyl cation and not the resting state complex. Since monomer pressure effects the branching rate, the equilibrium between the free Ni-alkyl and the resting state is believed to strongly influence the branching rate. Experimentally the relative magnitudes of the free energy barriers are ordered such that propagation < branching < termination.

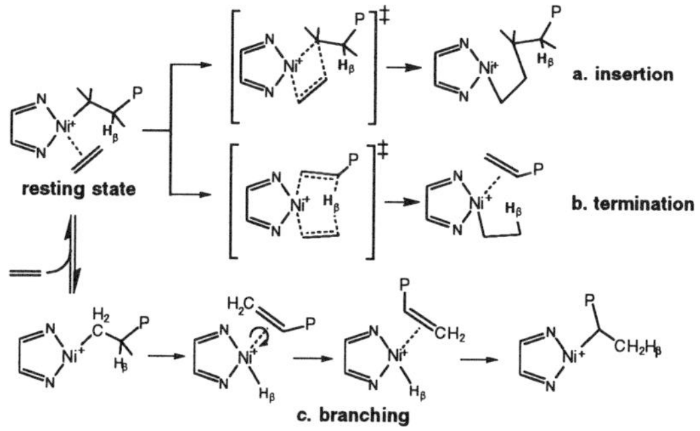

Figure 2. Proposed reaction mechanisms of (a) insertion, (b) chain termination and (c) chain branching mechanisms for the Brookhart Ni-diimine olefin polymerization catalyst.

Table 1. Comparison of Calculated and Experimental Barriers.

	reaction barriers (kJ/mol)		
	insertion	branching	termination
absolute:			
Pure QM[a] ($\Delta H^{\ddagger}$)	70.3	53.6	40.6
QM/MM ($\Delta H^{\ddagger}$)	55.3	64.0	77.9
Experimental[b] ($\Delta G^{\ddagger}$)	42-46	-	-
relative to insertion:			
QM/MM ($\Delta\Delta H^{\ddagger}$)	0.0	8.8	22.6
Experimental[c] ($\Delta\Delta G^{\ddagger}$)	0.0	5.4[d]	23.4[e]

[a]Reference 28. [b]Reference 29. [c]Reference 20. Polymerization of 1.6×10^{-6} mole of $(ArN=C(R)C(R)=NAr)Ni(CH_3)(OEt_2)]^+[B(3,5-C_6H_3(CF_3)_2)_4]^-$ where $Ar = 2,6-C_6H_3(i\text{-}Pr)_2$ and $R = Me$ in 100 mL of toluene at 0°C for 15 min. [d]NMR studies provide a ratio of 48 isomerization events per 500 insertions, assuming that all branches are methyl branches (methyl branches are experimentally observed to predominate). Applying Boltzmann statistics to this ratio at 273.15 K yields a $\Delta\Delta G$ of 5.4 kJ/mol. [e]The weight-average molecular weight, M_w, of 8.1×10^5 g/mol provides an estimate for the ratio of termination events to insertion events of 1:28900. Using Boltzmann statistics to this ratio gives a $\Delta\Delta G$ of 23.4 kJ/mol.

Table 1 compares the reaction barriers, $\Delta H^{\ddagger}$, for the propagation, branching and termination processes for both the pure QM model with no ligands modeled and the QM/MM model. Without the bulky ligands, the termination barrier is approximately 30 kJ/mol less than the insertion (propagation) barrier, suggesting that the system would be a poor polymerization catalyst. This is in agreement with experiment, where the Ni and Pd diimine systems without the bulky ligands are used as dimerization and oligermization catalysts.[30] When the bulky ligands are included, the termination and insertion barriers reverse their orders with a dramatic increase in the termination barrier. The termination transition state which is shown in Figure 3a, has both axial coordination sites of the Ni occupied. As proposed by Brookhart and coworkers[20,21] the bulky iso-propyl groups on the aryl rings act to block the axial coordination sites thereby dramatically increasing the termination barrier. In the insertion transition state which is sketched in Figure 3b the alkyl and olefin moieties lie in the coordination plane of the Ni center, removed from the bulky iso-propyl groups. In this case there is little steric hindrance to the insertion process. In fact the addition of the bulky aryl ligands actually reduces the insertion barrier. This stabilization of insertion transition state with bulky ligands in the QM/MM model compared to that without the bulky ligands results from two effects. First, the resting state structure is destabilized by the bulky ligands, thereby lowering the insertion barrier by increasing the energy of the precursor. Additionally, there is the relaxation of the orientation of the aryl rings relative to the Ni-diimine ring in the insertion transition state. Physically, there is a preference for the aryl rings and the diimine ring to be more parallel to maximize the π-bonding between the rings. When both axial coordination sites are occupied as in the resting state and the termination transition state, the aryl rings are forced to be perpendicular to the diimine ring. In the insertion transition state where the axial sites are empty, the aryl rings can rotate away from the perpendicular orientations as to enhance the π-bonding interaction between the rings, thereby stabilizing the insertion transition state. This concept helps rationalize the observed[20] increase in activity when the π-system of the diimine ring is extended. We conclude that the catalyst activity can be increased by enhancing the π-bonding interaction between the diimine rings and the aryl rings.

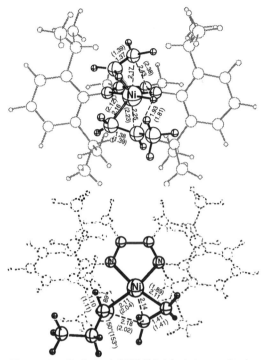

The chain branching mechanism originally proposed by Brookhart[20] involves discrete olefin hydride intermediate as sketched in Figure 2c. Our calculations implicate a concerted pathway since no stable olefin hydride complex could be located. (The optimized QM/MM chain branching transition state is not displayed.) With the hybrid model, the chain isomerization barrier is calculated to be 64 kJ/mol, which is only slightly increased from 53.6 kJ/mol in the pure QM model.

Table 1, reveals that the reaction barriers calculated from our combined QM/MM model are in good agreement with the experimentally determined free energy barriers, both in relative and absolute terms. This contrasts the results of the pure QM study where the bulky ligands were not modeled and even the order of the barrier heights is not reproduced. We note that there is an excellent agreement between the calculated and experimental chain termination barrier relative to the

Figure 3. Optimized QM/MM (a) chain termination transition state via β-hydrogen transfer to the monomer and (b) insertion transition state. MM atoms are ghosted for clarity, while the dummy hydrogen atoms are omitted. Parenthetic values refer to the same geometric parameter found in the corresponding pure QM geometry. The pure QM model does not contain the bulky ligands. Distances are in Ångstroms and angles in degrees.

insertion barrier. Specifically, the QM/MM relative barrier is $\Delta\Delta H^{\ddagger}=22.6$ kJ/mol whereas the value determined experimentally is $\Delta\Delta G^{\ddagger}=23.4$ kJ/mol. Our model therefore provides an accurate estimate of the polymer molecule weight, M_w. It is more difficult to directly compare the calculated barriers for chain branching with that of experiment because in our model we do not account for concentration effects. Figure 2c shows that the branching commences from the metal-alkyl complex and there is an equilibrium between with the olefin coordinated π-complex. Experimentally, the chain branching increases with decreased monomer pressure, suggesting that the equilibrium between the π-complex and the metal-alkyl is crucial To address this issue we are currently simulating the olefin capture process with the combined QM/MM *ab initio* molecular dynamics method.[31]

The QM/MM methodology has been successfully applied to examine the chemistry of the Brookhart Ni(II) diimine catalyst for which at least semi-quantitative results have been obtained. Moreover, we have demonstrated that the combined QM/MM method can be effectively applied to transition metal based catalytic processes in a detailed yet efficient manner.

3. *AB INITIO* MOLECULAR DYNAMICS SIMULATIONS

3.1 What is Molecular Dynamics?

Conventional electronic structure calculations can be classified as static simulations. In these calculations the nuclear positions are optimized to locate local minima and transition states on the potential surface at the zero temperature limit. This involves, for each nuclear geometry, converging the electronic structure in order to determine the energy and forces on the nuclei. This information is then used to move the nuclei to a more optimal geometry. In classical molecular dynamics the nuclei are allowed to move on the potential surface according to Newton's classical laws of motion (Eqn. 1) as to simulate nuclear motion at finite (non-zero) temperatures.

$$m_i \vec{a}_i = \vec{F}_i(t) = -\frac{\partial E_{TOT}}{\partial \vec{x}_i} \quad \text{i=1, 2..N}_{nuc} \quad (1)$$

The nuclear motion generated in a molecular dynamics simulation can be utilized for a variety of purposes. Stationary points can be optimized by simply applying friction to the nuclear motion, thereby causing the system to settle into a local minimum. The motion can also be used to sample configuration space as to construct a partition function from which properties can be derived rigorously from statistical mechanics. There are also global minimization schemes which utilize molecular dynamics, such as simulated annealing. In this section our intent is to briefly introduce the basic concepts involved with *ab initio* molecular dynamics and particularly those issues specific to our simulations of homogenous catalytic systems.

The motion of the nuclei in a MD simulation is determined by integrating Newton's equations of motion (Eqn 1). In other words, given the velocities and positions at a time t, one determines the same quantities with reasonable accuracy at a later time t+Δt using the calculated forces on the nuclei. With each new geometry that is generated, the forces on the nuclei have to be recalculated. In order to simulate molecular vibrations, the time step Δt must be at least a third smaller than the period of the fastest vibration. Thus in order to simulate C-H bond stretching which has a period of 0.01 psec, the Δt must be roughly 0.003 psec. Even a relatively short 10 ps simulation requires over 3300 time steps and therefore 3300 gradient calculations. (To put this in perspective, even the fastest enzyme catalyzed reactions have turnover periods of greater than 1000 picoseconds.) For this reason, most MD simulations are performed with a molecular mechanics force field. Their computational simplicity allows for extremely large simulations to be performed including protein folding processes. However, the empirical nature of molecular mechanics force fields has its limitations. For example, transition metal complexes which are of interest in our research are problematic because there is currently no force field general enough to handle all of the complicated bonding schemes available to these complexes. Moreover, chemical reactions cannot be simulated accurately (if at all) with a molecular mechanics force field. To overcome these problems *ab initio* or QM potentials can be used in conventional classical molecular dynamics. In other words, with the *ab initio* molecular dynamics (AIMD) method the forces are *not* determined from a molecular mechanics force field, but rather they are derived from a full electronic structure calculation at each time step. Although this is computationally demanding, it allows for bond breaking and forming processes to be simulated and it allows for molecular dynamics to be performed on systems where a molecular mechanics force field is not readily applicable.

3.2 Car-Parrinello *Ab Initio* Molecular Dynamics

Conventional AIMD involves moving the nuclei with the forces calculated from an electronic structure calculation. The electronic structure is normally described by a set of orthonormal molecular orbitals, φ_i, which are expanded in terms of a basis set, χ_k such that $\varphi_i = \sum_k c_{ik}\chi_k$. The optimal coefficients are solved variationally with the constraint that the molecular orbitals remain orthogonal. Generally this is done in a self consistent manner by the diagonalization of the Hamiltonian matrix or equivalent. An alternative method for determining the optimal coefficients draws analogy to nuclear dynamics which is used to optimize nuclear geometries. By assigning fictitious masses to the coefficients, fictitious dynamics can be performed on the coefficients which then move through electronic configuration space with forces given by the negative gradient of the electronic energy. The equivalent equations of motion are:

$$\mu \ddot{c}_{i,k} = -\frac{\partial E^{el}}{\partial c_{i,k}} - \sum_j \lambda_{i,j} c_{i,k} \langle \Psi_i | \Psi_k \rangle \qquad (2)$$

where μ is a fictitious mass, $\ddot{c}_{i,k}$ is the coefficient acceleration, and the last term corresponds to the constraint force imposed to maintain orthogonality. The coefficients move through electronic configuration space with a fictitious kinetic energy and by applying friction the coefficients can be steadily brought to settle into an optimal configuration.

In 1985 Car and Parrinello[32] developed a scheme by which to perform the nuclear dynamics and the electronic coefficient dynamics in parallel as to improve the efficiency of the AIMD method. In this way the electronic MD and nuclear MD equations are coupled:

$$\mu \ddot{c}_{i,k} = -\frac{\partial E}{\partial c_{i,k}} - \sum_j \lambda_{i,j} c_{i,k} \qquad m_I \ddot{x}_I = -\nabla E(x_I, c_{i,k}) \qquad (3)$$

Formally, the nuclear and electronic degrees of freedom are cast into a single, combined Lagrangian:

$$\mathcal{L} = \sum \mu_i \langle \dot{\Psi}_i | \dot{\Psi}_i \rangle + \tfrac{1}{2}\sum M_I \dot{R}_I^2 - E_{DFT}(|\Psi\rangle, R) + \sum_{i,j} \Lambda_{ij}(\langle \Psi_i | \Psi_j \rangle - \delta_{ij}) \qquad (4)$$

where the first two terms represent the kinetic energy of the wave function and nuclei, respectively, the third term is the potential energy and the last term accounts for the orthogonality constraint of the orbitals. The combined Car-Parrinello Lagrangian ensures that the propagated electronic configuration corresponds to the propagated nuclear positions. Although, the generated electronic configuration does not always correspond to the proper Born-Oppenheimer wavefunction, over time it generates a electronic structure that oscillates around it giving rise to stable molecular dynamics. The coupled Car-Parrinello dynamics, therefore, results in a speed up over conventional AIMD since the electronic wave function does not have to be converged at every time step, instead it only has to be propagated. The primary disadvantage of the Car-Parrinello MD scheme is that the electronic configuration oscillates about the Born-Oppenheimer wavefunction at a high frequency. Therefore, in order to generate stable molecular dynamics a very small time step must be used, usually an order of magnitude smaller than in conventional *ab initio* molecular dynamics.

Although other "first-principles" methods can be used, the Car-Parrinello coupled dynamics approach has mostly been implemented within the density functional framework with plane wave basis sets (as opposed to atom centered basis sets). Therefore, Car-Parrinello *ab initio* molecular dynamics generally refers only to this type of implementation. Applications of the Car-Parrinello AIMD method are concentrated in the area of condensed phase molecular physics.

The use of plane waves has its advantages in that the computational effort for the required integrations becomes minute on a per function basis. On the other hand, an enormous number of plane waves is required even when pseudo potentials are utilized to approximate the core. The number of plane waves required to accurately treat transition metals and first row elements becomes inhibitively large. This is clearly a problem in our research on homogenous catalysis since such systems almost always contain transition metals and the substrates are often made up of first row elements. Plane wave basis sets also introduce another problem when studying homogenous catalysts since periodic images are created automatically and therefore the simulation actually describes a periodic crystal. If non-periodic systems are simulated, then the cell size of the periodic systems must be sufficiently large that the wave functions of the images no longer overlap. This requires a vacuum region of approximately 5 Å between images. This is an issue because the computational effort increases with the cell size. If charged systems or systems with large dipole moments are simulated, the long range electrostatic interactions between periodic images will lead to artificial effects. Again, this is problematic in our research since many transition metal catalysts are charged species.

3.3 Projector Augmented Wave (PAW) Car-Parrinello AIMD

The Projector Augmented Wave (PAW) Car-Parrinello AIMD program developed by Peter Blöchl[34,35] overcomes the aforementioned problems such that AIMD simulations of transition metal complexes has become practical. PAW utilizes a full all-electron wavefunction with the frozen core approximation which allows both accurate and efficient treatment of all elements including first row and transition metal elements. In the PAW method, the plane waves are augmented with atomic based functions near the nuclei such that the rapidly oscillating nodal structure of the valence orbitals near the nuclei are properly represented. One can think of it as smoothly stitching in an atomic-like function into the plane waves such that the plane waves describe regions where the orbitals are smooth, allowing for rapid convergence of the plane waves. The plane wave expressions of PAW and those of the pseudopotential method are similar enough that the numerical techniques for the most computationally demanding operations are related and equally efficient. Therefore, PAW combines the computational advantages of using plane waves with the accuracy of all-electron schemes. The details of the implementation are described elsewhere.[33-35] To deal with charged systems, PAW has a charge isolation scheme to eliminate the spurious electrostatic interactions between periodic images. The charge isolation scheme involves fitting atomic point charges such that the electrostatic potential outside the molecule is reproduced (ESP fit). The ESP charges are then used to determine the electrostatic interaction between periodic images via Ewald sums. The spurious interactions between images is then subtracted. These features of the PAW AIMD package make it ideal to study transition metal catalysts at the *ab initio* molecular dynamics level.

3.4 Reaction Free energy barriers with AIMD

Reaction free energy barriers are routinely calculated from conventional static electronic structure calculations. Here, the excess free energy of the reactants and transition state can be determined by constructing a partition function from a frequency calculation and using a harmonic (normal mode) approximation. In most cases where the interactions are strong, the approximation is good. However, for processes where weak intermolecular forces dominate, the harmonic or quasi-harmonic approximation breaks down.[36] Alternatively, *ab initio* molecular dynamics simulations can be utilized to determine reaction free energy barriers. An MD simulation samples the available configuration space of the system as to produce a Boltzmann ensemble from which a partition function can be constructed and used to determine the free energy. However, finite MD simulations can only sample a restricted part of the total configuration space, namely the low energy region. Since estimates of the absolute free energy of a system requires a global sampling of the configuration space, only relative free energies can be calculated.

A number of special methodologies have been developed to calculate relative free energies. Since we are interested in reaction free energy barriers, the method we use in our research is derived from the method of thermodynamic integration.[37,38] Assuming we are sampling a canonical NVT ensemble the free energy difference, ΔA, between an initial state with $\lambda=0$ and a final state with $\lambda=1$, is given by Eqn 5.

$$\Delta A_{(0 \to 1)} = \int_0^1 \frac{\partial A(\lambda)}{\partial \lambda} d\lambda \quad (5)$$

Here the continuous parameter λ is such that the potential $E(\lambda)$ passes smoothly from initial to final states as λ is varied from 0 to 1. Since the free energy function can be expanded in terms of the partition function:

$$A(\lambda) = -kT \ln \left[\int \cdots \int e^{-\frac{E(X^N,\lambda)}{kT}} dX^N \right] \quad (6)$$

the relative free energy ΔA can be rewritten as:

$$\Delta A_{(0 \to 1)} = \int_0^1 \left(\int \cdots \int \frac{\partial E(X^N,\lambda)}{\partial \lambda} dX^N \bigg|_\lambda \right) \quad (7)$$

or

$$\Delta A_{(0 \to 1)} = \int_0^1 \left\langle \frac{\partial E(X^N,\lambda)}{\partial \lambda} \right\rangle_\lambda d\lambda \quad (8)$$

where the subscript λ represents an ensemble average at fixed λ. Since the free energy is a state function λ can represent any pathway, even non-physical pathways. However, if we choose λ to be a reaction coordinate as to represent a physical reaction path, this provides us with a means of determining an upper bound for a reaction free energy barrier by means of thermodynamic integration. The choice in reaction coordinate is important since a poorly chosen reaction coordinate will result in an unfavorable reaction path and potentially a significant over estimate of the barrier. The more the reaction coordinate resembles the intrinsic reaction coordinate (IRC) the potentially better the estimate. The reaction coordinate can be sampled with discrete values of λ on the interval from 0 to 1 or carried out in a continuous manner in what is termed a "slow growth" simulation[39] by

$$\Delta A = \sum_{i=1}^{N_{steps}} \left\langle \frac{\partial E(\lambda)}{\partial \lambda} \right\rangle_i \Delta \lambda_i \quad (9)$$

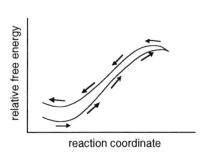

reaction coordinate

Figure 4 Schematic representation of the hysteresis in a slow growth free energy plot. The arrows designate the direction of the scan in terms of the reaction coordinate.

where i indexes the step number. Here the free energy difference becomes the integrated force on the reaction coordinate and can be thought of as the work necessary to change the system from the initial to final state. The discrete sampling resembles a linear transit calculation such that a series of simulations is set up corresponding to successive values of the reaction coordinate from the initial to final state. For each sample point, the dynamics must be run long enough to achieve an adequate ensemble average force on the fixed reaction coordinate. In a slow growth simulation[39] the reaction coordinate is continuously varied throughout the dynamics from the initial to the final state. Thus, in each time step the reaction coordinate is incrementally changed from that in the previous time step. Formally speaking the system is never properly equilibrated unless the change in the RC is infinitesimally small (reversible change). However, the smaller the rate of change the better the approximation. Since the RC is changed at each time step, the force on the reaction coordinate is biased depending on the direction in which the RC is varied. Therefore a forward and reverse scan of the RC is likely to give different results as depicted in Figure 4. This hysterisis as it is called is a direct consequence of the improper equilibration. Thus it is generally a good idea to perform both forward and reverse scans to reduce this error and to determine whether the rate of change of the reaction coordinate is appropriate. In other words, a slow growth simulation with virtually no hysteresis has its RC changed adequately slow, whereas a simulation with large hysteresis has its RC sampled too quickly. The advantage of the slow growth simulation is that the dynamics is not disrupted when the reaction coordinate is changed and hence the system only has to be thermally equilibrated once. On the other hand the method has the disadvantage that both the forward and reverse scans should be performed.

It should be noted that although the forces at each time step are determined from a full quantum mechanical electronic structure calculation, the dynamics itself is still classical. Therefore, quantum dynamical effects such as the tunneling are not included in the estimates of the reaction free energy barriers. Since the classical vibrational energy levels are continuous, $H_{vib} = \Sigma RT/2$ for all states, the zero point energy correction and ΔH_{vib} are also not included in the free energy barriers derived from the AIMD simulation.

3.5 Issues with Configurational Sampling

In order to properly sample the canonical ensemble, the system must be thermally equilibrated. Thermal equilibration often done by equilibrating the system at a particular temperature for a few picoseconds thereby ensuring that all vibrational modes are excited to an equal extent. Since we cannot afford to let the system thermally equilibrate for such a long period before we begin sampling, we take an alternative approach. To efficiently achieve a thermally equilibrated system, the nuclei are excited by a series of sinusoidal pulses. Each of the excitation vectors is chosen to be orthogonal to the already excited modes, thereby ensuring an evenly distributed thermal excitation.[40] Another requirement in order to properly generate a canonical ensemble is that the temperature be held constant. The most common procedure and the procedure we use is to couple the system to a heatbath by the Nosé thermostat method.[41] Applying a separate thermostat to both the wave function coefficients and the nuclei ensures a proper NVT ensemble is sampled.

Since the configurational averages in classical molecular dynamics do no depend on the masses of the nuclei,[42] a common technique to increase the sampling rate involves replacing the true masses with more convenient ones. Since nuclear velocities scale with $m^{-1/2}$, smaller masses move faster and therefore potentially sample configuration space faster. As a result, the masses of the heavy atoms can be scaled down in order to increase sampling. For example, we commonly rescale the masses of C, N and O in our simulations from 12, 14 and 16 amu, respectively, to 2 amu. There is a limit to the mass reduction, because at some point the nuclei move so fast that the time step has to be reduced. At this point there is no gain in reducing the masses further because if the time step has to be shortened, we have to perform more time steps to achieve the same amount of sampling. It is for this reason, we generally scale our hydrogen masses up from 1 amu to 1.5 amu or higher in order to use a larger time step. The "slow growth" AIMD method has been used in a number of studies of organimetallic reaction by Margl and coworkers[43-50].

4. COMBINED QM/MM*AB INITIO* MOLECULAR DYNAMICS

The combined QM/MM methodology can be easily embedded within the Car-Parrinello framework allowing for *ab initio* molecular dynamics simulations of extended systems to be performed efficiently. We have implemented the combined QM/MM methodology of Singh and Kollman[1] within the PAW code by extending the Car-Parrinello Lagrangian (Eqn. 4) to include the MM subsystem:

$$\mathcal{L} = \sum \mu_i \langle \dot{\Psi}_i | \dot{\Psi}_i \rangle + \tfrac{1}{2} \sum M_{I,QM} \dot{R}_{I,QM}^2 - E_{DFT}(|\Psi\rangle, R_{QM}) + \sum_{i,j} \Lambda_{ij} (\langle \Psi_i | \Psi_j \rangle - \delta_{ij})$$
$$+ \tfrac{1}{2} \sum M_{I,MM} \dot{R}_{I,MM}^2 - E_{MM}(R_{QM}, R_{MM}) \quad (10)$$

The last two terms in equation 10 are the kinetic energy of the MM nuclei and the potential energy derived from the MM force field. With the intent of validating the combined *ab initio* molecular dynamics Car-Parrinello QM/MM methodology we have applied the method to determine the free energy barrier of the hydrogen transfer to the monomer chain termination process (Figure 2b). As in the static QM/MM calculations presented in Section 2, the bulky R=Me and Ar=2,6-C_6H_3(i-Pr)$_2$ groups were treated by a molecular mechanics potential whereas the Ni-diimine core including the growing chain and monomer will be treated by a density functional potential. Full details of the simulation can be found in reference 57.

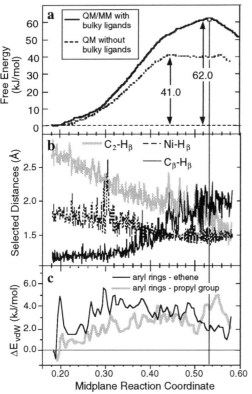

Plotted in Figure 5a is the free energy profile for the QM/MM AIMD simulation and the pure QM simulation without the bulky ligands. As expected there is a dramatic increase in the termination barrier when the bulky ligands are added. The estimated free energy barrier of 62 kJ/mol at 300 K is in good agreement with the experimental value[20] of ~67 kJ/mol at 272 K. Plotted in Figure 5b are various structural quantities that characterize the transfer process. Plotted in graph (c) is the relative molecular mechanics van der Waals energy for the interactions between the bulky aryl ligands and the active site propyl and ethene group. This plot reveals that the steric interaction between the active site moieties and the aryl ligands increases as the hydrogen transfer process occurs. This is a result of the expansion of the active site during the hydrogen transfer process. Displayed in Figure 6 is a snapshot structure of the slow growth simulation taken from the transition state region. The analogous transition state structure from a static QM/MM calculation is shown in Figure 3a. Again there is a strong similarity in the static and dynamic results.

Figure 5 Selected structural and energetic quantities as a function of the reaction coordinate from the combined QM/MM simulation. (a) Relative free energy (b): selected distances. (c): The molecular mechanics van der Waals interaction energies between the two bulky aryl groups and the active site ethene and propyl groups relative to the resting state value. The energies plotted have been "smoothed" and are the running average over 500 time steps.

The applicability of the combined QM/MM *ab initio* molecular dynamics approach to study large transition metal catalytic systems has been demonstrated. In applications where static QM/MM methods may be inadequate the QM/MM MD methodology shows particular promise. One such application is to reactions which have no enthalpic barrier on the static zero K potential surface. The capture of the monomer in the Brookhart catalyst system as shown in Figure 2 is one such case.[31] Another promising area of application, which we intend to explore in the future, is the simulation of solvent effects in chemical reactions. With the inclusion of electrostatic coupling into this method such that the electronic structure of the QM system (solute) can be polarized by electrostatic influences in the MM system(solvent), the influence of solvent can be approximated. Such coupling is currently being developed and validated in our lab.

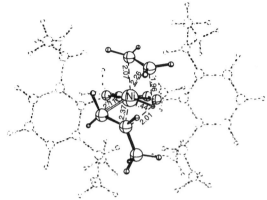

Figure 6. Snapshot structure extracted from the transition state region of the QM/MM AIMD simulation of the chain termination process in the Brookhart Ni-diimine catalyst. The snapshot structure is denoted by a vertical line in Figure 5. Conventions as in Figure 3.

Acknowledgment. The numerous contributions of Peter Blöchl of IBM Zurich are acknowledged. This work has been supported by the National Sciences and Engineering Research Council(NSERC) of Canada, as well as by the donors of the Petroleum Research Fund, administered by the American Chemical Society (ACS-PRF No. 31205-AC3) and by Novacor Research and Technology Corporation (NRTC) of Calgary. T.K.W. wishes to thank NSERC, the Alberta Heritage Scholarship Fund and the Izaak Walton Killam memorial foundation.

Literature Cited

(1)   Singh, U. C.; Kollman, P. A. *J. Comp. Chem.* **1986**, *7*, 718.
(2)   Field, M. J.; Bash, P. A.; Karplus, M. *J. Comp. Chem.* **1990**, *11*, 700.
(3)   Maseras, F.; Morokuma, K. *J. Comp. Chem.* **1995**, *16*, 1170.
(4)   Gao, J. In *Reviews in Computational Chemistry*; K. B. Lipkowitz and D. B. Boyd, Ed.; VCH: New York, 1996; Vol. 7.
(5)   Bakowies, D.; Thiel, W. *J. Phys. Chem.* **1996**, *100*, 10580.
(6)   Gao, J.; Xia, X. *Science* **1992**, *258*, 631.
(7)   Gao, J. *J. Phys. Chem.* **1992**, *96*, 537.
(8)   Warshel, A.; Levitt, M. *J. Mol. Biol.* **1976**, *103*, 227.
(9)   Théry, V.; Rinaldi, D.; Rivail, J.-L.; Maigret, B.; Ferenczy, G. G. *J. Comp. Chem.* **1994**, *15*, 269.
(10) Caldwell, J. W.; Kollman, P. A. *J. Am. Chem. Soc.* **1995**, *117*, 5179.
(11) Bakowies, D.; Thiel, W. *J. Comp. Chem.* **1996**, *17*, 87.
(12) Ho, L. L.; MacKerell-Jr, A. D.; Bash, P. A. *J. Phys. Chem.* **1996**, *100*, 4466.
(13) Hartsough, D. S.; Merz Jr., K. M. *J. Phys. Chem.* **1995**, *99*, 11266.
(14) Lyne, P. D.; Mulholland, A. J.; Richards, W. G. *J. Am. Chem. Soc.* **1995**, *117*, 11345.
(15) Harrison, M. J.; Burton, N. A.; Hillier, I. H.; Gould, I. R. *Chem. Commun.* **1996**, 2769.
(16) Bash, P. A.; Field, M. J.; Davenport, R.; Ringe, D.; Petsko, G.; Karplus, M. *Biochemistry* **1991**, *30*, 5826.

186

(17) Matsubara, T.; Maseras, F.; Koga, N.; Morokuma, K. *J. Phys. Chem.* **1996**, *100*, 2573.
(18) Deng, L.; Woo, T. K.; Cavallo, L.; Margl, P. M.; Ziegler, T. *J. Am. Chem. Soc.* **1997**, *119*, 6177.
(19) Haggin, J. in *Chemical and Engineering News*, February 5 1996; pp. 6.
(20) Johnson, L. K.; Killian, C. M.; Brookhart, M. *J. Am. Chem. Soc.* **1995**, *117*, 6414.
(21) Johnson, L. K.; Mecking, S.; Brookhart, M. *J. Am. Chem. Soc.* **1996**, *118*, 267.
(22) Cornell, W. D.; Cieplak, P.; Bayly, C. I.; Gould, I. R.; Jr., K. M. M.; Ferguson, D. M.; Spellmeyer, D. C.; Fox, T.; Caldwell, J. W.; Kollman, P. A. *J. Am. Chem. Soc.* **1995**, *117*, 5179.
(23) Baerends, E. J.; Ellis, D. E.; Ros, P. *Chem. Phys.* **1973**, *2*, 41.
(24) Baerends, E. J.; Ros, P. *Chem. Phys.* **1973**, *2*, 52.
(25) Becke, A. *Phys. Rev. A* **1988**, *38*, 3098.
(26) Perdew, J. P. *Phys. Rev. B* **1986**, *34*, 7406.
(27) Perdew, J. P. *Phys. Rev. B* **1986**, *33*, 8822.
(28) Deng, L.; Margl, P. M.; Ziegler, T. *J. Am. Chem. Soc.* **1997**, *119*, 1094.
(29) Professor Maurice Brookhart (Department of Chemistry, University of North Carolina at Chapel Hill). A private communication.
(30) Keim, W. *Angew. Chem., Int. Ed. Engl.* **1990**, *29*, 235.
(31) Woo, T. K.; Margl, P. M.; Deng, L.; Ziegler, T. in preparation.
(32) Car, R.; Parrinello, M. *Phys. Rev. Lett.* **1985**, *55*, 2471.
(33) Blöchl, P. E.; Margl, P. M.; Schwarz, K., *ACS Symposium Series 629: Chemical Applications of Density-Functional Theory.*; B. B. Laird; R. B. Ross and T. Ziegler Ed.; American Chemical Society Press: Washington DC, 1996; Vol. 629, pp 54.
(34) Blöchl, P. E. *J. Chem. Phys.* **1995**, *103*, 7422.
(35) Blöchl, P. E. *Phys. Rev. B* **1994**, *50*, 17953.
(36) Beveridge, D. L.; DiCapua, F. M. *Ann. Rev. Biophys. Chem.* **1989**, *18*, 431.
(37) Carter, E. A.; Ciccotti, G.; Hynes, J. T.; Kapral, R. *Chem. Phys. Lett.* **1989**, *156*, 472.
(38) Paci, E.; Ciccotti, G.; Ferrario, M.; Kapral, R. *Chem. Phys. Lett.* **1991**, *176*, 581.
(39) Straatsma, T. P.; Berendsen, H. J. C.; Postma, J. P. M. *J. Chem. Phys.* **1986**, *85*, 6720.
(40) Margl, P. unpublished work.
(41) Nosé, S. *Mol. Phys.* **1984**, *52*, 255.
(42) De-Raedt, B.; Sprik, M.; Klein, M. L. *J. Chem. Phys.* **1984**, *80*, 5719.
(43) Margl, P.; Blöchl, P.; Ziegler, T. *J. Am. Chem. Soc.* **1995**, *117*, 12625.
(44) Margl, P.; Blöchl, P.; Ziegler, T. *J. Am. Chem. Soc.* **1996**, *118*, 5412.
(45) Margl, P.; Lohrenz, J. C. W.; Blöchl, P.; Ziegler, T. *J. Am. Chem. Soc.* **1996**, *118*, 4434.
(46) Woo, T. K.; Margl, P. M.; Blöchl, P. E.; Ziegler, T. *Organometallics* **1997**, *16*, 3454.
(47) Woo, T. K.; Margl, P. M.; Lohrenz, J. C. W.; Blöchl, P. E.; Ziegler, T. *J. Am. Chem. Soc.* **1996**, *118*, 13021.
(48) Margl, P. M.; Schwarz, K.; Blöchl, P. E. *J. Chem. Phys.* **1995**, *103*, 683.
(49) Woo, T. K.; Margl, P. M.; Blöchl, P. E.; Ziegler, T. *J.Am.Chem.Soc.*, **1998**, 120,,2174
(50) Woo, T. K.; Margl, P. M.; Blöchl, P. E.; Ziegler, T. *J. Phys. Chem.. B* . **1997**, 101, 7877

Chapter 15

Theoretical Study of the Mechanism and Stereochemistry of Molybdenum Alkylidene Catalyzed Ring-Opening Metathesis Polymerization

Yun-Dong Wu and Zhi-Hui Peng

Department of Chemistry, Hong Kong University of Science and Technology, Clear Water Bay, Kowloon, Hong Kong, China

The mechanism and stereochemistry of molybdenum alkylidene catalyzed ring-opening metathesis polymerization have been investigated with model systems. Alkene significantly favors attacking on the CNO face of molybdenum alkylidene. All transition structures are in a distorted trigonal bipyramidal geometry, with the NH and one of the OR' groups axial. The alkene addition step is the rate determining step. The effects of OR' on the barrier of Mo=C bond rotation, the reactivity of alkylidene catalysts, and the stereo-preference of transition structures are discussed.

Alkylidene complexes of the type $M(CHR)(NAr)(OR')_2$ (M = W, Mo) are effective initiators for ring-opening metathesis polymerizations (ROMP) (1-4). These initiators catalyze polymerization of a variety of cyclic olefins, often in a living fashion. High stereospecificity and tacticity for the outcoming polymer can be achieved (5-10). Schrock et al. found that the reactivity of these alkylidenes towards olefins and the stereochemistry of the products are strongly influenced by the electronic properties of the alkoxide ligands (1-3). When R' is electron-donating (such as A, R' = t-Bu) and norbornadienes are the monomers, the reactivity of the catalyst is low and polymers are highly trans and highly syndiotactic (7); when R' is electron-withdrawing (such as B, R' = $CMe(CF_3)_2$), the reactivity of the catalyst is high and the polymers are highly cis and highly isotactic (8).

An important feature of the alkylidenes is the presence of syn and anti rotamers (11). Schrock et al. found that in most cases syn alkylidene is much more stable than anti ($K_{s/a} > 10^3$) (12). The interconversion of the syn and anti rotamers is relatively rapid when R' is an electron-donating group (A, $k_{s/a} = 1s^{-1}$), but relatively slow when R' is an electron-withdrawing group (B, $k_{s/a} = 10^{-5}$

s^{-1}). Kinetic studies with **B** as the catalyst have shown that the anti rotamer is at least 100 times more reactive than the syn rotamer (*12*).

Scheme 1

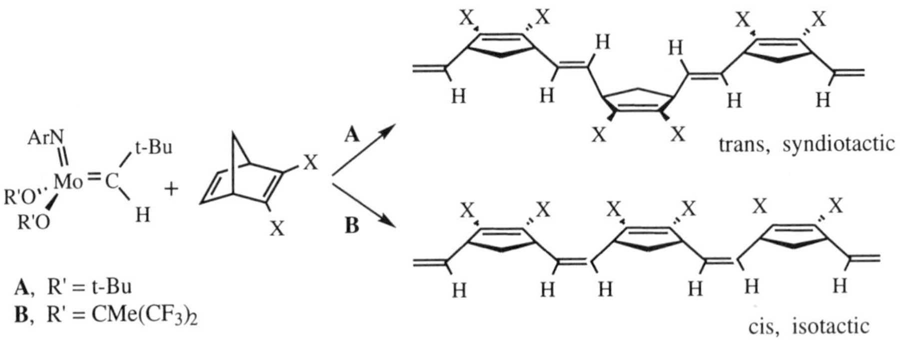

A, R' = t-Bu
B, R' = CMe(CF$_3$)$_2$

trans, syndiotactic

cis, isotactic

Based on detailed NMR studies of the metathesis reactions of the molybdenum alkylidenes with 2,3-bis(trifluoromethyl)norbornadiene (NBDF6) (*12*), Schrock *et al.* proposed that the trans-poly(NBDF6) prepared from Mo(NAr)(CHCMe$_2$Ph)(O-t-Bu)$_2$ (Ar = 2,6-i-Pr$_2$C$_6$H$_3$) was produced by the syn insertion of the monomer on the exo face of the unsubstituted double bond to the CNO face of the anti rotamer, since the polymerization is slow and the anti rotamer is kinetically accessible at room temperature. Because K$_{eq}$ ≈ 10^3, the anti rotamer actually must be at least 10^5 times more reactive than the syn rotamer in order to generate > 99% trans double bond. On the other hand, when Mo(NAr)(CHCMe$_2$Ph)[OCMe(CF$_3$)$_2$]$_2$ is used as the initiator, the polymerization is faster than the syn/anti isomerization, the syn rotamer is

Scheme 2

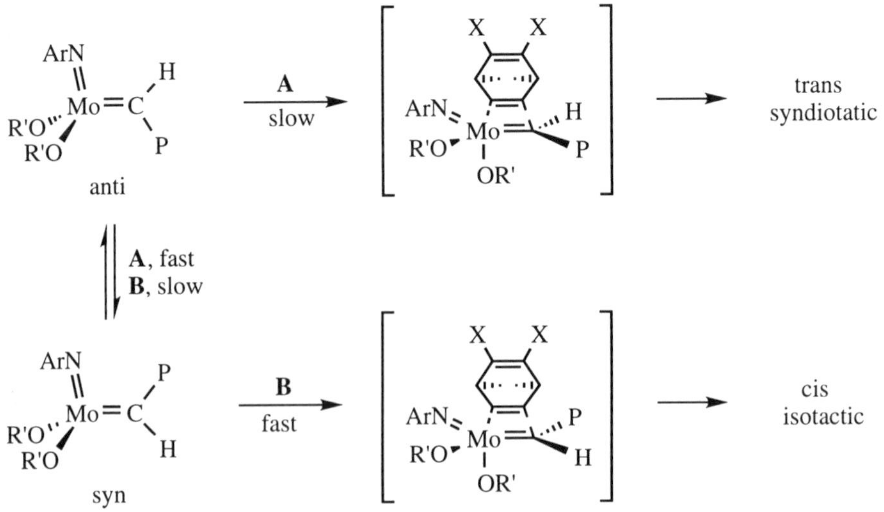

the active form and the syn insertion of the monomer to the CNO face of the syn rotamer generates the cis-poly(NBDF6). The proposal that syn and anti rotamer accessibility is the origin of the cis/trans structure of ROMP polymers has been further confirmed in a study of temperature effects (*13*). That is, when **B** is the initiator, an increase in reaction temperature decreases the percentage of cis-polymer.

In this Chapter, we present our recent theoretical studies on the reaction of ethene (*14*) and norbornadiene with $Mo(NH)(CHR)(OR')_2$ (R = H, Me; R' = Me, CF_3). These studies reveal detailed information about the alkene metathesis mechanism and geometrical and energetic preferences of transition structures. Our results fully support Schrock's explanation for the stereochemistry of ROMP.

Reaction of Alkylidenes with Ethene

All calculation were carried out with the Gaussian 94 program (*15*). In the following discussion, except when otherwise mentioned, all geometries and energies were calculated with the B3LYP nonlocal density functional approximation (*16-17*) using the HW3 basis set according to Frencking's definition (*18*), which was constructed by the contraction scheme [3311/2111/311] + ECP (*19*) on a 28-electron core for the molybdenum atom and the 6-31G* basis set for the other atoms. The enthalpies were calculated from the B3LYP/HWF (HWF adds a set of f-polarization functions (*20*) to the HW3 for the molybdenum atom and the 6-311G** basis set for the other atoms) energies on the B3LYP/HW3 geometries with thermal energy corrections derived from HF/3-21G frequency calculations.

Syn/Anti Preference of Alkylidenes. Figure 1 shows the calculated structures of syn (**1**) and anti (**2**) rotamers of $Mo(NH)(CHR)(OCH_3)_2$ and syn (**3**) and anti (**4**) rotamers of $Mo(NH)(CHR)(OCF_3)_2$ (R = Me, t-Bu). The calculated bond lengths and bond angles of these structures are in excellent agreement with experimental (*7*) and previous calculational results on similar compounds (*21-22*). The Mo–N–H angle in each structure is about 170°, indicating that the Mo–N has the character of a triple bond. The Mo–C–C angle ranges from 137° to 144° and the Mo–C–H angle ranges from 105° to 109° in the syn rotamers, quite close to the angles observed in the X-ray crystal structures of the molybdenum alkylidenes (*7*). On the other hand, the Mo–C–C and the Mo–C–H angles in the anti rotamers (**2**, **4**) range from 122° to 130° and 119° to 124°, respectively. This indicates that in the syn rotamer, there is an agostic interaction between the anti-C–H bond and the metal center while such an interaction is absent in the anti rotamers (*23-24*).

In agreement with experiments, the syn rotamer is more stable than the anti rotamer. The syn preference is about 2 kcal/mol for the methyl substituent and is increased to more than 3 kcal/mol for the tert-butyl substituent. Two factors are responsible for the syn preference: (1) the syn rotamer benefits from an agostic interaction that is absent in the anti rotamer; (2) the anti rotamer is destabilized by steric interactions between the alkylidene substituent and the alkoxyl groups, which point toward each other. This is the reason why the bulky tert-butyl group has a larger syn preference.

Alkylidene Rotational Barrier. The barrier to the syn/anti interconversion was evaluated with $Mo(NH)(CH_2)(OR')_2$ (R' = Me, CF_3). The transition structures correspond to about 90^0 rotation around the Mo=C bond. The calculated barrier to the Mo=CH_2 bond rotation for R' = Me, 10.1 kcal/mol, is somewhat lower than the value of 13.0 kcal/mol for $Mo(NH)(CH_2)(OH)_2$ calculated by Cundari with the MP2 method (*21b*). A barrier of about 13 kcal/mol is reported by Schrock *et al.* for the anti to syn isomerization of $Mo(NAr)(CHCMe_2Ph)(OCMe_3)_2$ (*12*). The calculated barrier for electron withdrawing R' = CF_3 is significantly increased to about 17 kcal/mol. This underlines the importance of the nature of the alkoxy group on the barrier to the syn/anti interconversion.

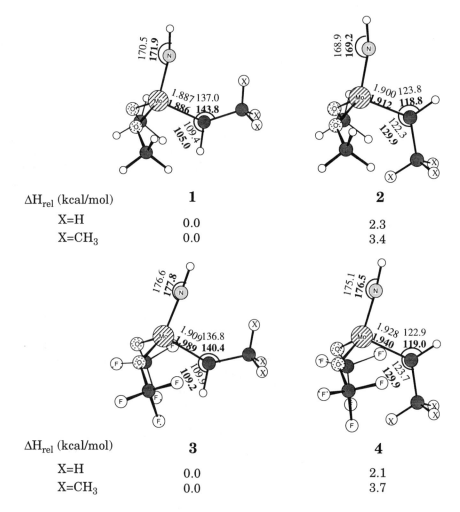

ΔH_{rel} (kcal/mol)	**1**	**2**
X=H	0.0	2.3
X=CH_3	0.0	3.4

ΔH_{rel} (kcal/mol)	**3**	**4**
X=H	0.0	2.1
X=CH_3	0.0	3.7

Figure 1. Syn and anti rotamers of molybdenum alkylidenes. The geometric parameters in bold are for tert-butyl alkylidene.

Transition structures of CNO and COO Face Cycloadditions. An important mechanistic aspect is the direction of olefin addition to the four-coordinate pseudotetrahedral catalysts. Two faces of cycloaddition are possible. One is the "CNO" face and the other is the "COO" face if the alkylidene rotates by 90⁰ before or during the process of olefin addition (*1c*, *7*). According to the structure of the base adduct of these catalysts (*25*) and other related reactions (*26*), Schrock *et al.* proposed that attack on the CNO face is most favorable (*1c*, *7*), with a transition structure in which the ring spans axial and equatorial sites with the nitrogen and one of the alkoxides taking the other two equatorial positions. Ziegler *et al.* proposed a similar transition structure in their theoretical study of the reaction of ethene with $Mo(O)(CH_2)Cl_2$ (*22*).

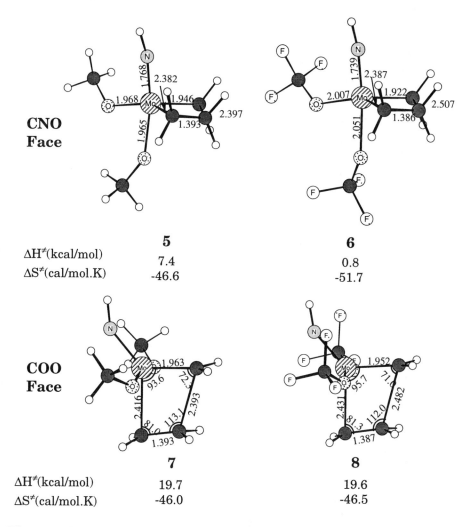

	5	**6**
$\Delta H^{\ddagger}$(kcal/mol)	7.4	0.8
$\Delta S^{\ddagger}$(cal/mol.K)	-46.6	-51.7

	7	**8**
$\Delta H^{\ddagger}$(kcal/mol)	19.7	19.6
$\Delta S^{\ddagger}$(cal/mol.K)	-46.0	-46.5

Figure 2. Transition structures of ethene addition to the CNO and COO faces of $Mo(NH)(CH_2)(OR')_2$ with the activation enthalpies and entropies.

Structures **5** and **6** are transition structures for CNO face addition of ethene to $Mo(NH)(CH_2)(OR')_2$. The electron-withdrawing property of the alkoxide ligands has little effect on the geometry of the transition structure. Thus, distinguishable from the earlier proposals, both structures are quite similar and are in a pseudo-TBP geometry with the NH and one of the OR' groups taking the axial positions. All efforts to locate a square-pyramidal (SP) transition structure failed; the metallacyclobutane ring is formed in the equatorial plane, just like the structures of the TBP tungstacyclobutanes which have been determined by X-ray crystal structure analysis (27).

In structures **5** and **6**, there is a significant Mo–C bond formation. The forming Mo–C bond lengths are about 2.38 Å, only about 0.3 Å longer than the Mo–C bond in the TBP structure of metallacyclobutanes (not shown). However, the metallacyclobutane rings are formed asynchronically, that is, the formation of the C–C bond lags behind the formation of the Mo–C bond. Thus, the C–C distances are about 0.8–0.9 Å longer than those in the metallacyclobutane products.

The calculated activation enthalpy for **5** is 7.4 kcal/mol. Bazan *et al.* reported an activation enthalpy of 6.6 kcal/mol for the reaction of W(NAr)(CH-t-Bu)(O-t-Bu)$_2$ with norbornadiene (7). The electron withdrawing alkoxide (OCF_3) lowers the activation energy significantly, and the calculated activation enthalpy for **6** is only 0.8 kcal/mol. Thus, the metal center behaves as an electrophile. The calculations indicate large activation entropies for the cycloadditions because of the partial formation of two bonds in the transition structures. The experimentally observed activation entropy for the reaction of W(NAr)(CH-t-Bu)(O-t-Bu)$_2$ with norbornadiene is about 40 eu (7).

Structures **7** and **8** are the "transition structures" of ethene addition to the COO face of $Mo(NH)(CH_2)(OCH_3)_2$ and $Mo(NH)(CH_2)(OCF_3)_2$. These structures were located using a Cs symmetry constraint with the Mo=CH$_2$ perpendicular to the Mo=NH. Because the perpendicular alkylidene is the transition structure for the Mo=C bond rotation, these "transition structures" both have two imaginary vibration of frequencies, corresponding to the Mo–C and C–C bond formation and the Mo=C bond rotation, respectively.

The calculated activation enthalpies for ethene attack on the COO face of the two Mo-alkylidenes are about 12.3 and 8.6 kcal/mol higher than those for the attack on the CNO face, respectively. The higher activation energy for the COO attack reflects the destabilization due to the 90^0 rotation of the Mo=C bond in the perpendicular structures. Therefore, it can be concluded that olefin addition to metal alkylidene takes place only on the CNO face.

Relative Reactivities of Syn and Anti Rotamers. Figure 3 gives the calculated activation enthalpies for the reactions of $Mo(NH)(CHMe)(OR')_2$ (R' = Me, CF$_3$) with ethene. The anti transition structure (**10**) is somewhat less stable than the syn transition structure. Since the anti rotamer is about 2 kcal/mol less stable than the syn rotamer in the reactant, the anti rotamer is actually predicted to be more reactive than the syn rotamer by about 1.8 kcal/mol for the alkylidene with electron-donating alkoxy (OCH_3) group and by about 1.1 kcal/mol with the electron-withdrawing alkoxy (OCF_3) group. This is qualitatively in accord with Schrock's experimental result. The agostic interaction, which benefits the syn rotamer, and the steric interaction, which destabilizes the anti rotamer, in the reactants nearly disappear in the transition structures. Therefore, we expect that the reactivity of the anti rotamer with respect to the syn rotamer is even larger when the substituent becomes more bulky.

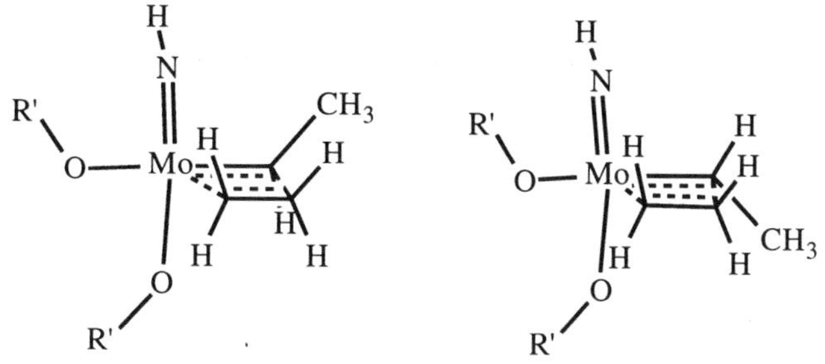

$\Delta H^{\neq}$ (kcal/mol)	**9**	**10**
R' = CH$_3$	6.7	4.9
R' = CF$_3$	0.4	-0.7

Figure 3. Calculated activation enthalpies of cycloaddition reactions of ethene with syn and anti rotamers of Mo(NH)(CHMe)(OR')$_2$.

Reaction of Norbornadiene with Alkylidenes

Due to the size of the systems, all geometries were optimized with the HF/HW3 method and the energies are at the same level except for those otherwise mentioned.

Syn and Anti Addition Transition Structures. Figure 4 shows the transition structures of the norbornadiene addition to the CNO face of Mo(NH)(CH$_2$)(OCH$_3$)$_2$ (**11, 12**) and Mo(NH)(CH$_2$)(OCF$_3$)$_2$ (**13, 14**). A major difference of these transition structures from those of ethene addition (Figure 2) is the larger distortion from ideal TBP geometry. That is, the N–Mo–O$_{ax}$ angle becomes smaller (152^0 in **11**, 144^0 in **12**, 160^0 in **13**, and 146^0 in **14**) compared to that in structures **5** (160^0) and **6** (171^0). There is also a distortion of about 10^0 between the reacting Mo=C and C=C bonds. The distortions in the syn norbornadiene structures (**11,13**) and the anti norbornadiene structures (**12,14**) are in opposite directions, in order to avoid the steric repulsion between the C$_7$ center of the norbornadiene and the NH group in the syn and the axial OR' group in the anti structures.

The calculated activation enthalpies are shown in Figure 4 (HF/HW3 level). They cannot be directly compared with those in Figure 2 for the addition of ethene, which are calculated with the B3LYP/HWF method. The Hartree–Fock method often significantly overestimates reaction activation energies *(28)*. With the HF/HW3 method, the calculated activation energies for structures **5** and **6** are 17.4 and 8.7 kcal/mol, respectively. We are more interested in the relative energies of the syn and anti structures (**11** vs. **12** and **13** vs. **14**), because these are relevant to the stereochemistry of metathesis reactions.

In both alkylidene cases, the syn arrangement of norbornadiene is more favorable than the anti arrangement. The calculated activation free energies of **11** and **13** are lower than those of **12** and **14** by 2.3 and 3.2 kcal/mol, respectively. The preference for the syn transition structures is apparently due to larger steric interactions in the anti transition structures involving the bridgehead of norbornadiene and the axial OR' group.

The result of this simplified model, which lacks the bulky substituents on the metal ligands, is in remarkable agreement with Schrock's explanation for the stereochemistry of ROMP. In the real catalysts, the flat aryl ring in the imido ligands can lie approximately in the same plane as $N/C_1/O_{ax}$ in the transition structure, and the steric interaction in the syn transition structure between the imido group and the bridgehead of the norbornadiene is thus minimized. On the other hand, the anti transition structure is expected to be further destabilized by the bulky axial alkoxide ligand.

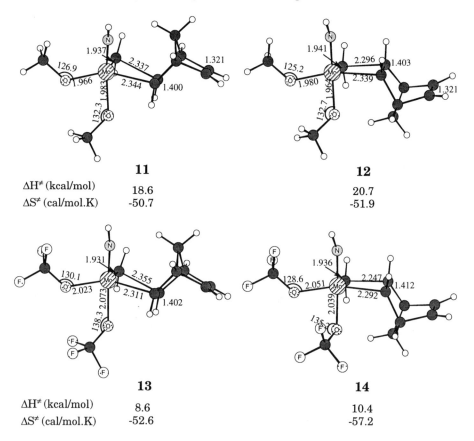

	11	**12**
$\Delta H^{\neq}$ (kcal/mol)	18.6	20.7
$\Delta S^{\neq}$ (cal/mol.K)	-50.7	-51.9

	13	**14**
$\Delta H^{\neq}$ (kcal/mol)	8.6	10.4
$\Delta S^{\neq}$ (cal/mol.K)	-52.6	-57.2

Figure 4. Transition structures of syn (**11,13**) and anti (**12,14**) cycloaddition reactions of norbornadiene with $Mo(NH)(CH_2)(OR')_2$ along with activation enthalpies and entropies.

Mechanism. Before discussing the stereochemistry of ring-opening metathesis, it is worth briefly describing the reaction mechanism. The metallacyclobutane, transition structures and products for ring-opening of metallacyclobutane have also been fully optimized. For the reaction of norbornadiene with $Mo(NH)(CH_2)(OCH_3)_2$, the results can be summarized as follows: (1) the initially formed metallacyclobutane is in a TBP geometry, which can be converted into a more stable SP geometry; (2) the transition structure for the ring opening is also in a TBP geometry. This supports Schrock's proposal that SP molybdacyclobutane must rearrange to the TBP structure before the ring-opening step can occur; (3) the ring-opening of metallacyclobutane is much more reactive than the elimination of norbornadiene. This is in agreement with Schrock's observation that the formation of metallacyclobutane is rarely reversible. Thus the stereochemistry of ROMP can be discussed based on the transition structures for metallacyclobutane formation. For the reaction of norbornadiene with $Mo(NH)(CH_2)(OCF_3)_2$, the initially formed TBP metallacyclobutane, which is more stable than the SP geometry, directly undergoes ring-opening with a small barrier.

Stereochemistry. Figure 5 shows the relative activation energies of the four possible transition structures for the reaction of norbornadiene with $Mo(NH)(CHMe)(OR')_2$. The two transition structures with syn-norbornadiene (**15** and **16**) are of similar stabilities, while the transition structures with anti-norbornadiene (**17** and **18**) are much less stable. Since the alkylidene methyl substituent has a small preference for the syn position (see Figure 3), structures **15** and **18** are somewhat destabilized by steric interactions. In polymerization reactions, the alkylidene substituents, t-Bu or CMe_2Ph in the initiator and isopropyl-equivalent in the polymerization steps, are sterically more bulky than methyl, and therefore **15** and **18** are further destabilized with respect to **16** and **17**.

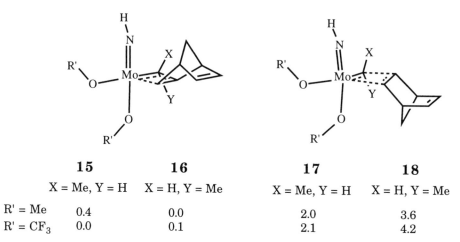

	15	**16**	**17**	**18**
	X = Me, Y = H	X = H, Y = Me	X = Me, Y = H	X = H, Y = Me
R' = Me	0.4	0.0	2.0	3.6
R' = CF_3	0.0	0.1	2.1	4.2

Figure 5. Calculated relative activation enthalpies (kcal/mol) of the cycloaddition reactions of norbornadiene with syn and anti rotamers of $Mo(NH)(CHMe)(OR')_2$ in syn (left) and anti (right) fashions.

The model reactions presented above allows a qualitative analysis of stereochemistry which is in full agreement with Schrock's rationalization shown in Scheme 2. Thus, (1) the syn approach of norbornadiene is more favorable for both syn and anti rotamers of alkylidene; (2) The outcome of polymer is dependent upon the relative rates of alkylidene isomerization and olefin metathesis; (3) if R' is strongly electron-withdrawing, the metathesis reaction is faster and an all-cis polymer is produced; (4) if R' is electron-donating, the alkylidene isomerization is faster and there is syn/anti equilibrium. Since the anti rotamer is more reactive, the formation of all-trans polymer becomes favorable.

In our future work, the steric effect of bulky ligand substituents will be accounted for by using the molecular mechanics force-field method or MM/QM method (29) so that the stereochemistry of ROMP can be predicted quantitatively.

References and Notes

1. (a) Schaverien, C. J.; Dewan, J. C.; Schrock, R. R. *J. Am. Chem. Soc.* **1986**, *108*, 2771. (b) Schrock, R. R.; Depue, R. T.; Fellmann, J. D.; Schaverien, C. J.; Dewan, J. C.; Liu, A. H. *J. Am. Chem. Soc.* **1988**, *110*, 1423. (c) Schrock, R. R.; Depue, R. T.; Feldman, J.; Yap, K. B.; Yang, D. C.; Davis, W. M.; Park, L.; DiMare, M.; Schofied, M.; Anhaus, J.; Walborsky, E.; Evitt, E.; Kruger, C.; Betz, P. *Organometallics* **1990**, *9*, 2262.

2. (a) Murdzek, J. S.; Schrock, R. R.; *Organometallics* **1987**, *6*, 1373. (b) Schrock, R. R.; Murdzek, J. S.; Bazan, G. S.; Robbins, J.; DiMare, M.; O'Regan, M. *J. Am. Chem. Soc.* **1990**, *112*, 3875.

3. (a) Feldman, J.; Schrock, R. R. *Prog. Inorg. Chem.* **1991** *39*, 1 (b) Schrock, R. R. *Polyhedron*, **1995**, *14*, 3177.

4. (a) Schrock, R. R. *Acc. Chem. Res.* **1990**, *23*, 158. (b) Schrock, R. R. *Pure & Appl. Chem.* **1994**, *66*, 1477.

5. Schrock, R. R. *Ring-Opening Polymerization*; Brunelle, D. J., Ed.; Hanser: Munich, **1993**, p 129.

6. (a) Schrock, R. R.; Feldman, J.; Grubbs, R. H.; Cannizzo, L. *Macromolecules* **1987**, *20*, 1169. (b) Murdzek, J. S.; Schrock, R. R. *Macromolecules* **1987**, *20*, 2640. (c) Schrock, R. R.; Krouse, S. A.; Knoll, K.; Feldman, J.; Murdzek, J. S.; Yang, D. C. *J. Mol. Cat.* **1988**, *46*, 243. (d) Bazan, G. C.; Schrock, R. R.; Cho, H.; Gibson, V. C. *Macromolecules* **1991** *24*, 4495.

7. Bazan, G. C.; Khosravi, E.; Schrock, R. R.; Feast, W. J.; Gibson, V. C.; O'Regan, M. B.; Thomas, J. K.; Davis, W. M. *J. Am. Chem. Soc.* **1990**, *112*, 8378.

8. Feast, W. J.; Gibson, V. C.; Marshall, E. L. *J. Chem. Soc. Chem. Commun.* **1992**, 1157.

9. McConville, D. H.; Wolf, J. R.; Schrock, R. R. *J. Am. Chem. Soc.* **1993**, *115*, 4413.

10. O'Dell, R.; McConville, D. H.; Hefmeister, G. E.; Schrock, R. R. *J. Am. Chem. Soc.* **1994**, *116*, 3414.

11. Ivin, K. J. *Olefin Metathesis*; Academic: New York, **1983**.

12. Oskam, J. H.; Schrock, R. R. *J. Am. Chem. Soc.* **1993**, *115*, 11831.

13. Schrock, R. R.; Lee, J.-K.; O'Dell, R.; Oskam, J. H. *Macromolecules* **1995**, *28*, 5933.
14. Wu, Y.-D.; Peng, Z.-H. *J. Am. Chem. Soc.* **1997**, *119*, 8043.
15. Gaussian 94 Revision B.3, Frisch, M. J.; Trucks, G. W.; Schlegel, H. B.; Gill, P. M. W.; Johnson, B. G.; Robb, M. A.; Cheeseman, J. R.; Keith, T. A.; Petersson, G. A.; Montgomery, J. A.; Raghavachari, K.; Al-Laham, M. A.; Zakrzewski, V. G.; Ortiz, J. V.; Foresman, J. B.; Cioslowski, J.; Stefanov B. B.; Nanayakkara, A.; Challacombe, M.; Peng, C. Y.; Ayala, P. Y.; Chen, W.; Wong, M. W.; Andres J. L.; Replogle, E. S.; Gomperts, R.; Martin R. L.; Fox, D. J.; Binkley, J. S.; Defrees, D. J.; Baker, J.; Stewart, J. P.; Head-Gordon, M.; Gonzalez, C.; Pople, J. A., Gaussian, Inc., Pittsburgh, PA, 1995.
16. Becke, A. D. *J. Chem. Phys.* **1993**, *98*, 5648.
17. (a) Lee, C.; Yang, W.; Parr. R. G. *Phys. Rev. B* **1988**, *37*, 785. (b) Miehlich, B.; Savin, A.; Stoll, H.; Preuss, H. *Chem. Phys. Lett.* **1989**, *90*, 5622.
18. (a) Jonas, V.; Frenking, G.; Reetz, M. T. *J. Comput. Chem.* **1992**, *13*, 919. (b) Jonas, V.; Frenking, G.; Reetz, M. T. *Organometallics* **1993**, *12*, 2111.
19. (a) Hay, P. J.; Wadt, W. R. *J. Chem. Phys.* **1985**, *82*, 299. (b) an additional d polarization function with $\xi_d = 0.09$ was added to valence basis set of Mo to form 311 contraction of d orbitals.
20. Ehlers, A. W.; Bohme, M.; Dapprich, S.; Gobbi, A.; Hollwarth, A.; Jonas, V.; Kohler, K. F.; Stegmann, R.; Veldkamp, A.; Frenking, G. *Chem. Phys. Lett.* **1993**, *208*, 111.
21. (a) Cundari, T. R.; Gordon, M. S. *J. Am. Chem. Soc.* **1991**, *113*, 5231. (b) Cundari, T. R.; Gordon, M. S. *Organometallics* **1992**, *11*, 55.
22. Folga, E.; Woo, T.; Ziegler, T. *Organometallics* **1993**, *12*, 325.
23. Schultz, A. J.; Brown, R. K.; Williams, J. M.; Schrock, R. R. *J. Am. Chem. Soc.* **1981**, *103*, 169.
24. (a) Brookhart, M.; Green, M. L. H.; *J. Organometal. Chem.* **1983**, *250*, 395. (b) Brookhart, M.; Green, M. L. H.; Wong, L. L. *Prog. Inorg. Chem.* **1988**, *36*, 1. (c) Hoffmann, R.; Jemmis, E.; Goddard, R. J. *J. Am. Chem. Soc.* **1980**, *102*, 7667. (d) Bai, Y.; Roesky, H.; Noltemeyer, M.; Witt, M. *Chem. Ber.* **1992**, *125*, 825. (e) Cummins, C. C.; van Duyne, G. D.; Schaller, C. P.; Wolczanski, P. T. *Organometallics* **1991**, *10*, 164.
25. Schrock, R. R.; Crowe, W. E.; Bazan, G. C.; DiMare M.; O'Regan, M. B.; Schofield, M. H. *Organometallics* **1991**, *10*, 1832.
26. Schrock, R. R.; Weinstock, I. A.; Horton, A. D.; Liu, A. H.; Schofield, M. H. *J. Am. Chem. Soc.* **1988**, *110*, 2686.
27. (a) Feldman, J.; Davis, W. M.; Thomas, J. K.; Schrock, R. R. *Organometallics* **1990**, *9*, 2535. (b) Schrock, R. R.; Depue, R. T.; Fellmann, J. D.; Schaverien, C. J.; Dewan, J. C.; Liu, A. H. *J. Am. Chem. Soc.* **1988**, *110*, 1423.
28. Houk, K. N.; Li, Y.; Evanseck, J. D. *Angew. Chem. Chem. Int. Ed. Engl.* **1992**, *31*, 682.
29. (a) Maseras, F.; Morokuma, K. *J. Comput. Chem.* **1995**, *16*, 1170. (b) Froese, R. D. J.; Musaev., D. G.; Morokuma, K. *J. Am. Chem. Soc.* **1998**, *120*, 1581.

Chapter 16

Theoretical Studies of the N_2 Cleavage by Three-Coordinate Group 6 Complexes ML_3

Djamaladdin G. Musaev[1], Qiang Cui, Mats Svensson, and Keiji Morokuma[1]

Cherry L. Emerson Center for Scientific Computation and Department of Chemistry, Emory University, Atlanta, GA 30322

The mechanism of the N_2 cleavage by three-coordinate group 6 complexes ML_3 where M=Cr, Mo and W for L=NH_2, and MoL_3 for L= H, Cl, NH_2 and OCH_3 have been studied with the hybrid density functional method. The first step of the reaction is found to be the coordination of N_2 to the metal atom of the quartet ground state ML_3 complex, **1**, to form doublet $(N_2)ML_3$, **2**, which, upon coupling with another reactant **1**, gives a stable intermediate **3** in the triplet ground state. From the triplet intermediate **3**, the reaction has to flip spin again and go over a substantial barrier at **4** to reach the product **5** in the singlet state. This barrier is found to be significant for Cr and Mo complexes, while it is only a few kcal/mol for W complexes. Having strongly π-donating ligands such as NH_2 and OCH_3 on the metal center decreases this rate-determining barrier. Therefore, we expect that W(III) complexes WL_3 with a strong π-donor ligand L would be more efficient in N_2 activation than its Cr(III) and Mo(III) analogs. The N-MoL_3 bond energy is calculated to be 146, 132, 155 and 154 kcal/mol for L=H, Cl, NH_2 and OCH_3, respectively, while the N-$Cr(NH_2)_3$ and N-$W(NH_2)_3$ binding energies are found to be 90 and 180 kcal/mol, respectively.

Activation of the N≡N triple bond and its chemical transformations are of a great interest from both fundamental and practical points of view, and have been a focus of extensive studies during the last several decades *(1,2)*. Perhaps the most famous reaction is the energy-intensive Haber-Bosch process *(2)*, which accounts for the production of millions of tons of NH_3 yearly from N_2 and H_2 molecules. While high pressure and temperature are necessary for this reaction, N_2 can also be fixed at atmospheric pressure and ambient temperature by certain types of bacteria to produce NH_3. The recent crystal structure of nitrogenase from *Azotobacter vinelandii* has provided insights into how these biological systems activate N_2 in an aqueous environment*(3)*.

Attempts to duplicate the Haber-Bosch process and the nitrogenase system in the

[1]Corresponding authors.

laboratory have not met with success. However, these studies have accumulated much accurate fundamental knowledge on the bonding modes and reactivity patterns of the N_2 molecule. In this aspect, the recent results of Cummins and coworkers *(4)* are extremely important. It has been shown *(4a)* that the strong N-N bond of N_2 and N_2O molecules can be cleaved by a three-coordinate Mo(III) complex, Mo(NRAr)$_3$, **1**, where R=C(CH$_3$)$_2$CH$_3$ and Ar=3,5-C$_6$H$_3$Me$_2$. The purification of red-orange **1** under an atmosphere of N_2 at -35°C led to an intensely purple colored solution (ethyl ether, 0.1 M), which gradually became golden on warming to 30°C and lost its paramagnetism. The ^{1}H NMR spectroscopy study confirmed that the final product is a terminal nitrido Mo(VI) complex, NMo(NRAr)$_3$, **5**. The proposed mechanism in Scheme 1 involves two different intermediates, **2** and **3**. Interestingly, it has been demonstrated (4d) that this process is accelerated in the presence of Chisholm's nitrido complex NMo(OR)$_3$, R=C(CH$_3$)$_3$.

However, these experiments raise several questions which can be elucidated by quantum chemical methods: i) Is the proposed mechanism reasonable from an electronic structural point of view; ii) what are the geometrical and electronic structures as well as the relative energies of the reactants, the assumed intermediates, transition states and products; and iii) how does the ease of the reaction depend on the nature of coligand L and metal M?

In order to answer the above mentioned questions and predict the factors controlling this fascinating reaction, we have carried out quantum chemical calculations of the model reaction:

$$2ML_3 \ (\mathbf{1}) + \ N_2 \ \rightarrow 2NML_3 \qquad (\mathrm{I})$$

where L = H, Cl, NH$_2$, OCH$_3$ and M= Cr, Mo and W. The preliminary results of the study have been published as a Communication previously *(5)* .

Calculation Procedure

Geometries and energetics of the stationary points on the potential energy surface of reaction (I) have been calculated with the hybrid density functional B3LYP (UB3LYP for open shell structures) method *(6)* in conjunction with the double-zeta (DZ) quality basis set lanl2dz *(7)*. For transition metal atoms and Cl we used Hay and Wadt's relativistic effective core potential (ECP) *(7)*. Normal-mode analyses have been performed for the transition states to verify their character. Energies all of the structures have been recalculated for the B3LYP/DZ optimized geometries at the B3LYP level with the DZP basis set, which includes the lanl2dz DZ basis set and polarization f_{Mo} and d_N- functions. All calculations were performed using the GAUSSIAN94 *(8)* and STOCKHOLM *(9)* (for the MCPF calculations) packages.

Previously, it has been shown that the hybrid density functional method, B3LYP, with valence double-z basis set gives reliable geometries. For example, it was shown *(5)* that the B3LYP/DZ optimized geometries of the model complex, (μ-N$_2$){Mo[N(HNCH$_2$CH$_2$)$_3$]}$_2$, agree very well with the experimental structure of the real system (μ-N$_2$){Mo[N(t-BuMe$_2$SiNCH$_2$CH$_2$)$_3$]}$_2$, confirming the reliability of the present method: N-N distance of 1.22Å (calculated) vs 1.20Å (X-ray), Mo-N(trans to N$_2$) distance of 2.30Å (calculated) vs 2.29Å (X-ray), and the other Mo-N distances of 1.97Å (calculated) vs 2.01Å (average, X-ray). However, the relative energies are normally more dependent on the basis sets and methods. Therefore, in order to test the accuracy of the computational approach B3LYP/DZ used in this paper, we have recalculated the relative energies of the potential energy surface of reaction (1) for M=Mo and L=H using more sophisticated methods, like CCSD(T) *(10)* and PCI-80 *(11)*, with the DZP basis set. In these calculations we used the B3LYP/DZ optimized geometries.

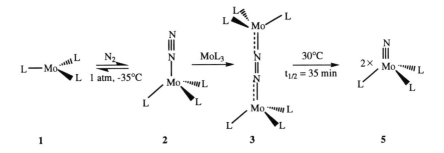

Scheme 1. Proposed (*4a*) sequence of reactions for the conversion of MoL$_3$ to NMoL$_3$ in presence of N$_2$.

Results and Discussion

In our discussions we will follow and examine the mechanism proposed by Cummins and coworkers *(4a)* presented in Scheme 1. The first step of this mechanism is coordination of N_2 to the metal center of ML_3, **1**, to form $(N_2)ML_3$, **2**, and then the second ML_3 complex attacks the N atom of the N_2 molecule of **2** and forms $L_3M(N_2)ML_3$, **3**, which dissociates into two nitrido complexes NML_3, **5**. The calculated geometries of the reactants, intermediates, transition states and products of these reactions are presented in Figure 1. The relative energies of various spin states of these species are presented in Table I and Figure 2. In Table II we present the results obtained at the different levels of theory for M=Mo, L=H.

At first, let us note that the B3LYP relative energies of reaction (I) calculated with the DZ and the DZP basis sets are quite different in absolute values, as seen in Table I, while the qualitative pictures derived from the two basis sets are very similar, especially with respect to the barrier heights for N-N activation. Therefore, below we will discuss only B3LYP/DZP results.

Since the ground states of the Cr, Mo and W atoms are $s^1d^5(^7S)$, with $s^2d^4(^5D)$ lying 23.2, 34.1 and 4.3 kcal/mol higher *(12)*, respectively, we may expect that the lower lying electronic states of the complex $M(III)L_3$, **1**, for M=Cr, Mo and W are quartet and doublet states with unpaired electrons in the d_{yz} (e"), d_{xz}(e") and $s(a_1')$ orbitals. The quartet $^4A_1'$ is found to be the ground state for all ML_3 complexes considered. However, the calculated energy gap between the first excited doublet $^2A_1'$ state and the ground $^4A_1'$ state, $\Delta E(^2A_1' - ^4A_1')$, depends on the nature of both the metal atom and the ligand L. As seen in Table I, for L=NH_2 $\Delta E(^4A_1' - ^2A_1')$ decreases via M=Mo(14.1kcal/mol) > Cr(13.2kcal/mol) > W(4.3kcal/mol), which correlates with the decrease in the $s^2d^4(^5D)$-$s^1d^5(^7S)$ energy gap of the free metal atom mentioned above. For a given metal atom M=Mo, $\Delta E(^2A_1' - ^4A_1')$ decreases via L=H(28.2kcal/mol) > Cl(14.7kcal/mol) > NH_2(14.1kcal/mol) > OCH_3(5.2kcal/mol). In other words, the π-donating ligand L decreases the doublet-quartet energy of the ML_3 complex, making the s^2d^4 electronic state of the metal more easily accessible.

The first step of the reaction, the coordination of the N_2 molecule to the ML_3 complex takes place without a barrier. As seen in Figure 1, N_2 coordinates end-on by one of the N atoms to give the complex $(N_2)ML_3$, **2**. The side-on complex is not energetically favorable and rearranges spontaneously to structure **2**. The N-N distance in **2** is not much longer than that in free N_2, indicating that the N≡N triple bond is not broken yet. Furthermore, the N-N bond is elongated significantly more in the doublet state of **2** than in its quartet state, indicating that the doublet state of ML_3 interacts with N_2 stronger than the quartet does, crosses with the quartet and gives the doublet ground state for **2** for all the systems except for M=Cr. Thus these systems have to make non-adiabatic transition from the quartet to the doublet, during this reaction step or after forming **2** in the quartet excited state. The stability of the doublet state is consistent with the calculated N_2-ML_3 bond energies, which may not be quantitatively very accurate because of the limited basis set, but should be accurate enough for comparison among different systems. As seen in Table I and Figure 2 the N_2-ML_3 bond energies are calculated to be 37.2, 23.4, 24.9, and 14.7kcal/mol for the doublet ML_3 and only 12.4, 5.2, 1.2, and 0.3 kcal/mol for the quartet ML_3, where M=Mo and L=H, Cl, NH_2 and OCH_3, respectively. These values are 19.4 and 6.7 kcal/mol for the doublet, and 0.7 and 1.7 kcal/mol for the quartet state with L=NH_2 and M=W and Cr, respectively. In other words the N_2-ML_3 bond energies calculated for the doublet states decrease for a given metal atom M=Mo via L=H > Cl > NH_2 > OCH_3, and for given L=NH_2 via Mo > W > Cr. These trends are consistent with the changes in the N-N bond length upon going from free N_2 to the complex $(N_2)MoL_3$.

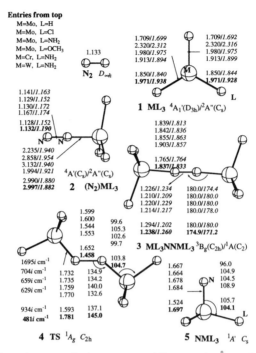

Entries from top
M=Mo, L=H
M=Mo, L=Cl
M=Mo, L=NH$_2$
M=Mo, L=OCH$_3$
M=Cr, L=NH$_2$
M=W, L=NH$_2$

Figure 1. Selected geometrical parameters (distances in Å, angles in deg.) of stationary points on the potential energy surface of the reaction: 2ML$_3$ + N$_2$ → 2NML$_3$, where L=H (first line), Cl (second line), NH$_2$ (third line), and OCH$_3$ (fourth line) for M=Mo, and L= NH$_2$, and M=Cr (fifth line) and W (sixth line). Numbers before and after a slash correspond to the high and low spin states, respectively. Symmetry labels are for L=NH$_2$.

Table I: The relative energetics (in kcal/mol) of the reaction 2MoL$_3$ + N$_2$ → 2 x NMoL$_3$ calculated at the B3LYP level using DZ and DZP basis sets.

Metal, M=	Mo						W		Cr	
Ligand, L=	Cl		NH$_2$		OCH$_3$		NH$_2$		NH$_2$	
Basis Set	DZ	DZP	DZ	DZP	DZ	DZP	DZ	DZP	DZ	DZP
Reactants, ML$_3$, 1										
Doublet	14.5	14.7	14.5	14.1	7.3	5.2	4.9	4.3	17.6	13.2
Quartet	0.0	0.0	0.0	0.0	0.0	0.0	0.0	0.0	0.0	0.0
(N$_2$)MoL$_3$, 2										
Doublet	-9.4	-8.7	-17.6	-10.8	-15.9	-9.5	-24.7	-15.1	7.1	6.5
Quartet	-5.3	-5.2	-2.5	-1.2	-0.7	-0.3	-2.3	-0.7	-3.0	-1.7
L$_3$Mo(N$_2$)MoL$_3$, 3										
Singlet, S	-27.3	-10.1	-47.5	-35.1	-32.1	-20.1	-72.0	-59.3	25.1	38.4
Triplet, T	-46.0	-33.3	-53.9	-41.6	-39.8	-27.0	-73.2	-60.7	2.7	21.1
ΔE(S-T)	18.7	23.2	6.4	6.5	7.7	6.9	1.2	1.4	22.4	17.3
Transition State, 4										
Singlet	-5.5	5.4	-33.1	-20.3	-20.0	-7.7	-68.4	-56.1	68.2	84.3
Activation Barrier										
	40.6	38.7	20.7	21.3	19.8	19.3	4.8	4.6	65.5	63.2
2 x NMoL$_3$, 5										
Singlet	-30.3	-27.2	-74.4	-69.0	-70.8	-66.6	-120.0	-114.4	40.0	52.0
D$_e$ (NML$_3$ → N + ML$_3$)										
	105.8	120.5	127.8	141.4	126.0	140.2	150.6	164.1	70.6	80.9
Best estimate[a]		132		155		154		180		90

a) see text for explanations.

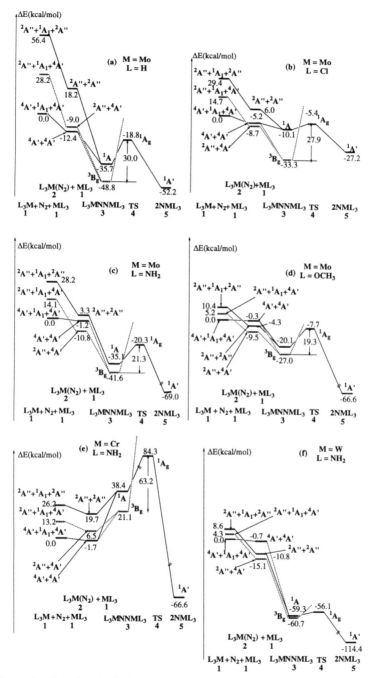

Figure 2. Calculated relative energies (in kcal/mol) of the stationary points on the potential energy surface of the reaction: $2ML_3 + N_2 \rightarrow 2NML_3$, for M=Mo and L= H (a), Cl (b), NH_2 (c), OCH_3 (d), as well as for M=Cr (e) and W (f) along with L=NH_2.

The next step of the reaction is coordination of the second ML_3 fragment to **2**, which is calculated to be an exothermic process, and occurs without a barrier, leading to the binuclear complex L_3MNNML_3, **3**, for all considered systems except M=Cr, which will be discussed in the next paragraph. The triplet complex **3** for M=Mo, being much more stable (7-23 kcal/mol) than the singlet, is considered as the resting stage of the reaction, and this paramagnetic intermediate should be experimentally detectable. For M=W (and L=NH_2), however, the singlet lies only 1.4 kcal/mol higher than the triplet, and non-adiabatic transition is likely to take place between the triplet and singlet in complex **3** and both spin states are likely to be populated in the resting stage of the reaction. The structure of **3** in Figure 1 indicates that this is a double end-on complex, with the N-N distance stretched by 0.08-0.13Å from the free N_2, suggesting that the N=N bond is now a double bond. The N-N bond stretch is larger for M=W than M=Mo. This correlates with the calculated exothermicity of the reaction $(N_2)MoL_3 + ML_3 \rightarrow L_3M(N_2)MoL_3$, which is several kcal/mol larger for M=W than Mo.

For M=Cr the reaction $(N_2)MoL_3 + ML_3 \rightarrow L_3M(N_2)MoL_3$ is endothermic by 20-40 kcal/mol and reaction (I) is unlikely to take place at moderate conditions. Therefore, below we will not discuss the results for M=Cr in detail, while we still present all the results in the corresponding tables and figures.

The N-N cleavage of complex **3** takes place through the singlet transition state **4**, which has an N-N bond distance of 1.46-1.60Å, corresponding qualitatively to an N-N single bond; the N=N π bond has been broken before the transition state. The N-N cleavage of triplet **3** leads to an excited state of the product and is energetically unfavorable. Thus the N-N cleavage from the resting state, triplet intermediate **3**, for M=Mo (for all considered L) requires a spin flip to the singlet before or during the cleavage process. The calculated barrier heights decrease via H > Cl > NH_2 > OCH_3, i. e. via increasing π-donating capability of the ligand L. On the other hand, the barrier for N-N cleavage in the singlet complex **3** for M=W is small (4.6 kcal/mol). Furthermore, since a substantial population is expected in singlet **3** for M=W, no spin flip is required for this reaction step. The overall reaction is in every case highly exothermic, which provides the thermodynamic driving force for the N_2 cleavage reaction studied here.

The entire reaction scheme presented here is smooth and reasonably consistent with the experimental findings (4a) for M=Mo and L=NRAr, **1**. The reaction from the quartet ground state of this ML_3 complex is likely to proceed by easy formation of the quartet intermediate **2**, which, upon coupling with another reactant **1**, gives a stable intermediate **3** with a triplet ground state. Since this state is not adiabatically connected to the quartet intermediate **2**, some non-adiabatic process would be required and this may slow down the reaction. From the triplet intermediate **3**, the reaction has to flip spin again to the singlet and go over a substantial barrier at **4** to reach the product **5**.

The present results show that the strongly π-donating ligands NH_2 and OCH_3 decrease the energy gap between the triplet ground state and the singlet excited state of the binuclear intermediate complex **3**. The decreased energy gap leads to a decrease in the barrier at the rate-determining transition state **4** in the singlet state, relative to the triplet resting state of **3**. The NH_2 and OCH_3 ligands also give the largest exothermicity for the overall reaction. The present results for W($NH_2)_3$ also indicate that the use of the W(III) complex, as compared with the Mo(III) complex used in the experiments, brings about very similar, even more profound effects. The smaller energy gap between the ground s^1d^5 and excited s^2d^4 states for the atomic W, compared to that for atomic Mo and Cr, puts the singlet state of **3** only slightly above the triplet ground state. Thus the rate-determining barrier at **4** is the smallest when M= W. Furthermore, since the reaction of this step will take place from the substantially populated singlet state of **3**, no spin flip is required, which would make the reaction more efficient. The exothermicity of this step and the entire reaction is largest when M= W. Therefore, we expect that W(III) complexes WL$_3$ with a strong π-donor ligand L would be more efficient in N$_2$ activation than the reported Mo(III) complex. This prediction is consistent with the fact that the CC bond in C$_2$ is cleaved more easily by WW compounds than by MoMo compounds (13). The Cr complex is not able to cleave the N≡N bond of N$_2$ molecule.

At end we would like to examine the reliability of the B3LYP/DZP method used for the energy calculations in this paper. Therefore we have performed single point HF, B3LYP, MCPF, CCSD(T) and PCI-80 energy calculations on the important structures of the reaction 2MoH$_3$ + N$_2$→ 2 x NMoH$_3$ (II) using B3LYP/DZ optimized geometries. The results obtained are presented in Table II. As seen from this table, the CCSD(T) and PCI-80 methods give, in general, similar results; the difference between the results produced by the two methods are within a few kcal/mol except for the thermodynamics of the overall reaction (II) and the N-N bond energy of N$_2$ molecule, which are calculated to be 14 and 16 kcal/mol, respectively, larger at the PCI-80 level than at the CCSD(T) level. This discrepancy can be explained in terms of the well documented basis set truncation effect (11). The B3LYP method used throughout this paper provides qualitatively the same picture as the more sophisticated PCI-80 method. The calculated energetics at the B3LYP and PCI-80 levels agree within a few kcal/mol for the reactant **1**, the quartet electronic state of **2**, and the singlet-triplet energy gap of **3**. These differences increase up to 10-13 kcal/mol for the N$_2$-MoH$_3$ (doublet) binding energy, for the energy of **3** relative to the reactants, as well as for the activation barrier. The exothermicity of the reaction (II) calculated at the B3LYP level is 27 kcal/mol smaller than that obtained at the PCI-80 level. Since the N-N bond of N$_2$ is broken and two N-MoH$_2$ bonds are formed during this reaction, the reasons for this discrepancy can be understood by analyzing the B3LYP and PCI-80 values of the N-N and N-MoH$_2$ bond energies. As seen from Table II, B3LYP and PCI-80 are in excellent agreement on the N-N dissociation energy of N$_2$ molecule. So, the reason for the above discrepancy between the B3LYP and PCI-80 results seems to be in the calculated N-MoH$_3$ bond energy. Indeed, as seen in Table II, the B3LYP/DZP method used in this paper underestimates the N-MoH$_3$ bond energy by 13.0 kcal/mol or by 10% in comparison with the PCI-80 method. Therefore, in Table I we have corrected our B3LYP/DZP bonding energies of N-ML$_3$ by 10%. Now, our best estimated value for the N-MoL$_3$ bond is calculated to be 146, 132, 155 and 154 kcal/mol for L=H, Cl, NH$_2$ and OCH$_3$, respectively, and 90 and 180 kcal/mol for N-M(NH$_2$)$_3$ for M=Cr and W, respectively.

Conclusions

One may draw the following conclusions from the above presented data:

1. The reaction (I) starts from the quartet ground state of the reactant ML$_3$ **1**, proceeds by easy formation of the doublet intermediate (N$_2$)ML$_3$ **2**, which, upon coupling with another reactant **1**, gives a stable intermediate L$_3$M(N$_2$)ML$_3$ **3** with a triplet ground state, from where it requires the flip a spin to the singlet state of **3**, and goes over a substantial barrier at **4** to reach the product **5**.

2. The increase of π-donating capability of the ligand L decreases the energy gap between the triplet ground and singlet excited states of complex **3**, and, consequently, decreases the barrier at the rate-determining transition state **4**.

3. The present results for W(NH$_2$)$_3$ indicate that the use of the W(III) complex brings about very similar, even more profound effects compared with the Mo(III) complex used in the experiments. When M=W, the rate-determining barrier at **4** is the smallest and the exothermicity of the entire reaction is largest. Furthermore, since the reaction **3** → **4** → **5** will take place from the substantially populated singlet state of **3**,

Table II: The atomization energy of N_2 molecule and relative energetics of the reaction $2MoH_3 + N_2 \rightarrow 2 \times NMoH_3$ calculated at different levels of theory.[a]

Elec. States	HF	B3LYP	MCPF	CCSD(T)	PCI-80
$D_e(N\equiv N)$ exp. = 225.94 ± 0.14 kcal/mol [b]					
	101.4	213.9	190.6	196.4	212.1
MoH_3, 1					
Doublet	38.5	28.2	33.0	33.2	33.2
Quartet	0.0	0.0	0.0	0.0	0.0
N_2MoH_3, 2					
Doublet	36.4	-9.0	0.7	-0.1	-1.1
Quartet	-0.9	-12.4	-12.5	-11.2	-12.0
$H_3Mo(N_2)MoH_3$, 3					
Singlet	9.3	-35.7	-16.3	-26.6	-26.1
Triplet	102.7	-48.8	-30.5	-41.8	-43.1
ΔE(Triplet-Singlet)	-6.4	13.1	14.2	15.3	16.9
Transition State, 4					
Singlet	154.7	-18.8	-10.5	-21.6	-25.7
Activ. Barrier	52.0	30.0	20.0	20.2	17.3
$2XNMoH_3$, 5					
singlet	91.8	-52.2	-50.4	-66.3	-79.9
$D_e(NMoH_3 \rightarrow N + MoH_3)$					
	-4.8	-133.0	-120.5	-131.4	-146.0

[a] All numbers are in kcal/mol. The energetics of the reaction are calculated relative to the ground quartet state of the $MoH_3 + N_2$. In these calculations we used DZP basis set.
[b] See: *CRC Handbook of Chemistry and Physics;* Lide, D. R.; Eds.; CRC Press, Roca Raton - Ann Arbor - Boston, **1992**.

no spin flip is required. Therefore, we expect that W(III) complexes WL_3 with a strong π-donor ligand L would be more efficient in N_2 activation than the reported Mo(III) complex.

Acknowledgments.

The authors express their gratitude to Prof. C. C. Cummins for intensive discussions on these and related results. The present research is in part supported by a grant (CHE96-27775) from the National Science Foundation.

References

1. (a) Fryzuk, M. D.; Love, J. B.; Rettig, S. J.; Young, V. G. *Science* **1997**, *275*, 1445. (b) Gambarotta, S. *J. Organomet. Chem.* **1995**, *500*, 117.
2. Ertl, G. in *Catalytic Ammonia Synthesis*; Jennings, J. R. Ed.; Plenium: New York, **1991**.
3. Chan, M. K. *et al. Science* **1993**, *260*, 792.
4. (a) Laplaza, C. E.; Odom, A. L.; Davis, W. M.; Cummins, C. C. *J. Am. Chem. Soc.* **1995**, *117*, 4999. *(b)* Laplaza, C. E.; Cummins, C. C. *Science* **1995**, *268*, 861. (c) Odom, A. L.; Cummins, C. C. *J. Am. Chem. Soc.* **1995**, *117*, 6613. (d) Laplaza, C. E.; Johnson, A. R.; Cummins, C. C. *J. Am. Chem. Soc.* **1996**, *118*, 709.
5. Cui, Q.; Musaev, D. G.; Svensson, M.; Sieber, S.; Morokuma, K. *J. Am. Chem. Soc.* **1995**, *117*, 12366.
6. (a) Becke, A. D. *Phys. Rev. A* **1988**, *38*, 3098. (b) Lee, C.; Yang, W.; Parr, R. G. *Phys. Rev. B* **1988**, *37*, 785. (c) Becke, A. D. *J. Chem. Phys.* **1993**, *98*, 5648.
7. For transition metals see (a) Hay, P. J.; Wadt, W. R. *J. Chem. Phys.* **1985**, *82*, 299. For Cl see (b) Hay, P. J.; Wadt, W. R. *J. Chem. Phys.* **1985**, *82*, 284.
8. (a)*GAUSSIAN-94*, Frisch, M. J.; Trucks, G. W.; Schlegel, H. B.; Gill, P. M. W.; Johnson, B. G.; Robb, M. A.; Cheesemen, J. R.; Keith, T. A.; Petersson, J. A.; Montgomery, J. A.; Raghavachari, K.; Al-Laham, M. A.; Zakrzewski, V. G.; Ortiz, J. V.; Foresman, J. B.; Cioslowski, J.; Stefanov, B. B.; Nanayakkara, A.; Challacombe, M.; Peng, C. Y.; Ayala, P. Y.; Chen, W.; Wong, M. W.; Andres, J. L.; Replogle, E. S.; Gomperts, R.; Martin, R. L.; Fox, D. J.; Binkley, J. S.; DeFrees, D. J.; Baker, J.; Stewart, J. J. P.; Head-Gordon, M.; Gonzales, C.; Pople, J. A.; Gaussian Inc., Pittsburgh, PA, **1995**. (b) Cui, Q.; Musaev, D. G.; Svensson, M.; Morokuma, K. *J. Phys. Chem.* **1996**, *100*, 10936.
9. STOCKHOLM is a general purpose quantum chemical set of programs written by Siegbahn, P. E. M.; Blomberg, M. R. A.; Pettersson, L. G. M.; Roos, B. O.; Almlof, J.
10. Bartlett, R. J. In *Modern Electronic Structure Theory,* Yarkony, D. R. Ed.; World Scientific: Singapore-New Jersey-London-Hong Kong, **1995**, *Part II*, 1047.
11. Siegbahn, P. E. M.; Blomberg, M. R. A.; Svensson, M. *Chem. Phys. Lett.* **1994**, *223*, 35.
12. *Atomic Energy Levels*; Moore, C. E.; Ed.; NSRD-NBS.; US Goverment Printing Office; Washington DC, **1971**, Vol. II and III.
13. Murdzek, J. S.; Schrock, R. R.; In *Carbyne Complexes*; Fischer, H.; Hofmann, P.; Keissl, F. R.; Schrock, R. R.; Schubert, U.; Wiess, K. Eds.;VHC: New York, **1988**, p. 147.

Chapter 17

Ethylene Polymerization by Zirconocene Catalysis

P. K. Das[1], D. W. Dockter[1], D. R. Fahey[1], D. E. Lauffer[1], G. D. Hawkins[2],
J. Li[2], T. Zhu[2], C. J. Cramer[2], Donald G. Truhlar[2], S. Dapprich[3],
R. D. J. Froese[3], M. C. Holthausen[3], Z. Liu[3], K. Mogi[3], S. Vyboishchikov[3],
D. G. Musaev[3], and K. Morokuma[3]

[1]Phillips Petroleum Company, 327 PL PRC, Bartlesville, OK 74004
[2]Department of Chemistry, University of Minnesota, 207 Pleasant Street S.E.,
Minneapolis, MN 55455
[3]Cherry L. Emerson Center for Scientific Computation and Department of Chemistry,
Emory University, Atlanta, GA 30322

The production of polyethylene by zirconocene catalysis is a multistep
process that includes initiation, propagation, and termination. Each of
these steps has a number of associated equilibrium and transition state
structures. These structures have been studied in the gas-phase
environment using density functional and integrated methods. We have
also examined the effects of solvation upon the energetics of the various
polymerization steps employing continuum and explicit representations
of the solvent (toluene). The reaction steps we have studied are
initiation, propagation, propylene and hexene incorporation, termination
by hydrogenolysis, termination by β-H transfer to the metal,
termination by β-H transfer to the monomer, and reactivation. The
solvation effect of toluene takes on special significance for the
initiation, termination by hydrogenolysis and by β-H transfer to the
metal, and reactivation steps.

Owing to their high commercial interest as catalysts for olefin polymerization, the
group IVA metallocenes have been the subject of many recent quantum-chemical
studies (1-29). For the most part, these studies have focused on the insertion of
ethylene into metallocenium cations (considered to be the active catalyst species).
Attention has also been given to termination steps in the polymerization scheme,
namely, those involving β-hydride transfer from the growing polymer chain to the
metal center and to the monomer in the monomer/metallocenium-cation complex. To
the best of our knowledge, hydrogenolysis has not been investigated as a termination
step in previous work, although hydrogen is a commonly used chain terminating agent

for molecular weight control. In addition, all of the previous computational work deals with the gas phase; the role of a solvent on the energetics of the various steps in the polymerization scheme is addressed here for the first time. In this paper, we present results from our computational studies of various relevant reaction steps involving two zirconocene catalyst systems, namely, Cp_2ZrR^+ (System I) and $[CpCH_2Cp]ZrR^+$ (System II) where R denotes H or an alkyl group. The reaction steps considered in this study are shown in Scheme A. While most of the results came from the application of full high level (full-HL) density functional methodology, in one case we have tested an integrated method (IMOMO) that uses high-level MO calculations for the reaction part and lower-level MO calculations for the spectator part of a molecule. Moreover, using two recently developed solvation models, we have computed the effect of toluene as a solvent on the reaction steps of Scheme A, as mediated by the two zirconocene catalysts.

Computation Methodologies

All full-HL computations (energetics, optimized geometries and analytical second derivatives) were performed using the B3LYP hybrid density functional theory method *(30)* that incorporates Becke's 3-parameter hybrid exchange *(31)* and Lee-Yang-Parr correlation functionals *(32,33)*. The full-HL calculations were carried out with the following basis sets *(34)*: LANL2DZ with a relativistic effective core potential (RECP) for Zr, LANL2DZ for the other atoms involved in the reaction center (H or alkyl groups bonded to Zr and coordinated olefins), and 3-21G for C and H atoms of the auxiliary ligands (L_1 and L_2, see Scheme A). Optimized equilibrium geometries and transition states (TSs) were obtained using non-redundant internal coordinates starting with Hessians calculated at the restricted Hartree-Fock (RHF) level. Analytical second derivatives *(35)* were used to obtain the frequencies for the enthalpies and free energies (scale factor = 1.0) and to verify minima and TSs.

The integrated method *(36-39)*, IMOMO, is similar to that of the integrated molecular orbital molecular mechanics (IMOMM) method that has been described earlier in the literature *(40,41)*, the difference being that the force-field treatment of the spectator portion of the molecule is replaced by a lower-level molecular orbital method. The notation IMOMO(HL:LL) will be used to designate an IMOMO calculation, where HL and LL denote the high-level and low-level MO methods, respectively. IMOMO HL energies and optimized geometries were calculated using the B3LYP method with the basis set described above; whereas the LL energies and geometries were obtained using the RHF method with the following basis sets *(34d-g)*: LANL1MB with RECP for Zr and LANL1MB for the other atoms. No frequencies were calculated at the IMOMO level. All full-HL and IMOMO calculations were performed using either GAUSSIAN 94 or the developmental version GAUSSIAN 95 *(42,43)*.

The two solvation models used in this work, namely, SM5.2R/MNDO/d and SM5.42R/BPW91/MIDI!(6D), are implemented in the codes, AMPAC5.4m1 *(44)* and DGAUSS4.0m1 *(45)*, respectively. They are based on extensions of the SM5 suite of solvation models *(46-49)* to organometallic complexes of zirconium. The SM5 suite is an extension of earlier SMx models *(50-52)* that were originally implemented in the AMSOL *(53)* program.

Scheme A: Polymerization Reaction Steps

Reaction steps **2-5** involving the catalyst systems Cp_2ZrR^+ (System I) and $[CpCH_2Cp]ZrR^+$ (System II), with R = H or an alkyl group, were studied.

1: Catalyst Activation (Not investigated in this study)

$$[L_1,L_2]ZrCl_2 + Cocatalyst \rightarrow [L_1,L_2]ZrCH_3^+ + \overline{Cocatalyst}$$

2: Initiation

$$[L_1,L_2]ZrCH_3^+ + CH_2=CH_2 \rightarrow \pi\text{-complex} \rightarrow TS \rightarrow$$
$$[L_1,L_2]ZrCH_2CH_2CH_3^+$$

3a, 3a': Propagation

$$[L_1,L_2]ZrCH_2CH_2CH_3^+ + CH_2=CH_2 \rightarrow \pi\text{-complex} \rightarrow TS \rightarrow$$
$$[L_1,L_2]ZrCH_2CH_2CH_2CH_2CH_3^+$$

3a: Ethylene approaching from the front side of the alkyl chain.
3a': Ethylene approaching from the back side of the alkyl chain.

3b: Comonomer incorporation

$$[L_1,L_2]ZrCH_2CH_2CH_3^+ + CH_2=CHR' \rightarrow \pi\text{-complex} \rightarrow TS \rightarrow$$
$$[L_1,L_2]ZrCH_2CH(R')CH_2CH_2CH_3^+$$

R' = CH_3 for propylene incorporation
R' = C_4H_9 for 1-hexene incorporation

4a, 4a' & 4a'': Termination by hydrogenolysis

$$[L_1,L_2]ZrR^+ + H_2 \rightarrow \pi\text{-complex} \rightarrow TS \rightarrow [L_1,L_2]ZrH(RH)^+ \rightarrow$$
$$[L_1,L_2]ZrH^+ + RH$$

4a: R = CH_3
4a': R = $CH_2CH_2CH_3$, H_2 approaching from the front side of the alkyl chain
4a'': R = $CH_2CH_2CH_3$, H_2 approaching from the back side of the alkyl chain

4b: Termination by β-H transfer to metal center

$$[L_1,L_2]ZrCH_2CH_2CH_3^+ \rightarrow TS \rightarrow [L_1,L_2]ZrH(CH_2=CHCH_3)^+ \rightarrow$$
$$[L_1,L_2]ZrH^+ + CH_2=CHCH_3$$

4c: Termination by β-H transfer to monomer in the complex

$$[L_1,L_2]ZrCH_2CH_2CH_3^+ + CH_2=CH_2 \rightarrow \pi\text{-complex} \rightarrow TS \rightarrow$$
$$[L_1,L_2]ZrCH_2CH_3(CH_2=CHCH_3)^+ \rightarrow [L_1,L_2]ZrCH_2CH_3^+ + CH_2=CHCH_3$$

5: Reactivation

$$[L_1,L_2]ZrH^+ + CH_2=CH_2 \rightarrow \pi\text{-complex} \rightarrow TS \rightarrow [L_1,L_2]ZrCH_2CH_3^+$$

Notations and Definitions

ΔE^c, ΔH^c, and ΔG^c: Energy, enthalpy, and free energy change for the <u>complexation</u> substep (applicable to all reaction steps above, except **4b**).

ΔE^r, ΔH^r, and ΔG^r: Energy, enthalpy, and free energy change for the <u>reaction</u> substep, measured from the complex except for **4b** where it is measured from the reactant cation.

$\Delta E^{\ddagger}$, $\Delta H^{\ddagger}$, and $\Delta G^{\ddagger}$: Activation energy, activation enthalpy, and activation free energy for the <u>reaction</u> substep, referenced from the π-complex except for **4b** where it is referenced from the reactant cation.

ΔE^s, ΔH^s, and ΔG^s: Energy, enthalpy, and free energy change for the <u>separation</u> substep (applicable to termination steps, **4a-a''**, **b**, **c**, above)

Born-Oppenheimer energies without ZPE will be denoted as E, and enthalpies and free energies for gas-phase processes at temperature, T, will be denoted as H_T and G_T, respectively. All solution-phase free energies correspond to a concentration of 1 mol L^{-1} and a temperature of 298 K and are denoted as G_S. Except for **4b**, each reaction step includes a reaction substep, r, that usually starts at the π-complex and ends at the product complex; the reaction substep for reaction step **4b** starts at the reactant cation and ends with the product complex. Also, each reaction step except for **4b** includes a complexation substep, c, preceding the reaction substep. All termination reaction steps, **4a-a''**, **b**, and **c**, include a separation substep, s, in which the product of the reaction substep separates.

The standard-state solvation energy, ΔG_S^0, of a solute molecule is computed as the sum of three terms, namely: (1) the gain, G_P, in electric polarization energy due to the polarization of the solvent; (2) the energy cost, ΔE_{EN}, for distorting the electronic/nuclear structure of the solute to be self-consistent with the polarized solvent; and (3) the contribution, G_{CDS}, to the free energy of solvation due to cavity formation, dispersion interactions, solvent structural changes, and other effects of the first solvation shell that differ from those included in G_P. Thus,

$$\Delta G_S^0 = \Delta E_{EN} + G_P + G_{CDS}. \tag{1}$$

The sum of G_P and ΔE_{EN} is determined by a self-consistent reaction field calculation (49, 54) and depends on the solvent dielectric constant. In the present work the nuclear relaxation part of ΔE_{EN} is not included explicitly, but rather it is included implicitly in the parameterization of the CDS terms. G_{CDS} is calculated from empirical atomic surface tension coefficients and the solvent-accessible surface area (SASA) of the solute. In particular G_{CDS} is a sum of atomic contributions of the form $\sigma_k A_k$, where σ_k is the surface tension of atom k and A_k is the exposed area of atom k.

Two models, SM5.2R/MNDO/d (48) and SM5.42R/BPW91/MIDI!(6D) (49), that were originally parameterized and tested against 2,135 experimental solvation free energy data for 275 neutral solutes, 49 ions, and 91 solvents (48,49) were adapted here to specifically treat zirconium compounds in toluene at room temperature. Since no experimental data were available for the free energy of solvation of any zirconium solute, an indirect route was used to develop the atomic surface tension coefficient (σ_{Zr}) needed for computing the contribution of the Zr atom to G_{CDS} in toluene. First, a free energy cycle was constructed as shown in Figure 1. In this cycle, M is a Zr-containing cation, T is a toluene molecule, C is a complex of M with a single toluene molecule, g stands for gas-phase, and s stands for toluene solution. $\Delta G_S^0(M)$, $\Delta G_S^0(T)$, and $\Delta G_S^0(C)$ are the solvation free energies in toluene for M, T, and C, respectively; and ΔG_T^c and ΔG_S^c are the complexation free energies in the gas phase and in solution, respectively. The key to the method is that ΔG_S^c should be zero because it simply corresponds to labeling the first-solvation-shell of toluene in two different ways, first as part of the solvent, then as part of the solute. This gives the following equation:

$$\Delta G_S^0(M) + \Delta G_S^0(T) = \Delta G_T^c + \Delta G_S^0(C). \tag{2}$$

In principle, both $\Delta G_S^0(M)$ and $\Delta G_S^0(C)$ contain the contributions to the CDS energies from Zr. An examination of three-dimensional space-filling models of the complexes as well as an estimate of the solvent-accessible surface area of Zr in the complexes indicate that the zirconium atom in the complex is totally buried, i.e., the Zr contribution to CDS should be zero for the complex; therefore, we chose the solvent radius large enough (1.7 Å) to insure that A_{Zr} is zero for four complexes of interest. (Note that a solvent radius of zero is used for computing A_k for a non-metallic

212

atom k in SM5.2R.) The atomic surface tension, σ_{Zr}, for Zr thus can be determined in the following way:

$$\sigma_{Zr} A_{Zr}(M) = \Delta G_T^c + \Delta G_S^0(C) - \Delta \widetilde{G}_S^0(M) - \Delta G_S^0(T), \qquad (3)$$

where $A_{Zr}(M)$ is the exposed surface area of Zr in M, and $\Delta \widetilde{G}_S^0(M)$ is the solvation free energy for M without a first-solvation shell contribution from Zr. The Zr atomic surface tensions for two specific solvation models described below were obtained using

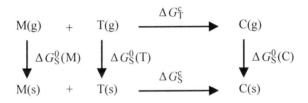

Figure 1. Free energy cycle used in developing the solvation models

Equation 3 and the gas-phase free energies for $Cp_2ZrCH_3^+$ and $Cp_2ZrCH_3(C_6H_5CH_3)^+$ (Figure 2A and Table I). These parameterizations were then tested using Equation 2 and the gas-phase free energies for the three zirconocenium cations, Cp_2ZrR^+ and the cation-toluene complexes, $Cp2ZrR(C_6H_5CH_3)^+$, with R = H, Cl, and $CH_2CH_2CH_3$.

Some important features of the two specific solvation models for Zr compounds in toluene (henceforth referred to as zirconium solvation models, ZSM1 and ZSM2) are given below.

ZSM1: Semiempirical method implemented with the MNDO/d parameters of Thiel et al. *(55-58)* except for the Zr α parameter that was optimized to 1.4 Å^{-1}.

Fixed (Rigid) solute geometries: We used optimized gas-phase geometries for the solute geometries. The contribution to E_{EN} from nuclear relaxation was absorbed in the CDS term.

Charge model: We used Mulliken charges (Class II charges) to compute G_P.

First-solvation-shell term: We used SM5-type surface tension functional forms *(46-49)* to compute G_{CDS}. For toluene, the surface tension coefficients were determined by the solvent properties, namely, index of refraction of 1.4961, hydrogen-bond acidity of 0.0, hydrogen-bond basicity of 0.14, and macroscopic surface tension of 40.2 cal mol^{-1} Å^{-2}.

ZSM2: DFT method implemented with the BPW91 *(59)* exchange-correlation functional; for Zr, DGAUSS-built-in valence basis sets PPC (atomic) and AP1 (fitting) were used with pseudopotentials. For the other atoms, the MIDI! basis set *(60)* was used with the original five-function spherical harmonic d set replaced by a six-function Cartesian d set.

Fixed (Rigid) solute geometries: We used the same geometries as for ZSM1. Charge model: We used Class IV CM2 charges (*61*), mapped from Löwdin population analysis, to compute G_P. In previous validation studies (*61*), the mapping was found to decrease the errors in dipole moments, typically by a factor of 3. Special attention was paid to obtaining realistic charges for aromatics.
First-solvation-shell term: We used parameters similar to those in ZSM1.

For the free energies of activation, adding the solvation energies at the gas-phase stationary points corresponds to the assumption of separable equilibrium solvation (*62*).

Results and Discussion

a. Specific Binding of Toluene to Zirconocenium Cations. In general, toluene may coordinate to the transition metal center of the zirconocenium cations in several different ways, i.e., via η^1(one C atom), η^2 (two C atoms), η^3 (three C atoms), η^6 (six C atoms), or its methyl-group. Two zirconocenium cation-toluene conformers, η^1 and pseudo-η^3, are shown in Figures 2A and 2B, respectively. We note also, although not

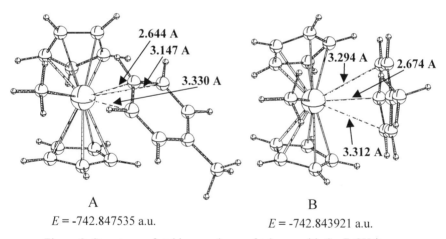

A
$E = -742.847535$ a.u.

B
$E = -742.843921$ a.u.

Figure 2. Structures of stable complexes of toluene with $Cp_2ZrCH_3^+$

shown in the figure, that the ethylene molecule is coordinated to the zirconocenium cations in a slightly asymmetric-η^2 manner. The difference between the coordination of the toluene and that of ethylene may be due to the relatively strong repulsive steric interactions of a toluene molecule with the auxiliary ligands. In Figure 2A, we show the lowest-energy conformer, having values of -22.1 and -10.3 kcal/mol for ΔH^c_{298} and ΔG^c_{298}, respectively. These numbers are only slightly larger (1-2 kcal/mol) than the corresponding values for ethylene coordination to the zirconocenium cations. Table I gives the complexation energies for Cp_2ZrR^+ with R= H, Cl, CH_3, and $CH_2CH_2CH_3$ and shows that toluene does bond strongly to zirconocenium cations. All of the energies listed in Table I correspond to those of the most stable conformers with structures similar to Figure 2A.

b. Gas-Phase Full-HL Calculations Ethylene Insertion. For the reaction steps in Scheme A, full-HL computations have been performed for the gas-phase stationary points on the potential energy surface. The classical energetics, i.e., the electronic contributions (including nuclear repulsion) without zero point energies (ZPE), are summarized in Table II. Overall, there are significant differences in energetics between

Table I. Full-HL gas-phase energetics (kcal/mol) of the complexation of toluene with zirconocenium cations.

Zirconocenium cation	ΔE^c	ΔH^c_{298}	ΔG^c_{298}
Cp_2ZrH^+	-27.8	-25.5	-14.9
Cp_2ZrCl^+	-25.0	-22.0	-12.0
$Cp_2ZrCH_3^+$	-24.7	-22.1	-10.3
$Cp_2ZrCH_2CH_2CH_3^+$			
β-agostic	-14.1	-11.4	0.1
almost non-agostic	-24.2	-21.6	-9.6

Table II. Energetics (kcal/mol) for the reaction steps involving $[CpCH_2Cp]ZrR^+$ and Cp_2ZrR^+

Reaction step[a]	$[CpCH_2Cp]ZrR^+$				Cp_2ZrR^+			
	ΔE^c	ΔE^r	$\Delta E^{\neq}$	ΔE^s	ΔE^c	ΔE^r	$\Delta E^{\neq}$	ΔE^s
2	-21.6	-7.7	7.1		-19.8	-8.1	6.5	
3a	-14.7	-9.7	6.0		-9.5	-9.9	3.5	
3a'	-14.7	-8.8	6.4		-10.8	-10.0	7.2	
3b	-16.9	-2.5	9.6		-13.8	-4.0	7.2	
4a	-7.8	-14.0	5.3	12.5	-7.8	-14.7	4.4	14.6
4a'	-6.6	-9.5	10.3	16.1	-3.9	-12.5	7.8	16.2
4a''	-8.5	-10.3	5.5	18.8	-6.0	-11.5	3.4	17.5
4b		10.9	12.0	26.9		12.5	12.3	25.0
4c	-14.7	-4.4	11.4	16.4	-9.5	-4.9	12.3	11.8
5	-24.7[b]	-15.6[b]	1.9[b]		-22.5[b]	-17.7[b]	1.6[b]	

[a]For numbering of steps and notation for energetics, see Scheme A.
[b]Ethylene is in a broad-side-on orientation.

the bridged and non-bridged zirconocene systems for the reaction steps under consideration. The reaction activation barriers ($\Delta E^{\neq}$) for propagation, hexene incorporation, and hydrogenolysis reaction steps are lower for the non-bridged system.

Since the zero point, thermal energetic, entropic, and solvation contributions to free energy changes play an important role in determining the energetics of the reaction steps, further examination of the reaction energetics will be deferred until solvent effects are discussed.

c. IMOMO(HL:LL) Calculations for Propylene Incorporation. IMOMO calculations were performed to test the adequacy of this method for modeling higher olefin insertion by using ethylene as a subsystem for the higher olefin homologues. The energies (ΔE) and geometries were calculated by IMOMO method for reaction step **3b**, where propylene was used as the comonomer. For the HL systems, we chose the homologous complexes $Cp_2Zr(C_2H_5)^+$ and $Cp_2Zr(C_4H_9)^+$ in which one methyl group in both the $Cp_2Zr(C_3H_7)^+$ and the propylene and two methyl groups in the π-complex, transition state, and products were replaced by hydrogen atoms. The reactants, intermediates, transition states, and products were calculated for a front side, 2,1-insertion, i.e., insertion by the addition of the metal alkyl to the coordinated propylene to give a secondary alkyl group attached to the metal. The potential energy surfaces and optimized geometries for this reaction, calculated at both the IMOMO and full-HL levels of theory, are presented in Figures 3 and 4, respectively.

Most of the IMOMO bond distances presented in Figure 4 are within 1% of those calculated using the full-HL method. In particular, the agostic interaction distances and carbon-zirconium bond distances shown in the figure are in good agreement with those optimized at full-HL level. The advantage of the IMOMO method is that it provides, at a lower cost, geometries comparable to those obtained from full-HL calculations. However, as Figure 3 shows, the differences in the energetics between the full-HL and the IMOMO calculations are as large as 3 kcal/mol. This suggests that caution should be used when applying this IMOMO approach to the calculation of energetics.

d. Solvation Effects. The modification of the gas-phase free energy changes by the solvent effect has been computed for all applicable substeps of the reaction steps in Scheme A, except step **1**. The data for the two zirconocene catalyst systems I and II are compiled in Table III, where we use the notation

$$\Delta\Delta G^x \equiv \Delta G_S^x - \Delta G_{298}^x, \quad x = c, \ r, \ s, \ or \ \neq. \tag{4}$$

The definitions of ΔG_S^x, ΔG_{298}^x, and the reaction substeps are given in Scheme A. Generally, the solvation models, ZSM1 and ZSM2, predict very similar solvation effects (to within ~1 kcal/mol) for the free energy changes of all reaction substeps, except for the hydrogenolysis reaction, where the unsigned differences in the $\Delta\Delta G^x$s between the two models for a given zirconium complex range from 1 to 4 kcal/mol.

Table IV shows enthalpy (ΔH_{298}^x) and free energy (ΔG_{298}^x, ΔG_S^x) changes for the reactions substeps involving systems I and II, respectively. To illustrate how the energetics change at the various stages of a reaction, energy diagrams for reaction steps **2**, **3a**, **4a''**, and **4b** involving $[CpCH_2Cp]ZrR^+$ are given in Figures 5-8. At this point in our gas-phase computational work, we have only identified saddle points for the reaction substep; the nature of the transition states (if any) for the complexation and

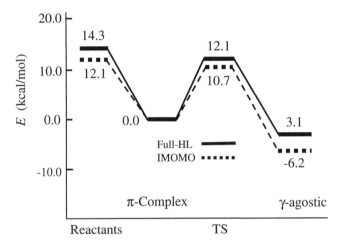

Figure 3. Energetics calculated at the IMOMO and full-HL levels for the propylene incorporation step, **3b**: $Cp_2ZrCH_2CH_2CH_3^+ + CH_2=CHCH_3 \rightarrow$ $Cp_2ZrCH_2CH(CH_3)CH_2CH_2CH_3^+$.

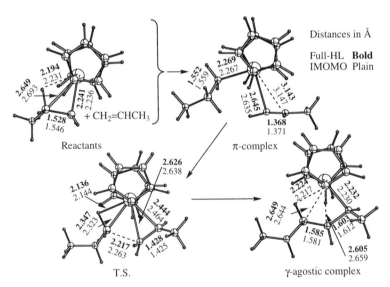

Figure 4. Optimized geometries of the reactants, intermediate, transition state, and product of the propylene incorporation reaction step, **3b**.

Table III. Solvent effects (kcal/mol) on free energy changes, upper entry computed by ZSM1 and lower entry computed by ZSM2.

Zirconocenium Cation	Reaction Step[a]	$\Delta\Delta G^c$	$\Delta\Delta G^r$	$\Delta\Delta G^{\neq}$	$\Delta\Delta G^s$
Cp$_2$ZrR$^+$	2	6.1	-0.1	-0.1	
		5.4	-0.4	0.5	
	3a	1.4	-0.6	-1.3	
		1.1	-0.5	-0.0	
	3a'	1.1	-0.4	0.1	
		1.1	-0.7	0.0	
	3b	1.7	0.6	0.9	
		2.2	0.5	0.9	
	4a	2.0	3.0	0.3	-9.6
		3.3	-0.4	0.3	-8.5
	4a'	-2.3	1.8	-1.0	-11.1
		-0.8	-1.6	-0.6	-10.0
	4a''	-3.3	3.1	1.1	-11.6
		-0.7	-1.0	0.1	-10.7
	4b		1.4	0.9	-11.3
			0.1	0.9	-10.5
	4c	1.4	0.2	0.0	-2.8
		1.1	0.1	0.2	-2.7
	5	10.1	-1.4	0.3	
		8.8	0.1	0.4	
[CpCH$_2$Cp]ZrR$^+$	2	9.9	-0.3	0.1	
		9.1	-0.4	0.6	
	3a	3.5	-0.5	-1.2	
		3.0	-0.6	-0.1	
	3a'	3.0	0.2	0.1	
		2.8	-0.2	0.1	
	3b	5.3	-0.6	-0.5	
		4.9	0.1	-0.1	
	4a	5.1	3.4	0.9	-14.9
		6.2	0.4	0.7	-13.7
	4a'	-0.3	2.1	-1.1	-17.0
		1.2	-1.2	-1.0	-15.8
	4a''	-1.1	2.9	1.3	-17.0
		1.0	-1.2	0.4	-15.7
	4b		3.2	1.4	-16.7
			2.3	1.8	-16.7
	4c	3.5	0.4	-0.1	-4.9
		3.0	0.2	0.1	-4.5
	5	15.4	-3.1	0.4	
		13.9	-2.0	0.8	

[a]For numbering of steps and notation for energetics, see Scheme A and Equation 4.

Table IV. Enthalpy and free energy changes in the gas phase and free energy changes in toluene for reaction steps involving Cp_2ZrR^+ (upper entry) and $[CpCH_2Cp]ZrR^+$ (lower entry).

Reaction Step[a,b]	ΔH^c_{298}	ΔG^c_{298}	ΔG^c_S	ΔH^t_{298}	ΔG^t_{298}	ΔG^t_S	$\Delta H^{\neq}_{298}$	$\Delta G^{\neq}_{298}$	$\Delta G^{\neq}_S$	ΔH^S_{298}	ΔG^S_{298}	ΔG^S_S
2	-18.1	-6.7	-0.6	-6.2	-3.4	-3.5	6.6	7.9	7.8			
	-20.2	-9.9	0.0	-5.8	-3.1	-3.4	7.1	10.1	10.2			
3a	-7.1	4.4	5.8	-9.0	-7.9	-8.5	2.9	4.5	3.2			
	-12.6	-1.2	2.3	-7.1	-5.3	-5.8	5.3	6.9	5.7			
3a'	-8.5	5.8	6.9	-8.9	-9.0	-9.4	7.7	9.6	9.7			
	-12.8	-1.0	1.0	-7.5	-6.8	-6.6	6.7	9.2	9.3			
3b	-11.3	0.8	2.5	-3.1	-2.4	-1.9	6.9	9.4	10.3			
	-15.0	-3.3	2.0	-1.3	0.5	-0.1	9.2	11.7	11.2			
4a	-5.9	1.4	3.4	-12.0	-11.5	-8.5	4.2	5.9	6.2	10.8	0.8	-8.9
	-6.2	1.1	6.2	-11.3	-11.6	-8.2	5.0	6.5	7.4	11.5	2.1	-12.8
4a'	-1.4	7.7	5.4	-10.3	-10.9	9.1	6.9	7.3	6.3	14.9	3.8	-7.3
	-4.6	3.2	2.9	-7.0	-9.5	-7.5	9.4	9.7	8.6	15.0	6.2	-10.8
4a''	-3.3	5.7	2.4	-9.3	-10.5	-7.4	2.7	3.3	4.4	16.0	4.4	-7.2
	-6.4	2.3	1.2	-8.0	-9.1	-6.2	4.9	5.1	6.4	17.7	6.6	-10.4
4b				10.9	11.5	12.9	9.6	10.5	11.4	23.3	10.7	-0.6
				9.2	8.4	11.6	9.4	9.5	10.9	25.5	13.9	-2.8
4c	-7.1	4.4	5.8	-4.9	-4.9	-4.7	9.9	12.6	12.6	9.5	-3.3	-6.1
	-12.6	-1.2	2.3	-4.5	-5.3	-4.9	8.9	11.4	11.3	14.2	3.5	-1.4
5	-20.8	-10.5	-0.4	-15.9	-15.1	-16.5	7.2	2.0	2.3			
	-23.3	-13.1	2.3	-13.8	-12.0	-15.1	1.3	2.5	2.9			

[a] Calculated by ZSM1. All data are in kcal/mol.

[b] For numbering of the steps and notation for energetics, see Scheme A.

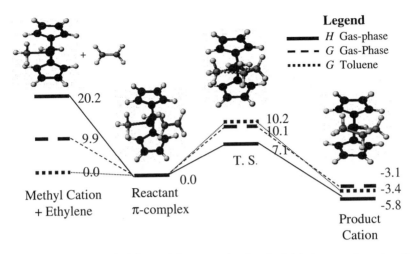

Figure 5. Relative enthalpies and free energies for the initiation reaction step, **2**, involving $[CpCH_2Cp]ZrCH_3^+ + C_2H_4$.

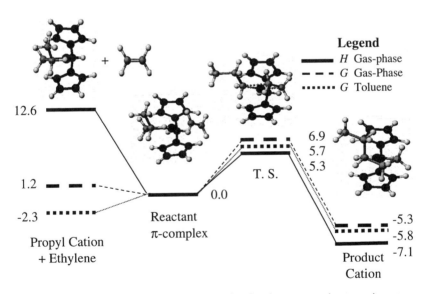

Figure 6. Relative enthalpies and free energies for the propagation reaction step, **3a**, involving $[CpCH_2Cp]ZrCH_2CH_2CH_3^+ + C_2H_4$.

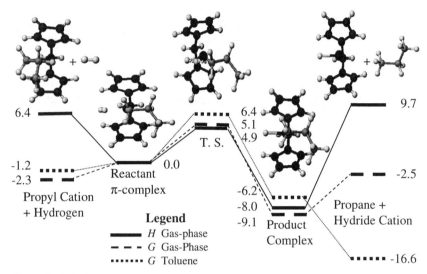

Figure 7. Relative enthalpies and free energies for the termination by hydrogenolysis reaction step, **4a**, involving [CpCH$_2$Cp]ZrCH$_2$CH$_2$CH$_3^+$ + H$_2$.

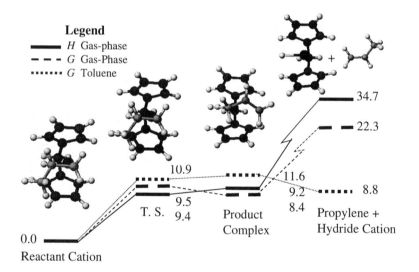

Figure 8. Relative enthalpies and free energies for the termination by β-H transfer to the metal center reaction step, **4b**, involving [CpCH$_2$Cp]ZrCH$_2$CH$_2$CH$_3^+$.

separation substeps is not clear. It is evident from the data in Table IV, as well as from Figures 5-8, that the greatest effect of the solvent is manifested in those substeps involving tri-coordinated zirconocenium cations, i.e., the complexation and separation substeps. The free energy changes and the activation free energies for the reaction substeps are changed to lesser extents in going from the gas phase to the solution phase. Also, we note that there is a significant difference in what is observed when one examines free energy changes rather than enthalpy changes.

The solvation effects are important in developing an understanding of the mechanisms involved in zirconcene-catalyzed ethylene polymerization. Specifically, solvation effects are pronounced in the initiation reaction step **2** (see Figure 5), the reactivation reaction step **5,** and the termination reaction steps **4a-a"** and **4b**, see Table III. Considering the case of the initiation reaction for [CpCH$_2$Cp]ZrR$^+$ without solvation, the transition state free energy is nearly the same (within 0.2 kcal/mol) as the free energy of the reactants, both referenced to the same energy level; but with solvation the former is 10 kcal/mol higher. For the termination reaction steps, solvation reduces the free energy changes for the separation into products by approximately 16 kcal/mol for reactions **4a-a"** and **4b**. Thus, predictions involving the energetics of polymerization mechanisms including these reactions are likely to depend significantly on solvation effects.

Solvation plays quite a different role in the free energies of activation for the two insertion steps, i.e., initiation and propagation. (Note that the free energy of activation for these reaction steps is $\Delta G^c + \Delta G^{\neq}$ if the reactants are in chemical equilibrium with the reactant complex.) The present gas-phase calculations (Figures 5 and 6) yield free energies of activation of 0.2 and 5.7 kcal/mol, respectively, for the initiation and propagation steps, referenced to separate reactants, implying that the latter provides a higher dynamical bottleneck. However, for a toluene solvent the calculations (also in Figures 5 and 6) yield 10.2 and 8.0 kcal/mol for corresponding free energies of activation for initiation and propagation, implying that the former provides a higher bottleneck and that the overall activation energy is about 10 kcal/mol. Thus the results of the condensed-phase calculations change our qualitative understanding of the polymer chain growth mechanism.

Summary and Conclusions

We have performed electronic structure and free energy of solvation calculations on two structurally distinct zirconocene catalyst systems. Comparing the two systems, significant differences are observed in the energetics at stationary points along the reaction paths for the various reaction steps involved in ethylene polymerization and hexene incorporation. Moreover, considering each zirconocene system separately, the changes in the gas-phase free energies and the solution-phase free energies differ significantly from each other. Two notable consequences for the catalyst systems studied are that the insertion reaction step is most likely the dynamical bottleneck for the polymer chain growth and that termination by β-H transfer to the metal center may indeed be a viable polymer chain termination process.

For large metallocene systems, the integrated method, IMOMO, shows promise for obtaining optimized geometries and possibly energies.

222

Acknowledgment

This work was performed in part under the support of the United States Department of Commerce, National Institute of Standards and Technology.

Literature Cited

1. Fujimoto, H.; Koga, N.; Fukui, K. *J. Am. Chem. Soc.* **1981**, *103*, 7452.
2. Fujimoto, H.; Yamasaki, T.; Mizutani, H.; Koga, N. *J. Am. Chem. Soc.* **1985**, *107*, 6157.
3. Fujimoto, H. *Acc. Chem. Res.* **1987**, *20*, 448.
4. Marynick, D. S.; Axe, F. U.; Hansen, L. M.; Jolly, C. A. *Topics in Physical Organometallic Chemistry;* Freund Publishing House: London, **1988**; Vol. 3, p. 37.
5. Shiga, A.; Kawamura, H.; Ebara, T.; Sasaki,, T.; Kikuzono, Y. *J. Organomet. Chem.* **1989**, *366*, 95.
6. Jolly, C. A.; Marynick, D. S. *J. Am. Chem. Soc.* **1989**, *111*, 7968.
7. Kawamura-Kuribayashi, H.; Koga, N.; Morokuma, K. *J. Am. Chem. Soc.* **1992**, *114*, 2359.
8. Kawamura-Kuribayashi, H.; Koga, N.; Morokuma, K. *J. Am. Chem. Soc.* **1992**, *114*, 8687.
9. Janiak, C. *J. Organomet. Chem.* **1993**, *63*, 452.
10. Bierwagen, E. P.; Bercaw, J. E.; Goddard, III, W. A. *J. Am. Chem. Soc.* **1994**, *116*, 1481.
11. Woo, T. K.; Fan, L.; Ziegler, T. *Organometallics* **1994**, *13*, 432.
12. Woo, T. K.; Fan, L.; Ziegler, T. *Organometallics* **1994**, *13*, 2252.
13. Axe, F. U.; Coffin, J. M. *J. Phys. Chem.* **1994**, *98*, 2567.
14. Meier, R. J.; Doremaele, G. H. J.; Iarlori, S.; Buda, F. *J. Am. Chem. Soc.* **1994**, *116*, 7274.
15. Weiss, H.; Ehrig, M.; Ahlrichs, R. *J. Am. Chem. Soc.* **1994**, *116*, 4919.
16. Sini, G.; Macgregor, S. A.; Eisenstein, O.; Teuben, J. H. *Organometallics* **1994**, *13*, 1049.
17. Fusco, R.; Longo, L. *Macromol. Theory Simul.* **1994**, *3*, 895.
18. Lohrenz, J. C. W.; Woo, T. K.; Ziegler, T. *J. Am. Chem. Soc.* **1995**, *117*, 12793.
19. Fan, L.; Harrison, D.; Woo, T. K.; Ziegler, T. *Organometallics* **1995**, *14*, 2018.
20. Lohrenz, J. C. W.; Woo, T. K.; Fan, L.; Ziegler, T. *J. Organomet. Chem.* **1995**, *497*, 91.
21. Fan, L.; Harrison, D.; Deng, L.; Woo, T. K.; Swehone, D.; Ziegler, T. *Can. J. Chem.* **1995**, *73*, 989.
22. Yoshida, T.; Koga, N.; Morokuma, K. *Organometallics* **1995**, *14*, 746.
23. Yoshida, T.; Koga, N.; Morokuma, K. *Organometallics* **1996**, *15*, 766.
24. Woo, T.; Margl, P. M.; Lohrenz, J. C. W.; Blöchl, P. E.; Ziegler, T. *J. Am. Chem. Soc.* **1996**, *118*, 4434.
25. Cruz, V. L.; Munoz-Escalona, A.; Martinez-Salazar, J. *Polymer* **1996**, *37*, 1663.
26. Margl, P.; Lohrenz, J. C. W.; Ziegler, T.; Blöchl, P. E. *J. Am. Chem. Soc.* **1996**, *118*, 4434.
27. Woo, T. K.; Margl, P. M.; Ziegler, T.; Blöchl, P. E. *Organometallics* **1997**, *16*, 3454.
28. Borve, K. J.; Jensen, V. R.; Karlsen, T.; Stovneng, J. A.; Swang, O. *J. Mol. Model.* **1997**, *3*, 193.
29. Fusco, R.; Longo, L.; Masi, F.; Garbassi, F. *Macromolecules* **1997**, *30*, 7673.
30. Stephens, P. J.; Devlin, F.J.; Chabalowski, C. F.; Frisch, M. J. *J. Phys. Chem.* **1994**, *98*, 11623.
31. Becke, A. D. *J. Chem. Phys.* **1993**, *98*, 5648.
32. Lee, C.; Yang, W.; Parr, R. G. *Phys. Rev. B* **1988**, *37*, 785.
33. Miehlich, B.; Savin, A.; Stoll, H.; Preuss, H. *Chem. Phys. Lett.* **1989**, *157*, 200.

34. (a) Binkley, J. S.; Pople, J. A.; Hehre, W.J. *J. Amer. Chem. Soc.* **1980**, *102*, 939.
(b) Gordon, M. S.; Binkley, J. S.; Pople, J. A.; Peitro, W. J.; Hehre, W. J. *J. Amer. Chem. Soc.* **1982** *104*, 2797.
(c) Pietro, W. J.; Francl, M. M.; Hehre, W. J.; Defrees, D. J.; Pople, J. A.; Brinkley, J. S. *J. Amer. Chem. Soc.* **1982**, *104*, 5039.
(d) Dunning, Jr. T. H.; Hay, P. J. in *Modern Theoretical Chemistry*, *Vol 3*; H. F. Schaefer, III, Ed.; Plenum: New York, **1976**; p.1.
(e) Hay, P. J.; Wadt, W. R. *J. Chem. Phys.* **1985**, *82*, 270.
(f) Wadt, W.R.; Hay, P.J. *J. Chem. Phys.* **1985**, *82*, 284.
(g) Hay, P.J.; Wadt. W. R. *J. Chem. Phys.* **1985**, *82*, 299.

35. Cui, Q.; Musaev, D. G.; Svensson, M.; Morokuma, K. *J. Phys. Chem.*, **1996**, *100*, 10936.

36. Humbel, S.; Sieber, S.; Morokuma, K. *J. Chem. Phys.* **1996**, *105*, 1959.

37. Svensson, M.; Humbel,S.; Morokuma, K. *J. Chem. Phys.* **1996**, *105*, 3654.

38. Coitiño, E.L.; Truhlar, D. G.; Morokuma, K. *Chem. Phys. Lett.* **1996**, *259*, 159.

39. Noland, M.; Coitiño, E.L.; Truhlar, D.G. *J. Phys. Chem. A* **1997**, *101*, 1193.

40. Matsabura, T.; Maseras, F.; Koga, N.; Morokuma, K. *J. Phys. Chem.* **1996**, *100*, 2573.

41. Maseras, F.; Morokuma, K. *J. Comput. Chem.* **1995**, *16*, 1170.

42. Frisch, M. J.; Trucks, G. W.; Schlegel, H. B.; Gill, P. M. W.; Johnson, B. G.; Robb, M. A.; Cheeseman, J. R.; Keith, T.; Petersson, G. A.; Montgomery, J. A.; Raghavachari, K.; Al-Laham, M. A.; Zakrzewski, V. G.; Ortiz, J. V.; Foresman, J. B.; Cioslowski, J.; Stefanov, B. B.; Nanayakkara, A.; Challacombe, M.; Peng, C. Y.; Ayala, P. Y.; Chen, W.; Wong, M. W.; Andres, J. L.; Replogle, E. S.; Gomperts, R.; Martin, R. L.; Fox, D. J.; Binkley, J. S.; Defrees, D. J.; Baker, J.; Stewart, J. P.; Head-Gordon, M.; Gonzalez, C.; Pople, J. A. *Gaussian 94, Revision C.3*, Gaussian, Inc., Pittsburgh PA, 1995.

43. Froese, R. J. D.; Morokuma, K. *Chem. Phys. Lett*, **1996**, *263*, 393; Dapprich, S.; Komaromi, I.; Byun, K. S.; Holthausen, M. C.; Morokuma, K.; Frisch, M. J. *J. Mol. Struct. (THEOCHEM)*, to be submitted.

44. Hawkins, G. D.; Liotard, D. A.; Cramer, C. J.; Truhlar, D. G. *AMPAC version 5.4ml*, University of Minnesota, Minneapolis, **1997**. This package is a set of modifications to *AMPAC version 5.4*, Semichem, Inc., Shawnee, Kansas to incorporate the SM5.2R solvation model for MNDO/d and other semiempirical methods.

45. Zhu, T.; Giesen, D. J.; Liotard, D. A.; Stahlberg, E. A.; Hawkins, G. D.; Chambers, C. C.; Cramer, C. J.; Truhlar, D. G. *DGSOL version 4.0ml*, University of Minnesota, Minneapolis, 1997. This package is a set of revisions to *DGAUSS version 4.0*, Oxford Molecular Group, UK to incorporate the SM5.42R solvation model for DFT. For a more complete description of *DGAUSS* see Andzelm, J.; Wimmer, E. *J. Chem. Phys.* **1992**, *96*, 1280.

46. Chambers, C. C.; Hawkins, D. G.; Cramer, C. J.; Truhlar, D. G. *J. Phys. Chem.* **1996**, *100*, 16385.

47. Giesen, D. J.; Cramer, C. J.; Truhlar, D. G. *J. Org. Chem.* **1996**, *61*, 8720.

48. Hawkins, D. G.; Cramer, C. J.; Truhlar, D. G. *J. Phys. Chem B.* **1997** *101*, 7147.

49. Zhu, T.; Li, J.; Hawkins, G. D.; Cramer, C. J.; Truhlar, D. G., to be published.

50. Cramer, C. J.; Truhlar, D. G. *J. Am. Chem. Soc.* **1991**, *113*, 8305.

51. Cramer, C. J.; Truhlar, D. G. *J. Comput.-Aided Molec. Des.* **1992**, *6*, 629.

52. Giesen, D. J.; Cramer, C. J.; Truhlar, D. G. *J. Phys. Chem.* **1995**, *99*, 7137.

53. Giesen, D. J.; Hawkins, G. D.; Chambers, C. C.; Lynch, G. C.; Rossi, I.; Storer, J. W.; Liotard, D. A.; Cramer, C. J.; Truhlar, D. G. *AMSOL version 6.1.1*, University of Minnesota, Minneapolis, **1997**.

54. Cramer, C. J.; Truhlar, D. G. in *Solvent Effects and Chemical Reactivity*; Tapia, O., Bertrán, J., Eds.; Kluwer: Dordrecht, **1996**; p. 1.

55. Thiel, W.; Voityuk, A. A. *Int. J. Quantum Chem.* **1992**, *44*, 807.

56. Thiel, W.; Voityuk, A. A. *Theor. Chim. Acta* **1996**, *93*, 315.

57. Thiel, W.; Voityuk, A. A. *J. Phys. Chem.* **1996**, *100*, 616.

58. Thiel, W., private communication of Zr parameters.
59. Becke, A. D. *Phys. Rev. A.* **1988**, *38,* 3088. Perdew, J. P.; Burke, K.; Wang, Y. *Phys. Rev. B* **1996**, *54,* 16533.
60. Easton, R. E.; Giesen, D. J.; Welch, A.; Cramer, C. J.; Truhlar, D. G. *Theor. Chim. Acta* **1996**, *93*, 281.
61. Li, J.; Zhu, T.; Cramer, C. J.; Truhlar, D. G. *J. Phys. Chem.* A, **1998**, *102*, 1820.
62. Chuang, Y.-Y.; Cramer, C. J.; Truhlar, D. G. *Int. J. Quantum Chem.*, to be published.

METALS AND METALLOIDS AS CATALYSTS

Chapter 18

Modeling Transition States for Selective Catalytic Hydrogenation Paths on Transition Metal Surfaces

Matthew Neurock and Venkataraman Pallassana

Department of Chemical Engineering, University of Virginia,
Charlottesville, VA 22901

Density functional theory (DFT) quantum chemical calculations have been used to analyze the reaction coordinate for β-hydride elimination of ethyl on Pd(111), as a model for general C-H bond activation and C=C bond hydrogenation. The DFT computed activation barrier of +69 kJ/mol is comparable to the activation energy of 40-57 kJ/mol measured experimentally by Kovacs and Solymosi [1]. The role of electron-withdrawing substituents, such as -OH and –F, on the structure and energetics of adsorption and selective hydrogenation for a series of different substituted ethylene intermediates were examined in an effort to construct structure-reactivity relationships. Strong electron-withdrawing substituents were found to reduce the adsorption energy of the di-σ binding mode. These substituents were also found to raise the activation barrier for β-hydride elimination of the corresponding β-substituted-ethyl intermediates. The reaction mechanism and transition state structures for various other C-H bond activation reactions are compared. The results indicate that there is a noticeable similarity between the transition state structures for various C-H bond activation reactions. This suggests that there may be a universal mechanism that governs a series of relevant selective hydrogenation reactions.

The selective hydrogenation of multifunctional molecules is important in the synthesis of fine chemicals and pharmaceutical intermediates. The ability to selectively hydrogenate specific olefin, aldehyde or ketone moieties within a given structure could have tremendous commercial relevance. In the area of fine chemical synthesis, many oxygenated intermediates are formed via the oxidation of hydrocarbons. Catalytic oxidation reactions are not easy to control and the natural tendency is to over-oxidize. Selective hydrogenation of specific moieties is often required to produce the desired product. The production of tetrahydrofuran, for example, first involves the oxidation of butane to maleic anhydride. Maleic anhydride is subsequently hydrogenated to THF by selectively saturating individual C=C and C=O moieties. In addition to site specificity, enantiomeric selectivity is also important and often critical in the design and synthesis of new drugs. For many pharmaceutical applications, the desired compound is optically active. One enantiomer may possess the requisite properties while the other may have detrimental side-effects in clinical use. The ability to selectively synthesize the appropriate enantiomeric isomer would be of great importance. A fundamental understanding of the mechanisms for selective hydrogenation would, therefore, be invaluable in developing new strategies for the targeted synthesis of chemical intermediates.

Model studies with α,β-unsaturated aldehydes on supported metal particles have been carried out to establish the mechanism for the selective hydrogenation of the C=C and C=O bonds, and to design new heterogeneous catalysts that can selectively saturate one specific moiety [2-11]. The ability to elucidate this chemistry under ultrahigh vacuum (UHV) conditions is difficult, because hydrogen will typically desorb before it will hydrogenate. To elucidate hydrogen addition to an adsorbed olefin, most of the surface science efforts have focused on analyzing the reverse reaction of the C-H bond activation of an adsorbed surface alkyl group. Surface alkyl groups are readily generated by decomposing the alkyl-halide reactant at low temperatures on the surface. By ramping the temperature, the products can be monitored to follow the C-H bond activation path.

In this work we use first-principle quantum chemical Density Functional Theory (DFT) to probe the hydrogenation of a series of single functional moieties in an effort to understand the effect of substituents on the chemisorption and surface reactivity of these intermediates. We examine the hydrogenation of a series of substituted-ethylene molecules as well as the hydrogenation of CO. A close examination of the reaction coordinates, indicate that there is an elementary C-H bond formation step that is relevant in a number of other chemical processes.

Modeling Transition States

Up until the last decade, the application of theory to modeling surface chemistry has primarily been used in a qualitative way. Pioneering efforts by Hoffmann [12,13], van Santen [14,15], Newns [16], Anderson, Nørskov [17,18], and numerous others, have shown how theory can provide tremendous insight into understanding the general concepts of bonding and how they relate to the governing catalytic mechanisms. The technological breakthroughs in quantum chemical methods and algorithms, that have occurred over the past decade, coupled with tremendous advances in CPU hardware, are now beginning to make it possible to also extract reliable energetic information for the chemisorption of small molecules on ideal systems, such as well-defined

228

organometallic clusters, zeolites and metal surfaces. The prediction of activation barriers on heterogeneous surfaces, however, is just beginning to emerge. Leading efforts by Ziegler [19], Siegbahn [20,21], Morokuma [22,23] and Goddard [24] for well-defined organometallic systems, van Santen [25] and Sauer [26] for zeolites, and Whitten [27], Nørskov [28,29], and van Santen [15] for metals, demonstrate the applicability of first-principle calculations toward computing activation barriers. Our focus here is on the application of first-principles methods to reactions on metal surfaces.

Preliminary efforts in the theoretical study of metals demonstrated its application to the dissociation of dimer molecules such as H_2, O_2, N_2, CO, and NO. The application of theory to elucidate the chemistry of more complex surface intermediates has only just begun. In a previous work, we described the application of DFT methods to the prediction of chemisorption properties as well as the reactivity of maleic anhydride, vinyl acetate as well as a host of other commercially relevant intermediates [30,31]. A critical analysis of these intermediates, as well as a series of other model species, will enable us to establish a set of substituent effects and to develop structure-reactivity relationships. Herein, we map out the reaction coordinate for the hydrogenation of the C=C bond of ethylene, isolate the transition state for this reaction, and examine a series of substituted ethylene species to establish the effect of electron-withdrawing groups.

Computational Details

Gradient corrected density functional theory (DFT) was used to compute all the structural and energetic results reported in this paper. The Vosko, Wilk and Nusair exchange-correlation functional was used within the local density approximation (LDA) [32]. Non-local gradient corrections of Becke (for exchange) and Perdew (for correlation) were explicitly incorporated in the exchange-correlation energy within each cycle of the self-consistent-field calculations [33-35]. The single-particle wavefunctions for the many-electron system are formed by a linear combination of an atomic orbital basis set. The basis sets used in our calculations are contracted Gaussian type functions of double zeta quality. For palladium, the core electrons are described by scalar-relativistically-corrected, frozen core pseudopotentials. The frozen core pseudopotential minimizes the CPU requirements for the self-consistent-field (SCF) calculations, as well as allows for an adequate inclusion of relativistic effects for the inner shell electrons. For all calculations, the SCF energy was converged to within 5.0×10^{-5} a.u. and the geometry was considered optimized when the gradient was less than 5.0×10^{-3} a.u./Å. Additional details on the implementation of DFT in the DGauss algorithm, which was used herein, can be obtained elsewhere [36,37].

In previous work, we have shown that constrained Pd(12,7) or Pd(12,6) clusters can be used to compute adsorption energies representative of that of the Pd(111) surface [30]. We allow for the complete relaxation of the adsorbate, while constraining the metal cluster to the bulk structure. Complete relaxation of the cluster can alter the energetics of adsorption by about 20-30 kJ/mol [30]. We have also performed periodic DFT-slab calculations on some of adsorption systems and have confirmed that cluster edge effects are negligible for our Pd_{18} and Pd_{19} cluster models. [30].

The binding and adsorption energies are computed as:

$$\Delta E_{Ads} = E_{Pd(12,7)\text{-}adsorbate} - E_{Pd(12,7)} - E_{Adsorbate}$$

where, $E_{Pd(12,7)\text{-}adsorbate}$ is the total energy at the optimized ground state of the adsorbate bound to the cluster, $E_{Pd(12,7)}$ is the total energy of the bare $Pd(12,7)$ cluster and $E_{adsorbate}$ is the total energy of the optimized adsorbate alone in the gas phase. Activation barriers for chemical reactions are computed by taking the differences in total energy for the structure at the transition state and the optimized reactant state.

Finite-sized metal clusters, such as that of palladium, can have spin multiplicities different from that of the lowest energy state of the bulk system. To properly account for such unpaired electrons in finite sized metal clusters, all our calculations are spin unrestricted. The most favorable spin multiplicity for the bare $Pd(12,7)$ cluster was determined to be the triplet. The spin multiplicities for all of the adsorbate/Pd19 systems described herein remained the same as that reported for the Pd_{19} cluster. The closed shell molecular adsorbates did not change the spin multiplicity of the system upon adsorption. Chemisorption of atomic hydrogen, however, lowered the ground state spin multiplicity of the $Pd(12,7)$ cluster to a doublet. By carefully analyzing the resulting orbital spectrum we chose a series of other possible symmetry occupations to help isolate the lowest energy state. A complete analysis of all possible states on a Pd_{19} cluster would be computationally infeasible. At 19 metal atoms, however, the energy differences between different states is very small.

Results and Discussion

A. C=C bond hydrogenation on Pd(111)

1) di-σ Adsorption on Pd(111) : Effect of Primary Substituents

Theoretical density functional calculations have shown that the energetically most favorable adsorption mode for ethylene on Pd(111) is di-σ [30,38,39]. UHV experimental studies, however, have identified both π and di-σ bound ethylene on single crystal Pd surfaces [40-42].

The optimized adsorption geometry along with its corresponding binding energy for ethylene on Pd(111), in both the π and di-σ chemisorption modes are depicted in Fig 1. Both cluster and periodic slab calculations were used to model this system. The adsorption energy computed from our Pd(12,6) cluster closely matches that of the periodic slab calculations, providing additional evidence that cluster edge effects for the Pd_{18} cluster model are very minor. The predicted adsorption energy of the di-σ

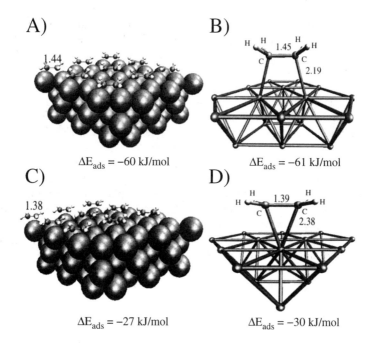

A) $\Delta E_{ads} = -60$ kJ/mol

B) $\Delta E_{ads} = -61$ kJ/mol

C) $\Delta E_{ads} = -27$ kJ/mol

D) $\Delta E_{ads} = -30$ kJ/mol

Figure 1. The di-σ and π adsorption modes of ethylene on Pd(111): Comparison of cluster and slab models for DFT.
A) Di-σ bound ethylene on Pd(111) 3 layer slab (√3 x √3 structure).
B) Ethylene adsorption on Pd(12,6) cluster in the di-σ mode.
C) π bound ethylene on Pd(111) 3 layer slab (√3 x √3 structure).
D) Ethylene adsorption on Pd(12,6) cluster in the π mode.

mode (-60 kJ/mol) is about 30 kJ/mol stronger than the π-adsorption mode in both the cluster and slab results. The DFT predicted di-σ adsorption energy for ethylene agrees well with reported experimental values [40-43]. Our results are also consistent with the theoretical values reported by Sautet [38].

The frontier orbitals for the di-σ adsorption mode show that there is electron-donation from the π orbitals of the adsorbate to the d orbitals of the metal and electron back-donation from the metal d orbitals to the π^* (LUMO) orbital of the adsorbate. This is illustrated in Fig 2. The synergy of these bonding interactions result in a strong dative bond between the adsorbate and the metal. While these interactions are also present for the π-adsorption complex of ethylene, they are weaker than that for di-σ adsorption.

To understand the correlation between adsorbate structure and chemisorption energy, we examined the effect of electron withdrawing substituents (X) on the $CH_2=CHX$ adsorption of ethylene in the di-σ mode (where X = -H, -OH and –F). The optimized geometry for each of these adsorbates on a Pd(12,7) cluster are shown in Fig 3. The vapor phase structures of the adsorbates were also optimized in order to compute the adsorption energies. The C=C bond distance for all of the vapor phase structures were between 1.34 and 1.35 Å. There is considerable elongation of the C=C bond distance upon adsorption in the di-σ mode, on account of rehybridization of the carbons from sp^2 to sp^3, as is seen in the adsorbate/cluster complexes depicted in Fig 3. Interestingly, the C=C bond distance for each of these chemisorbed intermediates is almost identical (1.45 Å) and appears to be independent of the nature of the substituent.

Since the substituents examined here do not directly interact with the surface, the changes in the chemisorption energy of the adsorbates can be attributed to the electronic interactions between the substituent and the C=C moiety. To help quantify the electron-withdrawing capability of the various substituents, charges were assigned to the substituent groups based on a simple Mulliken population analysis of the adsorbates in the vapor phase. A more rigorous accounting of charge would have likely demonstrated similar trends. The analysis of the charge on each substituent indicates that the fluorine is the most and hydrogen is the least electron withdrawing substituent. This is consistent with expectations based on the electron affinity of these species. The adsorption energies of substituted ethylene on a Pd(12,6) cluster and the Mulliken charges on the substituent are summarized in Table 1.

From Table 1, it is clear that an increase in the electron-withdrawing capability of the substituent (X) decreases the binding energy of the corresponding $CH_2=CHX$ species on Pd(111). We have been unable to find experimental adsorption energies for vinyl alcohol and fluoro-ethylene bound to Pd(111) through the C=C moiety. The reported adsorption energies for allyl alcohol (E_{ads} = -50 kJ/mol) and acrolein (E_{ads} = -50 kJ/mol) on Pd(111) are lower than that for ethylene, in spite of additional oxygen-surface interactions for these molecules [44]. The lowered binding strength through the C=C moiety in these unsaturated oxygenates is consistent with our predicted trends for electron-withdrawing groups. The observed trends in the chemisorption energy can

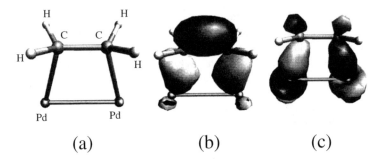

(a)   (b)   (c)

Figure 2.   Frontier orbital interactions for the di-σ adsorption mode of ethylene on Pd.
a) Optimized geometry for ethylene-palladium complex.
b) Molecular orbital corresponding to electron donation from ethylene π to metal d orbital.
c) Molecular orbital corresponding to electron back-donation from metal d orbital to ethylene π* orbital.

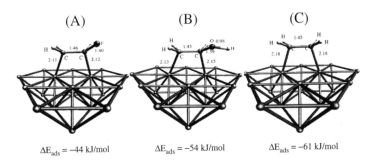

$\Delta E_{ads} = -44$ kJ/mol $\Delta E_{ads} = -54$ kJ/mol $\Delta E_{ads} = -61$ kJ/mol

Figure 3.   Non-local DFT optimized structures for different substituted ethylene molecules (CH_2=CHX) on Pd(111).
A) vinyl fluoride; B) vinyl alcohol and C) ethylene.
Reproduced with permission from reference 30.

Table 1. Effect of Direct Vinyl Substituents on the Adsorption Energy of Substituted Ethylene on Pd(111).

Adsorbate	Substituent	Charge on Substituent	Charge on the β-carbon	DFT computed Adsorption Energy kJ/mol
vinyl fluoride	-F	-0.210	+0.135	-44
vinyl alcohol	-OH	-0.029	-0.165	-54
ethylene	-H	0.098	-0.196	-61

be explained by the fact that the presence of strong electron withdrawing substituents on the carbon atom reduce the electron donation capability of the adsorbate and consequently decreases the chemisorption energy.

2) C-H Bond Activation of Substituted Ethyl on Pd(111)

The selective hydrogenation of C=C unsaturated compounds on group VIII metals, encompasses several important reactions in the chemical industry. Although each of these processes may have characteristics that are specific to the chemistry, the fundamental moiety that undergoes hydrogenation still remains the same. We, therefore, decided to examine the mechanism for hydrogenation of the C=C moiety in ethylene. However, the only fundamental data by which we can gauge our results are present for the reverse reaction, β-hydride elimination. This is due to the fact that probing hydrogenation reactions under UHV conditions is rather difficult. Instead, microscopic reversibility allows us to examine the back reaction of β-hydride elimination in an effort to understand hydrogenation. In this section, we explore the reaction coordinate for C-H bond formation in C=C bond hydrogenation through a detailed analysis of the reverse reaction of β-hydride elimination. Changes in the activation barrier and transition state geometry, due to electron-withdrawing substituents on the β-carbon atom, are also investigated.

The β-hydride elimination reaction proceeds through an agostic stretch of the C-H bond as the ethyl approaches the surface. Stretching the C-H bond lowers the energy of the σCH^* orbital of the ethyl group, allowing electron back-donation from the metal into this anti-bonding state. In addition to the C-H stretch, the C-C-Pd surface angle decreases which brings the "activated" CH_3 group closer to the surface. The combination of both of these processes act to further weaken the C-H σ bond, ultimately breaking it. The transition state for the reaction (refer figure 4) shows a long C-H bond (1.7 Å). The carbon and hydrogen of the activated C-H bond are coordinated to the central metal atom via a 3-center transition state complex. This is a metal atom insertion process. The hydrogen atom is stabilized by the neighboring metal atom, forming the 2-fold bridge site. The hydrogen atom subsequently migrates to the 3-fold hollow site, where it is most energetically stable (figure 4).

The reactant and product structures for β-hydride elimination of the different substituted ethyl ($-CH_2-CH_2X$) species (X = -F, -H and -CHO) were completely optimized on the fixed Pd(12,7) cluster. The transition state for the reaction was resolved by following the energy along a trial reaction coordinate. Since, for β-hydride elimination, the largest component of the reaction coordinate is the β C-H bond stretch, this internal mode was chosen as an approximate reaction coordinate in our transition state search procedure. The geometry of the adsorbate was optimized at a fixed C-H bond distance, to determine the lowest energy structure at that particular C-H bond stretch. This was repeated for selected points along the chosen reaction coordinate. The point of maximum energy along this trial reaction coordinate provides an approximate structure of the transition state and activation barrier for the reaction. To verify that our approximate transition state is close to the true transition state, we

(a)   (b)   (c)

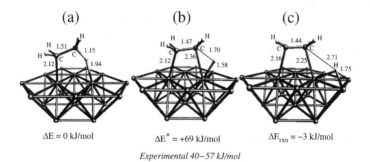

$\Delta E = 0$ kJ/mol $\Delta E^{*} = +69$ kJ/mol $\Delta E_{rxn} = -3$ kJ/mol

Experimental 40–57 kJ/mol

Figure 4. Reaction path for C-H bond activation of ethyl on Pd(111).
a) Optimized surface reactant (ethyl).
b) Transition state for β-hydride elimination of ethyl.
c) Optimized surface product (ethylene + atomic H).
Reproduced with permission from reference 30.

compute the vibrational frequencies for our predicted transition state geometry. Vibrational frequencies were computed on a Pd(7,0) fragment of the Pd(12,7) cluster so as to minimize the computational resource requirements. The frequency calculations led to the prediction of the imaginary modes that correspond to translation along the reaction coordinate. This confirmed that our optimization process brought us to within the vicinity of the true transition state for the reaction. Detailed transition state search calculations on the smaller Pd(7,0) cluster resulted in a geometry that was very similar to our "hand-optimized" transition state.

The transition state structures for the various substituted ethyl species, depicted in Fig 5, show remarkable similarity. The Pd-H bond distance at the transition state was found to be between 1.58-1.62 Å while the C-H bond distance was between 1.68-1.70 Å. This outstanding similarity between transition state structures for similar chemical reactions, if generic, presents an interesting opportunity in reaction coordinate analysis. The transition state for the reaction of one member of a homologous series can be used to estimate the structure and predict the properties of additional members of the series, without having to employ the detailed reaction coordinate analysis procedure.

As mentioned earlier, vibrational frequency calculations enabled us to confirm that we indeed isolated a saddle point on the potential energy surface. The normal mode eigenvectors corresponding to the reaction coordinate for ethyl β C-H bond breaking are shown in figure 6. The frequency corresponding to the reaction coordinate for β-hydride elimination of ethyl is -228 cm^{-1}. It is clear that the largest component to the reaction coordinate is the β C-H bond stretch. There is also a slight CH_2 bending element to the reaction coordinate, which corresponds to the change in hybridization of the β-carbon atom from sp^3 to sp^2.

The activation barriers for β-hydride elimination of the substituted-ethyl ($-CH_2-CH_2X$) species on a Pd(12,7) cluster are summarized in Table 2. The charges on the substituents (X) and the β-carbon atom are, again, based on Mulliken population analysis of the adsorbates in the vapor phase. The activation barrier for β-hydride elimination of ethyl is computed to be +69 kJ/mol. This is slightly higher than the experimental activation barrier of +40-57 kJ/mol measured by Kovacs and Solymosi for ethyl on the Pd(100) surface [1]. Since the surface metal atoms have a higher coordination number on the [111] surface, one would expect a slightly higher activation barrier for the β-hydride elimination process on this surface, which would agree more closely with our computed barrier. Temperature-programmed-desorption (TPD) studies of acrolein from the Pd(111) surface showed reaction-limited desorption of propanal at 340 K [44]. Based on this observation, the overall activation barrier for the hydrogenation reaction is estimated to be +85 kJ/mol. The DFT computed heat of reaction for C-H bond formation, in the hydrogenation of acrolein, is -7 kJ/mol [30]. If we assume that the C-H bond formation step is rate limiting, which is typically true for such reactions, then the DFT predicted activation barrier of +82 kJ/mol, for hydrogenation, is in good agreement with experiment [30,44].

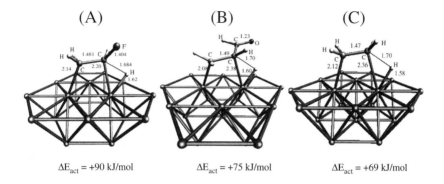

(A)   (B)   (C)

ΔE_{act} = +90 kJ/mol ΔE_{act} = +75 kJ/mol ΔE_{act} = +69 kJ/mol

Figure 5. Non-local DFT optimized transition state structures for β-hydride elimination of substituted ethyl (-CH$_2$-CH$_2$X) on Pd(111). A) vinyl fluoride; B) acrolein and C) ethylene.

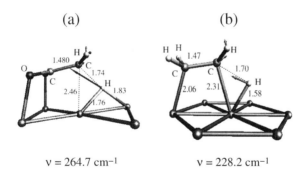

(a)   (b)

ν = 264.7 cm^{-1} ν = 228.2 cm^{-1}

Figure 6. Normal mode eigenvectors corresponding to the reaction coordinate for β-hydride elimination.
a) acetate and b) ethylene.
Reproduced with permission from reference 30.

Table 2. Substituent Effects on β-hydride Elimination of β-substituted Ethyl on Pd(111).

Adsorbate	Substituent	Charge on Substituent	Charge on β-carbon	DFT computed Activation Energy kJ/mol	Experimental Activation Energy kJ/mol
vinyl fluoride	-F	-0.210	+0.135	+90	-
acrolein	-CHO	-0.108	-0.165	+75	+77*
ethylene	-H	0.098	-0.196	+69	+40 to +57§

§ from [1]
*estimated from TPD data of Davis and Barteau [44]

It is evident, from Table 2, that the presence of electron-withdrawing substituents raise the activation barrier for the β-hydride elimination reaction. Strong electron withdrawing groups decrease the electron density on the β-carbon atom, making the process of β-hydride elimination (i.e. $H^{\delta+}$ transfer to metal) more difficult. This is in consonance with the conclusions of Gellman and co-workers, who observed that substitution of fluorine at the γ-carbon of the propyl group increases the barrier for β-hydride elimination over copper, and associated it with the decreased electron density on the β-carbon atom [45]. They found that the barrier increased by 20-30 kJ/mol per fluorine substituent. If we neglect the differences between Cu and Pd, this is in general agreement with our results where there is a 20 kJ/mol increase in the addition of a single fluorine substituent.

B. Comparison of Transition States for Different C-H Bond Activation Reactions

In the previous section, we demonstrated the noticeable similarities between the transition state structures for β-hydride elimination of ethyl species, with different β-carbon substituents. The trends in the activation barriers were also rationalized based on the electron-withdrawing nature of the substituent groups. Comparison of the β-hydride elimination reaction with other C-H bond activation reactions reported in the literature, such as methane activation on Ni, or acetate C-H bond activation, show outstanding congruence in the reaction pathway and structure of the transition state. To test this assumption and illustrate the resemblance, the transition state structures and activation barriers for different C-H bond activation reactions are compared in this section. For multifunctional adsorbates with C=C and C=O moieties, analyzing the differences between C=C and C=O hydrogenation mechanisms is crucial in elucidating the key factors that control hydrogenation selectivity. In an initial effort to understand the differences in hydrogenating these moieties, we examine the mechanism for CO hydrogenation and compare it with that for C=C hydrogenation.

DFT optimized transition state structures for the C-H bond activation processes of additional hydrocarbon molecules are depicted in Fig 7. C-H bond activation of acetate is believed to be an important precursor in its decomposition to CO_2 on metal surfaces such as Pd(111). By performing detailed reaction coordinate calculations, we have isolated the reaction pathway for this C-H bond activation process [30,31]. The activation barrier for the C-H bond breaking reaction is computed to be about +115 kJ/mol, which is considerably higher than that of ethyl on Pd(111) at +69 kJ/mol. The acetate group is initially bound perpendicular to the Pd(111) surface in a di-σ conformation. These results are consistent with experimental evidence from HREELS measurements [46]. The reaction pathway first involves the tilting of the acetate group away from the surface normal and in a direction perpendicular to the initial plane of the COO group. When the terminal CH_3 group is close enough to the surface, the mechanism is quite similar to ethyl C-H bond activation. The reaction coordinate again involves a C-H bond stretch coupled with the bending mode to bring the CH_3 group toward the surface [30,31]. The transition state is late along the C-H bond stretch coordinate. The C-H bond distance at the transition state (1.76 Å) is slightly longer than that found for ethyl (1.7 Å) on Pd(111). The dissociating hydrogen

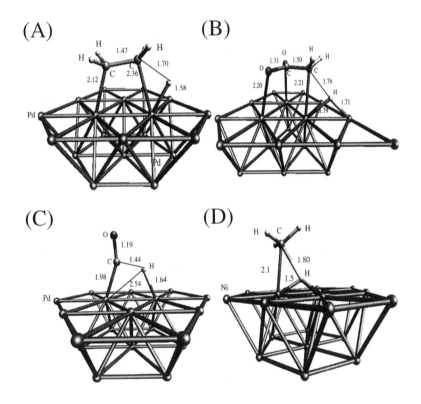

Figure 7.   Comparison of transition states for different C-H bond activation
reactions.
A) ethyl; B) acetate; C) formyl and D) methane.

atom is observed to be close to the 2-fold bridge site, similar to that for ethyl, at the transition state. If the CO_2^- group is viewed as an electron withdrawing substituent, it is easily envisioned why the C-H bond activation process has higher activation energy for acetate, as compared to ethyl. There is also an additional energy expense associated with the significant tilting of the acetate species away from the surface normal. A notable distortion about the O atoms occurs to tilt the acetate species towards the surface. In addition, there is a repulsive interaction associated with bringing a methyl group towards the surface. This repulsive interaction remains high until there is enough energy for electrons from the surface to be donated to the σ_{CH}^* orbital. The predicted activation barrier of +115 kJ/mol for C-H bond activation of acetate is a little higher than the overall barrier of +85 kJ/mol measured by Davis and Barteau for the decomposition of acetate [46].

C-H bond activation of methane is another example where the structure of the transition state is likely to be similar to that of ethyl and acetate. Methane activation over supported Ni particles has received considerable attention, due to its relevance in steam reforming. The reaction has been studied extensively on single metal atoms [47-49], well-defined clusters [50-53] and extended slabs [54]. The transition state for the reaction is shown in figure 7. The reaction pathway for C-H bond activation of methane on Ni is very similar to that of ethyl on Pd(111) and involves metal insertion into the C-H bond. The transition state is late along the C-H stretch coordinate and occurs at a C-H bond distance of 1.4-1.8 Å. The metal-H bonding is also evident at the transition state with metal-H bond distances of 1.7-1.8 Å. The C-H bond activation barrier for ethane or methane (+121 kJ/mol) [54] is considerably higher than that of ethyl on Pd(111). This is possibly due to the additional stability provided by the CH_2 moiety, that anchors the ethyl species to the surface [30].

The hydrogenation of CO to form oxygenate intermediates occurs readily over supported Pd clusters. The initial step of CO hydorgenation to form the surface formyl intermediate has been speculated to be the rate determining in this chemistry. The reaction path for CO hydrogenation involves a CO and hydride migration. Surface CO and atomic hydrogen react either over one or two metal atom centers. The preferred path is shown in Fig 7C. A closer analysis of this reaction indicates that the reverse reaction involves metal insertion into the CH bond. The mobility of the CO intermediate enables the HCO complex to easily shift to a di-σ like intermediate bridge complex where the CO group is stabilized by interacting with one metal atom center while the atomic hydrogen can stabilize by interacting with an adjacent metal atom. This helps to lower the barrier. The structure itself is quite similar, but now involves a four-center transition state where both metal atoms contribute to lowering the barrier. Despite these minor changes the mechanism still follows the basic pattern of an agostic C-H bond stretch and the rotation of the hydrocarbon intermediate species toward the surface to help stabilize the C-H bond activation.

242

Conclusions

The selective hydrogenation of C=C and C=O unsaturated adsorbates finds numerous applications in the fine-chemicals synthesis industry. Understanding the mechanistic pathways for selective hydrogenation is crucial in the design of effective catalysts for specific applications. In this paper, we have detailed the reaction pathway for the β-hydride elimination of ethyl, which is the microscopic reverse reaction of ethylene hydrogenation. The mechanism involves a coupled C-H stretch and a bending of the C-C-Pd angle. The computed activation barrier of +69 kJ/mol for the β-hydride elimination reaction on Pd(111) is comparable to the experimentally measured reaction barrier of 40-57 kJ/mol [1]. By studying the effect of substituents on the C=C bond adsorption and ethyl β-hydride elimination, we have speculated a generalized fragment based approach in analyzing activation barriers for geometrically similar reactions. In general, strongly electron-withdrawing substituents tend to raise the activation barrier for β-hydride elimination of β-substituted ethyl species. For instance, the activation barrier for β-hydride elimination of $-CH_2-CH_2F$ species on Pd(111) is about 20 kJ/mol higher than that for ethyl. Electron-withdrawing groups also lower the energy of adsorption through the C=C bond for substituted-ethylene. For example, the adsorption energy of vinyl fluoride is about 15 kJ/mol lower than that of ethylene. Finally, we have compared the reaction mechanism and transition state structure for ethyl β-hydride elimination with other C-H bond activation reactions, reported previously in the literature. Comparison of transition states for different kinds of C-H bond formation reactions, brings out the inherent similarities in the reaction mechanism and the structure of the transition state. The mechanism for all the C-H bond activation reactions show metal insertion into the C-H bond and a late transition state along the C-H bond stretch, similar to ethyl β-hydride elimination. These observations seem to suggest an intercomparable mechanism governing a large number of C-H bond activation processes relevant to the chemical process industry.

Acknowledgments

The authors would like to thank Prof. Jens Nørskov, Bjørk Hammer and Lars Hansen from the Center for Atomic Scale Materials Physics, Technical University of Denmark, for help with the use of their plane wave pseudopotential program and Dr.George Coulston (DuPont Chemical Company) for helpful discussions. The DuPont Chemical Company and NSF (Career Award CTS-9702762) are also acknowledged for financial support.

References

1.   Kovacs, I. and F. Solymosi, *J. Phys. Chem.*,**97**: p. 11056-11063,**1993**
2.   Sen, B. and M.A. Vannice, *J. Catal.*,**113**: p. 59,**1988**
3.   Makouangou, R.M., et al., *Ind. Eng. Chem. Res.*,**33**: p. 1881,**1994**
4.   Vannice, M.A. and B. Sen, *J. Catal.*,**115**: p. 65,**1989**
5.   Beccat, P., et al., *J. Catal.*,**126**: p. 451,**1990**
6.   Simonik, a.J.C.B., *Coll. Czechos. Chem. Comm.*,**37**: p. 353,**1972**

7.   Birchem, C.M.P., Y. Berthier, and G. Cordier, *J. Catal.*,**146**: p. 503,**1994**
8.   Satagopan, V. and S.B. Chandalia, *J. Chem. Tech. Biotechnol.*,**59**: p. 257,**1994**
9.   Yoshikawa, K.a.Y.I., *J. Mol. Catal. : A Chem.*,**100**,**1995**
10.   Jenck, J. and J.E. Germain, *J. Catal.*,**65**: p. 141,**1980**
11.   Paseka, I., *J. Catal.*,**121**: p. 349,**1990**
12.   Hoffmann, R., *Ang. Chem. Ent. Ed. Eng.*,**21**: p. 711,**1982**
13.   Hoffmann, R., *Solids and Surfaces, A chemist's view of bonding in extended surfaces*, **1988**: VCH.
14.   van Santen, R.A., *Theoretical Heterogeneous Catalysis*. World Scientific Lecture Notes in Chemistry. Vol. 5. **1991**: World Scientific Publishing Company, Pvt. Ltd.
15.   van Santen, R.A. and M. Neurock, *Catal. Rev.*,**37**(4): p. 557,**1995**
16.   Newns, D.M., *Phys. Rev. B.*,**178**: p. 1123,**1969**
17.   Norskov, J. and B.I. Lundqvist, *Phys. Rev. B.*,**1979**
18.   Norskov, J., et al., *Phys. Rev. Lett.*,**46**: p. 257
19.   Ziegler, T., *Chem. Rev.*,**91**: p. 651-667,**1991**
20.   Siegbahn, P.E.M., *J. Phys. Chem.*,**97**: p. 9096-9102,**1993**
21.   Carroll, J.J., et al., *J. Phys. Chem.*,**99**: p. 14388-14396,**1995**
22.   Cui, Q. and K. Morokuma, *J. Chem. Phys.*,**108**(10): p. 4021-4030,**1998**
23.   Froese, R.D.J., et al., *J. Am. Chem. Soc.*,**119**(31): p. 7190-7196,**1997**
24.   Low, J.J. and W.A.Goddard III, *J. Am. Chem. Soc.*,**106**: p. 8321-8322,**1984**
25.   van Santen, R.A. and G.J. Kramer, *Chem. Rev.*,**95**: p. 637-660,**1995**
26.   Sauer, J., in *Modeling of Structure and Reactivity in Zeolites*, C.R.A. Catlow, Editor, **1992**, Academic Press, New York.
27.   Yang, H., et al., *Surf. Sci.*,**277**: p. L95.**1992**
28.   Hammer, B. and J.K. Norskov, *Nature*,**376**: p. 238,**1995**
29.   Hammer, B. and J.K. Norskov, *Theory of Adsorption and Surface Reactions*, in *Chemisorption and Reactivity on Supported Clusters and Thin Films*, M.L. R and G. Pacchioni, Editors. **1997**, Kluwer Academic Publishers, Netherlands. p. 285-351.
30.   Neurock, M., *Catalytic Surface Reaction Pathways and Energetics from First principles*. in *Studies in Surface Science and Catalysis*. **1997**: Elsevier Science.
31.   Neurock, M., *Applied Catalysis A: General*,**160**: p. 169-184,**1997**
32.   Vosko, S.J., L. Wilk, and M. Nusair, *Can. J. Phys.*,**58**: p. 1200-1211,**1980**
33.   Becke, A.D., *Phys. Rev. A*,**38**: p. 3098,**1988**
34.   Becke, A., *ACS Symp. Series*,**394**: p. 165,**1989**
35.   Perdew, J.P., *Phys. Rev. B*,**33**: p. 8822,**1986**
36.   Andzelm, J., *DGauss: Density Functional - Gaussian approach. Implementation and Applications*, in *Density Functional Methods in Chemistry*, J. Labanowski and J. Andzelm, Editors. **1991**, Springer-Verlag New York, Inc. p. 155-169.
37.   Andzelm, J. and E. Wimmer, *J. Chem. Phys.*,**96**(2): p. 1280-1303,**1992**
38.   Sautet, P. and J.F. Paul, *Catal. Lett.*,**9**: p. 245,**1991**
39.   Maurice, V. and C. Minot, *J. Phys. Chem.*,**94**: p. 8579,**1990**
40.   Stuve, E.M. and R.J. Madix, *J. Phys. Chem.*,**89**: p. 105-112,**1985**
41.   Nishijima, M., et al., *J. Chem. Phys.*,**90**(9): p. 5114-5127,**1989**

244

42.	Gates, J.A. and L.L. Kesmodel, *Surface Science*,**120**: p. L461-L467,**1982**
43.	Ratajczykowa, I. and I. Szymerska, *Chem. Phys. Lett.*,**96**(2): p. 243-246,**1983**
44.	Davis, J.L. and M.A. Barteau, *Journal of Molecular Catalysis*,**77**: p.109-124,**1992**
45.	Forbes, J.G. and A.J. Gellman, *J. Am. Chem. Soc.*,**115**: p. 6277,**1993**
46.	Davis, J.L. and M.A. Barteau, *Langmuir*,**256**: p. 50,**1989**
47.	Siegbahn, P.E., M.R.A. Blomberg, and M. Svensson, *J. Am. Chem. Soc.*,**115**: p. 4191,**1993**
48.	Low, J.J. and W.A.Goddard III, *J. Am. Chem. Soc.*,**106**: p. 8321,**1984**
49.	Swang, O., K. Faegri, and O. Gropen, *J. Phys. Chem.*,**98**: p. 3006,**1994**
50.	Burghgraef, H., A.P.J. Jansen, and R.A.v. Santen, *Chem. Phys.*,**177**: p. 407,**1993**
51.	Burghgraef, H., A.P.J. Jansen, and R.A. van Santen, *J. Chem. Phys.*,**101**: p. 11012,**1994**
52.	Burghgraef, H., A.P.J. Jansen, and R.A.van Santen, *Surf. Sci.*,**1995**
53.	Yang, H. and J.L. Whitten, *J. Am. Chem. Soc.*,**113**: p. 6442, **1991**
54.	Kratzer, P., B. Hammer, and J.K. Norskov, *J. Chem. Phys. (submitted)*,**1996**

Chapter 19

Dissociation of N_2, NO, and CO on Transition Metal Surfaces

M. Mavrikakis[1], L. B. Hansen[1], J. J. Mortensen[1], B. Hammer[2], and J. K. Nørskov[1]

[1]Center for Atomic-Scale Materials Physics, Department of Physics, Technical University of Denmark, DK-2800 Lyngby, Denmark
[2]Institute of Physics, Aalborg University, DK-9220 Aalborg, Denmark

Using density functional theory we study the dissociation of N_2, NO, and CO on transition metal surfaces. We discuss an efficient method to locate the minimum energy path and the transition state, and review recent calculations using this method to determine the transition state for dissociation of N_2 on Ru(0001) and NO on Pd(111), Pd(211), and Rh(111) surfaces. We also show how steps and adsorbed alkali metal atoms can significantly decrease the dissociation barrier. Finally, trends in the properties of the transition state for N_2, NO and CO dissociation on transition metals are discussed in some detail.

1 Introduction

Density functional theory (DFT) with a non-local description of exchange and correlation has reached a level of efficiency and accuracy that it can be applied meaningfully to the study of chemical reactions on metal surfaces [1, 2]. Current computational demands of DFT calculations allow for a realistic modeling of the surface, typically by a large cluster or an extended slab [3]. The accuracy is still not comparable to thermal energies, but has reached a point where reaction mechanisms can be determined, and where variations in binding energies and activation energies from one system to the next can be calculated with reasonable accuracy. DFT calculations are therefore developing into an important tool for improving our understanding of surface reactions, and in conjunction with experiment for designing new surfaces with improved catalytic properties [4].

Reactions at surfaces often involve a combination of bond breaking and bond making events leading to changes in the configuration of the adsorbed molecules and displacements of the surface atoms. Therefore, finding the transition state separating reactants from products usually involves searching for a saddle point in a high-dimensional space. In spite of the large progress in computational speed, the DFT method is still computationally demanding. As a result, developing efficient methods for finding transition states is a very important goal. Having determined the transition state of a reaction path, it is equally important to extract from the calculation the key characteristics of the reaction to build general models which could be used for making predictions regarding other similar reactions, instead of having to perform large scale calculations for each new system of interest.

In the present contribution, we review some recent progress made in understanding the dissociation of three simple diatomic molecules (CO, N_2 and NO) on transition metal surfaces. There are two main parts of the paper; in the first part, we give some details of the DFT calculations and the method we apply to locate the transition state for a surface reaction. In the second part, we further elaborate on the interpretation of the geometry and energetics of the activated complexes.

2 Methods applied for determining the transition state of surface reactions

2.1 DFT calculations

In the present work density-functional theory is used for the calculation of total energies [5]. Metallic surfaces are described using the slab approach. Typically between three and nine layers of metal are repeated periodically in a super cell geometry with about 10 Å of vacuum between the slabs. We consider the close packed (111) surface of Pd and Rh and the similar (0001) surface of Ru. To discuss the effect of steps on surface reactivity we also consider the Pd(211) surface which consists of (111) terraces and (100) steps. Calculations for the Pd(211) surface are done in a (2×1) surface unit cell with two metal atoms per cell along the step edge. Calculations for the (111) and (0001) surfaces are done in (2x2), (2x3), (2x4), or $\begin{pmatrix} 2 & -2 \\ 0 & 3 \end{pmatrix}$ surface unit cells. For each system, the adsorption is allowed on only one of the two surfaces exposed and the electrostatic potential is adjusted accordingly [6]. The top one to two (for the fcc(111) and hcp(0001)) or four (for the fcc(211)) surface layers are relaxed unless otherwise stated.

The ionic cores are described by soft (for: Pd and Ru) or ultra-soft (for Rh) pseudopotentials [7, 8] and the Kohn-Sham one-electron valence states are expanded in a basis of plane waves with kinetic energies below 40 Ry (soft) or 30 Ry (ultra-soft). The exchange-correlation energy and potential are described by the generalized gradient approximation (GGA-II) expression, using the PW91 functional self-consistently [9, 10]. The self-consistent GGA-II density is determined by iterative diagonalization of the Kohn-Sham Hamiltonian, Fermi-population of the Kohn-Sham states ($k_B T$=0.1 eV), and Pulay mixing of the resulting electronic density [11]. All total energies have been extrapolated to $k_B T = 0$ eV, following the procedure described in [12].

2.2 Minimum Energy reaction path and the Transition State

Having calculated the binding energies for adsorption of reactants and products using total energy calculations as described above, one can then proceed with the second part of the calculation focusing on the determination of the minimum energy reaction path. Any chemical reaction can be viewed as a transition from one minimum (the reactants) to another minimum (the products) on the potential energy surface describing the energy of all the atoms in the system as a function of their coordinates. In the following we denote a point in this configuration space by a vector containing all the coordinates of the M atoms in the system:

$$\vec{\Gamma} \;=\; (\vec{R}_1, \vec{R}_2, ..., \vec{R}_M) \tag{1}$$
$$=\; (x_1, y_1, z_1, x_2, y_2, z_2, ..., x_M, y_M, z_M). \tag{2}$$

Every point along the minimum energy path has the property of being the lowest energy point in all degrees of freedom except for the one along the path. The highest-energy point along the minimum energy path represents the classical barrier height for the transformation described by that path. It is in general difficult to define algorithms for finding the minimum energy path [13, 14, 15]. The energy must be minimized relaxing all coordinates of the problem except one, the local direction of the path. For our purposes, we use a method proposed by Ulitsky and Elber [16] which can be described as a simplified version of the more general Nudged Elastic Band method of Mills, Jónsson and Schenter [13].

According to this method, an initial guess for the path has to be made first. We make such an initial guess for the reaction path by linearly interpolating between the initial and the final state ($\vec{\Gamma}_I$ and $\vec{\Gamma}_F$), as it is illustrated for the case of N_2 dissociation on Ru(0001) in Fig. 1. The figure shows the (z, b) plane spanned by the height z of the N_2 molecule above the Ru surface and the N–N bond length b. The two end points represent the initial molecularly adsorbed state of N_2 (I) and the final atomically chemisorbed state (F), which is closer to the surface and has a much longer N–N "bond" length. The straight line connecting the two end points represents the initial guess which we use to start our calculations. It is important to notice that in our calculations not only the degrees of freedom described by z and b, but all six N_2 coordinates and all degrees of freedom of the metal atoms in the top two layers of the slab are considered. Although the path found with this method slightly depends on the choice of coordinates, stationary points along the path, such as the transition state, are independent of such choices.

In the implementation of this method, only a finite number of points (images) along the path are used. The richer the potential energy landscape being probed, the larger the number of images one has to include in the calculation. The

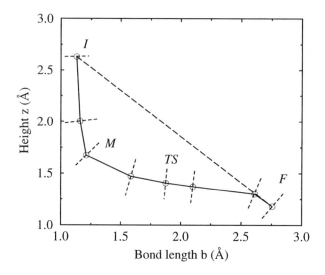

Figure 1: Illustration of the method for finding the minimum energy reaction path. Both the initial guess for the minimum energy reaction path and the final minimum energy reaction path for N_2 dissociation on Ru(0001), as determined by using the Ulitsky-Elber [16] method, are shown. z represents the height of the center of mass of the molecule over the Ru surface plane, whereas b represents the N–N bond length. The initial guess for the path is shown by the dashed line connecting the initial (I) and final (F) state. Minimizing the energy perpendicularly to the local path, leads to the progressive development of the minimum energy path (solid line). The eight small circles correspond to the eight images used for this calculation. M and TS denote a metastable and the transition state respectively (see also Fig. 2). The constraint is defined locally as the hyperplane perpendicular to the line connecting the two neighboring points on the path, and is updated consistently as the calculation evolves. The hyperplanes for the final path are sketched at the images used. [Figure adapted from ref. [17], copyright: Academic Press].

coordinates of each image are first relaxed towards lower energy within the hyperplane normal to the path. This is accomplished by moving each image only along the force perpendicular to the path. Often it is necessary to add new images along the path; this can be done at any time during the course of the calculation. For the dissociation of N_2 on Ru(0001), as shown in Fig. 1, a total of eight such images were necessary to capture the main configurations along the reaction pathway.

To make sure that the minimum energy path is found, the hyper-plane constraints have to be updated as the path evolves. In general, vector $\vec{r}_{n+1} - \vec{r}_{n-1}$ is defined as the normal to the hyper-plane passing through the image number n. As a result, the local constraint corresponding to each image is updated under the influence of the motion of the path at its neighboring images, and consequently the local constraint is the outcome of the motion of the whole path.

The final minimum energy path, as shown in Fig. 1, is remarkably different from the initial guess for the path. Notice that the final constraints (shown dashed) are not perpendicular to the initial guess for the path. Overall, this method has been proven to be very robust even for treating rather complicated reaction paths, where simple intuitive approaches usually fail [13].

Within the above framework of describing a reaction path, it is natural to define a reaction coordinate s as the norm of the total change in all coordinates along the path, starting at the initial state $\vec{\Gamma}_I$:

$$s = \int_{\vec{\Gamma}_I}^{\vec{\Gamma}} |d\vec{\Gamma}|. \tag{3}$$

For a finite number of images along the path one can use the discretized version of Eq.3:

$$s_n = \sum_{i=1}^{n-1} |\vec{\Gamma}_{i+1} - \vec{\Gamma}_i|. \tag{4}$$

In concluding this section, we should emphasize that the method described here is unbiased, since knowledge of the initial (reactant) and final (product) states of the reaction is the only input required for the calculation. Other advantages of this method include the fact that only one constraint is imposed during relaxation and that relaxations of the substrate can be easily taken into account.

3 Properties of the transition state

We now discuss some examples where this methodology has been followed to determine the transition state for dissociation of both homonuclear and heteronuclear diatomic molecules. We first consider the dissociation of N_2 and NO in some detail. Then, we attempt to extract some general common characteristics emerging from these calculations and compare these with results pertaining to CO dissociation.

3.1 N_2 dissociation

Figure 2 shows the calculated energy of the $N_2/Ru(0001)$ system, relative to molecular N_2 and the clean $Ru(0001)$ surface, as a function of the dissociation reaction coordinate. Dissociation of N_2 on $Ru(0001)$ is calculated to have a significant barrier [17], which is in good qualitative agreement with recent experiments [18, 19].

The picture emerging from Fig. 2 is the following. As the molecule approaches the surface, it is first attracted towards a molecularly adsorbed state (I). This molecular state is perpendicular to the surface. Further along the reaction coordinate, the molecule follows a trajectory of sliding on the surface and simultaneously

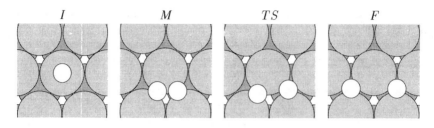

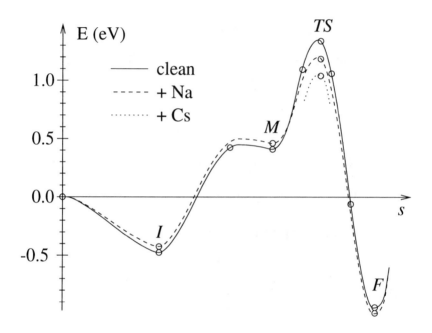

Figure 2: Top: Snapshots along the minimum energy path for N_2 dissociation on a clean Ru(0001) surface. Successive images shown here represent the initial state (I), a metastable state (M), the transition state (TS), and the final state (F) of the dissociation process. Bottom: The energy (E) along the minimum energy path is plotted as a function of the reaction coordinate (S). Energy is referenced to the sum of the total energies of the corresponding (clean or promoted) surface and gas-phase N_2 molecule (see refs. [17, 20] for details).

bending towards the surface until a new metastable molecularly adsorbed state is found (M). Continuing along the reaction coordinate, the molecule becomes increasingly stretched whereas the energy of the system keeps increasing until the transition state (TS) is reached. The transition state is characterized by a very stretched "molecule". At the transition state, one of the N atoms is already close to its final position in a three-fold site, whereas the other N atom is coordinated to two ruthenium atoms, nearly in a "bridge-like" configuration. Finally, the two N atoms end up in hcp sites. The energy corresponding to the final configuration of the dissociation products, as shown in Fig. 2, is higher than the lowest possible energy state of the two N atoms, simply because they are still close enough to strongly repel each other (they "share" one Ru atom). Subsequent diffusion of the N atoms to adjacent hcp sites can further decrease the energy of the system to -1.54 eV.

The effect of coadsorbing Na or Cs in the vicinity of the dissociating N_2 molecule has also been included in Fig. 2 [20]. Both Na and Cs clearly decrease the barrier for N_2 dissociation on Ru(0001), in good agreement with experiments suggesting that alkalis promote ammonia synthesis over Ru catalysts and that Cs is the best promoter among the two [21].

3.2 NO dissociation

The dissociation of NO on Pd(111), Pd(211), and Rh(111) [22, 23] is considered next. Figure 3 summarizes the results we got by applying the method described in section 2.2 in order to determine the transition state corresponding to each one of these three systems. Data shown for NO/Rh(111) represent calculations [23] where only the adsorbates' degrees of freedom were allowed to relax, whereas for both NO/Pd(111) and NO/Pd(211) surface relaxation effects have been included [22]. Since the main focus of this contribution is the location and characteristics of the transition state, Fig. 3 does not show the whole reaction path, which can be found elsewhere [22, 23], but simply illustrates initial, final, and transition states of the corresponding minimum energy paths.

NO dissociation on Pd(211) was studied for four alternative final states, whereas the initial molecularly adsorbed state was kept fixed at the most preferred three-fold site near the step-edge. We found that the path involving NO dissociation parallel to the step, characterized by the same local geometry as on the (111) surface, has the lowest barrier, whereas paths involving NO dissociation via molecular configurations which are perpendicular to the step have a considerably higher barrier [22].

At low coverages, NO prefers to occupy the fcc site on a Pd(111) surface, with a calculated binding energy of the upright molecular state of 2.24 eV. The most energetically favorable final state for NO dissociation on Pd(111) has both N and O atoms on adjacent fcc sites. Our calculations predict that NO dissociation is much more facile at the step than on the flat (111) terrace, in excellent agreement with experiment [24, 25].

Finally, NO preferentially adsorbs on hcp sites of Rh(111) at low coverages, with a calculated binding energy of ca. 2.38 eV. The final state of the minimum energy NO dissociation path on Rh(111) finds both N and O atoms in adjacent hcp sites. The barrier to dissociation calculated without surface relaxation is considerably lower than the barrier on either Pd surfaces, which is in good agreement

252

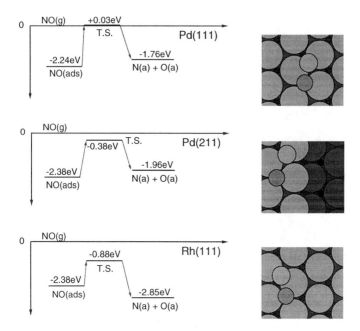

Figure 3: Energies of the molecularly adsorbed state, the transition state and the final dissociated state for NO dissociation on Pd(111), Pd(211), and Rh(111). Energies are referenced to the sum of the total energies of the corresponding clean surface and a gas-phase NO molecule. The geometry of the transition state is shown to the right, in a top view of the surface. Each panel spans more than one unit cells, and adatoms corresponding to neighboring unit cells have been removed for clarity. Two different shades are used for surface atoms in the Pd(211) panel to indicate the two terraces defining the step. Small dark and light circles represent the N and O adatoms respectively. Notice the nearly "center-bridge" configuration of the transition state. Adatoms in the final dissociated state of NO prefer hcp sites on Rh(111), but fcc sites on Pd(111) and Pd(211).

with available experimental data [26, 27]. The low barrier for NO dissociation is among the main reasons for rhodium being used for the catalytic reduction of NO in automobile catalytic converters [28]. As shown in Fig. 3, the dissociation channel is the main path, since the alternative path to the gas phase has a much higher barrier (1.5 eV vs 2.38 eV).

4 Discussion

In this section we further discuss similarities and differences characterizing the dissociation barriers and geometries of the corresponding transition states for small diatomic molecules.

From the calculations presented in the previous section for both N_2 on Ru and NO on Pd and Rh surfaces, one can easily conclude that the geometry of the transition state to dissociation is very similar for both molecules and on all three metals examined. In all cases studied here, the transition state involves one adatom close to a three-fold site (center), and the second adatom close to a two-fold site (bridge). Even the stepped Pd(211) surface gives a transition state with similar geometry. We have found that these similarities extend to the case of CO dissociation on the close packed surfaces of Ni, Pt and Ru as well [29, 30]. Furthermore, work on H_2 dissociation on Ni(111) [15] suggests a similar geometry for the transition state (center-bridge configuration), although not exactly the same with the configuration we are calculating for NO, CO, and N_2 dissociation. However, the small difference observed in the transition state geometry can be attributed to the corresponding difference in the final dissociated state having the two H-atoms closer to each other, compared to the dissociation products of NO, CO and N_2. This is a direct consequence of the smaller repulsion between two hydrogen atoms, compared to the repulsion between N, C and O atoms. Presumably, the center-bridge geometry of the transition state is a general result characterizing a hexagonal local geometry of the surface, whereas the transition state geometry could be substantially different for the dissociation of the same small diatomics on other facets of the same metals (see [3], for an example on a (100) surface).

In particular, one can view the transition state configuration as a result of a compromise between the energy cost of stretching the molecule and the energy gain obtained by the adsorption of the reaction products. The cost of the molecular stretching will dominate at small intra-molecular bond lengths, but as the bond becomes longer this term saturates and the adsorption energy of the atomic components of the almost dissociated molecule takes over. The atomic adsorption energy term can be expected to be more attractive, the more surface atoms the transition state complex is coordinated to. At the same time, one should take into account that "sharing" of surface atoms leads to adsorbate-adsorbate repulsion through the covalent interaction of the adsorbates with the surface d-bands on the "shared" surface atoms as discussed in Ref. [17]. With this distribution of energy, the transition state geometry reflects the balance between keeping the bond length small, while allowing the dissociating atoms (or radicals for larger molecules) to interact with as many surface atoms as possible. A simple consequence of this picture is that molecules similar in size and composition experience similar transition states on transition metal surfaces with similar local geometries.

In all cases studied here, and from the results of similar theoretical studies performed by others (see for example: NO dissociation on Cu(111) [31]), one can

254

Table 1: Calculated values for the total energy of the transition state (E_{TS}), the difference between E_{TS} and the total energy of the molecularly adsorbed state (E_{MA}), i.e. the barrier to dissociation, and the difference between E_{TS} and the total energy of the final dissociated state (E_F). E_{TS} is referenced to the sum of the total energies of the respective clean surface and the gas-phase molecule. Calculations for the NO/Pd(111) and NO/Pd(211) systems included surface relaxations [22]; the rest of the data correspond to calculations with the metal atoms fixed at their bulk truncated positions [23, 29].

System	E_{TS} (eV)	E_{TS}-E_{MA} (eV)	E_{TS}-E_F (eV)
CO/Ni(111)	+1.40	+2.80	+1.80
CO/Pt(111)	+2.90	+4.40	+2.00
NO/Pd(111)	+0.03	+2.27	+1.79
NO/Pd(211)	-0.38	+2.00	+1.58
NO/Rh(111)	-0.88	+1.50	+1.97

conclude that the transition state to dissociation of NO, CO and N$_2$ on transition and noble metals is much more final-state-like than initial-state-like. In other words, it is more reasonable to think of the transition state as being two atomic adsorbates interacting repulsively with each other rather than as a stretched molecule. This picture is further supported by the data collected in Table 1, where it is clear that the variation in the last column is much smaller than the variation in the middle column, suggesting that the difference between the energy of the transition state and the energy of the initial state (E_{TS}-E_{MA}) varies much less than the corresponding difference between the energy of the transition and the final state (E_{TS}-E_F). Therefore, the energy of the transition state follows the energy of the final state much more closely than it does with the energy of the initial state. Moreover, the large variations observed in the transition state energy from one metal to the next, and from one facet to the next for the same metal, can be attributed mainly to variations in the adsorption energies of the chemisorbed atoms alone. This conclusion certainly holds for both NO and CO dissociation over different metals and facets, for which we have performed detailed calculations [29, 30].

The final question arising then, is what determines the variation in the atomic chemisorption energies. Based on the above arguments, one can expect that factors determining atomic chemisorption bond strength will largely determine variations in the transition state energies and thus dictate trends in the activation energy for dissociation. We have suggested that these variations are determined from the interaction of the adsorbate valence states with the metal d-states [30, 32, 33, 34, 35]. There are two components to this interaction, and three resulting rules of thumb dictating the final atomic chemisorption bond strength:

1. A Pauli repulsion proportional to the overlap matrix element squared (V_{ad}^2) between the adsorbate state(s) and the metal-d states (rule-I).

2. An attraction due to the formation of bonding (and anti-bonding) states between the adsorbate and metal-d states. The magnitude of this term is determined by a) the overlap matrix element V_{ad}, b) the position of the adsorbate state(s) and the d states relative to the Fermi level, and c) the filling of the bonding and anti-bonding states. The general rule is that the closer the d states are to the Fermi level, the stronger the interaction (rule-II)); also, the fewer the d electrons the metal has, the fewer the anti-bonding states which will be filled and the stronger the interaction (rule-III)).

In comparing 4d metals like Pd and Rh, the variation in the coupling matrix element is modest, and the main difference in chemisorption strength comes from the fact that Rh has one d electron less than Pd and according to rule-III has a stronger net interaction with N and O (there will be fewer filled adsorbate – metal-d anti-bonding states). This explains why rhodium is a better NO dissociation catalyst. Structure sensitivity of NO dissociation can be similarly rationalized within the same framework of ideas: the d states at a step are generally higher in energy (closer to the Fermi level) [22], making step atoms on Pd(211) more active for NO dissociation compared to terrace (111)-type Pd atoms [rule-II]. Hence, the smallest NO dissociation barrier on Pd(211) compared to Pd(111).

Comparing the transition state for dissociation of CO on Ni and Pt, on the other hand, the two metals belonging to the same group of the periodic table have their d-shells almost filled. Therefore, nearly all anti-bonding adsorbate – metal-d states are filled, and as a result the repulsive term (1) above dominates over the attractive term (2). The Pt 5d states are much more extended than the Ni 3d states, yielding a V_{ad}^2 for Pt larger than the corresponding for Ni [rule-I]. This makes C and O bind weaker on Pt, pushing the CO dissociation transition state higher in energy. A direct consequence is that Pt does not dissociate CO [36, 37, 38] whereas Ni does [39, 40].

The barrier for dissociation on a metal surface is determined by the properties of the metal at the surface. The reactivity of a given metal surface can be modified by changing the position of its d states relative to the Fermi level. We saw above that this can be achieved by changing the coordination number of the surface atoms at a step. Changes in metal surface reactivity can also be accomplished by alloying the metal into another metal surface [41]. Adding promoters or poisons to a surface represents an alternative route to modifying a surface's chemical activity. A simple demonstration of this effect was given above, where the addition of Na or Cs to a Ru surface substantially decreased the barrier for N_2 dissociation.

Finally, we briefly discuss the origin of the alkali promotion effect [20]. In Fig. 4 we have plotted the change in the transition state energy (ΔE_{TS}) as a function of the quantity

$$\Delta E_{dip} = -\varepsilon \mu \qquad (5)$$

256

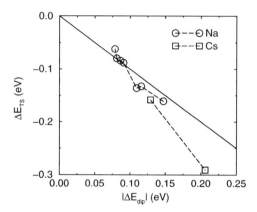

Figure 4: Interaction energy between alkali (Na, Cs) atoms and the transition state for N_2 dissociation on Ru(0001) as a function of ΔE_{dip}. Circles (for Na) and squares (for Cs) correspond to different positioning of the alkali atoms on the surface, relative to the transition state. Lines are drawn to guide the eye. For details, see Ref.[20].

which is the electrostatic interaction between a dipole, with electric dipole moment μ, and an electric field ε. For μ in Eq. 5 we have taken the calculated dipole moment of the transition state complex in the absence of the alkali:

$$\mu = \int d\vec{r} \Delta \rho_A(\vec{r}) z \tag{6}$$

We calculate a value of μ=-0.13 eÅ for the transition state to N_2 dissociation outside a Ru(0001) surface. The electric field ε is determined from the alkali-induced electrostatic potential $\Delta \phi_{alkali} = \phi_{alkali/Ru} - \phi_{Ru}$. We use the maximum field outside the surface as a measure of ε rather than attempt to find the local field at the height of the adsorbate.

Fig. 4 suggests that there is a very good correlation between ΔE_{TS} and ΔE_{dip}. Moreover, at higher values of ΔE_{dip} there is a trend that ΔE_{TS} varies even stronger with ΔE_{dip}. This happens simply because the dipole-electric field interaction remains proportional to the field strength for small field strength, while for stronger fields polarization effects introduce an extra attractive term, which is second order in the field strength [42]. The correlation elucidated in Fig. 4 strongly supports that the direct electrostatic interaction between the adsorbed alkalis and the transition state complex dominates the effect of promotion [20].

Acknowledgments

MM gratefully acknowledges financial support from EU, through a Marie-Curie Fellowship (contract ERBFMBICT #961691). The present work was in part financed by The Danish Research Councils through The Center for Surface Reactivity and grant #9501775. The Center for Atomic-scale Materials Physics is sponsored by the Danish National Research Foundation.

References

[1] G. A. Somorjai, *Introduction to Surface Chemistry and Catalysis*, (John Wiley, New York, 1994).

[2] R. I. Masel, *Principles of adsorption and reaction on solid surfaces*, (Wiley-Interscience, New York, 1996).

[3] See e. g. R. A. van Santen and M. Neurock, *Catal. Rev.-Sci. Eng.*, 1995, **37**, 557, or Ref. [2].

[4] F. Besenbacher, I. Chorkendorff, B. S. Clausen, B. Hammer, A. M. Molenbroek, J. K. Nørskov, I. Stensgaard, *Science*, 1998, **279**, 1913.

[5] M. C. Payne, M. P. Teter, D. C. Allan, T. A. Arias and J. D. Joannopoulos, *Rev. Mod. Phys.*, 1992 **64**, 1045.

[6] J. Neugebauer and M. Scheffler, *Phys. Rev. B*, 1992, **46**, 16067.

[7] N. Troullier and J. L. Martins, *Phys. Rev. Lett.*, 1991 **43**, 1993.

[8] D. H. Vanderbilt, *Phys. Rev. B*, 1990, **41**, 7892.

[9] J. P. Perdew, J. A. Chevary, S. H. Vosko, K. A. Jackson, M. R. Pederson, D. J. Singh, and C. Fiolhais, *Phys. Rev. B*, 1992, **46**, 6671.

[10] J. A. White and D. M. Bird, *Phys. Rev. B*, 1994, **50**, 4954.

[11] G. Kresse and J. Forthmüller, *Comput. Mat. Sci.*, 1996, **6**, 15.

[12] M. Methfessel and A. T. Paxton, *Phys. Rev. B*, 1989, **40**, 3616.

[13] G. Mills, H. Jonsson, G. K. Schenter, *Surf. Sci.*, 1995, **324**, 305.

[14] T. N. Truong, D. G. Truhlar, and B. C. Garrett, *J. Phys. Chem.*, 1989, **93**, 8227.

[15] T. N. Truong, and D. G. Truhlar, *J. Phys. Chem.*, 1990, **94**, 8262.

[16] A. Ulitsky and R. Elber, *J. Chem. Phys.*, 1990, **92**, 1510.

[17] J. J. Mortensen, Y. Morikawa, B. Hammer, and J. K. Nørskov, *J. Catal.*, 1997 **169**, 85.

[18] H. Dietrich, P. Geng, K. Jacobi, and G. Ertl, *J. Chem. Phys.*, 1996, **104**, 375.

[19] L. Romm, G. Katz, R. Kosloff, and M. Asscher, *J. Phys. Chem. B*, 1997, **101**, 2213.

[20] J. J. Mortensen, B. Hammer, J. K. Nørskov, *Phys. Rev. Lett.*, 1998, **80**, 4333.

[21] K. I. Aika, T. Takano, S. Murata, *J. Catal.*, 1992, **136**, 126; A. Ozaki and K. Aika, in *"Catalysis: Science and Technology"*, edited by J. R. Anderson and M. Boudart (Springer Verlag, Berlin,1981), Vol. 1.

[22] B. Hammer, *Faraday Discuss. Chem. Soc., submitted*.

[23] M. Mavrikakis, L. B. Hansen, B. Hammer, and J. K. Nørskov, *in preparation*.

[24] Q. Gao, R. D. Ramsier, H. Neergaard Waltenburg, and J. T. Yates, Jr., *J. Am. Chem. Soc.*, 1994, **116**, 3901.

[25] R. D. Ramsier, Q. Gao, H. Neergaard Waltenburg, K. W. Lee, O. W. Nooij, L. Lefferts, and J. T. Yates, Jr., *Surf. Sci.*, 1994, **320**, 209.

258

[26] H. J. Borg, J. Reijerse, R. A. van Santen, J. W. Niemantsverdriet, *J. Chem. Phys.*, 1994, **101**, 10052.

[27] T. W. Root, G. B. Fisher, L. D. Schmidt, *J. Chem. Phys.*, 1986, **85**, 4679.

[28] K. C. Taylor, *Catal. Rev.*, 1993, **35**, 457.

[29] Y. Morikawa, J. J. Mortensen, B. Hammer, and J. K. Nørskov, *Surf. Sci.*, 1997, **386**, 67.

[30] M. Mavrikakis, B. Hammer, and J. K. Nørskov, *in preparation*.

[31] M. Neurock, R. A. van Santen, W. Biemolt, and A. P. J. Jansen, *J. Am. Chem. Soc.*, 1994, **116**, 6860.

[32] B. Hammer and J. K. Nørskov in *"Chemisorption and Reactivity on Supported Clusters and Thin Films"*, 285-351, R. M. Lambert and G. Pacchioni (Eds.), (Kluwer Academic Publishers, The Netherlands, 1997).

[33] S. Holloway, B. I. Lundqvist, and J. K. Nørskov, *Proc. Int. Congress on Catalysis*, Berlin 1984, Vol. 4, p.85.

[34] J. K. Nørskov, *Rep. Prog. Phys.*, 1990, **53**, 1253.

[35] B. Hammer and J. K. Nørskov, *Nature*, 1995, **376**, 238.

[36] D. F. Olegtree, M. A. Van Hove, G. A. Somorjai, *Surf. Sci.*, 1986, **173**, 351.

[37] H. Steininger, S. Lehwald, H. Ibach, *Surf. Sci.*, 1982, **123**, 264.

[38] P. R. Norton, J. W. Goodale, E. B. Selkirk, *Surf. Sci.*, 1979, **83**, 189.

[39] D. W. Goodman, R. D. Kelley, T. E. Madey, J. M. White, *J. Catal.*, 1980, **64**, 479.

[40] R. D. Kelley, D. W. Goodman, *Surf. Sci.*, 1982, **123**, L743.

[41] A. Ruban, B. Hammer, P. Stoltze, H. L. Skriver, J. K. Nørskov, *J. Mol. Catal. A*, 1997, **115**, 421.

[42] J. K. Nørskov, S. Holloway, N. D. Lang, *Surf. Sci.*, 1984, **137**, 65.

Chapter 20

A Theoretical Study of the Mechanism of the Adsorptive Decomposition of Nitrous Oxide on Copper

Peter Wolohan[1], William J. Welsh[1], Robert Mark Friedman[2], and Jerry R. Ebner[2]

[1]Department of Chemistry and Center for Molecular Electronics, University of Missouri, St. Louis, MO 63121
[2]Monsanto Corporate Research, Monsanto Company, St. Louis, MO 63167

We present results from a Density Functional Theory (DFT) study of the mechanism of the dissociative adsorption of nitrous oxide on the principal low index crystal planes of copper (<100> and <110>) by applying the generalized gradient approximation (GGA) to three-dimensional supercell representations of these crystal planes. We have studied the "molecular adsorption" of N_2O on these surfaces and the site and structure of the "atomic" oxygen product. The transition state of the reaction mechanism has been located using the four-fold hollow site as the primary catalytic site on both planes. The calculated activation energy to dissociation is 0.67 eV lower on the Cu<110> surface than on the Cu<100> surface. Furthermore, adsorption on the Cu<110> surface results in a greater weakening of the N=O bond of N_2O. The present calculations indicate that adsorption of N_2O in this site on the Cu<110> surface is extremely favorable, which is consistent with the highly reactive nature of this surface.

Copper, often highly dispersed on a variety of supports, is widely used as a catalyst in many industrially important catalytic processes including, for example, methanol synthesis, the water gas shift reaction, and the oxidative dehydrogenation of alcohols. One method for quantification of the free-copper surface area of supported and unsupported copper catalysts uses adsorptive decomposition of N_2O as described by Dell el al. (1). In this procedure, the number of exposed surface atoms of Cu is estimated from the amount of N_2 released in the adsorptive decomposition of nitrous oxide according to the reaction:

$$N_2O_{(g)} + 2Cu_{(s)} = N_{2(g)} + -(Cu-O-Cu)-_{(ads)} \qquad (1)$$

Numerous analytical techniques have been published for monitoring the course of this reaction and the completion of the monolayer (2-7). General agreement exists regarding the stoichiometry of the chemisorbed layer: $Cu(s)_2O_{ads}$, corresponding to full coverage [$\theta = 1$]. Many investigations of this reaction have been reported including a recent detailed study by G. Sankar et al *(8)* who employed a specially designed *in-situ* reaction cell that also accommodated both X-ray absorption and X-ray diffraction measurements. These workers identified two preferred configurations for the chemisorbed atomic oxygen culminating from N_2O dissociation: (1) a two-fold long bridge site in which the oxygen atom bridges two Cu atoms

separated by a distance of 3.62 Å, and (2) chemisorption of the atomic oxygen in the four-fold hollow site. The two-fold long bridge site is unique to the Cu<110> surface, which is known to be the most reactive of the three principal low-index surfaces: <100>, <110>, and <111> *(9-10)*. This site exhibits a primary Cu-O distance of 1.87 Å and a secondary Cu-O distance of 3.30 Å. Adsorption at the alternative four-fold hollow site gives rise to a primary Cu-O distance of 2.23 Å. Comparable adsorption sites have been studied and discussed for O_{ads} on the various crystal planes of copper generated by reaction of the metal with O_2 *(11-14)*. Still, many questions remain unanswered with respect to the N_2O reaction, including: What is the mechanism of the adsorptive decomposition of N_2O on copper? What is the activation energy for dissociation on the principal crystallographic planes and at the different adsorption sites? What is the transition state of the mechanism and is it independent of the crystallographic plane? What are the implications of the structural specificity for the general application of the dissociative adsorption for quantification of the catalytic active?

The bonding of a molecule to an extended metal surface or to the surface of a supported metal crystallite is a fundamental issue. Although the bonding within a molecular species or within a metal has been well studied, many aspects regarding the bonding of a molecule with a metal surface such as that of a catalyst remain unclear and largely unexplored beyond the cluster model. *Ab initio* theoretical techniques allow us to study such important chemical processes at a level not possible with available experimental techniques. Density functional theory *(15-21)* (DFT), in particular the gradient corrected DFT (GC-DFT) procedures allow us to calculate the properties of interacting species.

Computational Methods

One particular implementation of DFT is that used in the quantum chemistry program CASTEP *(22)* which allows one to study the interaction of a discrete molecular species with a periodic representation of the extended metal surface *(23-26)*. In a chemical system, the molecular orbitals are eigenstates of the system. In order to describe the bonding and chemistry of the molecule, it is essential to understand such properties as the shape, symmetry and energy of these eigenstates. Charge densities of the eigenstates of a system are calculated as

$$\rho_{total}(r) = \sum_{i_{band}} \sum_{k_i} \rho_{i_{band},k_j}(r) \qquad (2)$$

$$\rho_{total}(r) = \sum_{i_{band}} \sum_{k_i} \left| \psi_{i_{band},k_j}(r) \right|^2 \qquad (3)$$

where i_{band} is the band number, k_j is the k-point, and the summations run over all occupied bands and sampled k-points, respectively. The term $\psi_{i_{band},k_j}(r)$ represents a Bloch state that is an eigenstate of the extended periodic system. The charge densities of all the sampled Bloch states, ρ_{i_{band},k_j}, then give a representation of the molecular orbitals of the system being studied.

In the present study, the adsorption of N_2O and atomic O were modeled on both the Cu<110> surface (the most reactive for dissociation) and Cu<100> surface (much less reactive for dissociation) using three-dimensional cell representations that contained three atomic layers of the copper catalyst. For the Cu<110> surface, the O atom was placed above the two-fold long bridge site at a Cu-O distance of 1.87 Å and above the four-fold hollow site at a Cu-O distance of 2.23 Å in accordance with those observed by Sankar *et al.* *(8)* (Figure 1). The former distance compares favorably with 1.84 Å for Cu-O in the two-fold long bridge site as determined from angular dependent surface extended X-ray absorption fine structure (SEXAFS) of (2x1) O on Cu<110> *(11)*. In the absence of experimental geometric data with regard to positioning the N_2O molecule, it was placed at the same distance given above for the O atom in each case. Similar placement of the N_2O molecule and atomic oxygen with respect to their O atoms allowed interpretation of differences in calculated properties between the two systems as due to perturbative effects arising from differences in structure between the intact N_2O molecule and atomic oxygen. In particular, their similar placement enables comparison of the electron density distributions relative to the adsorptive decomposition reaction and, thus, facilitates an understanding of the adsorptive mechanism.

Similar geometric data from experiment were again to our knowledge unavailable in the case of N_2O on the Cu<100> surface, although this is not surprising given the less reactive nature of the Cu<100> surface. As a result, suitable Cu-O distances were estimated using molecular mechanics (MM) to carry out an extensive study of N_2O adsorption on both copper surfaces and adsorption sites using much larger representations of the extended Cu surface. Values for the equilibrium adsorbed Cu-O distance given by these MM calculations were 1.78 Å on the two-fold site and 2.11 Å on the four-fold hollow site of the Cu<100> surface. We used our MM-calculated distances for both the N_2O and $O_{(ads)}$ for consistency with calculations performed on the Cu<110> surface. These results can be compared with data from an extensive array of studies on the dissociative adsorption of O_2 on the Cu<100> surface *(12-14)*. The chemisorption of O at a four-fold hollow site on the Cu<100> surface yielded Cu-O distances of $\approx$ 1.9 Å from multiple scattering (MS) analysis of near-edge X-ray absorption fine structure (NEXAFS) *(12)*, 1.94 Å from surface extended X-ray absorption fine structure

262

(a)

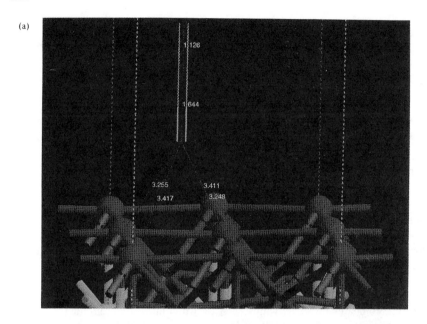

(b)

Figure 1. Ball and stick illustration of the calculated geometry of the transition state of the N_2O adsorptive decomposition mechanism when N_2O is adsorbed in (a) the four-fold hollow site on Cu⟨110⟩ surface and (b) the four-fold hollow site on Cu⟨100⟩ surface. The following color scheme is employed: blue = nitrogen, red = oxygen and green = copper. For color, see the color insert.

(SEXAFS) *(13)*, and 1.81 Å from angle resolved core-level X-ray photoemission *(14)*.

Throughout the DFT-GGA calculations, the ground state electronic configurations for all species (i.e., gaseous N_2O, N_2, and atomic O; free copper surfaces; and metal-adsorbate systems) were calculated by applying a type of Car-Parrinello *(27)* approach utilizing a conjugate gradient minimization scheme to locate directly the electronic ground states. As is commonly adopted in this formalism *(28)*, we employed a plane-wave basis set and a GGA exchange-correlation functional based on that proposed by Perdew and Wang *(29)*. An energy cutoff, which relates to the resolution of the basis set, was set at 800 eV together with a k-point sampling of 0.100 Å and a termination criterion of 2×10^{-6} eV energy gain per atom between subsequent steps in the band calculation for the electronic calculation.

After calculating the single point energy of each system, further calculations were performed specifically on the four-fold hollow sites to determine the equilibrium geometry of the adsorbed species. A BFGS minimization routine was employed allowing atomic motion to occur in all of the atoms of the supercell except in the bottom layer of copper atoms, in order to replicate the effect of the bulk. The criteria selected in the minimization scheme are as follows: Root-Mean-Square (RMS) Stress of < 1.00 GPa, RMS force of < 0.100 eV/Å, and RMS displacement < 0.010 Å.

Results and Discussion

As a preliminary study, the various levels of theory of DFT in the program CASTEP were benchmarked for accuracy and consistency. We did this by calculating the equilibrium geometry and electronic properties of the N_2O molecule at the various levels of theory. The results, presented in Table I, show that the GGA "fine" parameters yielded the most accurate geometric properties of the N_2O molecule compared to experiment. As a result, these program specifications were used in all subsequent calculations. Preliminary calculations that included construction of the metal surfaces and adsorption of the intact N_2O molecule at frozen geometries for the different are presented elsewhere *(30)*.

A summary of the ground state energetics of the supercell representations calculated using single point energy calculations is presented in Table II. The adsorption energy for N_2O is exothermic on the two-fold site of the Cu<100> surface although the corresponding dissociation energy is endothermic. The dissociation energy (D_e) for N_2O is defined in the present study as D_e = [E(copper surface-oxygen complex) + $E(N_{2(g)})$] - [$E(N_2O_{(g)})$ + E(copper surface)]. Adsorption of N_2O on the four-fold hollow site of Cu<100> and on both the two-fold site and four-fold hollow site of Cu<110> is, as expected, endothermic as the aforementioned distances correspond to the adsorption of atomic oxygen and not to the intact N_2O molecule. One general observation from these results is that, for either surface, N_2O adsorption is more favorable on the two-fold site than on the corresponding four-fold hollow site.

The calculated adsorption energy of atomic oxygen, which was held fixed at a distance corresponding to the experimental value on the Cu<110> surface, is

Table I. Comparison of results from CASTEP calculations on the N_2O molecule

| | LDA | | GGA | | |
	Medium	Coarse	Medium	Fine	Exp.
r_e (NN)	1.241	1.353	1.252	1.150	1.127
r_e (NO)	1.090	1.269	1.233	1.096	1.185
E_T	-946	-884	-929	-973	*

All bond lengths in Å and energies in eV. Coarse, medium and fine refer to the cutoff energy of the plane-wave basis set, where the accuracy of the basis set increases from coarse to fine.

Table II. Comparison of results from GGA "fine" single point energy calculations on N_2O and atomic oxygen adsorbed in the four-fold hollow and two-fold bridge sites on the Cu<110> and Cu<100> surfaces

| | $E_{surface}$ | N_2O | | O | | D_e | |
		2f	4f	2f	4f	2f	4f
Cu<110> surface	-13891.0	0.52	3.33	-4.53	-6.78	1.97	-0.28
Cu<100> surface	-13890.9	-2.59	4.84	-4.65	-4.90	1.85	1.60

Energies in eV. 2f refers to two-fold site and 4f to four-fold site.

exothermic for both surfaces and both sites considered here. In contrast to that calculated for N_2O, adsorption of atomic oxygen on a four-fold site on either surface is more favorable than on the corresponding two-fold site. In particular, the adsorption of atomic oxygen is most favorable in the four-fold hollow site of the Cu<110> surface. The large magnitude of this favorable adsorption of atomic oxygen in the four-fold site on the Cu<110> surface accounts for the overall exothermic value of the calculated dissociation energy on this surface. This favorable adsorption of atomic oxygen is consistent with the experimental evidence, i.e., that the Cu<110> surface is significantly more reactive than the alternative Cu<100> surface for the adsorptive decomposition of N_2O. The same calculated dissociation energies are endothermic for the two-fold site on the Cu<110> surface and for both the two-fold site and four-fold site on the Cu<100> surface.

Given the preference for adsorption in the four-fold hollow site, the respective equilibrium geometries of N_2O and atomic O adsorbed in the four-fold hollow site on both the Cu<100> and Cu<110> surfaces were calculated. The calculated energetics of binding and dissociation in the four-fold hollow site on each surface is presented in Table III, while a summary of the key geometric parameters from these calculations is given in Table IV. It is seen that the adsorption of both N_2O and atomic O in the four-fold site is more favorable on the Cu<110> surface than on the alternative Cu<100> surface. Only in the case of N_2O in the four-fold hollow site of the Cu<100> surface is adsorption endothermic. As can be seen from Table III, adsorption of atomic O is strongly preferred energetically over N_2O. The magnitude of this energetic preference is increased as a result of performing the energy minimization calculations. Interestingly, the dissociation energy D_e for N_2O in the four-fold site on the Cu<110> surface became substantially more exothermic (from -0.28 eV in Table II to -2.15 eV in Table III) as a result of energy minimization while the corresponding energy on the Cu<100> surface changed only slightly (from 1.60 eV in Table II to 1.29 eV in Table III).

Optimization of the model system to a stress-free representation was made difficult to achieve by the presence of an unusual hysteresis effect in the model systems (Table IV). In particular, the stresses in the Cu<100> model system proved difficult to reduce. During the optimization, the hysteresis was characterized most obviously in the two Cu<110> models by oscillation in the gamma lattice angle between $90°$ and $83°$ coupled with oscillation of the two Cu-O distances. The current three-layer model of the copper surface, in which only the bottom layer is held fixed, may not be sufficient to replicate the effect of the copper bulk and thus anchor the cell parameters. A possible solution would be to include at least another layer of fixed copper atoms.

Considering the geometric parameters, the calculated value of 2.30 Å for the equilibrium Cu-O distance associated with the adsorption of atomic O on the Cu<110> surface is in good agreement with the corresponding experimental value (2.23 Å) (8). The value (2.09 Å) calculated for this distance on the Cu<100> surface is much shorter and in good agreement with value of 2.11 Å estimated by our preliminary molecular mechanics calculations (Table IV). In contrast to atomic O, N_2O is only weakly adsorbed as evidenced by the large values of the Cu-O distance (3.25 and 3.41 Å). Nevertheless, the calculated binding energy is very

Table III. Comparison of properties of N_2O and atomic oxygen adsorbed in four-fold hollow site on the Cu<110> and Cu<100> surfaces, as calculated from GGA "fine" optimizations

	Free Surface		Free N_2O		N_2O_{ads}	O_{ads}	D_e
	E_{Fermi}	|e⁻|	E_{LUMO}	|e⁻|	4f	4f	4f
Cu<110>	-3.20	0.08	-1.02	0.00	-1.47	-8.90	-2.15
Cu<100>	-3.78	1.00	-1.02	0.00	7.23	-5.46	1.29

Energies in eV. 4f refers to four-fold site, |e⁻| refers to electron occupancy, E_{Fermi} refers to the Fermi energy of surface and E_{LUMO} refers to energy of LUMO.

Table IV. Calculated GGA "fine" geometric properties of N_2O and atomic oxygen adsorbed in four-fold hollow sites on the Cu<110> and Cu<100> surfaces

	r_0			Lattice Parameters				RMS
	Cu-O	N=O	N=N	a_0	α	β	γ	Stress
Cu<110>	-	-	-	3.603	90.0	90.0	85.69	0.72
Cu<100>	-	-	-	4.050	90.0	90.3	90.0	0.97
O_{abs} Cu<110>	2.011, 2.289	-	-	3.340	90.1	90.0	83.4	0.62
O_{abs} Cu<100>	2.088, 1.966	-	-	3.879	90.0	90.0	90.5	0.98
N_2O_{abs} Cu<110>	3.251, 3.414	1.091	1.157	3.816	90.1	90.0	87.2	0.57
N_2O_{abs} Cu<100>	2.659, 2.524	1.121	1.153	4.222	90.0	90.2	89.1	0.98

All bond lengths are in Å. r_0 refers to equilibrium bond length. RMS Stress is the residual stress on the cell which is used as a convergence criteria in the optimization scheme.

favorable ($E_{binding}$ = -34.73 eV). In addition, the N=O bond has compressed slightly (from r_o = 1.096 Å for free N_2O to r_o = 1.091 Å for bound N_2O) indicating strengthening while the N=N bond has correspondingly lengthened (from r_o = 1.150 Å for free N_2O to r_o = 1.157 Å for bound N_2O) indicating weakening. An significant finding of this study is the observation that adsorption of N_2O is more favorable on the structurally open Cu<110> surface compared with the Cu<100> surface, although at longer Cu-O distances. This finding may explain the more reactive nature of the Cu<110> surface to the adsorptive decomposition of N_2O. Another finding is the extensive degree of breathing of the four-fold adsorption site on both surfaces when N_2O or O are adsorbed, except when atomic O is adsorbed on the Cu<110> surface (see equilibrium lattice constants, Table IV).

Further evidence of the more reactive nature of the Cu<110> for the adsorptive decomposition of N_2O was obtained by carrying out a linear search of the potential energy surface to locate the transition state of the reaction on both the surfaces. This linear search was facilitated in each case by extending the N=O bond of the N_2O adsorbate in increments of 12% of the calculated equilibrium N=O bond length for free N_2O (Table I) while reducing the corresponding N=N bond in increments of 12% of the calculated equilibrium N≡N bond length of free N_2 (r_o = 1.101 Å from calculations and 1.098 Å from experiment *(30)*). This procedure was continued until the N=O bond length exceeded twice that calculated for the equilibrium N=O bond of free N_2O and the N=N bond reached that calculated for free N_2. The possibility of a NO transition state was discounted since there is no experimental evidence for such a transition state in the adsorptive decomposition of N_2O on copper *(2, 8-10)*. Figure 1 illustrates the geometry of the transition state located on each surface, while Figure 2 presents a schematic drawn to scale which depicts the calculated energetics of the complete reaction coordinate of the N_2O decomposition reaction from our studies.

Inspection of the geometry of the transition states reveals that, in the four-fold site, the N=N bond is shortened (i.e., strengthened) while the N=O bond is lengthened (i.e., weakened) to a greater extent on the Cu<110> surface than on the Cu<100> surface and at a lower cost in energy to achieve this strained conformation. The calculated activation energy for dissociation of N_2O in the four-fold hollow site on the Cu<100> surface is 2.80 eV while the corresponding energy on the more reactive Cu<110> surface is 2.40 eV. In addition, the energetics associated with the decomposition reaction in the four-fold site is mainly downhill for the Cu<110> surface but sharply uphill for the Cu<100> surface.

Figure 3 illustrates the highest occupied band (HB) calculated from the energy-minimized structure of N_2O adsorbed in a four-fold hollow site on the Cu<100>and Cu<110> surfaces, respectively. This HB associated with the N_2O-Cu<100> complex quite clearly involves the interaction of the metal surface with the LUMO of N_2O, constructed from the antibonding overlap of the $2p_x$ or $2p_y$ atomic orbitals of the nitrogen and oxygen atoms of N_2O to form two degenerate antibonding orbitals of π symmetry. This eigenstate thus represents the overlap of the principal frontier orbitals of the two interacting species. Further analysis of the two HBs for the N_2O-metal complex with respect to energetics (HB = -2.86 eV and HB+1 = -3.49 eV) and occupancy (HB = 2.00 |e-| and HB+1 = 2.00 |e-|) reveals that energy is expended to transfer charge density from the metal surface to the

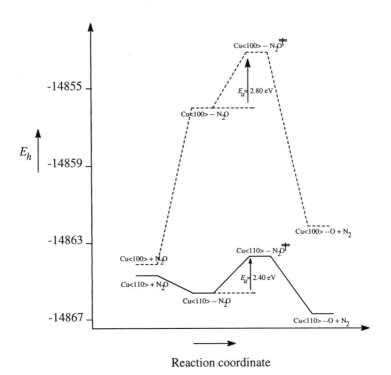

Figure 2. Schematic of the energetics of the various species along the reaction coordinate of the adsorptive decomposition mechanism of N_2O in the four-fold hollow site on the Cu<110> and Cu<100> surfaces.

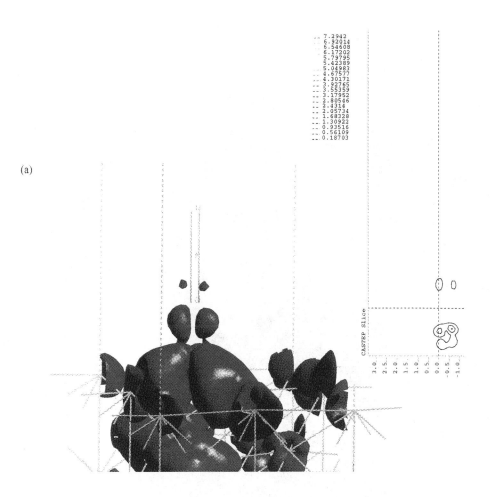

(a)

Figure 3. Volumetric illustration of the shape and symmetry of the calculated highest occupied band (HB) of a N_2O adsorbate in the four-fold hollow site (a) on the Cu⟨110⟩ surface and (b) on the Cu⟨100⟩ surface. The contour plot in the upper right-hand corner is taken from a slice of the charge density through the LUMO orbital in the x-direction. For color, see the color insert.

(b)

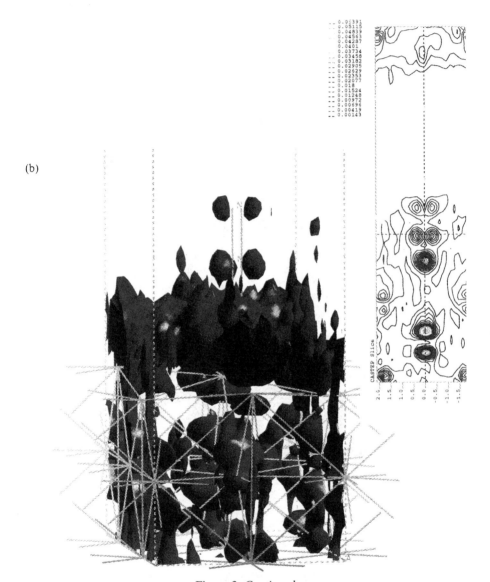

Figure 3. *Continued.*

antibonding orbital of the N_2O (Table III). This observation is further corroborated by the significant weakening of the N=O bond ($r_o = 1.121$ Å) and to a lesser extent the N=N bond ($r_o = 1.153$ Å) of the N_2O adsorbate which had started at the experimental distance for the N=O bond in gas-phase N_2O (Table I).

The same type of frontier orbital interaction is more difficult to identify for the corresponding orbital of the N_2O-Cu<110> complex. The present calculations indicate that only the atomic orbitals of the O atom in the N_2O adsorbate contribute to the HB of the complex. This conclusion is consistent with the calculated geometry which shows that the N_2O adsorbate is too distant from the Cu<110> surface to develop a significant interaction. Indeed, the N=O bond of the adsorbed system is only slightly contracted in the energy-minimized model ($r_o = 1.091$ Å) relative to its starting point corresponding to the experimental distance for N=O in gas-phase N_2O (Table I). It thus appears that the primary effect of the Cu<110> surface is to perturb the geometry of the N_2O adsorbate in preparation for the exothermic process of dissociation (Table III).

Conclusions

From the present DFT study of the mechanism of the adsorptive decomposition of N_2O on copper, we can conclude the following:

- N_2O adsorption is energetically more favorable in the two-fold sites than in the corresponding four-fold sites, particularly adsorption in the short two-fold site on the Cu<100> surface which is exothermic at the distances modeled in the single point calculations.
- Adsorption of atomic oxygen, placed at the experimental distances for the Cu<110> surface and at the estimated distances for the Cu<100> surface, is exothermic in all cases. Adsorption is energetically more favorable in the four-fold sites than in the corresponding two-fold sites, particularly the four-fold site on the Cu<110> surface which is the most energetically favorable of the sites investigated. This finding is not supported by the experimental results of Sankar et al. *(8)* in which the long two-fold bridge is assigned as the location of the product O from the N_2O decomposition. However, it is important to remember that the experiments were carried out at saturation whereas the present calculations relate to very low coverage without inclusion of adsorbate-adsorbate interactions *(32)*. These effects are the focus of future work. At the same time, the binding energy of atomic O calculated in the present work (-4.90 eV from the single-point calculations and -5.46 eV from the optimized results) is consistent with estimates of 5-6 eV reported elsewhere.[33]
- Dissociation of N_2O was calculated to be exothermic only in the four-fold hollow site on the Cu<110> surface which is observed as the most reactive surface by far. Upon optimization of the supercell models, these conclusions are confirmed but the magnitude of the difference as a ratio of dissociation energies is reduced.

- The calculated activation energy for dissociation of N_2O in the four-fold hollow site on the Cu<100> surface is 2.80 eV while the corresponding energy on the Cu<110> surface is 2.40 eV.

- In the transition state of the dissociation process in the four-fold site, the N=N bond is strengthened and the N=O bond is weakened to a greater extent on the Cu<110> surface than on the Cu<100> surface despite costing less energy to achieve this strained conformation.

- The adsorptive decomposition of N_2O on copper is initiated by reduction of the N_2O whereby electron density is transferred from the metal surface to the LUMO of the adsorbate molecule. However, this interaction must be energetically favorable to occur.

Further studies using the DFT dynamics capabilities of CASTEP are being carried out using these optimized copper supercells. Besides accounting for the effects of temperature, these extended studies will allow us to explore the mechanism of the dissociative adsorption of N_2O on the Cu <100> and <110> principal crystallographic planes. They should provide further evidence of the more reactive nature of the Cu<110> surface.

Acknowledgments

PW and WJW wish to thank the Monsanto Company for funding of this research.

Literature Cited

(1) Chinchen, G. C.; Hay, C. M.; Vanderveil H. D.; Waugh, K. C. J. *Catalysis* **1987**, *79*, 103.

(2) Dell, R, M.; Stone, S.; Tiley, P. F. *Trans. Faraday Soc.* **1953**, *49*, 159.

(3) Scholten, J. J. F.; Konvalinka, J. A. *Trans Faraday Soc.* **1969**, *65*, 2465.

(4) Dvorak, B.; Pasek, J. *J. Catal.* **1970**, *18*, 108.

(5) Evans, J. W.; Wainwright, M. S.; Bridgewater, A.; Young, D. J. *Appl. Catal.* **1983**, *7*, 75.

(6) Chinchen, G. C.; Hay, C. M.; Vandervell, H. D.; Waugh, K. C. J. *Catal.* **1987**, *103*, 79.

(7) Giamello, E.; Fubini, B.; Lauro, P.; Bossi, A. *J. Catal.* **1984**, *87*, 443.

(8) Sankar, G.; Thomas, J. M.; Waller, D.; Couves, J. W.; Catlow, c. R. A.; Greaves, G. N. *J. Phys. Chem.* **1992**, *96*, 7485.

(9) Arlow, J. S.; Woodruff, D. P. *Surf. Sci.* **1985**, *157*, 327.

(10) Arlow, J. S.; Woodruff, D. P. *Surf. Sci.* **1985**, *162*, 310.

(11) Döbler,, U.; Baberschke, K.; Haase, J.; Pushmann, A. *Phys. Rev. Lett.* **1984**, *52*, 1437.

(12) Vvedensky, D. D.; Pendry, J. B.; Döbler, U.; Baberschke, K. *Phys. Rev. B* **1987**, *35*, 7756.

(13) Döbler, U.; Baberschke, K.; Stöhr, J.; Outka, D. A. *Phys. Rev. B* **1985**, *31*, 2532.

(14) Kono, S.; Goldberg, S. M.; Hall, N. F. T.; Fadley, C. S. *Phys. Rev. B* **1980**, *22*, 6085.

(15) Hohenberg, P.; Kohn, W. *Phys. Rev. B* **1964**, *136*, 864.

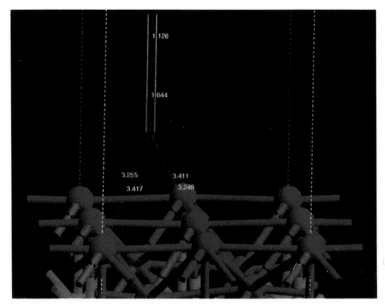

(a)

1 126

1 644

3.255   3.411
3.417   3.248

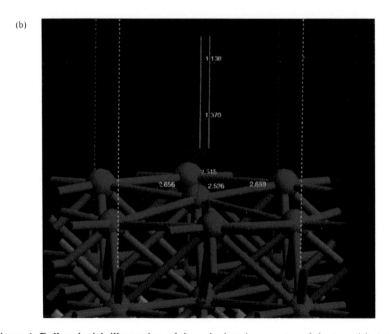

(b)

1 138

1 370

2.515
2.656   2.526   2.659

Figure 1. Ball and stick illustration of the calculated geometry of the transition state of the N₂O adsorptive decomposition mechanism when N₂O is adsorbed in (a) the four-fold hollow site on Cu⟨110⟩ surface and (b) the four-fold hollow site on Cu⟨100⟩ surface. The following color scheme is employed: blue = nitrogen, red = oxygen and green = copper.

(a)

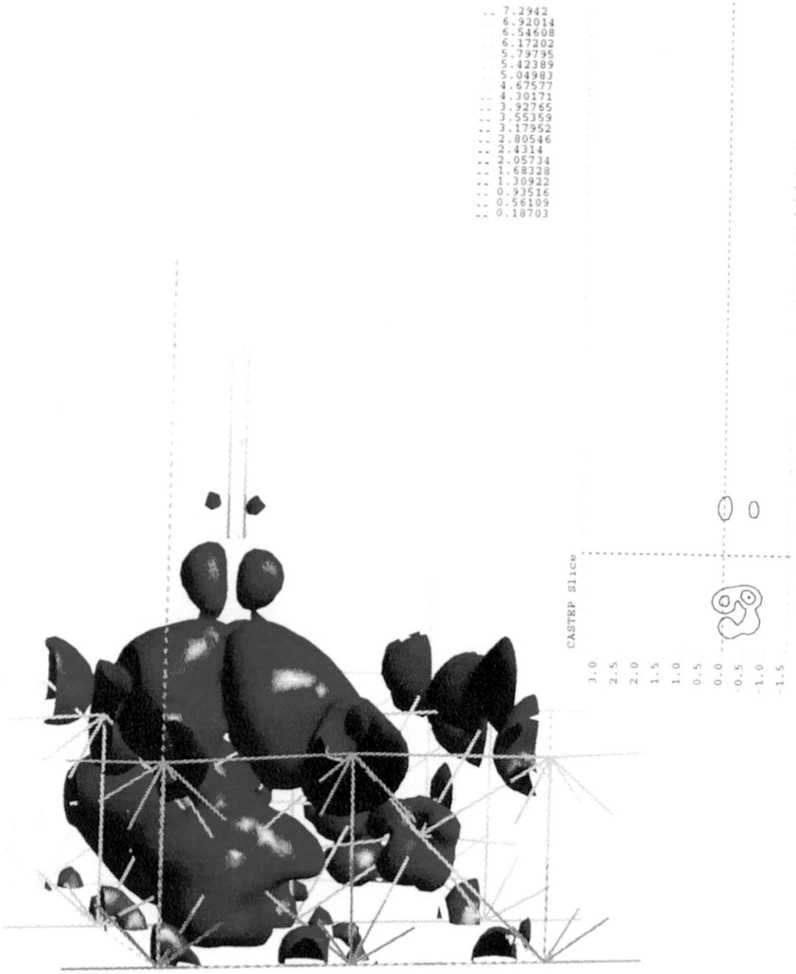

```
..  7.2942
    6.92014
    6.54608
    6.17202
    5.79795
    5.42389
    5.04983
    4.67577
..  4.30171
..  3.92765
..  3.55359
..  3.17952
..  2.80546
..  2.4314
..  2.05734
..  1.68328
..  1.30922
..  0.93516
..  0.56109
..  0.18703
```

Figure 3. Volumetric illustration of the shape and symmetry of the calculated highest occupied band (HB) of a N_2O adsorbate in the four-fold hollow site (a) on the Cu⟨110⟩ surface and (b) on the Cu⟨100⟩ surface. The contour plot in the upper right-hand corner is taken from a slice of the charge density through the LUMO orbital in the x-direction.

(b)

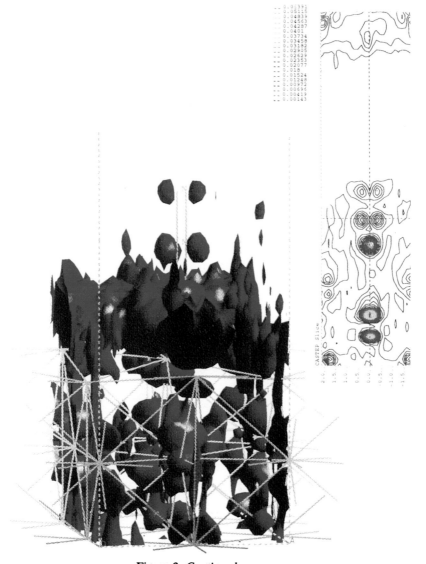

Figure 3. *Continued.*

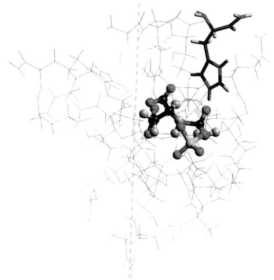

Figure 4. Active site model of ATCase with bounded transition state (rendered as sticks and balls). H134 is highlighted.

(16) Kohn, W.; Sham, L. J. *Phys. Rev. A* **1965**, *140*, 1133.

(17) Parr, R. G.; Yang, W. *Density-Functional Theory of Atoms and Molecules*; Oxford University Press: New York, 1989.

(18) Jones, R. O.; Gunnarsson, O. *Rev. Mod. Phys.* **1989**, *61*, 689.

(19) *Density Functional Methods in Chemistry*; Labanowski, J. K.; Andzelm, J., Eds.; Springer-Verlag: New York, 1991.

(20) Ziegler, T. *Chem. Rev.* **1991**, *91*, 651.

(21) Wigner, E. P. *Trans. Faraday Soc.* **1938**, *34*, 678.

(22) Cambridge serial total energy package, University of Cambridge, Madingley Road, Cambridge, United Kingdom.

(23) White, J. A.; Bird, D. M.; Payne, M. C. *Phys. Rev. B* **1996**, *53*, 1667.

(24) Owen, J. H. G.; Bowler, D. R.; Goringe, C. M.; Miki, K.; Briggs, G. A. D. *Phys. Rev. B* **1996**, *54*, 14153.

(25) Milman, V.; Pennycook, S. J.; Jesson, D. E. *Thin Solid Films* **1996**, *272*, 375.

(26) Kantorovich, L. N.; Gillan, M. J.; White, J. A. *J. Chem. Soc.* **1996**, *92*, 2075.

(27) Car, R.; Parrinello, M. *Phys. Rev. Lett.* **1985**, *55*, 2471.

(28) Lin, J. S.; Qteish, A.; Payne, M. C.; Heine, V. *Phys. Rev. B* **1993**, *47*, 4174.

(29) Perdew, J. P.; Wang, L. *Phys Rev. B* **1991**, *40*, 3425.

(30) Wolohan, P.; Welsh, W. J.; Friedman, R. M.; Ebner, J. R. (unpublished work).

(31) *The Surface Chemistry of Metals and Semiconductors*; Gatos, H. C., Ed.; John Wiley and Sons, Inc.: New York, 1959.

(32) *Chemistry and Physics of Solid Surfaces*, Vanselow, R., Ed.; CRC Press: Boca Raton, Florida, 1979, Vol. 2.

(33) Madhaven, P. V.; Newton, M. D. *J. Chem. Phys.* **1987**, *86*, 4030, and references cited therein.

Chapter 21

Theoretical Studies of Ethyl to Ethylene Conversion on Nickel and Platinum

J. L. Whitten and H. Yang

Department of Chemistry, North Carolina State University, Box 8201, Raleigh, NC 27695–8201

Ethylene production from ethyl adsorbed on nickel and platinum surfaces is investigated using first-principles theory. Calculations are based on an embedded cluster formalism that permits an accurate determination of energies and adsorbate structure. The key issue is the nature of the activation barrier associated with the transfer of a beta hydrogen to the metal surface. The minimum energy pathway is found to require an initial repulsive interaction with the surface in which the adsorbed ethyl fragment is tilted until there is a slight penetration of the electron density of the metal surface by the ethyl beta hydrogen. The H atom can then be transferred from this more energetic configuration with an overall activation energy of 17 kcal/mol for Ni(100) and 14 kcal/mol for Pt(100). Differences between nickel and platinum are discussed.

1. Introduction

The goal of this research is the development and application of theoretical techniques that will provide a molecular level understanding of surface processes, especially energetics, adsorbate structure and reaction mechanisms. The work relates to two broad subject areas: catalytic processes on metals and properties of electronic materials. First-principles theoretical methods are employed to obtain an accurate description of these systems (1,2).

The difficulty in treating surface reactions using first-principles theory is that there are conflicting demands on the theory. At the surface, the treatment must be accurate enough to describe surface-adsorbate bonds and energy changes accompanying chemical reactions (generally, this is most readily achieved if systems are small); while, for a metal, a large number of atoms is required to describe conduction and charge transfer processes (in this case methods for treating large symmetric systems are most appropriate). The objective of the embedding theory employed in this work is to balance the accuracy and size aspects of the problem (2).

The present paper deals with catalytic reactions leading to ethylene formation on nickel and platinum. The ethylene study which involves consideration of α and β elimination pathways in the reaction of ethane with nickel is motivated by the short contact time catalytic studies of Schmidt and coworkers who observed a high degree of selectivity in several transition metal systems (3,4). In the formation of ethylene, the key issue is the nature of the activation barrier associated with the transfer of a beta hydrogen to the metal surface. In the present work, the minimum energy pathway for beta hydrogen transfer is shown to involve an initial repulsive interaction with the surface in which the adsorbed ethyl fragment is tilted until there is a slight penetration of the electron density of the metal surface by the ethyl beta hydrogen. Dissociation of CH then becomes more facile compared to H transfer from the equilibrium configuration of adsorbed ethyl.

2. Theory

In order to describe reactions at a metal surface the theoretical treatment must be adequate to deal with the complexities of chemical bond formation or dissociation and accompanying effects of charge transfer, polarization and electronic correlation. *Ab initio* configuration interaction theory provides an attractive way to proceed only if systems are of manageable size.

The purpose of the embedding approach adopted in the present work is to organize the theoretical treatment in such a manner that an accurate many-electron treatment of the adsorbate/surface portion of a system can be carried out while coupling this region to the bulk lattice (2). Localized orbitals extracted from a treatment of the clean surface are used to define an electronic subspace encompassing the adsorbate and neighboring surface atoms.

Calculations are carried out for the full electrostatic Hamiltonian of the system (except for core electron potentials), and wavefunctions are constructed by self-consistent-field (SCF) and multi-reference configuration interaction (CI) expansions,

$$\psi = \Sigma_k \lambda_k \det (\chi_1^{k} \chi_2^{k} \ldots \chi_n^{k}).$$

In work to date, surfaces have been modeled by a large cluster of atoms and the system subsequently reduced to a smaller effective size by the embedding procedure. In the treatment of transition metals, the first stage deals only with the most delocalized part of the electronic system, the s,p band; d functions on surface atoms are added after a surface s,p electronic subspace is defined. SCF calculations are performed on the s band of the initial cluster and the resulting occupied orbitals are localized by a unitary transformation based on the maximization of exchange interactions with bulk atomic orbitals (2). Final electronic wavefunctions, including the adsorbate, are constructed by configuration interaction, and the coupling of the local electronic subspace and adsorbate to the bulk lattice electrons, $\{\phi_j, j = 1, m\}$, is represented by the modified Hamiltonian,

$$H = \Sigma_i^{N}(-1/2\nabla_i^2 + \Sigma_k^{Q} - Z_k/r_{ik}) + \Sigma^{N}_{i<j} 1/r_{ij} + \Sigma_i^{N} V_i^{eff}$$

where

$$< a(1) \,|\, V_1^{eff} \,|\, b(1) > \; = \; < a(1)\, b(1) \; | \; 1/r_{12} \; | \; \rho(2)>$$
$$- \; < a(1)\, b(2) \; | \; 1/r_{12} \; | \; \gamma(1,2)> \; + \; \Sigma_m \, \lambda_m \, < a|Q_m><Q_m|b>$$

and ρ, γ and Q_m denote densities, exchange functions and atomic orbitals derived from $\{\phi_j, j = 1, m\}$, respectively. Numerous applications to adsorption on transition metal surfaces have been reported and details of the procedure are reported in references (1,2).

For platinum, relativistic atomic solutions of Okada and Matsuoka (5) are used to construct valence orbitals and core electron potentials. These basis functions are of the form

$$\Phi_{p\lambda\alpha} = N \, r^k \exp(-a_p r^2) \, \chi_{\lambda\alpha}(\theta\phi) \qquad \text{(large component spinors)}$$

$$X_{p\lambda\alpha} = \sigma \cdot p \, \Phi_{p\lambda\alpha} \qquad \text{(small component spinors)}$$

In forming valence atomic orbitals and in representing Coulomb and exchange potentials of core electrons acting on the valence basis, it is a good approximation to neglect the contribution of the small component spinors. In the present work, this assumption is made and large component expansions, after renormalization, are used to form core densities, core density matrices and valence orbitals. Atomic orbitals are represented as follows:

$$\psi = \Sigma_p c_p \, f_p(r) \, Y(\theta,\phi) \quad \text{where} \quad f_p(r) = N \, r^k \exp(-a_p r^2).$$

Defining the exact core density as

$$\rho = \Sigma_m \, |\psi_m(1)|^2$$

where the sum is over all core states, an approximate density

$$\rho' = \Sigma_p w_p \, (2b_p/\pi)^{3/4} \exp(-b_p r^2)$$

is obtained by minimization of $< \rho(1) - \rho'(1) \; | \; 1/r_{12} \; | \; \rho(2) - \rho'(2)>$. Core-valence exchange is expressed in terms of a core density matrix,

$$\gamma = \Sigma_m \, \psi_m(1)\psi_m(2)$$

approximated by $\Sigma_m \, \psi'_m(1)\psi'_m(2)$ where the ψ' are smaller gaussian expansions determined by maximum overlap with the exact core orbitals.

Computations based on the above density and exchange expansions can be quite time consuming. Fortunately the summations over spherical harmonic, or the corresponding real, atomic orbitals can be written much more compactly to produce very easily evaluated expansions.

If we consider atomic orbitals of the form, $F_k = Y_k \, G(r)$, where, for example,

$$s_1 = G(r) \qquad\qquad p_1 = x \, G(r)$$
$$d_1 = \sqrt{3} \, xz \, G(r) \qquad\qquad f_1 = 5/2 \, (x^3 - 3/5 \, r^2 x) \, G(r)$$
$$\cdots$$

it is possible to sum over the complete set of angular components for each shell. Defining

$$L = \; \Sigma_k \, Y_k(1) Y_k(2) \qquad \text{(relative to origin M)}$$

it can be shown that

$$L_s = 1$$

$$L_p = x_1 x_2 + y_1 y_2 + z_1 z_2$$

$$L_d = 3/2 \, (x_1 x_2 + y_1 y_2 + z_1 z_2)^2 - \tfrac{1}{2} \, r_1^2 \, r_2^2$$

$$L_f = 5/2 \, (\, x_1 x_2 + y_1 y_2 + z_1 z_2 \,)^3 - 3/2 \, (x_1 x_2 + y_1 y_2 + z_1 z_2) \, r_1^2 \, r_2^2$$

for s, p, d and f shells. In addition to these exchange expressions, relationships for core densities used to generate core potentials acting on valence electrons are obtained by setting $2 = 1$. Projectors, $\Sigma_k \, \langle a(1)|F_k(1)\rangle \langle F_k(1)|b\,(1)\rangle$, follow immediately from evaluation of $\langle a\,(1) \mid L \, G(1)G(2) \mid b(2)\rangle$ and then separating variables. The above relationships make the direct evaluation of core potentials, exchange and Phillips-Kleinman projectors quite tractable. In this work, relativistic contributions are incorporated via the Pauli operators, $-1/8 \, \alpha_f^2 \, \Sigma_i \, p_i^4$ (mass-velocity) and $1/8 \, \alpha_f^2 \, \Sigma_i \, \nabla_i^2 V$ (Darwin). Spin-orbit contributions are not included in the Hamiltonian, although spin-orbit interactions are present in the determination of the relativistic orbitals of Okada and Matsuoka.

3. Ethane Conversion to Ethylene on Nickel and Platinum

Effective utilization of alkanes to produce desirable products requires highly selective pathways that overcome the difficulty of activating CH bonds. A great deal of research has been carried out on the optimization of catalytic processes to achieve the desired selectivity at low cost. Our previous contributions to this subject have involved an accurate theoretical treatment of the reaction of methane with nickel, including the determination of minimum energy reaction pathways and activation barriers for CH_4 dissociation on Ni(111), and a study of the activation of methane over a Fe/ Ni(111) alloy surface (1).

During the past several years, Schmidt and coworkers have studied the production of olefins by oxidative dehydrogenation of ethane and other alkanes on gauze-like supports coated with Pt, Rh or Pd by very short contact time, millisecond, processes

(3,4). On Pt, the selectivity for C_2H_4 is found to be greater than 70% and the conversion of C_2H_6 exceeds 80%. The most probable initiation step is thought to be the oxidative dehydrogenation of C_2H_6 by reaction with adsorbed oxygen atoms,

$$C_2H_6 + O \text{ (ads)} \rightarrow C_2H_5 \text{ (ads)} + OH \text{ (ads)}$$

or possibly with adsorbed hydroxyl

$$C_2H_6 + OH \text{ (ads)} \rightarrow C_2H_5 \text{ (ads)} + H_2O \text{ (ads)}$$

Following C_2H_5 adsorption, gaseous ethylene can be formed by elimination of a β-hydrogen:

$$C_2H_5 \text{ (ads)} + OH \text{ (ads)} \rightarrow C_2H_4 \text{ (gas)} + H_2O \text{ (ads)}$$

or

$$C_2H_5 \text{ (ads)} + O \text{ (ads)} \rightarrow C_2H_4 \text{ (gas)} + OH \text{(ads)}$$

or possibly by direct interaction of H with the surface. Elimination of an α-hydrogen leads to adsorbed ethylidene (-CHCH$_3$) and eventually to undesired products. How these various pathways compete, through their different activation energies and probabilities of reaction, is the subject of the present theoretical study.

The interaction of ethyl groups with transition metals has been examined experimentally under ultrahigh vacuum conditions at low temperature (100~ 400 K) by temperature-programmed desorption (TPD), x-ray photoelectron spectroscopy, high-resolution electron energy loss spectroscopy, and secondary ion mass spectroscopy *(6-9)*. The first hydrogen is apparently abstracted from the ethyl by β-elimination to form C_2H_4. The C_2H_4 may desorb or react on the surface to form ethylidyne (-CCH$_3$). In TPD studies, C_2H_4 was shown to desorb at temperatures of 170 K; thus, at high temperature, one would expect C_2H_4 to desorb immediately after being formed on the metal surface, thereby averting ethylidyne formation. These and other experimental studies suggest that the more noble metals tend to favor β-elimination rather than α-elimination and cracking to carbon, but these metals are catalytically less active than nickel.

In order to investigate the hierarchy in selectivity, high quality first-principles calculations on the adsorption of ethyl (-CH$_2$CH$_3$) on nickel and platinum surfaces are being carried out and the energetics of β-hydrogen elimination by direct interaction with the surface

$$CH_2CH_3 \text{ (ads)} \rightarrow C_2H_4 \text{ (ads)} + H \text{ (ads)}$$

calculated. In this paper, we report the results of calculations on Ni(100) and Pt(100). The goal is to understand the α and β-hydrogen elimination mechanisms by calculation of the reaction energetics, activation barriers and geometric features of the pathways.

Insight into the reaction mechanism can be obtained by concentrating on two key questions: a) the energetics of adsorbed ethyl as it tilts toward the surface and b) the energetics and geometry of ethylene coadsorbed with hydrogen. We begin with a consideration of ethylene on the surface. When ethylene adsorbs on Ni(100), several adsorption sites are plausible. Figure 1 gives the energies for planar and nonplanar ethylene at four-fold, bridge and other sites on Ni(100).

The classical di-σ configuration (C-C axis above a Ni-Ni bond) with the HCH plane of nuclei tilted away from the surface by 20° is found to be strongly bound, E_{ads} = 8.9 kcal/mol. However, the di-σ configuration at the four-fold site is slightly more stable, E_{ads} = 9.8 kcal/mol. While the latter result is perhaps surprising, the opportunity to form one bond with a pair of surface Ni atoms (a three-center bond) can have advantages over forming a bond with a single Ni atom since there is a different pattern of rupture of surface Ni-Ni bonds. At all sites studied, going from a planar to nonplanar ethylene geometry lowers the energy by ~ 3 kcal/mol. The ethylene C-C bond lengthens from the gas phase value of 1.35 Å to 1.43 Å. Ethylene on Pt(100) is

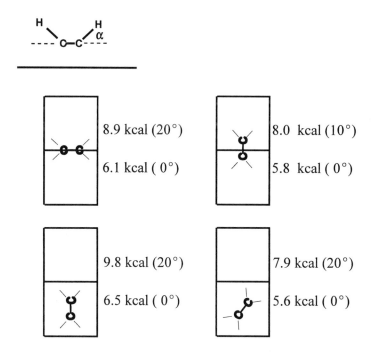

Figure 1   Ethylene adsorption at several sites on Ni(100). Surface nickel atoms are located at the vertices of the boxes shown in the figure. Energies from embedded cluster calculations are reported for planar and nonplanar geometries.   For the nonplanar geometry, the HCH tilt angle is 20° from the horizontal. Unless otherwise indicated, the C-surface distance is 2.35 Å and the C-C distance is 1.38 Å for the planar cases, and 2.28 Å and 1.43 Å, respectively, for the nonplanar cases. These values correspond to the equilibrium distances for the most favorable adsorption site.

qualitatively similar. Figure 2 shows the energy variation with HCH angle for ethylene adsorbed at a four-fold site on Pt(100) with H coadsorbed at the center of an adjacent four-fold site. The angle variation is for an asymmetric location of ethylene in which one C is above the center of the four-fold site, corresponding to the product formed immediately after the transfer of H from ethyl adsorbed at a four-fold site to the surface. Also shown in the figure is the energy lowering that occurs when ethylene is moved so that the midpoint of the C-C bond is directly above the four-fold site. The C-C bond length for ethylene at the asymmetric location is shorter than that obtained for ethylene at the center of the four-fold site without H coadsorbed. The value is much shorter than the C-C bond length in adsorbed ethyl which is typical of that for a C-C single bond, 1.53 Å. This observation is an important factor in explaining what geometry changes must accompany H transfer to the surface to produce ethylene from adsorbed ethyl.

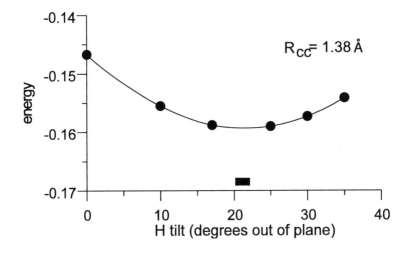

Figure 2. Ethylene coadsorbed with H on Pt (100). The energy variation is for an out-of-plane tilting of ethylene hydrogens for ethylene adsorbed on Pt(100) with one C atom above the four-fold site and H coadsorbed above the center of an adjacent four-fold site. The point labeled (■) shows the energy when ethylene in its optimum geometry is shifted so that the center of the C-C bond is directly above the center of the four-fold site. The energy is in a.u. (1.a.u. = 27.21 eV).

Next we consider adsorbed ethyl. The calculations show that the C-C bond length remains close to that of a typical single bond. Figure 3 shows that tilting ethyl toward the surface gives an equilibrium geometry corresponding to a $50°$ tilt from perpendicular for ethyl/Ni(100). This tilt angle is primarily a consequence of the nearly tetrahedral orientation of bonds around the α carbon. On tilting further, the graph shows a sharp increase in energy as a hydrogen on the β carbon comes too close to the surface. For a tilt angle of $90°$, the β hydrogen would be within 0.8 Å of the surface which is much too close for a hydrogen that remains bonded to another atom. For tilt angles of 60° and 90°, other symbols in the figure denote the change in energy when the distance from the α carbon to the surface is increased. At 60°, the energy increases slightly as the C-surface distance is increased from its equilibrium value. At 90°, the energy decreases as the C-surface distance is increased until the stabilization gained by moving the β hydrogen away from the surface is matched by the destabilization from the increased C-surface distance.

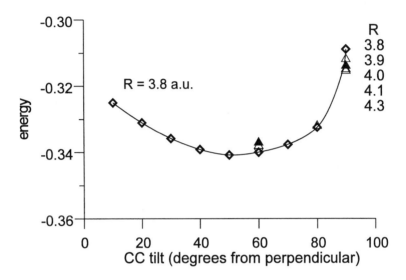

Figure 3. Energy change on C_2H_5 tilt on Ni(100) . Ethyl is adsorbed on Ni(100) at the center of a four-fold site. The equilibrium geometry corresponds to a 50° tilt from the surface normal. The sharp rise in energy at larger tilt angles is due to repulsion between a β hydrogen and the surface. Symbols at 60° and 90° indicate changes in energy when the distance from the α carbon to the surface is increased from 3.8 a.u. (2.01 Å). The energy is in a.u. (1.a.u. = 27.21 eV).

Figure 4 reports the variation in energy for ethyl tilt on Pt(100) in which the ethyl C is bonded at the center of the four-fold site. Since the Pt-Pt nearest neighbor distance is much longer than the Ni-Ni distance, both ethyl and adsorbed H are closer to the surface in their equilibrium geometries. However, the energy variation on tilting ethyl is very similar to that for Ni(100). Once H begins to penetrate the electron density of the surface, for example for a tilt greater than 70°, the energy increases sharply. The optimum tilt angle is 60° sightly larger than the value for Ni(100). For tilt angles of 60° and 80°, other symbols in the figure denote the change in energy when the distance from the α carbon to the surface is increased. At 60°, the energy increases slightly as the C-surface distance is increased from its equilibrium value. At 80°, the energy decreases considerably as the C-surface distance is increased until the stabilization gained by moving the β hydrogen away from the surface is matched by the destabilization from the increased distance of the α carbon from the surface.

We now explore possible transition states for beta H elimination. If we consider ethyl on Ni(100), tilted 50°, the distance from the β H to the surface is 2.44 Å which is much larger than the equilibrium distance for a completely dissociated H of 1.0 Å. If the C-H bond is simply stretched in order to move H toward the surface, the energy increases sharply. The same is found for ethyl on Pt(100): transferring H from ethyl adsorbed at its equilibrium geometry is a very energetic process. Although the d-electrons assist in preferentially stabilizing the transition state, their bonding is insufficient to allow H to be transferred through space from the beta C.

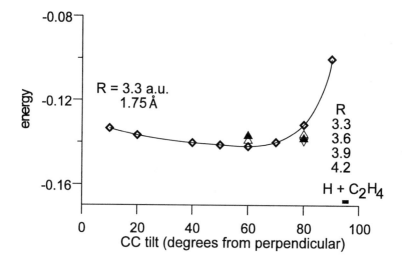

Figure 4.   Energy change on C_2H_5 tilt on Pt(100). Ethyl is adsorbed on Pt(100) at the center of a four-fold site. The equilibrium geometry corresponds to a 60° tilt from the surface normal.   The sharp rise in energy at larger tilt angles is due to repulsion between a β hydrogen and the surface.   Symbols at 60° and 80° indicate changes in energy when the distance from the α carbon to the surface is increased from 3.3 a.u. (1.75 Å) to the values shown in the inset.   The total energy of H and ethylene coadsorbed at the centers of adjacent four-fold sites is indicated. The energy is in a.u. (1.a.u. = 27.21 eV).

Of course, it is an oversimplification to consider only the CH stretch reaction coordinate. In order to calculate the energy barrier accurately, it is necessary to allow simultaneously a relaxation of the resulting - CH_2 - CH_2 geometry as H is transferred to the surface. As noted earlier, one of the principal coordinates affecting the energy is the C-C distance where a large decrease in energy is expected in going from a single to double bond distance; likewise, angle variations accompanying the transition from tetrahedrally bonded C to that corresponding to ethylene with tilted hydrogens are expected to be important.

The present work explores only a few dissociation pathways, but one feature of the (100) surface seems clear: the dissociation is facilitated if the H and ethylene products are in different four-fold cells on the surface. In all of the pathways investigated, CH bond stretch occurs across a metal-metal bridge. Equally promising are pathways in which CH bond stretch and dissociation take place across a single surface atom site, and such possibilities will be considered in future studies.

After carrying out rather extensive geometry optimizations, it turns out that there is a simple way to describe the motion leading to the transition state for H transfer. The key point is not to focus on H transfer from the equilibrium geometry of adsorbed ethyl but rather to allow ethyl to vibrate into or otherwise penetrate the electron density of the metal before transfer of hydrogen. Specifically, by tilting ethyl beyond its equilibrium geometry to $80°$, a more energetic configuration is created that lies ~ 5 kcal/mol above that of the equilibrium geometry. Hydrogen is now closer to the surface (at 1.32 Å on Ni and 1.06 Å on Pt) and CH bond stretch is found to occur with a much smaller increase in energy. This is true for ethyl on both Ni(100) and Pt(100). The key steps in such a reaction pathway are summarized in Tables 1 and 2 for nickel and

Table 1. Ethylene formation from ethyl adsorbed on Ni(100). The energy increase for key steps in the lowest energy reaction pathway for beta hydrogen transfer from ethyl to the surface are shown. The perpendicular distances of the alpha carbon and beta hydrogen from the surface plane of nuclei are denoted by R_C and R_H, respectively. The ΔE value is the increase in energy compared to the previous geometry; the first entry, 5.5 kcal/mol, is relative to ethyl in its equilibrium geometry on the surface.

	ΔE (kcal/mol)
Ethyl bends toward surface ($\alpha = 80°$) slightly penetrating repulsive region $R_{CC} = 1.53$ Å, $R_C = 2.01$ Å $R_{CH} = 1.09$ Å, $R_H = 1.32$ Å	5.5
CH stretch 0.26 Å (24%) $R_{CH} = 1.36$ Å, $R_H = 1.09$ Å, $\Delta CC = -0.05$ Å	5.8
CH stretch 0.43 Å (40%) $R_{CH} = 1.52$ Å, $R_H = 1.09$ Å HCH angles optimized	4.7
CH stretch 0.64 Å (58%) $R_{CH} = 1.73$ Å, $R_H = 0.95$ Å	0.8
Overall activation barrier (sum of steps)	16.8

platinum surfaces, respectively, and motions leading to the transition state for beta hydrogen transfer for Pt(100) are depicted in Figure 5. On both surfaces, the HCH angles remain nearly tetrahedral in the early stages of CH stretch, e.g., up to an increase in CH bond length of 24% from the equilibrium value of 1.09 Å. For the reaction on Pt(100), as shown in Table 2, most of the barrier is found to occur during the early stages of CH stretch. Between CH bond length increases of 24% and 40%, HCH angle variations begin to be important.

The calculations reveal a significant difference between Ni and Pt, however. For ethyl on Pt(100) it is found that as the CH bond is stretched the emerging ethylene and H separate by ethylene sliding toward a more favorable site on the surface whereas for Ni(100) the H transfer occurs with less motion of the ethylene fragment. Table 2 shows a decrease in the intermediate and final contributions to the barrier height for Pt due to this effect. More work is currently underway in an effort to understand such subtle distinctions of the dissociation process on these two metals.

Table 2. Ethylene formation from ethyl adsorbed on Pt(100). The energy increase for key steps in the lowest energy reaction pathway for beta hydrogen transfer from ethyl to the surface are shown. The ΔE value is the increase in energy compared to the previous geometry; the first entry, 5.0 kcal/mol, is relative to ethyl in its equilibrium geometry on the surface. The perpendicular distances of the alpha carbon and beta hydrogen from the surface plane of nuclei are denoted by R_C and R_H, respectively. The lateral shift moves ethylene toward the center of the four-fold site increasing its distance from the H being transferred to the surface.

	ΔE (kcal/mol)
Ethyl bends toward surface ($\alpha = 80°$) slightly penetrating repulsive region $R_{CC} = 1.53$ Å, $R_C = 1.75$ Å $R_{CH} = 1.09$Å, $R_H = 1.06$ Å	5.0
CH stretch 0.26 Å (24%) $R_{CH} = 1.36$ Å, $R_H = 0.82$ Å $\Delta CC = -0.05$ Å	8.1
CH stretch 0.55 Å (40%) $R_{CH} = 1.53$ Å, $R_H = 0.67$ Å 0.21Å lateral shift of C_2H_4	1.0
CH stretch 0.66 Å (53%) $R_{CH} = 1.68$ Å, $R_H = 0.48$ Å 0.21Å lateral shift of C_2H_4	-0.1
Overall activation barrier (sum of $\Delta E > 0$ steps)	14.1

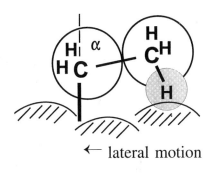

CH stretch

penetration of density

← lateral motion

Figure 5. Depiction of the pathway leading to the transition state for the transfer of a beta hydrogen from adsorbed ethyl to the surface of Pt(100). Ethyl is tilted beyond its equilibrium geometry until one of the beta hydrogens penetrates the repulsive region of the metal electron density corresponding to a 5 kcal/mol energy increase for a tilt angle $\alpha = 80°$. CH stretch is accompanied by a lateral separation of the emerging ethylene and adsorbed H. Contributions to the overall activation barrier of 14 kcal/mol are indicated at points along the reaction pathway in Table 2.

Acknowledgment

Support of this research by the U.S. Dept. of Energy is gratefully acknowledged.

Literature Cited

(1) Whitten, J. L.; Yang, H. *Surf. Sci. Reports* **1996**, *24*, 55..
(2) Whitten, J. L.; Yang, H. *Int. J. Quantum Chem.* **1995**, *29*, 41.
(3) Huff, M.; Schmidt, L. D. *J. Phys. Chem.* **1993**, *97*, 11815; Goetsch, D. A.; Schmidt, L. D. *Science* **1996**, *271*, 1561.
(4) Huff, M.; Torniainen, P. M.; Schmidt, L. D. *Catal. Today* **1994**, *21*, 113.
(5) Okada, S.; Matsuoka, O. *J. Chem. Phys.* **1989**, *91*, 4183.
(6) Zaera, F. *J. Am. Chem. Soc.* **1989**, *111*, 8744; *J. Phys. Chem.* **1990**, *94*, 8350.
(7) Jenk, C. J.; Chiang, C. M.; Bent, B. E. *J. Am. Chem. Soc.* **1991**, *113*, 6308.
(8) Liu, Z. M.; Zhou, X. L.; Buchanan, D. A.; White, J. M. *Chem. Ind.* **1994**, *53*, 521.
(9) Zhou, X. L.; Solymosi, F.; Blass, P. M.; Cannon, K. C.; White, J. M. *Surf. Sci.*, **1989**, *219*, 294.

Chapter 22

Electrostatic Stabilization of the Transition-State by a Solid Catalyst: Dissociative Chemisorption of NH₃ on the Stepped Si(111) Surface

Krisztina Kádas[1] and Gábor Náray-Szabó[2]

[1]Department of Theoretical Physics, Institute of Physics, Technical University of Budapest, H-1521 Budapest, Hungary
[2]Department of Theoretical Chemistry, Loránd, Eötvös University, Pázmány Péter st. 2, H-1117 Budapest, Hungary

The dissociative chemisorption of NH₃ on the stepped silicon (111) surface was studied by means of semiempirical quantum chemical calculations based on molecular orbital theory. The activation energy of the dissociation for different geometrical arrangements were examined and compared to that of the gas phase ammonia molecule. It was stressed that electrostatics plays an important role in the reduction of the energy of the transition state.

1. Introduction

During a catalytic process the activation free energy of a system decreases via energetic or entropy stabilization of the transition-state complex or via alteration of the reaction pathway of the non-catalytic process. There exist various effects determining the stabilization of the transition-state complex. Perhaps the most important is the electrostatic interaction with the catalyst that may be effective in homogeneous, heterogeneous and enzyme catalysis, as well *(1)*. Several types of crystal surfaces are known to be good catalysts and this may be due to their strongly different atomic arrangement and electronic properties as compared to the bulk which results in specific interactions with the reacting species.

In our previous papers *(2,3)* we have shown that the increased reactivity of the reconstructed Si(111) surface is mainly due to its specific electrostatic properties. The molecular electrostatic potential (MEP) changes extremely fast on the surface providing a large gradient, i.e. large molecular electrostatic field (MEF). Therefore we suggested that the MEF can be applied as a simple reactivity map for the silicon surface; experimentally located adsorption sites are correctly predicted by MEF maps. We have also shown *(4)* that the magnitude of the MEF considerably increases in the vicinity of surface defects, which is in good agreement with the experimental observation that steps and kinks are preferred catalytic sites *(5)*.

In this paper we report on a preliminary study of the dissociative chemisorption of ammonia on the stepped Si(111) surface, a simple model for heterogeneous catalysis. We provide some evidence for our hypothesis, namely that the transition state of the reaction is stabilized by the electrostatic field emerging in the vicinity of the surface. Though our results are promising further, more detailed, calculations are necessary to support this idea in a sufficient way.

2. Electrostatic Catalysis

We define electrostatic catalysis as a process where electrostatic interactions stabilize the transition-state complex as compared to the initial state. In the following we mention some examples. One is the acceleration of 1, 5-hydride shift in cyclopentadiene upon the influence of a lithium cation (6). Water and other polar solvents may influence the activation energy of reactions involving polar or ionic reaction partners and stabilize structures with higher dipole moments as compared to the gas phase (7). The activation energy decreases in aqueous solution if the dipole moment of the transition-state complex is larger than that of the initial structure formed by the reactants. In the opposite case, when the dipole moment is larger for the reactants, the solvent can stabilize them to a larger extent than the transition state does, thus the reaction becomes slower in water than in the gas phase, an electrostatic "anticatalysis" takes place.

Enzymatic processes are also good examples of electrostatic catalysis, since one of the main factors (beyond entropy and general acid–base catalysis), that are widely accepted to contribute to enzyme catalysis is electrostatics. Enzymes can stabilize polar species, like ion pairs or $(- + -)$ charge distributions, more effectively than water, because they have appropriately oriented dipoles, whereas water dipoles are randomised by outer solvation shells interacting with bulk solvent. While the water dipoles have to reorient in order to stabilize the transition–state charges, the enzyme dipoles are already pre–oriented toward the transition–state charge distribution (8,9). Therefore enzymes can act as "supersolvents" that are designed to stabilize transition states, relative to the corresponding states in aqueous solution.

Heterogeneous catalytic processes can also be of electrostatic nature. A widely known example for this is catalysis inside microporous materials, e.g. zeolite pores. The MEP changes remarkably fast on the surface of such pores, which means that its gradient, i.e. the MEF is large. It has been accepted that the catalytic effect of zeolites is mainly due to their electrostatic properties. On the one hand, the strong electrostatic field within the pores polarizes covalent bonds, promoting them for fission, and stabilizes ion pairs. On the other hand, the large negative MEP inside the pores stabilizes cations (10,11,12). This latter effect prefers e.g. the formation of protonated species and may result in stabilization of the corresponding transition states.

Several examples can be found for electrostatic heterogeneous catalysis among the reactions at gas–crystal surfaces. Here either the surface reconstruction or surface defects can lead to atomic structures which differ remarkably from the ideal bulk crystal structure, providing specific electrostatic properties, and giving rise to catalytic effects. In our previous papers (2,3) we examined the role of electrostatics in the reactivity of the reconstructed Si (111)–(7x7) surface. We have found that the MEF has the largest values above rest atoms, center and corner adatoms, respectively, and it is larger above the faulted than the unfaulted subunit of the (7x7) unit cell. We have shown, that the MEF maps calculated in a plane lying above the surface changes parallel with the binding energy maps obtained for the combined Si surface and ammonia system. Therefore we drew the conclusion that MEF maps, as simple reactivity maps, can be used to predict the experimental reactivity order. In another study (4), we have shown that electrostatics plays a determining role also in the enhanced reactivity observed near surface defects on the silicon (111) surface.

In the following we will discuss an example of electrostatic catalysis, the dissociative chemisorption of ammonia on the stepped Si (111) surface, in some more detail.

3. Models and Methods

Ultrahigh vacuum cleavage of the Si(111) surface at low temperature results in a (2x1) structure. It is described by the π–bonded chain reconstruction model (Pandey chain) (13). This model provides the best agreement with experimental results, e.g. low–energy electron diffraction (14), electron–energy–loss spectroscopy (15), STM (16), photoemission (17), ion scattering (18), optical absorption (19), and gives the lowest total energy (20).

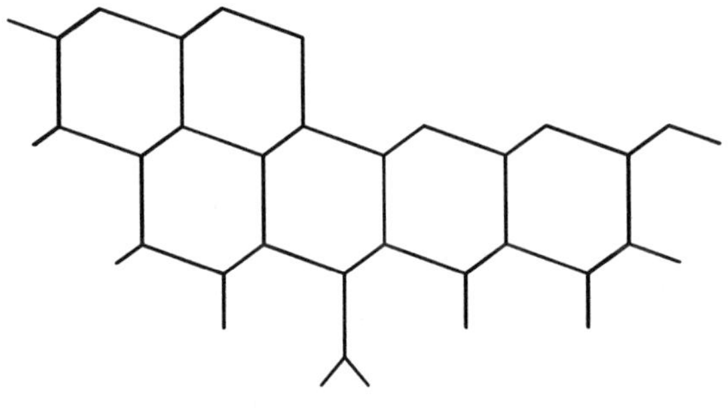

Figure 1: Unreconstructed surface step model (side view) showing also the terminating hydrogen atoms.

The unreconstructed Si(111) surface with a surface step was modelled by a finite six–layer slab of ideal crystal structure. This thickness has been proven to represent reliably the real surface *(2,3)*. Dangling bonds at the cluster boundaries were terminated by hydrogen atoms pointing towards the bulk. The Si–H bond distance was set to 1.43 Å. The surface step was created by removing appropriate atoms from the first and second layers (Figure 1). Our unreconstructed surface model contains 80 Si and 56 H atoms. In this case a $=SiH_2$ group has been introduced into the model pointing toward the bulk in order to obtain a closed shell cluster.

The reconstructed surface step model was built on the basis of a low–energy electron diffraction experiment published by Himpsel et al. *(14)*. The surface step was obtained after removing appropriate atoms from the first and second layer and retaining the reconstruction of the topmost double layer also near the surface step, in the third and fourth layers (Figure 2). This model consists of 97 Si and 72 H atoms. A part of this model, displayed by thick lines on Figure 2, will be used to represent selected steps of the chemisorption process later.

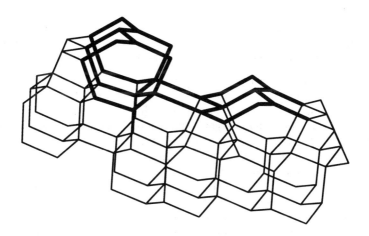

Figure 2: Reconstructed surface step model without displaying terminating H atoms. The part of the model cluster denoted by thick lines will be used to represent reaction steps later.

The semiempirical NDDO method with AM1 parametrization *(21)* has been used to calculate the optimal geometries of the models. Semiempirical calculations have been performed by the MOPAC program package *(22)* on HP 9000/809 K–100 and IBM 9076 SP2 workstations. Typical running times were 14–17 hours (CPU time).

4. Results and Discussion

In our previous paper *(4)* we calculated the MEF maps both for the unreconstructed and reconstructed surface step models. As compared to the ideal surface, where the MEF varies between 0.4 and 1.8 V/nm, the field increases significantly near surface steps reaching values of 8 V/nm in certain regions. The largest MEF increase was observed above surface atoms in the close vicinity of the surface step. The calculated MEF increase was found larger for the reconstructed, than for the unreconstructed step.

The initial structural arrangement of the "surface step model + ammonia molecule" system was taken from a previous geometry optimization *(4)* both for the unreconstructed and reconstructed surface step models. In both cases the NH_3 molecule is situated in a region with increased MEF near the surface step. The Si–N distance is 1.86 Å for the unreconstructed and 1.90 Å for the reconstructed surface step model. The hydrogen atoms of the ammonia molecule point towards the gas phase. The N–H bond distances are near 1.01 Å.

To model the dissociative chemisorption, one of the N–H bonds was elongated by 0.2 Å at each step. At every elongated distance, atoms of the ammonia molecule were relaxed so that the elongated N–H bond length was kept fixed.

Figure 3 shows the total energy curves of the reconstructed surface step model as a function of the elongated N–H distance. The solid line curve belongs to a model in which the H pointing towards the surface step dissociates from the NH_3 molecule. Points A, B and C will be used to represent steps of the chemisorption process later. The short dashed line curve shows the total energy of a model in which the H of the elongated N–H bond points away from the surface step. The long dashed line curve displays the results obtained for the gas phase ammonia molecule. As Figure 3. shows, the activation energy of the dissociation significantly decreases above the silicon surface as compared to that of the gas phase NH_3 molecule, meaning, that the surface stabilizes the transition–state complex. The activation energies are slightly different for the H pointing towards or away from the surface step (217.1 and 202.5 kJ/mol, respectively). The smaller activation energy obtained in the latter case indicates that the dissociation is easier for this geometrical arrangement.

For the unreconstructed surface step model the activation energies are 136.4 and 148.8 kJ/mol for the H pointing towards and away from the surface step, respectively. The dissociation is energetically favourable for the H pointing towards the surface step. The activation energy is smaller for the unreconstructed surface step than for the reconstructed one by 80.7 kJ/mol for the H pointing towards the surface step, and by 53.7 kJ/mol for the other hydrogen.

We suggest that electrostatic interaction plays an important role in the stabilization of the transition–state complex in the dissociative chemisorption of NH_3 on the stepped silicon surface.

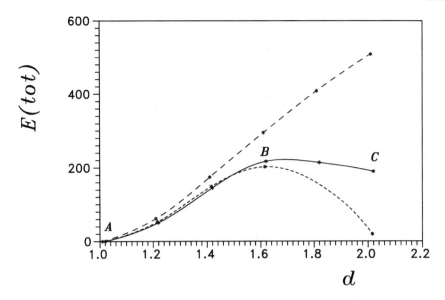

Figure 3: Total energies (in kJ/mol) as a function of the elongated N–H bond distance (Å) obtained for the reconstructed surface step model. Energies are related to that of the "surface step + NH$_3$" system in its initial geometry. The solid and short dashed line curves belong to the models in which the H of the elongated N–H bond points towards and away from the surface step, respectively. The long dashed line curve shows the total energy of the gas phase ammonia molecule.

It is not so simple to estimate this effect since both in the initial and transition states covalent bonds are formed with surface silicon atoms and charge transfer between the reactant and the surface takes place. A rough estimate can be obtained, however, by considering the electrostatic potential of the unperturbed surface model at the atoms of the dissociating ammonia molecule. The ammonia-surface interaction energy is then obtained as $E_{int} = \sum V(i)q(i)$ where $V(i)$ and $q(i)$ are the surface potential and Mulliken net charge on atom i. According to our semiempirical AM1 calculations we obtained 61.1 and -22.4 kJ/mol for E_{int} in the initial and transition states, respectively. Thus, while the initial state is destabilized, the transition state is stabilized by the electrostatic field of the stepped silicon surface resulting in the reduction of the barrier height. Studies to provide further evidence for the electrostatic nature of catalytic rate acceleration are underway. Though, as we have shown, electrostatics is important, it is also possible that other, e.g. dynamical, effects play a crucial role *(23)*.

Figure 4. displays snapshots of the dissociative chemisorption. Only the part of the model cluster – drawn by thick lines in Figure 2. – represents the surface. Figure 4.A, B and C correspond to point A, B and C on Figure 3, respectively. Figure

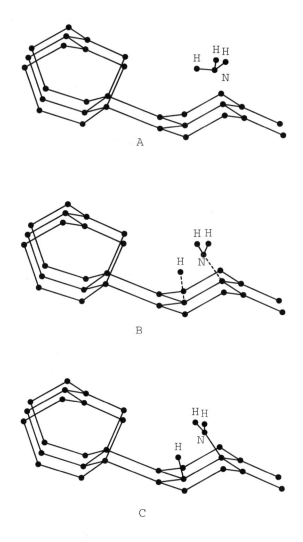

Figure 4: Snapshots of the dissociative chemisorption of NH_3 on the stepped Si(111) surface. Panel (A) shows the initial position, while panel (B) displays the system in its transition state, where the elongated N–H bond is already dissociated. The dissociation species (H and NH_2) are bound to surface silicon atoms on panel (C).

4.A shows the initial position with normal N–H bond distances. Figure 4.B belongs to the transition–state complex at maximal total energy, where the elongated N–H bond is already dissociated. At the end of the chemisorption process the dissociation species (H and NH_2) form bonds with surface Si atoms (Figure 4.C).

5. Conclusions

We examined the dissociative chemisorption of NH_3 on the stepped silicon surface by means of semiempirical quantum chemical calculations. For both the unreconstructed and reconstructed surface step models the surface stabilizes the transition–state complex. We suspect that the main effect governing this stabilization is the electrostatic interaction. The chemisorption process was discussed in detail.

Acknowledgements

This work was supported by National Scientific Research Fund (OTKA) under grant numbers D 23456 and T 019387.

Literature Cited

(1) Náray–Szabó, G. in *Encyclopedia of Computational Chemistry*, P.v.R. Schleyer ed., Wiley, to be published.

(2) Kádas, K.; Farkas, Ö.; Náray–Szabó, G. *ACH Models Chem.* **1995,** *132*, 125.

(3) Kádas, K.; Kugler, S.; Náray–Szabó, G. *J. Phys. Chem.* **1996,** *100*, 8462.

(4) Kádas, K.; Náray–Szabó, G. *J. Mol. Struct. THEOCHEM*, in press.

(5) Somorjai, G.A. *Chemistry in Two Dimensions: Surfaces*, Cornell University Press: Ithaca, 1981.

(6) Jiao, H.; Schleyer, P.v.R. *J. Am. Chem. Soc.* **1995,** *117*, 11529.

(7) Jorgensen, W.L. *Solvent as Catalyst - Computational Studies of Organic Reactions in Solution*, Lecture No. 188 at the 215th National Meeting of the ACS, Computers in Chemistry Division, Dallas, TX, 1998.

(8) Warshel, A. *Proc. Natl. Acad. Sci. U.S.A.* **1978,** *75*, 5250.

(9) Rao, S.N.; Singh, U.C.; Bash, P.A.; Kollman, P.A. *Nature (London)* **1987,** *328*, 551.

(10) White, J.C.; Nicholas, J.B.; Hess, A.C. *J. Phys. Chem.* **1997,** *101*, 590.

(11) Ángyán, J.G.; Ferenczy, G.; Nagy, P.; Náray–Szabó, G. *Collect. Czech. Chem. Commun.* **1988,** *53,* 2308.

(12) Ferenczy, G.; Ángyán, J.G. *J. Chem. Soc. Faraday Trans.* **1990,** *86,* 3461.

(13) Pandey, K.C. *Phys. Rev. Lett.* **1981,** *47,* 1913.

(14) Himpsel, F.J.; Marcus, P.M.; Tromp, R.; Batra, Inder P.; Cook, M.R.; Jona, F.; Liu, H. *Phys. Rev. B* **1984,** *30,* 2257.

(15) Matz, R.; Lüth, H.; Ritz, A. *Solid State Commun.* **1983,** *46,* 343.

(16) Stroscio, J.A.; Feenstra, R.M.; Fein, A.P. *J. Vac. Sci. Technol. A* **1987,** *5,* 838.

(17) Himpsel, F.J.; Heimann, P.; Eastman, D.E. *Phys. Rev. B* **1981,** *24,* 2023.

(18) Tromp, R.M.; Smit, L.; van der Veen, J.E. *Phys. Rev. Lett.* **1983,** *51,* 1972.

(19) Chiaradia, P.; Cricenti, A.; Selci, S.; Chiarotti, G. *Phys. Rev. Lett.* **1984,** *52,* 1145.

(20) Pandey, K.C. *Phys. Rev. Lett.* **1982,** *49,* 223.

(21) Dewar, M.J.S; Zoebisch, E.G.; Healy, E.F.; Stewart, J.J.P. *J. Am. Chem Soc.* **1985,** *107,* 3902.

(22) Stewart, J.J.P. *QCPE Bull.* **1989,** *9,* 10. QCPE Program 455, MOPAC Version 6.0.

(23) Stich, I. *Surface Science* **1996,** *368,* 152.

CATALYSIS BY METAL OXIDES AND ZEOLITES

Chapter 23

Kinetic Theory and Transition State Simulation of Dynamics in Zeolites

Chandra Saravanan[1], Fabien Jousse[1], and Scott M. Auerbach[1-3]

[1]Departments of Chemistry and [2]Chemical Engineering, University of Massachusetts, Amherst, MA 01003

We have developed and applied modeling techniques specialized for infrequent events to study the transport of benzene in the zeolite faujasite, focusing on the microscopic factors that control short and long range mobility. We describe the first exact site-to-site flux correlation function calculation for a non-spherical molecule inside a zeolite. We find that transition state theory is qualitatively correct for some but not all jumps. We outline a recently developed analytical theory indicating which transition states control diffusion in Na-Y zeolite. Our new theory predicts self-diffusion coefficients in qualitative agreement with pulsed field gradient NMR, and in qualitative disagreement with tracer zero-length column data.

The transport properties of adsorbed molecules play a central role in catalytic and separation processes that take place within zeolite cavities. Although significant effort has been devoted to understanding diffusion in zeolites (1,2), several basic questions persist: Can transition state theory and flux correlation function theory (3) be used to model complicated molecular jump processes in zeolites? Do these simulations need to allow for framework and molecular flexibility? How does the competition between energy and entropy modify the rates of these jump processes? What are the fundamental interactions that control the temperature and pressure dependence of diffusion in zeolites? In the present article, we begin to address these issues by modeling benzene jump diffusion in faujasite type zeolites.

Significant effort has been devoted to modeling benzene adsorption and

---

[3]Corresponding author.

diffusion in the faujasites Na-X and Na-Y (4–20), motivated by persistent discrepancies among different experimental probes of mobility (21–23). Pulsed field gradient NMR diffusivities for benzene in Na-X decrease monotonically with loading (21), while tracer zero-length column data *increase* monotonically with loading (23). Addressing this discrepancy with theory and simulation will provide deep understanding of the microscopic physics essential to these transport phenomena. Our new theory predicts self-diffusion coefficients in qualitative agreement with pulsed field gradient NMR, and in qualitative disagreement with tracer zero-length column data.

Benzene Jump Dynamics in Na-Y Zeolite

We model benzene self-diffusion in Na-Y by replacing the zeolite with a three dimensional lattice of binding sites. Benzene has two predominant binding sites in Na-Y (Si:Al=2.0). In the primary site, denoted as S_{II}, benzene is facially coordinated to a zeolite 6-ring, *ca.* 2.7 Å above a Na cation (24). In the secondary site, denoted as W, benzene is centered in the 12-ring window separating adjacent cages, *ca.* 5.3 Å from the S_{II} site (24). Figure 1 shows schematic adsorption sites and jumps for benzene in Na-Y. The lattice of benzene binding sites in Na-Y contains four tetrahedrally arranged S_{II} sites and four tetrahedrally arranged, doubly shared W sites per cage. A saturation coverage of *ca.* 6 molecules per cage is found for benzene in Na-Y, corresponding to occupation of all S_{II} and W sites.

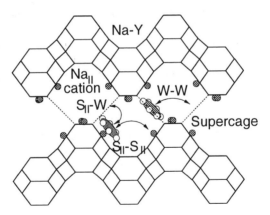

Figure 1. Schematic sites and jumps for benzene in Na-Y. Adapted from Ref. (20).

Since benzene becomes trapped with long residence times at cationic sites, transition state theory (TST) and flux correlation function (FCF) theory (3) must be used to calculate site-to-site jump rate coefficients. The exact rate coefficient for a jump from site i to site j takes the standard form:

$$k_{i \to j} = k_{i \to j}^{\text{TST}} \times f_{ij}(\tau), \tag{1}$$

where $k_{i \to j}^{\text{TST}}$ is the TST rate coefficient given by:

$$k_{i \to j}^{\text{TST}} = \frac{1}{2} \left(\frac{2k_B T}{\pi m} \right)^{1/2} \frac{Q^{\ddagger}}{Q_i}. \tag{2}$$

In Equation (1), $f_{ij}(\tau)$ is the normalized reactive flux correlation function evaluated at the plateau time, i.e. the typical time required for relaxation from the transition state, which is usually much less than $1/k_{i \to j}$. Please see Ref. (18) for computational details. Although previous TST and FCF calculations for jumps of spherical species in zeolites have been reported (25–28), modeling jumps of non-spherical species such as benzene is much more complicated due to angle-dependent dividing surfaces (16–18,29–31). We begin to address this by defining a planar dividing region of spatial width ε, which can encompass some dividing surface curvature, and by using FCF dynamics to correct TST for an inaccurately defined dividing surface. Our calculations below provide the first numerically exact site-to-site rate coefficients for non-spherical molecules in zeolites.

We applied Voter's Monte Carlo displacement vector approach (32) to the calculation of the partition function ratio in Equation (2). We generalized this method to the 6-dimensional case of a rigid, non-spherical molecule as described in Ref. (18). All TST and FCF calculations were carried out with our forcefield for aromatics in zeolites (9), which includes electrostatic and Lennard-Jones interactions between benzene and Na-Y. For the case of a single benzene molecule in Na-Y, we found the fast multipole method (33) to be the most efficient for calculating electrostatic interactions; while for the other cases considered below, the standard Ewald approach was the best. The simulation cell consisted of 640 zeolite atoms and one benzene molecule, under cubic periodic boundary conditions with a repeat distance of 24.8 Å. The Si:Al ratio of 2.0 requires 64 Na atoms in the simulation cell, assumed to occupy fully the I′ sites in the β cages and the II sites in the supercages. We used the "average T-site" approach, wherein the 128 Si and 64 Al atoms are replaced by 192 identical "T-atoms," each with properties intermediate between that of Si and Al. All calculations reported below were carried out on an IBM RS/6000 with a 604e 200 MHz PowerPC processor, capable of calculating ca. 20 host–guest energies per CPU second.

Rigid Benzene in a Rigid Framework. We begin by calculating the $S_{II} \to S_{II}$, $S_{II} \to W$, $W \to S_{II}$ and $W \to W$ rate coefficients keeping both benzene and Na-Y rigid. A typical TST calculation for this system required 10 CPU hours, averaging over 10^5–10^6 configurations. The corresponding FCF dynamical correction typically required 48 additional CPU hours, consisting of 2000 trajectories, each lasting ca. 2 ps. In addition to the usual checks for statistical convergence of $k_{i \to j}^{\text{TST}}$ and $f_{ij}(\tau)$, we confirmed that the TST rate coefficients obey detailed balance for the $S_{II} \rightleftharpoons W$ equilibrium, by calculating independently $K_{\text{eq}}(S_{II} \to W)$ using Voter's jump vector method. In addition, we confirmed explicitly that the TST rate coefficients depended upon the dividing surface location; while the FCF rates were independent of the dividing surface location.

The results of these calculations are shown in Table I. Several remarks can be made about the data in Table I. First, the $S_{II} \rightleftharpoons W$ rate coefficients do indeed obey detailed balance, when compared to the simulated equilibrium coefficient. This equilibrium coefficient energetically favors the S_{II} site for its strong π−cation interaction, but entropically favors the W site for its greater flexibility; a trend that is mirrored by the rate coefficients. In all cases the activation energies from FCF calculations, denoted "Corr. function" in Table I, agree well with those from minimum energy path calculations, denoted "MEP" in Table I. TST overestimates the FCF rate coefficients by a factor of *ca.* 2 for all jumps beginning and/or ending at an S_{II} site, but gives a *qualitatively* wrong activation energy for the W→W jump. The prefactor for this jump, which is reasonably well approximated by TST, gives us a clue why TST is so bad in this case. Since the FCF W→W prefactor is nearly an order of magnitude smaller than that for the W→S_{II} jump, there is likely to be a strong entropic bottleneck in the former case. This can arise from either a tight transition state, which TST should be able to handle, or from other final states that lie close to the W→W dividing surface, which TST *cannot* treat accurately because of its blindness to the eventual fate of dividing surface flux. Figure 1 shows that the W→W path crosses right through the S_{II}→S_{II} path, suggesting that most of the flux through the W→W dividing surface relaxes at an S_{II} site. Most of these W→W dividing surface configurations have nothing to do with actual W→W jumps, but do have energies slightly higher than the W site energy, explaining the very small TST activation energy for this jump.

Table I. Activation Energies and Prefactors for Benzene in Na-Y

	Activation Energy (kJ mol^{-1})			Arrhenius prefactors	
	MEP	TST	Corr. function	TST	Corr. function
$k(W{\to}S_{II})$	16	17.0 ± 0.1	16.4 ± 0.3	$2.7\ 10^{12}$ s^{-1}	$1.1\ 10^{12}$ s^{-1}
$k(W{\to}W)$	18	1.1 ± 0.5	15.1 ± 4.0	$6.0\ 10^{11}$ s^{-1}	$2.4\ 10^{11}$ s^{-1}
$k(S_{II}{\to}W)$	41	44.8 ± 0.1	44.4 ± 0.1	$1.6\ 10^{13}$ s^{-1}	$0.8\ 10^{13}$ s^{-1}
$k(S_{II}{\to}S_{II})$	35	37.4 ± 0.1	36.8 ± 0.3	$1.6\ 10^{13}$ s^{-1}	$0.8\ 10^{13}$ s^{-1}
$K_{eq}(S_{II}{\to}W)$	25	28.0 ± 0.2		7.1	

SOURCE: Adapted from Ref. (*18*).

Flexible Benzene in a Flexible Framework. In the previous section we held benzene and Na-Y rigid to test the feasibility of performing TST and FCF calculations, and to obtain initial estimates of the jump rate coefficients. Here we relax these constraints to determine how flexibility affects these rates. Since these new calculations can be rather computationally intensive, we attempted to diagnose which host–guest vibrational couplings would be most important.

Towards this end, we performed several constant $-NVE$ molecular dynamics (MD) calculations for benzene in Na-Y, using our previously published force-field for Na-Y framework dynamics (9). We used MD to calculate power spectra, i.e. Fourier transforms of velocity autocorrelation functions (34), keeping various parts of the system rigid. Data were collected during 200,000 1 fs steps, after initial equilibration periods of 20,000 steps, all at approximately 300 K.

Figure 2a shows the power spectrum of Na at site II in Na-Y (see Figure 1 for Na(II) position), in the absence (solid line) and presence (dashed line) of benzene. Figure 2a suggests that the vibrational frequency of Na(II) is blue-shifted upon adsorption of benzene, as benzene pulls Na(II) up an anharmonic vibrational potential. Figure 2b shows the power spectrum of benzene's center of mass (COM) at the S_{II} site in Na-Y, with full framework flexibility (solid line), with only the nearby Na(II) flexible (dashed line), and with a rigid framework (dots). Several remarks can be made about Figure 2b. The peaks near 13–17 cm^{-1} correspond to COM motion parallel to the local Na-Y surface, whereas the higher frequency peaks arise from vibrations normal to the surface. Only the normal vibrations are perturbed from framework flexibility. These normal vibrations are significantly red-shifted by nearby Na(II) vibrations, and are further slightly red-shifted by the remaining lattice vibrations. Indeed, roughly 80% of the total red shift arises from Na(II) motion. These power spectra suggest qualitatively which zeolite vibrations $must$ be included in TST/FCF calculations, namely the nearby Na(II); and suggest quantitatively how these coupled vibrations might affect the jump prefactor in a harmonic picture of adsorption. In particular, the counterbalancing blue-shifted Na(II) vibration and red-shifted benzene normal vibration foreshadows a partial cancellation, potentially giving a flexible-lattice jump prefactor nearly equal to the rigid lattice one.

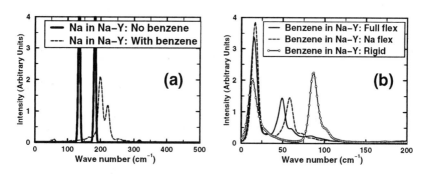

Figure 2. Power spectra for (a) Na(II) and (b) benzene's COM.

In order to quantify these effects, we performed several TST and FCF calculations, allowing for benzene internal flexibility and/or Na vibration in Na-Y zeolite. We focused on the $S_{II} \rightarrow S_{II}$, $S_{II} \rightarrow W$ and $W \rightarrow S_{II}$ jumps, since TST agreed rather well with FCF for these jumps as discussed above, allowing us to use TST for these much more demanding calculations. Where possible, FCF calculations were performed for comparison with TST (not shown in Table II), generally giving agreement comparable to that shown in Table I. We considered three new combinations of flexibility: flexible benzene and rigid zeolite (denoted "Flex B–Rigid Z" in Table II), flexible benzene and vibrating nearby Na(II) ("Flex B–Na(II)"), and flexible benzene and all Na cations vibrating ("Flex B–Na(all)"). A typical "Flex B–Na(all)" TST calculation required 35 CPU hours on an IBM PowerPC.

The results in Table II demonstrate a remarkable insensitivity to the space of coordinates included. It is difficult to extract clear trends from Table II, except the fact that jumping from an S_{II} site is *slightly* faster when the nearby Na(II) is allowed to move. Even the $W{\rightarrow}S_{II}$ barrier decreases when Na(II) is flexible, presumably because Na(II) approaches and stabilizes the $W{\rightarrow}S_{II}$ transition state. The prefactors in Table II also appear relatively insensitive to the space of coordinates included, and show no clear trend. This presumably arises from the competition between the Na(II) and benzene COM vibrational frequencies discussed in Figure 2.

One conclusion from our work is that site-to-site jump dynamics in zeolites are well described by TST when the initial or final sites involve relatively deep potential mimina. Another conclusion is that molecular jump dynamics in a large pore zeolite is well described by including only a small number of degrees of freedom. Completely different conclusions might be drawn from modeling the jump dynamics of molecules in small pore zeolites, where window breathing modes can play a crucial role in transition state stabilization.

Table II. Effect of Benzene and Na Flexibility on Jumps in Na-Y

	Activation Energy (kJ mol^{-1})			Prefactors (10^{13} s^{-1})		
	$S_{II}{\rightarrow}S_{II}$	$S_{II}{\rightarrow}W$	$W{\rightarrow}S_{II}$	$S_{II}{\rightarrow}S_{II}$	$S_{II}{\rightarrow}W$	$W{\rightarrow}S_{II}$
Rigid B–Z	37.4 ± 0.1	44.8 ± 0.1	17.0 ± 0.1	1.6	1.6	0.27
Flex B–Rigid Z	36.7 ± 0.1	44.6 ± 0.4	16.5 ± 0.3	1.4	2.8	0.28
Flex B–Na(II)	36.7 ± 0.4	43.3 ± 0.3	15.2 ± 0.5	2.6	1.9	0.17
Flex B–Na(all)	—	44.0 ± 0.3	—	—	2.1	—

Benzene Diffusion Theory in Na-Y Zeolite

In order to make contact with macroscopic transport measurements (*1,2*), we must relate our site-to-site jump rate coefficients with quantities such as the self-diffusion coefficient. In the following sections, we describe such a connection based on the mean field dynamics of cage-to-cage transition state motion.

Transition State Picture of Diffusion. We can simplify diffusion in cage-type zeolites by imagining that—although hops really take place among actual potential minima—long range motion involves jumps from one "cage site" to an adjacent "cage site" (14). A random walk through Na-Y reduces to hopping on the tetrahedral lattice of cages. In what follows, the W and S_{II} sites are denoted sites 1 and 2, respectively.

We have previously shown that for self-diffusion in cage-type zeolites at finite fractional loadings, θ, the diffusion coefficient is given by: $D_\theta \cong \frac{1}{6} k_\theta a_\theta^2$, where a_θ is the mean intercage jump length ($a_\theta \cong 11$ Å for Na-Y) and $1/k_\theta$ is the mean cage residence time (14). Furthermore, we have determined that $k_\theta = \kappa \cdot k_1 \cdot P_1$, where $P_1 = [1+K_{eq}(1{\rightarrow}2)]^{-1}$ is the probability of occupying a W site, $\langle \tau_1 \rangle = 1/k_1$ is the mean W site residence time, and κ is the transmission coefficient for cage-to-cage motion (14). Our theory thus provides a picture of cage-to-cage motion involving transition state theory ($k_1 \cdot P_1$) with dynamical corrections (κ). This finding is significant since it shows that kinetics needs to be considered only for jumps originating at threshold sites between cages. It is reasonable to expect that $\kappa \cong \frac{1}{2}$ for all but the highest loadings. We also expect that P_1 will increase with loading, and that k_1 will decrease with loading. Below we determine the loading dependencies of κ, k_1 and P_1 using mean field theory (35) and kinetic Monte Carlo simulations. Our results elucidate how the balance between k_1 and P_1 controls the resulting concentration dependence of the self-diffusion coefficient.

Parabolic Jump Model. We have determined P_1, the probability of occupying a W site, using standard Ising lattice theory (35). In order to develop viable theory and simulation strategies for modeling k_1, the total rate of leaving a W site, we need to account for blocking of target sites and adsorbate–adsorbate interactions that modify jump activation energies. In order to account for such effects, we have generalized a model that relates binding energies to transition state energies used previously by Hood *et al.* (36) and also used by us for predicting mobilities in zeolites (10). We assume that the minimum energy hopping path connecting adjacent sorption sites is characterized by intersecting parabolas, shown in Figure 3, with the site-to-site transition state located at the intersection point. The mathematical details of this model will be reported in a forthcoming paper (20). We have performed several many-body reactive-flux correlation function calculations (35) that give barriers in qualitative agreement with the parabolic jump model. A more detailed test of this method will be reported shortly (37).

Mean Field Ising Theory. As discussed above, $\kappa = \frac{1}{2}$ in mean field theory, and standard Ising mean field theory can be used to obtain P_1. The parabolic jump model for k_1 is also amenable to mean field theory. Assuming that fluctuations in the pre-exponentials can be ignored and that activation energies are Gaussian-distributed, we have that $\langle k_{i \rightarrow j} \rangle \cong \nu_{i \rightarrow j} \langle e^{-\beta E_a(i,j)} \rangle = \nu_{i \rightarrow j} e^{-\beta \langle E_a(i,j) \rangle} e^{-\beta^2 \sigma_a^2(i,j)/2}$, where $\sigma_a^2(i,j)$ is the variance of the Gaussian distribution of activation energies, i.e. $\sigma_a^2(i,j) = \langle [E_a(i,j) - \langle E_a(i,j) \rangle]^2 \rangle = \langle [E_a(i,j)]^2 \rangle - \langle E_a(i,j) \rangle^2$. These quantities can all be obtained analytically using the parabolic jump model.

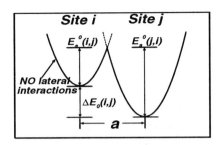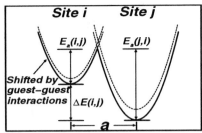

Figure 3. Relating binding energies to modified TS energies. Adapted from Ref. (*20*).

Kinetic Monte Carlo Simulation. To determine the accuracy of our mean field treatment, we perform kinetic Monte Carlo (KMC) simulations (*15*) on benzene in Na-Y, using the parabolic jump model described above. A hop is made every KMC step and the system clock is updated with variable time steps (*38*). The details of our implementation are reported elsewhere (*15*). We use time averages to calculate equilibrium values of θ_1, θ_2, κ, P_1, k_1, k_θ, a_θ and D_θ. Below we focus on comparing theory and simulation for k_θ, the cage-to-cage rate coefficient.

For a given temperature and loading, the following parameters are required by our new model to calculate the self-diffusion coefficient: $E_a^{(0)}(i,j)$, $\nu_{i\to j}$, a_{ij} and J_{ij} for $i,j = 1,2$. These are the infinite dilution jump activation energies and pre-exponential factors (*cf.* Tables I and II), jump lengths and Ising nearest neighbor interactions for each site pair. The jump lengths can be deduced from structural data (*24*), and fall in the range *ca.* 5–9 Å. The Ising coupling can be obtained from the second virial coefficient of the heat of adsorption (*39*), yielding *ca.* -3 kJ mol^{-1}. Since the diffusivity is especially sensitive to activation energies, we must recognize that our calculated barriers in Tables I and II may not be the most accurate of all available data. We first regard these barriers as flexible parameters, to determine in the most general sense what loading dependencies are consistent with our model.

Figure 4a shows that three "diffusion isotherm" types emerge. We see in Figure 4a excellent qualitative agreement between theory (lines) and simulation (dots). Theory consistently overestimates simulated diffusivities because mean field theory neglects correlation effects that make $\kappa < \frac{1}{2}$ for finite loadings. These diffusion isotherm types differ in the coverage that gives the maximum diffusivity: $\theta_{\max} = 0$ is defined as type I, $\theta_{\max} \in (0,0.5]$ is type II, and $\theta_{\max} \in (0.5,1]$ is type III. Defining the parameter $\chi \equiv \beta[E_a^{(0)}(2,1) - E_a^{(0)}(1,2)]$, we find that type I typically arises from $\chi < 1$, type II from $\chi \sim 1$, and type III from $\chi > 1$. This suggests that when the S_{II} and W sites are nearly degenerate, i.e. $\chi \lesssim 1$, the coverage dependence of P_1 is weak, and hence k_θ and D_θ are dominated by the decreasing coverage dependence of k_1. Alternatively, when $\chi \gtrsim 1$, the enhancement of P_1 at higher loadings dominates the diffusivity until $\theta_1 \sim \theta_2$, at which point the decreasing k_1 begins to dominate. We have also compared the temperature dependencies generated from theory and simulation for various loadings. We have found again that our theory gives excellent qualitative agreement with simulation.

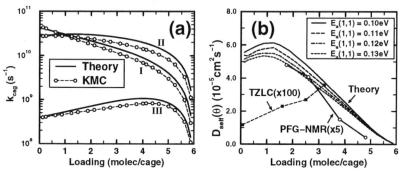

Figure 4. (a) Possible diffusion isotherms, (b) benzene in Na-Y. Adapted from Ref. (*20*).

Barriers consistent with energetic data for benzene in Na-X (*5*), the system for which experiments disagree, are $E_a(2 \rightarrow 1) \cong 25$ kJ mol^{-1} and $E_a(1 \rightarrow 2) \cong 10$ kJ mol^{-1}. Figure 4b shows the resulting diffusion isotherms for various values of $E_a(1 \rightarrow 1)$ at $T = 468$ K, compared to pulsed field gradient NMR data (*21*) (uniformly scaled by a factor of 5) and tracer zero-length column data (*23*) (uniformly scaled by a factor of 100) at the same temperature. Our model predicts that benzene in Na-X has a type II diffusion isotherm, in qualitative agreement with the pulsed field gradient NMR results. Benzene in Na-Y has a significantly larger value of χ than in Na-X (*10*), as seen from the activation energies in Tables I and II. Thus, our model predicts that benzene in Na-X will have a type II diffusion isotherm, while benzene in Na-Y will have a type III diffusion isotherm.

Conclusions. We have described a powerful combination of transition state simulation and analytical theory of activated self-diffusion in zeolites with adsorbate–adsorbate interactions. Our new theory, which is based on mean field dynamics of cage-to-cage motion, gives excellent qualitative agreement with kinetic Monte Carlo simulations for a wide variety of system parameters. Moreover, our theory provides deep understanding of the microscopic physics essential to these transport phenomena. Our new theory predicts self-diffusion coefficients for benzene in Na-X in qualitative agreement with pulsed field gradient NMR, and in qualitative disagreement with tracer zero-length column data. We anticipate that this combination of transition state simulation and mean field theory can be used to complement experimental data for a wide variety of other host–guest transport systems.

Acknowledgments. We acknowledge support from NSF grants CHE-9625735 and CHE-9616019, and from Molecular Simulations, Inc. Acknowledgment is made to the donors of the Petroleum Research Fund, administered by the American Chemical Society, under grant ACS-PRF 30853-G5.

Literature Cited

(1) Kärger, J.; Ruthven, D. M. *Diffusion in Zeolites and Other Microporous Solids;* John Wiley & Sons: New York, 1992.

(2) Chen, N. Y.; Degnan, Jr., T. F.; Smith, C. M. *Molecular Transport and Reaction in Zeolites;* VCH Publishers: New York, 1994.

(3) Chandler, D. *J. Chem. Phys.* **1978**, *68*, 2959.

(4) Demontis, P.; Yashonath, S.; Klein, M. L. *J. Phys. Chem.* **1989**, *93*, 5016.

(5) Bull, L. M.; Henson, N. J.; Cheetham, A. K.; Newsam, J. M.; Heyes, S. J. *J. Phys. Chem.* **1993**, *97*, 11776.

(6) Uytterhoeven, L.; Dompas, D.; Mortier, W. J. *J. Chem. Soc., Faraday Trans.* **1992**, *88*, 2753.

(7) Klein, H.; Kirschhock, C.; Fuess, H. *J. Phys. Chem.* **1994**, *98*, 12345.

(8) O'Malley, P. J.; Braithwaite, C. J. *Zeolites* **1995**, *15*, 198.

(9) Auerbach, S. M.; Henson, N. J.; Cheetham, A. K.; Metiu, H. I. *J. Phys. Chem.* **1995**, *99*, 10600.

(10) Auerbach, S. M.; Bull, L. M.; Henson, N. J.; Metiu, H. I.; Cheetham, A. K. *J. Phys. Chem.* **1996**, *100*, 5923.

(11) Auerbach, S. M.; Metiu, H. I. *J. Chem. Phys.* **1996**, *105*, 3753.

(12) Auerbach, S. M.; Metiu, H. I. *J. Chem. Phys.* **1997**, *106*, 2893.

(13) Auerbach, S. M. *J. Chem. Phys.* **1997**, *106*, 7810.

(14) Saravanan, C.; Auerbach, S. M. *J. Chem. Phys.* **1997**, *107*, 8120.

(15) Saravanan, C.; Auerbach, S. M. *J. Chem. Phys.* **1997**, *107*, 8132.

(16) Mosell, T.; Schrimpf, G.; Brickmann, J. *J. Phys. Chem. B* **1997**, *101*, 9476.

(17) Mosell, T.; Schrimpf, G.; Brickmann, J. *J. Phys. Chem. B* **1997**, *101*, 9485.

(18) Jousse, F.; Auerbach, S. M. *J. Chem. Phys.* **1997**, *107*, 9629.

(19) Saravanan, C.; Jousse, F.; Auerbach, S. M. *J. Chem. Phys.* **1998**, *108*, 2162.

(20) Saravanan, C.; Jousse, F.; Auerbach, S. M. submitted, **1998**.

(21) Germanus, A.; Kärger, J.; Pfeifer, H.; Samulevic, N. N.; Zdanov, S. P. *Zeolites* **1985**, *5*, 91.

(22) Shen, D. M.; Rees, L. V. C. *Zeolites* **1991**, *11*, 666.

(23) Brandani, S.; Xu, Z.; Ruthven, D. *Microporous Materials* **1996**, *7*, 323.

(24) Fitch, A. N.; Jobic, H.; Renouprez, A. *J. Phys. Chem.* **1986**, *90*, 1311.

(25) Klein, H.; Fuess, H.; Schrimpf, G. *J. Phys. Chem.* **1996**, *100*, 11101.

(26) June, R. L.; Bell, A. T.; Theodorou, D. N. *J. Phys. Chem.* **1991**, *95*, 8866.

306

(27) Mosell, T.; Schrimpf, G.; Hahn, C.; Brickmann, J. *J. Phys. Chem.* **1996**, *100*, 4571.

(28) Mosell, T.; Schrimpf, G.; Brickmann, J. *J. Phys. Chem.* **1996**, *100*, 4582.

(29) Snurr, R. Q.; Bell, A. T.; Theodorou, D. N. *J. Phys. Chem.* **1994**, *98*, 11948.

(30) Maginn, E. J.; Bell, A. T.; Theodorou, D. N. *J. Phys. Chem.* **1996**, *100*, 7155.

(31) Jousse, F.; Leherte, L.; Vercauteren, D. P. *J. Phys. Chem. B* **1997**, *101*, 4717.

(32) Voter, A. *J. Chem. Phys.* **1985**, *82*, 1890.

(33) Greengard, L.; Rokhlin, V. *J. Comp. Phys.* **1987**, *73*, 325.

(34) Allen, M. F.; Tildesley, D. J. *Computer Simulation of Liquids;* Oxford Science Publications: Oxford, 1987.

(35) Chandler, D. *Introduction to Modern Statistical Mechanics;* Oxford University Press: New York, 1987.

(36) Hood, E. S.; Toby, B. H.; Weinberg, W. H. *Phys. Rev. Lett.* **1985**, *55*, 2437.

(37) Jousse, F.; Auerbach, S. M. in preparation, **1998**.

(38) Maksym, P. *Semicond. Sci. Technol.* **1988**, *3*, 594.

(39) Barthomeuf, D.; Ha, B. H. *J. Chem. Soc., Faraday Trans.* **1973**, *69*, 2158.

Chapter 24

Alkylation and Transalkylation Reactions of Aromatics

S. R. Blaszkowski[1] and R. A. van Santen[2]

[1]DSM Research, P.O. Box 18, 6160 MD Geleen, Netherlands
[2]Schuit Institute of Catalysis, Laboratory for Inorganic Chemistry and Catalysis, Eindhoven University of Technology, P.O. Box 513, 5600 MB Eindhoven, Netherlands

Density Functional Theory calculations have been performed to analyse the reaction energy of solid acid catalyzed methyl transfer reactions. Different mechanistic routes for the alkylation of benzene and toluene by methanol have been compared. An associative reaction path via an intermediate complex of methanol and benzene or toluene is found to be the preferred route. The activation energy is 123 and $\cong$120 kJ/mol for benzene and toluene, respectively. A methoxy-mediated path involves very high activation barriers compared to the associative route. However coadsorbated water gives a large reduction of the activation energy for this reaction. Different mechanisms for toluene transalkylation: involving biphenyl methane as an intermediate, directly via methyl transfer, and methoxy mediated have been compared. When the reaction proceeds via a biphenyl methane intermediate the preferred route is the one where the reaction chain of elementary reactions is propagated via hydride transfer. The limiting step is the initial dehydrogenation, with an activation energy of +277 kJ/mol, which is present only in the very first step of the reaction chain. In the following steps, instead of initial dehydrogenation is replaced by proton assisted cracking of biphenyl methane becomes the step with the highest activation barrier. The direct mechanisms via methyl transfer or via intermediate methoxy do present activation barriers that are lower than the dehydrogenation step but higher than via biphenyl methane/hydride transfer mediated reaction. For small pore zeolites, where large molecules like biphenyl methane cannot be formed, they should be considered as optional routes for the transalkylation reaction.

The discovery that the medium-pore zeolite ZSM-5 can alkylate or disproportionate mono-substituted benzene compounds to achieve with nearly 100% selectivity para-substituted xylene has generated significant fundamental (*1-9*) as well as practical interest (*10*). Here we will present an analysis of the possible intermediates and elementary reaction steps leading to (trans)alkylation of aromatics. In the zeolite catalyzed alkylation of benzene or toluene a first question is whether the carbon-carbon bond formation proceed via an associative reaction mechanism (*11*) or in consecutive reaction steps. Consecutive reaction steps require the formation of intermediate methoxy species, generated by dissociation of methanol or toluene. Associative reactions proceed via coadsorption of the corresponding molecules and no intermediate methoxy formation is necessary. Until some time ago it was generally accepted that the alkylation of aromatics proceeds via intermediate methoxy formation (*4-6,12*). However recently Ivanova and Corma (*13*) by using C^{13} NMR have shown that an associative reaction path is more likely. Just as for alkylation reactions, transalkylation reactions can be proposed to occur according to a consecutive reaction scheme via intermediate methoxy formation or according to an associative reaction mechanism with direct CH_3 transfer. Additionally for the transalkylation a third option is the formation of a biphenyl methane intermediate. This reaction route appears to dominate (*1,2,7*). As we will show in the latter case adsorbed benzyl-cation intermediates have to be proposed and the reaction has to be considered a reaction chain propagated by hydride transfer. For both reactions, alkylation and transalkylation of aromatics, there is now ample evidence that in the zeolite not a primary reaction step (*4,11*) but consecutive reactions and diffusional limitations result in the high selectivity for p-xylene (*14*). Isomerisation has also been found to proceed via a bimolecular route similar to the transalkylations.

Here we will analyze several reaction paths and their corresponding reaction energy diagrams based on the results of DFT calculations. The zeolite will be represented by a 1T-atom cluster. Earlier we successfully demonstrated the use of this approach to analyze dimethyl ether and carbon-carbon bond formation mechanisms (*11*). There are two limitations to the use of the cluster approximation to study proton activated reactions in zeolites. Firstly the acidity of the cluster can be different from the acidity of the real zeolitic site and it may depend on cluster size. Secondly no information on the size or shape of the cavities is provided for by the cluster calculations. Elsewhere we have extensively discussed the effect of cluster size on computed transition state energies (*15*). Deprotonation of the zeolite requires very high activation energies (*15*). Nevertheless, the breaking of the proton-cluster bond is compensated for by a co-interacting change in the electrostatic interaction between positively charged transition state and negatively charged cluster, making the energy involved in protonation reactions to become much lower. Also by comparing computed transition states energies using different clusters one can obtain converged values using the Brønsted-Polanyi (*15*) relation between activation energies and reaction energies of reactions with the same reaction-mechanism. According to this relation there is a linear relationship between activation energy and reaction energy. The latter is in our case proportional to the deprotonation energy. We have shown earlier (*15*) that for cracking reaction this lowers the activation energies of transition states on a 1TH cluster by 25 kJ/mol more than it would in a low aluminum contents zeolite.

Clearly when the micropores of the zeolites are smaller in size than the transition states considered in the model calculations, the size of the micropore will prevent their formation. For the ZSM-5 zeolites this implies that the bimolecular reactions will only occur at the channel cross sections. Embedding procedures using empirical potentials will have to be used to actually analyze the increases in transition state energies that occur in narrow pores. The use of Car-Parinello approaches (16) that enable full consideration of the three dimensional structures of zeolites offers no solution. This is because the quantum mechanical method applied is also based on the electronic Density Functional Theory, that cannot be properly used to compute the weak Van der Waals interactions that dominate the interaction of the adsorbate with the zeolite cage.

Cluster calculations give relevant results when applied to a system where steric constraints do not play an important role and the only dominant effect of the zeolite cage is a weak stabilization of the complex by Van der Waals interactions. The reference state for the reacting molecule within the cluster approximation is not the molecule in the gas-phase and empty zeolite, but the molecule (or molecules) physically adsorbed in the micropores of the zeolite (17). The cluster approximation implies that the attractive Van der Waals interaction-energies do not change during reaction. For a first order reaction with a proton activated elementary reaction step as rate limiting step this leads to the following relation between apparent and true activation energy (18):

$$E_{ads}(app) = E_{act}(true) - (1-\theta) E_{ads}$$

E_{ads} is the absolute energy value of reactant adsorption and θ is its coverage. The energies predicted by the cluster calculations are the true activation energies. In view of the relatively high adsorption energies of aromatic molecules in medium pore zeolites, ≈ 80 kJ/mol (19), the corrections to the apparent activation energies due to adsorption are considerable. This is even more the case for bimolecular reactions, with a more complex dependence of the apparent activation energy on adsorption energies.

After a short summary of the computational details, we will start the results section with a discussion of benzene and toluene methylation by reaction with methanol. Addition of a surface methoxy species is compared with associative methylation from methanol. The addition of methoxy to benzene proceeds via a transition state close to the transition state obtained earlier by Beck et al. (20) for H/D exchange between deuterated zeolite and benzene. A characteristic of such transition states is that several oxygen atoms of the protonic sites are involved. Protonation is to be considered as Lewis base/Brønsted acid assisted reaction. In the second part of the results section we will compare the different transalkylation routes.

Computational details

All calculations in this work are based on Density Functional Theory, DFT (21). The Dgauss program (versions 2.1 and 3.0) part of the UniChem package from Cray research Inc. (22) was used. The local density approximation (LDA) with the exchange-correlation potential given in the form parameterized by Vosko et al. (23) is used to obtain transition state geometries. The LDA without non-local correction has been found to be inadequate for the calculation of accurate binding energies for reactions which involve hydrogen transfer (24,25). Because of this, non-local exchange and correlation corrections (NL) due to Becke (26) and Perdew (27), respectively, are included to the final total LDA energy. This level of correction is found to be excellent for binding energies.

The basis sets are of double-zeta quality and include polarization functions for all non-hydrogen atoms, DZPV (28). They were optimized for use in density functional calculations in order to minimize the basis set superposition error (29), BSSE, as has been demonstrated by Radzio et al. (30) in studies of the Cr_2 molecule. A second set of basis functions, the fitting basis set (31) is used to expand the electron density in a set of single particle Gaussian-type functions. Total LDA energy gradients and second derivates are computed analytically (32). No symmetry constrains have been used in the optimisation of any of the studied structures.

Several different approaches can be used to represent the zeolitic/adsorbate system: i) the cluster approach, used in the present work, ii) to embed the cluster in a set of point charges chosen to reproduce the bulk potential (33) and, more recently, iii) first principle calculations where the full zeolite structure is calculated using periodic boundary conditions (17,34). We commented on the limitations and also use of our approach in the introductory part of this paper.

The molecular systems considered are the zeolitic cluster interacting with methanol and/or aromatic molecules as benzene, toluene and xylene or other reaction intermediate. The single $HOHAl(OH)_3$ cluster has been used to represent the acidic zeolite. The main reason for choosing this small cluster is the very large size of the organic molecule what results in a very large system to be optimized. The transition states obtained are of course very useful as initial structure when computer resources become available for calculations on larger systems. Elsewhere (15) we have shown that the estimated error in the computed interaction energies due to the DFT approximation is of the order of 25 kJ/mol.

Results and discussion

Alkylation reactions. For the alkylation reaction Table 1 summarizes elementary reaction steps and their respective activation energies. A comparison is made between the associative reaction mechanism and via intermediate methoxy species. In Figures 1a, 1b, and 1c the transition states of three different benzene alkylation mechanisms are shown. The transition states for toluene alkylation are similar, except that an extra methyl group appears in the o-, m- or p- positions.

The associative reaction path of the proton assisted reaction between methanol and benzene resulting in toluene (the transition state is shown in Figure 1a) presents an activation barrier of 123 kJ/mol. Alkylation by surface methoxy is compared in the presence (Figure 1b) and absence (Figure 1c) of coadsorbed water. The table shows that in presence of water the alkylation is much easier. The conclusion is that in the absence of water the direct associative reaction path is preferred. This agrees with the recent work of Ivanova et al. (13). However the presence of water makes the activation energies of both reaction mechanisms, associative and via surface methoxy (water

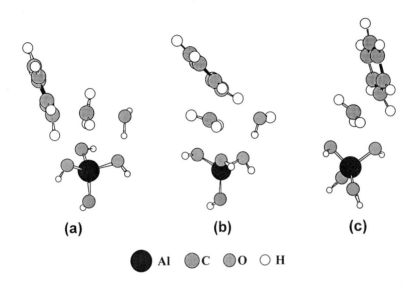

(a)   (b)   (c)

● Al   ◉ C   ◉ O   ○ H

Figure 1. Transition state for (a) direct methylation of benzene by methanol (b) reaction of the surface methoxy and benzene, and (c) water assisted reaction of surface methoxy and benzene. Note the nearly 180° angle between $O\text{-}C(CH_3)\text{-}C(C_6H_6)$.

assisted) very close. This means that in the consecutive reaction steps the aromatic molecule reacts with methoxy species generated by dissociative adsorption of methanol.

Table 1. Activation energies for alkylation of benzene and toluene[1]

figure	initial state	final state		E_{act} (kJ/mol)
1a	benzene + methanol	toluene		123
1b	benzene + methoxy + water	toluene + water	forward	133
			backward	180
1c	benzene + methoxy	toluene	forward	192
			backward	240
(1a)	toluene + methanol	o-xylene	118	
(1b)	toluene + methoxy + water	o-xylene + water	123	
(1c)	toluene + methoxy	o-xylene	183	
(1a)	toluene + methanol	m-xylene	121	
(1b)	toluene + methoxy + water	m-xylene + water	128	
(1c)	toluene + methoxy	m-xylene	190	
(1a)	toluene + methanol	p-xylene	119	
(1b)	toluene + methoxy + water	p-xylene + water	124	
(1c)	toluene + methoxy	p-xylene	183	

[1] The transitions states for toluene alkylation are analogous to benzene alkylation, expect by the presence of a methyl group in the positions o-, m-, or p-. The corresponding benzene figure numbers are shown in parenthesis.

The activation energy of methoxy formation from methanol has been computed to be equals to 212 kJ/mol (*11,35*). With coadsorbed H_2O this value is reduced to 160 kJ/mol (*11*). The lower reaction barrier for reaction of methoxy in the presence coadsorbed water is due to the nearly 180° angle between cluster oxygen methylcarbenium ion intermediate and benzene carbon atom in this particular transition state. The scaffolding effect that causes this has been noted first by Sinclair and Catlow (*36*) and is also operational on water promoted dissociative adsorption of methanol (*11*). For completeness Table 1 also collects the activation energies for the alkylation of toluene. As expected for an electrophylic substitution reaction, toluene is more reactive than benzene. The same preference for associative direct alkylation with methanol as for benzene is found. Ortho alkylation is slightly preferred over para and meta alkylation has the lowest rate. This agrees with the observed (*37,38*) preference for initial formation of ortho-xylene. As discussed in the introductory part, the p-xylene will be the main product result of consecutive equilibration and diffusion reactions, as also confirmed by Monte Carlo calculations (*7*).

The transalkylation reactions. Three reaction mechanisms for transalkylation can be distinguished:

1. Direct associative transalkylation:

$$\text{Ph–}CH_3 + \text{Ph–}R \xrightarrow{\text{HZ}} \text{Ph} + H_3C\text{–Ph–}R \qquad (1)$$

2. Indirect, methoxy mediated transalkylation:

$$\text{Ph–}CH_3 + HZ \longrightarrow \text{Ph} + CH_3\text{-}Z \qquad (2a)$$

$$H_3C\text{-}Z + \text{Ph–}R \longrightarrow H_3C\text{–Ph–}R + HZ \qquad (2b)$$

3. Reaction chain, via biphenyl methane:

Dehydrogenation:

$$\text{Ph–}CH_3 + HZ \longrightarrow \text{Ph–}CH_2\text{-}Z + H_2 \qquad (3a)$$

Oligomerization:

$$\text{Ph–}CH_2\text{-}Z + \text{Ph–}R \longrightarrow \text{Ph–}CH_2\text{–Ph–}R + HZ \qquad (3b)$$

Cracking:

$$\text{Ph–}CH_2\text{–Ph–}R + HZ \longrightarrow R\text{–Ph–}CH_2\text{-}Z + \text{Ph} \qquad (3c)$$

Hydride Transfer:

$$R\text{–Ph–}CH_2\text{-}Z + \text{Ph–}CH_3 \longrightarrow R\text{–Ph–}CH_3 + R\text{–Ph–}CH_2\text{-}Z \qquad (3d)$$

The direct associative transalkylation in which protonation of one aromatic molecule induces methyl transfer to the other is presented in equation 1a and the corresponding transition state is shown in Figure 2. The computed activation energy for this reaction is given in Table 2. The value of 252 kJ/mol of this associative

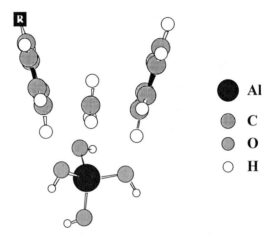

Figure 2. Transition state for direct methyl transfer. In the calculation R=H.

reaction step is slightly higher than via generation of a methoxy intermediate from toluene according to equations 2a and 2b. The activation energy step for methoxy formation from toluene[*] is 240 kJ/mol as shown in Table 1 (backwards reaction) and the reaction of the surface methoxy and toluene giving xylene requires 183 kJ/mol (Table 1). The methoxy formation is thus the limiting step for this reaction. However, one should keep in mind that despite such reaction does not occur in presence of co-adsorbed water, in the case it would be present in the reaction (humidity) methoxy formation decreases to 180 kJ/mol.

Table 2. Activation energies for transalkylation of toluene* (non methoxy intermediated)

equation	initial state	final state	E_{act} (kJ/mol)	
	Direct methyl transfer (without biphenyl methane formation)			
1	toluene + toluene[*]	xylene[*] + benzene	252	
	Reaction via biphenyl methane (dimer)			
3a	toluene	benzyl$_{ads}$ + H$_2$	forward	277
			backward	210
3b	benzyl$_{ads}$ + toluene[*]	dimer[*]	175	
3c	dimer[*]	benzyl[*]$_{ads}$ + benzene	197	
3d	benzyl[*]$_{ads}$ + toluene	benzyl$_{ads}$ + xylene[*]	179	

* In the calculations, in place of toluene, a benzene molecule has been used, which is computationally less intensive. Table 1 shows differences of the order of 5-10 kJ/mol for benzene and toluene alkylation.

Another possibility to consider is route 3 (presented in equations 3a-d and Figure 3) that proceeds via intermediate formation of biphenyl methane (*1,2,7*) The reaction is initiated by dehydrogenation of a toluene molecule so as to give a benzyl cation as presented in equation 3a. The benzyl cation then forms biphenyl methane (equation 3b) and the CH$_2$ unit is transferred in a consecutive cleavage reaction (cracking shown in equation 3c). Finally product formation and reaction chain propagation occurs via hydride transfer, as proposed by Guisnet et al. (*14*). Table 2 summarizes the activation energies of the respective elementary reaction steps. Whereas the initial generation of the benzyl cation (the dehydrogenation step) has the highest activation energy (277 kJ/mol) of all the elementary reaction steps considered so far, in order to maintain the reaction chain one proceeds via a series of elementary reaction steps that circumvents the dehydrogenation step. The cleavage of biphenyl

---

[*] Table 1 shows that the energy differences between the variuos alkylation mechanisms for benzene and toluene are about 10 kJ/mol. The conclusion is that the methoxy formation from xylene (not calculated) should be circa 10 kJ/mol lower ($\cong$ 230 kJ/mol) than for toluene (240 kJ/mol).

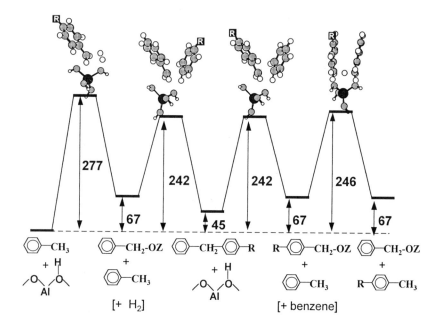

Figure 3. The reaction energy diagram for the transalkylation reaction via biphenyl methane (energy unit kJ/mol). Reaction intermediates are sketched at the bottom of the figure. The corresponding transition states are shown on top of the barriers. In the calculations R=H.

methane now has the highest activation energy: 197 kJ/mol. This value is considerably lower than the activation energies for the elementary reaction steps of the associative mechanism as well as the methoxy intermediated mechanism as long as co-adsorbed water is absent. Hence one concludes that in the absence of water and as long as the zeolite cavities are large enough so that biphenyl methane formation can occur, reaction route 3 is preferred. Note that the continuation of the process via hydrogenation as proposed by Jacobs et al. (*39*) is not probable since it involves much to high activation energies. This seems to agree with the conclusion of deuterium labeled experiments by Chen et al. (*1*) and Anderson et al. (*8*). Another possibility is the direct reaction of biphenyl methane to xylene plus benzene promoted by a hydrogen molecule (see the transition state in Figure 4). The activation barrier involved is relatively high, 240 kJ/mol, and the continuation processes involves the dehydrogenation step, making this route not really probable.

Transalkylation can also occur in more narrow pore zeolites in which biphenyl methane formation is suppressed. Then the mechanism via methoxy surface (equations 2a and 2b) will take over. Unless water is added to the reaction mixture the activation energy will be higher in this case than for a larger pore zeolite, where biphenyl methane can be formed. It cannot be excluded that other molecules may play a similar promoting role as water.

Conclusion

The Density Functional studies of zeolite catalyzed (trans)alkylation reactions show that analysis of the transition states of elementary reaction steps enables discrimination of preference for several different reaction paths. Although the clusters to represent the acidic zeolite site used here are quite small, specially when considering the size of the molecules involved in the reactions, the results are quite satisfactory. Also they can be used in the future as starting point for calculations involving a larger zeolite cluster or embedded systems.

Acknowledgment

S.R. Blaszkowski thanks CNPq (Brazil) for the scholarship. Computational resources were supplied by NCF (The Netherlands) under project SC-417.

Literature Cited

1. Chen, N.Y.; Reading, W.W.; Dwyer, F.G. *J. Am. Chem. Soc.* **1979**, *101*, 6783.
2. (a) Gnep, N.S.; Tejudo, J.; Guisnet, M. *Bull. Soc. Chim. Fr.* **1982**, *1-2, 1-5.* (b) Olson, D.H.; Haag, W.O. *ACS Symposium Series* **1984**, *248*, 275.
3. Guisnet, M. *Stud. Surf. Sci. Catal.* **1985**, *20*, 283.
4. (a) Mirth, G.; Lercher, J.A. *J. Catal.* **1994**, *147*, 199. (b) Mirth, G.; Lercher, J.A. *J. Catal.* **1991**, *132*, 244.
5. Rakogy, J.; Romotanski, T. *Zeolites* **1993**, *13*, 256.

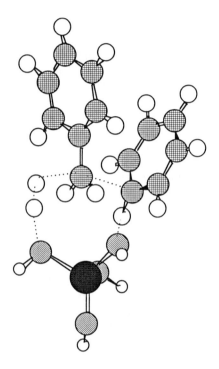

Figure 4. Transition state for the direct reaction of biphenyl methane giving xylene and benzene promoted by a hydrogen molecule.

6. (a) Corma, A.; Sastre, G.; Viruela, P. *Stud. Surf. Sci. Catal.* **1994**, *84*, 2171. (b) Corma, A.; Sastre, G.; Viruela, P. *J. Mol. Catal. A* **1995**, *100*, 75. (c) Corma, A.; Llopsis, F.; Viruela, P.; Zicovich-Wilson, C. *J. Am. Chem. Soc.* **1994**, *116*, 134.

7. Wang, J.G.; Li, Y.W.; Chen, S.Y.; Peng, S.Y. *Zeolites* **1995**, *15*, 286.

8. Anderson, J.R.; Dong, C.N.; Chang, T.F.; Western, R.J. *J. Catal.* **1991**, *127*, 113.

9. (a) Das, J.; Baht, Y. S.; Halgeri, A.B. *Ind. Eng. Chem. Res.* **1994**, *33*, 246. (b) Keading, W.W.; Chor, C.; Young, L.B.; Wemstein, B.; Butta, S.D. *J. Catal.* **1981**, *67*, 159. (c) Beltrame, P.; Beltrane, P.L.; Cartini, P.; Forni, L.; Zuretti, G. *Zeolites* **1985**, *5*, 400. (d) Blaskar, G.V.; Do, D.D. *Ind. Eng. Chem. Res.* **1990**, *29*, 355.

10. Chen, N.Y.; Garwood, W.E.; Dwyer, F.G. *Shape selective catalysis in industrial applications;* M. Dekker: New York, 1989; pp 212.

11. (a) Blaszkowski, S.R.; Santen, R.A. van *J. Am. Chem. Soc.* **1996**, *118*, 5152. (b) Blaszkowski, S.R.; Santen, R.A. van *J. Phys.Chem.* **1997**, *101*, 2292.

12. (a) Chen, N.Y. *J. Catal.* **1988**, *114*. (b) Smirniotis, P.G.; Ruckenstein, E. *Ind. Eng. Chem. Res.* **1995**, *34*, 1517.

13. Ivanova, I. I.; Corma, A. *J. Phys. Chem. B* **1997**, *101*, 547.

14. Morin, S.; Gnep, N. S.; Guisnet, M. *J. Catal.* **1996**, *159*, 296.

15. Santen, R.A. van *Catalysis Today* **1997**, *30*, 377.

16. Santen, R.A. van *J. Mol. Catal.* **1996**, *A1 309*, 405.

17. (a) Shah, R.; Gale, J.G.; Payne, M.C., *J. Phys. Chem.* **1996**, *100*, 11688. (b) Shah, R.; Payne, M.C.; Lee, M.H.; Gale J.D., *Science* **1996**, *271*, 1395.

18. (a) Santen, R.A. van; Niemantsverdriet, J.W. *Chemical kinetics and Catalysis;* Plenum: New York; 1995. (b) Talu, O.; Guo, G.J.; Taghurst, D.T. *J. Phys. Chem.* **1993**, *93* 7294.

19. (a) Thamm, H., *Zeolites* **1987**, *7*, 341. (b) Wu, P.; Debebe, A; Ma, Y.H., *Zeolites* **1983**, *3*, 118. (c) Trikoyinnis, J.G.; Wei, J. *Chem. Eng. Sci.* **1991**, *40*, 155.

20. (a) Beck, L.W.; Xu, T; Nicholas, J.B.; Haw, J.F. *J. Am. Chem. Soc.* **1995**, *117*, 11594. (b) Santen, R.A. van ; Kramer, G.J. *Chem. Rev.* **1996**, *95*, 637.

21. (a) Hohenberg, P.; Kohn, W. *Phys. Rev. B* **1964**, *136*, 864. (b) Kohn, W.; Sham, L.J. *Phys. Rev. A* **1965**, *140*, 1133.

22. Andzelm, J.; Wimmer, E. J. *Chem. Phys.* **1992**, *96*, 1280.

23. Vosko, S.H.; Wilk, L.; Nusair, M. *Can. J. Phys.* **1980**, *58*, 1200.

24. Fan, L.; Ziegler, T., *J. Am. Chem. Soc.* **1992**, *114*, 10890.

25. Blaszkowski, S.R.; Jansen, A.P.J.; Nascimento, M.A.C.; Santen, R.A. van *J. Phys. Chem.* **1994**, *98*, 12938.

26. Becke, A.D. *Phys. Rev. A* **1988**, *38*, 3098.

27. Perdew, J.P. *Phys. Rev. B* **1986**, *33*, 8822.

28. Godbout, N.; Andzelm, J.; Wimmer, E.; Salahub, D.R. *Can. J. Chem.* **1992**, 70, 560.

29. Sauer, J. *Chem. Rev.* **1989**, *89*, 199.

30. Radzio, E.; Andzelm, J.; Salahub, D.R. *J. Comp. Chem.* **1985**, *6*, 553.

31. Andzelm, J.; Russo, N.; Salahub, D.R., *Chem. Phys. Lett.* **1987**, *142*, 169

32. (a) Schlegel, H.B. *Ab Initio Methods in Quantum Chemistry*; John Wiley & Sons: New York; **1987**. (b) Head, J.D.; Zerner, M.C. *Advances in Quantum Chemistry* **1989**, *20*, 239.

33. Greatbanks, S.P.; Hillier, I.H.; Burton, N.A.; Sherwood, P. *J. Chem. Phys.* **1996**, *105*,3770.

34. Nusterer, E.; Blöchl, P.E.; Schwarz, K. *Ang. Chem., Int. Ed. Engl.* **1996**, *35*, 175.

35. Blaszkowski, S.R.; Santen, R.A. van *J. Phys. Chem.* **1995**, *99*, 11728.

36. Sinclair, P.E.; Catlow, C.R.A. *J. Chem. Soc., Faraday Trans.* **1996**, *92*, 2099.

37. Allen, R.M.; Yates, L.I., *J. Am. Chem. Soc.* **1961**, *83*, 2799.
38. Young, L.B.; Butter, S.A.; Keuding, W.W. *J. Catal.* **1982**, *76*, 418.
39. (a) Jacobs, P.A.; Martens, J.A. in *Introduction to Zeolite Science and Practice*; Bekkum, H. van; Flanigen, E.M.; Jansen, J.C., Eds.; Elsevier: Amsterdam; 1990. (b) Jacobs, P.A.; Martens, J.A. *Stud. Surf. Sci. and Catalysis* **1991**, *50*, 445.

Chapter 25

Density Functional Study on the Transition State of Methane Activation over Ion-Exchanged ZSM-5

Yusuke Ueda, Hirotaka Tsuruya, Tomonori Kanougi, Yasunori Oumi, Momoji Kubo, Abhijit Chatterjee, Kazuo Teraishi, Ewa Broclawik, and Akira Miyamoto

Department of Materials Chemistry, Graduate School of Engineering, Tohoku University, Aoba-ku, Sendai 980-8579, Japan

In order to investigate the methane dissociation over Ga-ZSM-5, the transition state of the reaction pathway was determined by density functional calculations (DFT). We revealed that the activation energy is 25.8 kcal/mol and that the chemisorption state in which CH_3 is attached to the Ga^{3+} ion and H forms a hydroxyl group with the extraframework oxygen is the main product. We also investigated the influence of the cluster size on the transition state of methane dissociation. It was shown that $[Al(OH)_4]^-$ can express the trend and is adequate for use to model the zeolitic structure. The transition states of the methane dissociation reaction over Al-ZSM-5 and In-ZSM-5 were also investigated.

1. Introduction

The main role of the catalyst is to lower the activation energy of the reaction by stabilizing the transition state. However, as the life time of the transition state is of pico seconds order, only a limited experimental research has so far been performed. Recently, with the development of the femtosecond pulsed laser (FPL), it has become possible to observe transition states experimentally *(1,2)*. However, complicated systems such as inhomogeneous catalytic reactions cannot be analyzed by FPL. On the other hand, computational technique to analyze the transition state is already established, and is routinely used to investigate the reaction pathway. Therefore, it would be the suitable method to study the catalytic reaction mechanism.

Being responsible for the acid rain which destroy the global environment, nitrogen oxides must be removed from the automobile emission. The cation-exchanged zeolitic catalysts such as Cu- *(3)*, Co- *(4,5)*, Mn- *(6)*, Ni- *(6)*, Pd- *(7)*, Ga- *(8,9)*, In- *(10)*, Ce- *(11)*, and H-ZSM-5 *(12)* as well as Cu- *(13)* and Co-mordenite *(6)* and Cu-SAPO-34 *(14)* have been actively investigated as catalysts for the selective reduction of nitrogen monoxide using hydrocarbons as reductants under the excess oxygen condition. In particular, as methane is present in the exhaust gas, catalysts which are active in NOx decomposition with methane as reductant are desirable. Ga- and In- exchanged ZSM-5, introduced by Yogo and co-workers *(8,10)* are one of such catalysts and have been studied extensively. Methane is notoriously an inactive agent

(15) and only the TPD (temperature programmed desorption) investigation has so far been carried out to study the dissociative mechanism over Ga-ZSM-5 (16). However, consensus has already been obtained that the dissociative adsorption of methane is the rate-determining step in many other heterogeneous catalytic processes such as steam reforming, alkylation of hydrocarbons and combustion (17). In this study, in order to investigate the reaction path of methane dissociation over Ga-ZSM-5, the transition state and activation energy were determined by density functional calculations (DFT).

Two possible routes for the methane dissociation over Ga-ZSM-5 are considered here. The first route leads to a methyl attached to the Ga^{3+} ion and a hydrogen forming a hydroxyl group with the extraframework oxygen (route 1). The second one corresponds to the reverse combination, i.e. H is attached to Ga^{3+} ion and CH_3 is connected to the extraframework oxygen to make a methoxyl group (route 2). We already revealed by DFT calculations that the product of the former route is preferred in view of energy stabilization. Both reaction routes yield highly stable products in comparison to the methane physisorbed systems (18). Based on this result, we extended our investigations to fully describe the reaction pathway of methane dissociation in the present work. For this purpose, transition state search calculations were performed and the structures, molecular orbitals and the activation energies were analyzed.

First of all, in order to clarify which reaction path is more feasible, either route 1 or route 2, the activation energies of both routes were compared to confirm the actual dissociation process (transition state of route 1 was already reported in ref. (19)). Secondly, in order to validate our choice of the cluster model, influence of cluster size on the transition state of the methane dissociation was investigated. Finally, because activity, selectivity and durability depend on the exchange cations present in ZSM-5, the catalytic performance of three exchange cations, namely, Ga, In and Al was investigated. The same calculations were thus performed on the physisorption state, chemisorption state and transition state of methane in Al-ZSM-5 and In-ZSM-5.

2. Method and model

Although the crystal structure was determined by X-ray diffraction study (20), the physical and chemical state of the active gallium species in H-ZSM-5 zeolites has not been unequivocally established. There exists, however, strong evidence that Ga is present in a highly dispersed monomeric form, coordinated to basic oxygen within zeolite channels (21). In a reducing atmosphere it could be a reduced hydride moiety, while in the presence of excess oxygen an oxidized $[GaO]^+$ unit may be proposed. Actually $[GaO]^+$ species in Ga-ZSM-5 are suggested by some experiments (22,23). Furthermore, earlier quantum chemical studies have reported that T12 is the energetically favorable site for the incorporation of aluminum (24). Therefore in the present study $[GaO]^+$ was assumed to be the active site in the framework of ZSM-5 whose T12 site was substituted by aluminum.

MD calculation was performed under periodic boundary condition to determine the conformation of Ga-exchanged site within the zeolitic lattice (Figure 1a). A single AlO_4 tetrahedron which represents the active site, was extracted as the cluster

(a)

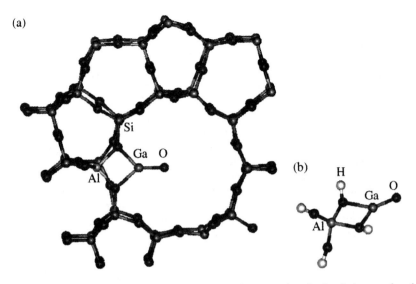

(b)

Figure 1   (a) Structure of Ga-ZSM-5 framework obtained by molecular dynamics, and (b) the [Al(OH)$_4$]GaO model cluster of the Ga active center in Ga-ZSM5.

model for quantum chemical calculations, where adjacent silicons were replaced by hydrogens and charge deficiency was compensated by [GaO]$^+$ cationic species (Figure 1b). This cluster model is the same as that used in our earlier works (18,19).

Of this model, zeolitic framework including terminal hydrogen atoms were fixed and only [GaO]$^+$ unit and reacting methane were optimized. Our preliminary calculation at LDA level using large cluster model (vide infra) revealed that relaxation of framework near Al site (Al(OSi)$_4$ moiety) upon the dissociative adsorption of methane stabilizes the system by 7.0kcal/mol. Since only relative energies are discussed, relaxation effect should be even less than this value. Transition state search was also conducted within the same space of freedom by locating the point where the first derivative of the energy vanishes while the second derivative (Hessian) matrix has only one negative eigenvalue.

MD calculations were carried out with the MXDORTO program developed by Kawamura (25) for the determination of the structure of the zeolite framework. The Verlet algorithm (26) was used for the calculation of atomic motions, while the Ewald method (27) was applied for the calculation of electrostatic interactions. DFT calculations were performed by solving Kohn-Sham equation self-consistently (28) as implemented in the DMol program (29-31). We employed local density approximation (LDA) with Vosko-Wilk-Nusair (VWN) functional (32) for geometry optimization. In order to calculate accurate energies, we applied the Beck-Lee-Yang-Parr (BLYP) (33) nonlocal functional as a correction. Double numerical with polarization functions (DNP) basis set was used.

3. Result and Discussion

3. 1 Reaction pathway and transition state for methane dissociation over Ga-ZSM-5.

As described in introduction, we considered two chemisorption models. The transition state of methane activation process (route 1) was calculated by DFT following the procedure described in ref. *(19)*. Our cluster model is strictly based on the MD result, while in our earlier work Cs symmetry was applied in order to reduce the computational cost. In the first step the physisorbed state was calculated. This system may be described as the complete methane molecule interacting weakly with the Ga cation. From now on we will call the hydrogen which is abstracted from methane molecule *dissociative hydrogen*. We selected the distance between the dissociative hydrogen and the extraframework oxygen as the approximate reaction coordinate. In the second step the energy profile was calculated with respect to this coordinate keeping all other parameters fixed in order to approximately locate the transition state region. The energy maximum in this profile was found at the O-H and C-H distance being 1.6 and 1.8 Å, respectively. This geometry was then taken as the initial coordinate in the next step of calculation in which the full transition state optimization was performed.

Transition state structure was confirmed by the analysis of the eigenvector of Hessian which corresponds to the negative eigenvalue. The normal coordinate consists almost exclusively of the in-plane movement of the dissociative hydrogen between its neighboring carbon and oxygen atoms, supplemented by the Ga-O stretch, which also supports our choice of the approximate reaction coordinate.

Transition state of the route 2 was calculated by the similar procedure. The situation is more complicated for route 2 as only one bond length can not represent the reaction coordinate. Therefore in this route we varied two bond distances, namely, between the dissociative hydrogen and carbon and between carbon and extraframework oxygen, and drew an energy profile. The energy maximum in this profile was found at the O-C and C-H distance of 2.0 and 1.5 Å, respectively. Also in this case, the structure at energy maximum was employed as the initial state for the transition state search.

Figure 2 shows the reaction pathway and the transition state for methane dissociation. The adsorption energy was defined by eq. 1:

$$E_{adsorption} = E_{host+guest} - (E_{host} + E_{guest}) \qquad (1)$$

Here $E_{host+guest}$, E_{host}, and E_{guest} denote the total energy of the system including both the host zeolite cluster and the incorporated guest molecules, the bare zeolite cluster, and the adsorbate molecule in the gas phase, respectively. As seen from the figure, the adsorption energy of the methane at transition state and chemisorption state of route 1 are +28.6 kcal/mol and −39.8 kcal/mol, respectively. These values are slightly different from what are reported in ref. *(19)*, as in our earlier work, the geometry of Ga-ZSM-5 cluster was different from the present model.

The adsorption energy of the methane at transition state and chemisorption state of route 2 are +68.6 kcal/mol and −20.9 kcal/mol respectively. From the comparison of the adsorption energy, it was found that the activation energy of route 1 is about 40 kcal/mol lower than that of route 2. As a result, we can confirm that route 1 is the main reaction path, where hydrogen atom is abstracted by extraframework oxygen attached to Ga atom.

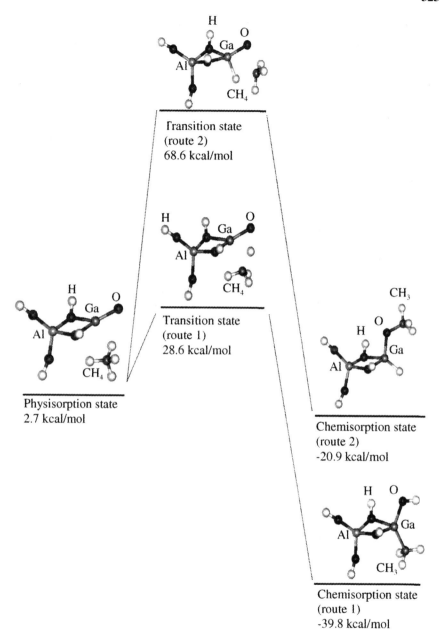

Figure 2 Reaction pathway and transition state of methane dissociation in Ga-ZSM-5.

Table 1   Geometrical parameters of transition states.

		Transition state of route 1	Transition state of route 2
C-Dissociative H	(Å)	1.35	1.63
Ga-Extraframework O	(Å)	1.72	1.73
Extraframework O-Dissociative H	(Å)	1.46	-
Ga-Dissociative H	(Å)	-	1.64
Ga-C	(Å)	2.21	-
Extraframework O-C	(Å)	-	2.10

The geometrical parameters at transition state of both routes are listed in Table 1. One can see that the C-H distance of route 1 is smaller than that of route 2. It means that in route 1 cleaving C-H bond and forming O-H bond reaches the balancing point earlier than in the case of route 2 due to the strong O-H interaction. It may be considered that the interaction of H with O is more favorable than the interaction of H with Ga because O and Ga are electron-acceptor and electron-donor, respectively. As a result, activation energy of route 1 is lower than that of route 2.

3.2 Effect of cluster size. One of the problems that arises when target is large is the choice of the model. It is still computationally very difficult to include the entire zeolite into the quantum chemical calculation. In order to study efficiently, the model should be as small as possible but large enough to express the phenomena of interest. Thus we investigated the influence of cluster size on the activation energy and the structure of transition state so as to examine if the cluster size is adequate.

The large cluster model shown in Figure 3 was constructed by replacing the terminating hydrogen of small cluster model with $Si(OH)_3$, where Si and O atoms were placed at the position obtained by MD. The same procedure as was performed with the small model was applied to the large model to obtain the transition state.

Table 2 shows the adsorption energy of methane on the large cluster model and small cluster model along with the activation energy for methane dissociation. As one can see from the table, the difference in adsorption energy and activation energy between large and small cluster models are less than 2 kcal/mol. Table 3 shows the geometrical parameters at the transition state of two models. The notation of the atoms are given in Figure 4. It can be noted that the difference in bond distance is as small as 0.01~0.03 Å between small and large clusters. Table 4 shows the Mulliken charge on constituting atoms at the transition state along with HOMO energy levels. Atomic charges on framework oxygens significantly differ between small and large cluster models as framework oxygen of small cluster model is terminated by hydrogen atom. However, the difference in charge on the Ga atom and extraframework oxygen are small. The energy level of HOMO of the large cluster model is 0.7 eV higher than that of the small cluster model. It may thus be concluded that the electronic state of the active site is not affected significantly by the Si atoms and that $[Al(OH)_4]$ is adequate for use to model the zeolitic structure.

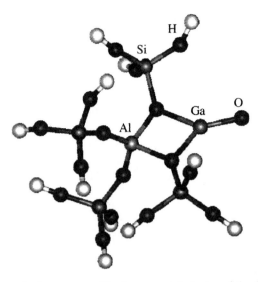

Figure 3   Geometrical structure of the large model cluster of the Ga active center in Ga-ZSM5, [AlO$_4$(Si(OH)$_3$)$_4$]GaO.

Table 2   Physical and dissociative adsorption energy of methane in Ga-ZSM5 along with the activation energy for methane dissociation.

	Adsorption energy [kcal/mol]			Activation energy [kcal/mol]
	Physisorption	Transition state	Chemisorption	
Large cluster	1.0	27.1	-40.0	26.1
Small cluster	2.7	28.6	-39.8	25.9

3.3 Effect of exchanged cations. We prepared Al-ZSM-5 and In-ZSM-5 cluster models by the same procedure as the case for Ga-ZSM-5 small model. Physisorbed, chemisorbed and transition states corresponding to route 1 over these cluster models were investigated similarly Table 5 shows the adsorption energies of methane on Al-, Ga- and In-ZSM-5 along with the activation energy for methane dissociation. The order of adsorption energy is Ga < In < Al. The activation energies over Ga- and In-ZSM-5 are almost the same, whereas that over Al-ZSM-5 is considerably smaller than the former two.

Table 3 shows the geometrical parameters of the transition state on Al-, Ga- and In-ZSM-5. The distance between C-exchanged cation is related to the ionic radius. Correlation between activation energy and C-H / O-H distances can also be found. It is considered that the activation energy for methane dissociation is proportional to the

Table 3   Geometrical parameters at transition state of Ga-, Al-, In-ZSM5

	Ga-ZSM5 large	Ga-ZSM5 small	Al-ZSM5	In-ZSM5
Cation-O (Å) 10-11*	1.72	1.72	1.64	1.94
O-H (Å) 11-16*	1.45	1.46	1.52	1.41
C-H (Å) 12-16*	1.34	1.35	1.32	1.37
Cation-C (Å) 10-12*	2.24	2.21	2.12	2.41
Angle HCH (degree) 16-12-13*	128.99	132.27	130.16	131.12
Angle HCH (degree) 16-12-15*	109.31	104.11	96.58	97.30
Angle HCH (degree) 16-12-14*	94.28	97.59	106.84	107.24
Dihedral angle(degree) 10,11-16,12*	0.16	3.56	-4.17	-1.12
Dihedral angle(degree) 10,11-12,13*	28.26	11.73	-18.28	-20.70

*See Figure 4 for atomic number.

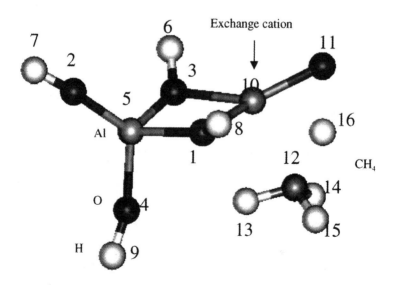

Figure 4   Structure of transition states.

Table 4   Mulliken charge on constituting atoms at the transition state of methane
dissociation in Ga-ZSM-5 along with HOMO energy levels.

			Small cluster	Large cluster
HOMO (eV)			-4.96	-4.25
Mulliken charge				
Framework oxygen	O	1*	-0.98	-0.90
	O	2*	-0.83	-0.70
	O	3*	-0.99	-0.89
	O	4*	-0.85	-0.71
Aluminum	Al	5*	1.05	1.17
Exchange cation	Ga	10*	1.23	1.35
Extraframework oxygen	O	11*	-0.80	-0.82
Methane	C	12*	-1.34	-1.38
	H	13*	0.34	0.39
	H	14*	0.39	0.40
	H	15*	0.42	0.39
Dissociative hydrogen	H	16*	0.26	0.26
	total	-	0.07	0.06

*See Figure 4 for atomic number.

distance of C-H (inversely proportional to O-H distance), which will be discussed in the next section.

Table 6 shows the Mulliken charge on the constituting atoms of each models. In Al-ZSM-5, methane bares positive charge as a net, which suggests that electron transfer from methane molecule to Al-ZSM-5 took place. Due to the electrostatic interaction between methane and Al-ZSM-5 cluster, the transition state is more stabilized than in In- or Ga-ZSM-5. This may be the reason why Al-ZSM-5 shows relatively small activation energy.

In our earlier work, it was revealed by DFT calculation that Al-ZSM-5 shows strong affinity to H_2O compared with Ga- and In-ZSM-5 (34). Although the activity of Al-ZSM-5 has not been found experimentally, our DFT calculation suggests that under the highly dry condition, Al-ZSM-5 would show higher activity compared with Ga- or In-ZSM-5.

3.4 Basicity and catalytic performance. In order to rationalize the catalytic performance of cation exchanged ZSM-5 in methane activation reaction, the results are further analyzed. First of all, the atomic charges given in Table 6 indicate that dissociative hydrogen is positively charged, hence is protonic. Therefore the basicity of the catalysts is expected to be important in this reaction.

Furthermore, a close look at the structural parameters given in Table 3 suggests the correlation between the activation energy and   C-H / O-H distances at   transition

Table 5   Physical and dissociative adsorption energy of methane in Al-ZSM-5, Ga-ZSM-5 and In-ZSM-5 along with the activation energy for methane dissociation.

	Adsorption energy [kcal/mol]			Activation energy [kcal/mol]
	Physisorption	Transition state	Chemisorption	
Al-ZSM-5	0.4	14.7	-44.2	14.3
Ga-ZSM-5	2.7	28.6	-39.8	25.9
In-ZSM-5	1.8	27.9	-41.6	26.1

Table 6 Mulliken charge on constituting atoms at the transition state for methane dissociation in Ga-, Al- and In-ZSM-5.

			Ga-ZSM-5	Al-ZSM-5	In-ZSM-5
Mulliken charge Framework oxygen	O	1*	-0.98	-0.92	-0.98
	O	2*	-0.84	-0.83	-0.84
	O	3*	-0.99	-0.95	-1.00
	O	4*	-0.85	-0.85	-0.85
Aluminium	Al	5*	1.05	1.11	1.07
Terminating hydrogen	H	6*	0.59	0.60	0.56
	H	7*	0.47	0.46	0.46
	H	8*	0.58	0.58	0.55
	H	9*	0.46	0.46	0.46
	total	-	-0.51	-0.34	-0.57
Exchange cation	Ga,Al,In	10*	1.23	0.73	1.40
Extraframework oxygen	O	11*	-0.80	-0.75	-0.79
	total	-	0.43	-0.02	0.61
Methane	C	12*	-1.34	-1.27	-1.35
	H	13*	0.34	0.42	0.40
	H	14*	0.38	0.44	0.30
	H	15*	0.42	0.38	0.37
Dissociative hydrogen	H	16*	0.26	0.38	0.21
	total	-	0.06	0.35	-0.07

*See Figure 4 for atomic number.

state when Al-, Ga- and In-ZSM-5 are compared. The schematic diagram of the energy profile as a function of the reaction coordinate is shown in Figure 5. Here, the reaction coordinate corresponds to the position of proton transferred from methane to extraframework oxygen, and right and left energy curves denote the dissociation of proton from reactant (CH) and product (OH), respectively. Proton dissociation curve from methane is assumed to be constant regardless to the environment at this qualitative approximation, while that from extraframework oxygen depends on its basicity. Conceptually, the points at which right and left curves intersect correspond to the transition state. This figure suggests that if the minimum of the left curve is deeper, i.e. the basicity of oxygen is stronger, the interaction appears at shorter C-H / longer O-H distance, and the activation energy is also lower.

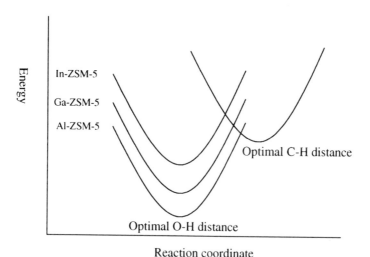

Figure 5   The schematic diagram of the energy profile as a function of the reaction coordinates.

4. Conclusion

We performed density functional calculations to investigate the methane dissociation over Ga-ZSM-5, and determined the transition state of the reaction pathway. The activation energies of reaction along route 1 and route 2 are 25.9 kcal/mol and 65.9 kcal/mol respectively, and the adsorption energies of the chemisorption products are -39.8 kcal/mol and -20.9 kcal/mol respectively. From these results, it was confirmed that the route 1 is the main pathway.

We also investigated the influence of cluster size on the transition state for the methane dissociation. Geometrically as well as energetically, the difference between large and small models was very small. It was shown that [Al(OH)$_4$] can express the trend and is adequate for use to study the zeolitic structure.

The transition states of the methane dissociation over Al- and In-ZSM-5 were also investigated. Calculated activation energy over Al-ZSM-5 is considerably smaller

332

than that over Ga- and In-ZSM-5. This suggests that Al-ZSM-5 shows higher activity compared with Ga- and In-ZSM-5 under highly dry condition.

References

(1)   Motzkus, M.; Pedersen, S.; Zewail, A. H., *J. Phys. Chem.* **1996**, 100, 5620.
(2)   Zewail, A. H., *J. Phys. Chem.* **1996**, 100, 12701.
(3)   Sato, S.; Yu-u, Y.; Yahiro, H.; Mizuno, N.; Iwamoto, M. *Appl. Catal.* **1991**, 701, L1.
(4)   Kishida, M.; Tachi, T.; Yamashita, H.; Miyadera, H. *Shokubai* **1992**, 34, 148.
(5)   Li, Y.; Armor, J. N. *Appl. Catal.*, B **1992**, 1, L31.
(6)   Li, Y.; Armor, J. N. *Appl. Catal.*, B **1993**, 2, 239.
(7)   Nishizawa, V.; Misono, M. *Chem. Lett.* **1993**, 1295.
(8)   Yogo, K.; Ihara, M.; Terasaki, I.; Kukuchi, E. *Chem. Lett.* **1993**, 229.
(9)   Li, Y.; Armor, J. N. *J. Catal.* **1994**, 145, 1.
(10)   Kikuchi, E.; Yogo, K. *Catal. Today* **1994**, 22, 73.
(11)   Misono, M.; Kondo, K. *Chem. Lett.* **1991**, 1001.
(12)   Kintaichi, Y.; Hamada, H.; Tabata, M.; Sasaki, M.; Ito, T. *Catal. Lett.* **1990**, 6, 239.
(13)   Mabilon, G.; Durand, D. *Catal. Today* **1993**, 17, 285.
(14)   Ishihara, T.; Kagawa, M.; Miauhara, Y.; Takita, Y. *Chem. Lett.* **1992**, 2119.
(15)   Zhang, X.; Walters, A. B.; Vannice, M. A. *J. Catal.* **1994**, 146, 568.
(16)   Tabata, T.; Kokitsu, M.; Okada, O. *Catal. Lett.* **1994**, 25, 393.
(17)   Introduction to Zeolite Science and Practice, edited by H. van Bekkum, E. M. Flanigen, J. C. Jensen, Studies in Surface Science and Catalysis 58 (Elsevier, Amsterdam, **1991**); Zeolites and Related Microporous Materials: State of the Art 1994, edited by J. Weitkamp, H. G. Korge, K. Pfeifer, W. W. Holderich, Studies in Surface Science and Catalysis 84 (Elsevier, Amsterdam, **1994**).
(18)   Himei, H.; Yamadaya, M.; Kubo, M.; Vetrivel, R.; Broclawik, E.; Miyamoto, A., *J. Phys. Chem.* **1995**, 99, 12461.
(19)   Broclawik, E.; Himei, H.; Yamadaya, M.; Kubo, M.; Miyamoto, A., *J. Chem. Phys.* **1995**, 103, 2102.
(20)   Flanigen, E. M.; Bennet, J. M.; Grose, R. W.; Cohen, J. P.; Patton, R. L.; Kirchner, R.M.; Smith, *J. V. Nature* **1978**, 271, 512.
(21)   Meitzner, G. D.; Iglesia, E.; Baumgartner, J. E.; Huang, E. S. *J. Catal.* **1993**, 140, 209.
(22)   Meitzner, G. D.; Iglesia, E.; Baumgartner, J. E.; Huang, E. S., *J. Catal.*, **1993**, 140, 209.
(23)   Kwak, B. S.; Sachtler, W. M. H. *J. Catal.* **1993**, 141, 729.
(24)   Derouane, E. G.; Fripiat, J. G. *Zeolites* **1985**, 5, 165.
(25)   Kawamura, K. in Molecular Dynamics Simulation; Yonezawa, F., Ed.; Springer-Verlag: Berlin, **1992**; p 88
(26)   Verlet, L. *Phys. Rev.* **1967**, 159, 98.
(27)   Ewald, P. P. *Annu. Phys.* **1921**, 64, 253.
(28)   Kohn, W.; Sham, L. J. *Phys Rev.* **1965**, A 140, 1133.
(29)   DMol version 2.3.5 San Diego: MSI, **1993**.
(30)   Delley, B. *J. Phys. Chem.* **1990**, 92, 508.
(31)   Delley, B. *J. Phys. Chem.* **1991**, 94, 7245.
(32)   Vosko, S. H.; Wilk, L.; Nusair, M. *Can. J. Phys.* **1980**, 58, 1200.
(33)   (a) Becke,A. *J. Chem. Phys.* **1988**, 88, 2547.; (b) Lee, C.; Yang, W.; Parr, R. G. *Phys. Rev.* **1988**, B37, 786.
(34)   Kanougi, T.; Tsuruya, H.; Oumi, Y.; Chatterjee, A.; Fahmi, A.; Kubo, M.; Miyamoto, A. *Appl. Surf. Sci.*, in Press.

Chapter 26

First Principles Study of the Activation of Methane on Defects of Heteropolyanion Structures: A simple Way to Model Oxide Surfaces

J.-F. Paul and M. Fournier

Laboratoire de catalyse homogène et hétèrogène, CNRS URA 402, batiment C3, 59655 Villeneuve d'Ascq Cedex, France

The activation of the CH bond in hydrocarbon molecules and in particular in methane is a subject of great importance from the industrial point of view as well as the fundamental one. Some molecular oxide species such as the heteropolyanions (HPA) are able to break these strong bonds in ODH (oxydeshydrogenation) reactions, but the mechanism is still under debate. We have performed a DFT study on some Linqvist ($Mo_6O_{19}^{2-}$) HPA in order to enlighten their electronic properties. These HPA also represent good models of oxide surfaces without the problem of the unsaturated bonds. In a second step, we have studied some possible active sites on the Linqvist molecule, and the ability of these sites to activate the CH bond of a methane molecule

Due to their interesting acid and redox properties, the polyoxometallates (iso or hetero) are now very extensively used as catalysts. They also provide a good basis for the molecular design of mixed oxide catalysts. The catalytic function of heteropolycompounds (HPC) has attracted much attention in the last two decades (1,2). and there are already some industrial processes which use HPCs (3,4). The major reason for this increasing interest on that kind of oxocomplexes is probably their good ability to act as strong Brönsted acids and powerful oxidants, exhibiting fast and reversible multielectron exchanges (5,6) under rather mild conditions. Solid heteropolyacids (HPA) possess a discrete ionic structure exhibiting an extremely high proton mobility (7). Numerous reviews have been devoted to the various aspects of HPA catalysis in either homogeneous or heterogeneous systems (8,9).

One of the promising challenges in the field of heterogeneous selective oxidation is the dehydrogenation oxidative reaction (ODH) of organic compounds and in particular of alkane molecules (10,11). The methane oxidation remains a very strong challenge. Some previous works reported the use of HPA as an efficient and selective material for the ODH process. The reaction seems to proceed through a Mars and Van Kreveulen mechanism, which, in the stationary state, implies the consumption of the oxygen atoms from the lattice (i.e. the polyoxoanion oxygen atoms) leading to the formation of oxygen vacancies (12).

Our purpose is to check the capability for the optimized structure to accept that kind of perturbation and the capability of the defect structure to activate a C-H bond.

Calculation Methods

All the calculations of this study have been made using the density functional program ADF (13). The basis set for all the atoms include triple z Slater functions for the valence orbitals and a frozen core approximation for the inner shell electrons. This core includes the electrons up to 3d for the Mo, 1s one for C and O and 2p for P. We have added polarized orbitals for all the elements, including H atom. For the calculation at the LDA level, we use the VWN formalism (14). The effect of gradient corrections on exchange and correlation are included in the approximations introduced by Perdew and Wang (PW91) (15). In order to reduce the computational effort, these gradient corrections are only computed after the self-consistence convergence of the electronic cycles. The accuracy parameter ACCINT has always been set to 5 giving reliable results. All the structures are optimized at the same level of accuracy.

Results and Discussions.

The Keggin structure.

The dodecamolybdophosphate ($PMo_{12}O_{40}^{3-}$) is a heteropolyanion (HPA) with several isomeric structures (5). The Keggin structure (figure 1) is the most symmetrical. This high level of symmetry (T_d) allows the full geometrical optimization of the molecule despite the large number of atoms, including the exchange and correlation effect, which are important for the transition metal atoms.

There are four types of oxygen atoms in the structure. The Oa oxygen atoms are bound to the phosphorus atom, and so they are not accessible to the incoming molecules. The most important for the reactivity of the HPA are the two bridge type atoms (Ob, Oc) between two Mo and the top one (Ot) which is bound only to one Mo by a double bond. Ob, Oc and Ot atoms are located on the exterior shell of the Keggin anion.

The experimental geometry of the Keggin structure with various cations (Na^+, H^+, Cs^+) is available from X ray or neutron diffraction techniques. The calculated and experimental distances and angles are summerized in the Table I. The agreement between them is good, except for the Ot oxygen (1.72 vs. 1.67). This kind of error is typical for this type of DFT calculation which overestimates distances somewhat. The constraints on the other parameters due to the T_d symmetry correct this defect. It seems that the basis sets with polarization orbitals are large enough to describe the geometric and electronic properties of the anion.

Table I. Distances and angles in the Keggin HPA. The experimental values are taken in ref. 20.

	Calculated values	Experimental values
P-Oa	1.569Å	1.50-1.55Å
Mo-Oa	2.437 Å	2.43-2.47Å
Mo-Ob	1.939 Å	1.91-1.92Å
Mo-Oc	1.927 Å	1.90-1.93Å
Mo-Ot	1.727 Å	1.62-1.70Å
Mo-Ob-Mo	124.3°	126-129°
Mo-Oc-Mo	152.4°	151-152°

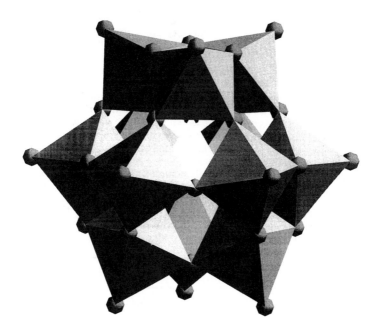

Figure 1. Scheme of a Keggin HPA.

In order to understand the chemical behavior of the HPA, we study the HOMO and LUMO orbitals. The HOMO and LUMO ones have an E symmetry and so are two-fold degenerated. These results are in agreement with previous LDA calculations (17). The HOMO orbital is mainly centered on the bridge oxygen atoms and is quite totally nonbonding. Most interestingly, the LUMO is localized on the twelve Mo atoms, corresponding to the d orbital in the plane of the four bridging oxygen atoms. This orbital is also a nonbonding one. So it should be possible to introduce four electrons in this orbital with small variations of the Mo-O distances. This prediction is consistent with the experimental data (18). The HPA are good oxidizing agents.

It is also possible to compare the HOMO LUMO gap with the results of Yamazoe et. al.. They have found a gap of 1.60 eV but at the LDA level for an non-optimized structure. We calculate a value of 1.91 eV. In order to test the influence of the electrostatic array in the crystal, we have computed the electronic properties for various electrostatic fields. We construct three point charge distributions for the first cationic shell. In the first one, a +3 charge is placed in the 111 direction, in the second the same point charge is located in the 100 direction. The distance between the charge and the P atom is fixed to 7Å. The point charge distribution is transformed to have T_d symmetry. The last distribution is the mean of the two previous ones. The effect is an overall shift of the energy levels with very small variation of the gap (Table II) and of the Mulliken charge of the various atom types (Table III). The electrostatic field is screened by the exterior oxygen atoms, which reduce the influence of the crystal field on the structure. This effect is also increased by the size of the Keggin anion and by the high symmetry of the point charge distribution. These three converging phenomena are responsible for the weak influence of the crystal field on the HOMO-LUMO gap as the effect on all orbitals centered on the Mo atoms and exterior O atoms is similar.

Table II. Variation of the HOMO-LUMO gap with the point charge distribution. The first column represented the computation without point charge distribution, the three last ones are in the order of the text.

Charge Distribution	1	2	3	4
Gap (eV)	1.914	1.938	1.987	1.962

Table III. Variation of the Mulliken charges with the point charge distribution. Same organization than for the table 2

Charge Distribution	1	2	3	4
P	1.87	1.88	1.87	1.88
Mo	2.38	2.36	2.35	2.35
Oa	-0.88	-0.88	-0.88	-0.88
Ob	-0.90	-0.91	-0.90	-0.90
Oc	-0.93	-0.94	-0.93	-0.94
Ot	-0.67	-0.63	-0.63	-0.63

The charge on the Mo atoms is 2.87 electron, which is to be compared with a formal oxidation number of 6 according to the ionic model. The bond between the atoms in the structure is also a covalent one. In order to describe accurately the behavior of the electron in the 4d orbitals of the Mo, it is necessary to include the gradient corrections.

The chemistry of the Keggin $PMo_{12}O_{40}^{3-}$ anion is well described by the calculation. For example the most nucleophilic oxygen atoms should be the bridge one. The HOMO is localized on these atoms. Furthermore, the Mulliken charges on the bridge (Ob, Oc) atoms are greater than the values on the Ot atoms.

Despite the good description of the molecule, we cannot test the reactivity of the HPA on the Keggin models. The CPU time for one electronic step is about 3 hours, within the Td symmetry. We have tested some calculations on systems with lower symmetry but the calculation did not converge in ten hours on a Cray C90. In order to investigate the chemistry and more precisely the activation of methane on heteropolyanion catalysts, we turn our attention to Linqvist systems ($Mo_6O_{19}^{2-}$) which are more simple from the computational point of view.

The Linqvist structure

The number of the characterized HPA is very important and is increasing each year (19). Most of them are composed of more than 50 heavy atoms. This size is a problem for ab initio methods when we include no symmetry in the calculation. The isopolyanions can be much smaller. The so-call Linqvist structure is constructed with six MoO6 octahedra. The Mo atoms build an octahedra around the central oxygen atom (figure 2). This model seems accurate not only for the HPA, but also for the bulk oxide MoO_3 which is constituted of similar building units.

The geometrical parameters are close to those of the Keggin HPA (Table IV). Then the Linqvist structure seems to be a good manageable model for the HPA structure. The calculated distances are also in good agreement with the experimental data (Mo-Mo: 3.274 vs. 3.28). The point group is C1, even if all the Mo-Oa distances are very close. It is confirmed by the calculation, the O_h symmetry leads to two imaginary frequencies. From the electronical point of view, the differences between the two structures are more important. The Mulliken charges are summarized in table 4. The Mo atoms are less charged in the $Mo_6O_{19}^{2-}$ ion, while the charges on the Oc ones decrease slightly. On the contrary, there are only small variations on the Ot atoms. Those three effects show that the bonds in the Linqvist structure are more covalent than in the Keggin polyanion.

Table IV. Geometric and electronic parameters of the Linqvist isopolyanion. For simplicity, we have supposed an octahedral geometry.

	Distance (Å) or angle (degree)
Mo-Oa	2.327
Mo-Oc	1.725
Mo-Ot	1.931
Mo-Mo	3.274
Mo-c-Mo	116.0
Atom	Mülliken Charge
Mo	2.213
Oa	-1.06
Oc	-0.85
Ot	-0.67

The orbitals show also some variation between the two HPAs. Let us focus on the HOMO and LUMO. The HOMO has an important component on the central oxygen atom, so the contribution on the bridge oxygen decreases. In contrast to the Keggin

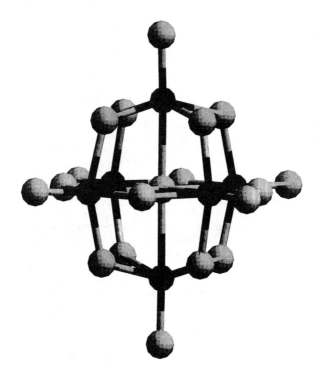

Figure 2. Scheme of the Linqvist structure.

structure, the LUMO is an antibonding one. The weight on the Mo 4d orbital and the Ob are close. This is in agreement with the experimental facts. Indeed, the Linqvist structure is more difficult to reduce than the Keggin anion. The Linqvist structure is not perfect model, but it will give reliable hints concerning the reactivity of the HPAs. The comparison with other clusters (21) which mimic the surface of MoO_3 indicated that the bonding seems to be more ionic in the Linqvist anion than in other possible clusters. For example the Mo charge goes from +2.23 in $Mo_6O_{19}^{2-}$, to +1.3 in $Mo_2O_9H_{10}$. The differences may also be induced by the saturation of the dangling bond by hydrogen atoms and by the small size of the model.

Nucleophilicity of the Linqvist anion

The description of the geometric and electronic parameters of the isopolyanion seems to be accurate. In a second step we have tested the ability of our model to describe the reactivity of the HPA and of the MoO_3 surface. So we study the acidic properties of the anion.

Taking into account the experimental conditions involved in the Linqvist structure condensation, the anion should be a strong acid in aqueous solution. The two kinds of oxygen (Oc, Ot) atoms, as in the Keggin anion, should have different acid behavior. The bridge ones should correspond to the stronger acid, considering the electrostatic potential and the distribution of the HOMO. We have compared the bond strength for the two sites. The binding energies of the proton on an Oc is 15.63 eV while it is 14.99 eV on the Ot oxygen atom The Oc-H distance is 0.962Å close to the free OH species (0.969Å) and close to the calculated distance on small Mo clusters (21). The addition of the proton did not change the geometry of the structure. Even the charge on the Mo and the oxygen do not change (for Mo 2.23 in $Mo_9O_{19}H^-$ vs. 2.21 in $Mo_9O_{19}^{2-}$). The relaxation of the anion is very small. The perturbation due to the H atoms does not influence the whole structure. The Linqvist anion should be a good model to test the reactivity of the HPAs toward the proton. These calculations confirm that, on the Keggin structure, the bridge oxygen atoms represent the acidic site.

Activation of the Linqvist anion.

Much experimental evidence (22) suggest that the mechanism of C-H bond activation during the oxydehydrogenation reaction of saturated hydrocarbon involves the creation of oxygen vacancy sites on the oxide surface or in the HPA structure. We have tested the possibility of creating such vacancies in a Linqvist anion, despite its small number of atoms. The most probable mechanism involves the departure of one of the exterior oxygen atoms, most probably the Ob one which is the most reactive for a nucleophilic attack. In this case, the active site is a vacancy on the anion.

The optimized geometry of the vacnacy structure Mo_6O_{18} is presented in figure 3. The parts of the anion far from the defect are only slightly modified. The Mo-Ob and Mo-Ot distances are respectively 1.91 and 1.69 Å. In the perfect structure, those distances are 1.95 and 1.72 Å. The main geometrical perturbation is reduced to the nearest neighbors of the vacancy. In spite of its small size, the reconstruction is very important. The two Mo atoms try to restore their initial coordination and move away from the vacancy while the Oa oxygen relaxes into. The Mo-Oa distance shows great variations, from 2.10 to 2.31 Å. This reconstruction reduces the energy of the structure by 1.99 eV. It seems difficult to study the creation of the vacancy without including those huge effects in the model. The total variation energy associated with the reaction $Mo_6O_{19}H_2 \Leftrightarrow Mo_6O_{18} + H_2O$ is +14.75 eV. This value is similar to the energy of abstraction of OH^- from the $Mo_2O_{11}H_{10}$ cluster used

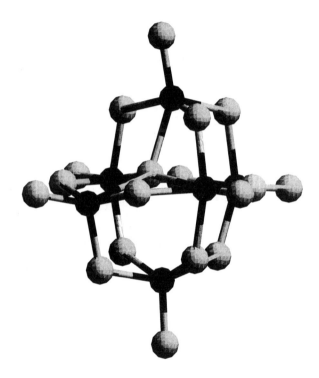

Figure 3. Geometry of the vacancy anion.

by M. Witko and K. Hermann, who found a value of 10.70 eV for the abstraction of an OH⁻ from their cluster. For the same reaction, we compute a value of 9.67 eV, taking into account the relaxation energy

The spreading of the electronic perturbation over all the anion is another parameter that will indicate the ability of the model to mimic the creation of vacancy in more important structures (Keggin HPA and MoO_3 oxide). As for the geometrical perturbation, the part of anion far from the defect is only slightly perturbed. The charges on the Oc atoms are now -0.83, which represents a variation of 0.02 electron with respect to the Linqvist structure. Clearly, the reconstructed part is subject to much more important variations. The Ot oxygen atoms now have a charge of –0.56 and the bond between them and the Mo are more covalent. It seems that the Linqvist anion is a good model to study the activation of a surface. The Ot atoms are used to restore the valence and the charge of the Mo (2.21 vs. 2.22), while the Ob oxygens seem to localize the perturbation to the nearest atoms and so dissociate the rest of the system from the perturbation.

Table V. Mulliken charges for the vacancy anion. Mo1 was previously bound to the missing Oc. Mo2 are also around the vacancy. Mo3 are on the other part of the Linqvist strucute. Oc1 are located around the vacancy. Oc2 are in trans-position of the vacancy with respect to Mo1. Oc3 are on the other part of the anion.

Atom	Mülliken Charge
Mo1	2.19
Mo2	2.31
Mo3	2.23
Oa	-1.06
Oc1	-0.82
Oc2	-0.80
Oc3	-0.83
Ot	-0.56

As the number of atoms and electrons change, the shape of the HOMO and LUMO is greatly modified (figure 4a and 4b). The HOMO is now localized all around the vacancy. This will allow a great overlap of the orbital with the incoming molecule. The LUMO is centered on the less perturbed part of the Linqvist. We have considered the activation of the Ot oxygen; even if they are less probable candidates for the creation of the active site. The Mo-Ot bond is shorter than Mo-Oc and the oxygen atoms are also less charged. This optimized structure is less stable from 1.57 eV, and the abstraction of the OH⁻ fragment is more difficult by 0.93 eV. Taking into account this difference, we have focused the last point of the study on the abstraction of the Oc oxygen atom. In order to test the possible catalytic activity of the vacancy, we have studied the activation of methane on the hexamolybdate Linqvist anion.

Activation of the methane on a vacancy.

The HPA have been proved to be good catalysts for the activation of the CH bond in the ODH of propane and isobutane (10). The reaction seems to proceed through a Mars and Van Kreveulen. We have studied the ability of a vacancy to activate methane. We have postulated a mechanism with a hydride species. In effect the Keggin anion is able to adsorb dihydrogen (22). Firstly, we have investigated the adsorption of a methane molecule. Secondly, we have tried to determine the transition state and the activation energy for the C-H activation.

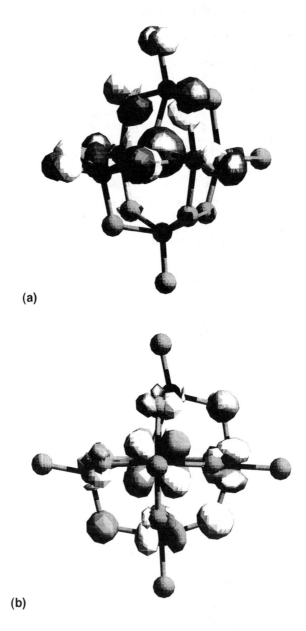

(a)

(b)

Figure 4. HOMO (a) and LUMO (b) orbitals of Mo_6O_{18}.

Firstly we have computed the most stable geometry which can be described as an H atom bound to one Mo previously coordinated with the abstracted Oc atom. The Mo-H distance is 1.7 Å. The CH_3 fragment is adsorbed on the Oc atoms. The geometry is represented in figure 5. The charge on the H atom is -0.33 electron. The charge on the CH_3 moiety is +0.38. Consequently, the net charge of the Mo_6O_{18} fragment is close to 0. The distances are all quite similar to the distances in the Mo_6O_{18} fragment. The electronic charges are also conserved during this dissociative adsorption except for the Mo bond with the H atoms. This Mo charge is equal to 2.15 while the charge on all the other Mo atoms is 2.25. The H atom restores the initial valence of the metallic atom, which then move toward its initial position.

From the energetic point of view, the reaction $Mo_6O_{18} + CH_4 \Leftrightarrow Mo_6O_{18}CH_3H$ is endothermic. The computed value is $\Delta E = -0.79 eV$.

Secondly, we have looked for the transition states. We have chosen the C-H distance and the distance between the middle of the C-H bond and the Oa atom as reaction coordinates. The final point of the computation has the geometry of the previous fragment. All the other parameters have been optimized according to this reaction coordinate. We have verified that the transition state found in this study has one imaginary frequency. The geometry of the transition state is close to the final geometry with a long C-H distance (2.5 Å) (figure 6). The activation energy for the reaction is 1.52 eV. At the LDA level, this energy is reduced to 1.35 eV. This reflect the general tendency of LDA to underestimate the activation energies. In order to check to influence of the electrostatic field of the surface, we have introduce point charge in the half space The electrostatic field reduce the activation energy to 1.40 eV. This is only a rough estimation because the point charge distribution will overestimate the surface electrostatic flied. The value for the C-H activation in the gas phase is close to 4 eV. The Linqvist HPA acts as a catalyst even if the remaining activation energy is quite large.

Preliminary experimental studies seems to show that the Linqvist structure is not very active in the ODH reaction while the Keggin HPA can perform the reaction. The chosen model demonstrates that the Linqvist is indeed not yet the perfect catalyst. In further studies, changing some properties of the model could lead to the improvement of the experimental conditions. The various parameters could include the nature of the metal, the oxidation properties and even the shape or the size of the considered HPA.

Conclusion.

In this study, we have tested the possibility to use the Linqvist isopolyanion as a model of the HPA or even of the surface of MoO_3. The calculations on the HPA are, at the present time, only possible when the full symmetry of the ideal anion is used. In this case, the optimized geometry is in very good agreement with the X ray diffraction measurements. We can predict that the more nucleophilic sites are the Ob and the Oc ones. The Linqvist isopolyanion is a more simple structure than the HPA. It seems to represent a good compromise in order to compute electronic and geometric properties of the catalyst. The three systems are built with octahedral MoO_6 units. The nucleophilic properties of the Linqvist anion confirm the prediction made for the HPA.

In order to improve our knowledge of the catalytic properties of the molybdenum oxide, we have tested the ability of the $Mo_6O_{19}^{2-}$ to create vacancy site, which is supposed to be active in catalytic mechanisms. The electronic and geometric properties of the vacancy fragments show that the perturbation due to the abstraction of an oxygen atom are localized in a very small region. This vacancy can activate the C-H bond of a methane molecule. The activation energy is still relatively high: 1.52

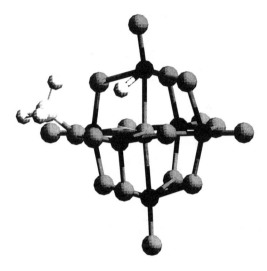

Figure 5. Methane adsorbate on Mo_6O_{18}.

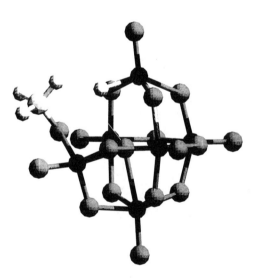

Figure 6. Scheme of the transition state.

eV. We will check various parameters such as the mechanism or the metallic center, in order to find structures with lower activation energies and to garner clues for the construction a better catalyst.

Acknowledgement.

The authors thank I.D.R.I.S. (Institut du développement et des ressources en informatique scientifique) for a computer time allocation under the project number 980882.

References

1. Hill, C.L.; Prosser-McCartha, C.M. *Coord. Chem. Rev.* **1995**, 143, 407.
2. Kozhevnikov, I.V. *Catal. Rev. Sc. Eng.* **1995**, 37, 311.
3. Misono M.; Nojiri N. *Appl. Catal.* **1993**, 64, 1.
4. Nojiri, N.; Misono, M. *Appl. Catal.* **1990**, 93,103.
5. Pope, M. T.; *Heteropoly and Isopoly oxometallates*, Springer-Verlag: Berlin, Germany, 1983.
6. Souchay, P; *Ions Minéraux Condensés*, Masson: Paris, France, 1969.
7. Nakamura, O.; Kodama, T.; Ogino, I.; Miyake, Y. Japanese Patent JP 51106694, 1976, *Chem. Abstr.* **1976**, 86, 19085.
8. Kozhevnikov, I. V. *Chem. Rev.* **1998**, 98, 171 and references therein.
9. Mizuno, N.; Misono, M. *Chem. Rev.*, **1998**, 98, 199 and references therein.
10. Cavani, F.; Trifiro, F. *Catal. Today,* **1997**, 36, 431.
11. Moffat, J.B. *Appl. Catal. A.*, **1996**, 1246, 65.
12. Ernst,V.; Barbaux, Y. ; Courtine, P. *Catalysis Today* **1987**, 1, 167 and reference therein
13. Guerra, C. F; et al. *Parallelisation of the Amsterdam Density Functional Program*, Methodes and Tehchniques in Computational Chemistry, Clementi, E. Corongiu, G.: Cagliari, Italy, 1995, Vol. METECC-95
14. Vosko, S.H.; Wilk, L.; Nusair, M. *Can. J. Phys.* **1980**, 58, 1200
15. Perdew. J.P. et al. *Phys. Rev. B*, **1992**, 46, 6671
16. Boeyens, J. C.; Mac Dougal, G. J.; Van R. Smit, J. *J. Solid State Chem.*, **1976**, 18, 191.
17. Taketa, H.; Katsuki, S.; Eguchi, K.; Seiyama, T.; Yamazoe, N. *J. Phys. Chem.*, **1986**, 90, 2959.
18. Souchay,P.; Massart, R.; Hervé, G. *Rev. Polarogr.* **1967**, 14, 270.
19. Baker, L.C.W.; Glick, D.C. *Chem.Rev.* **1998**, 98, 3.
20. Allcock, R.; Bissel, E. C.; Shawl; E. T. *Inorg Chem.*, **1973**, 12, 2963
21. Hermann, K.; Michalak, A.; Witko, M. *Catalysis Today*, **1996**, 32, 321.
22. Jalowiecki-Duhamel, L.; Monnier, A.; Barbaux, Y. *J. Catal.* (in press).

Chapter 27

Determination of Transition State Structures Using Large Scale Ab-Initio Techniques

E. Sandré[1,4], M. C. Payne[1], I. Stich[2], and J. D. Gale[3]

[1]TCM Group, Cavendish Laboratory, Madingley Road, Cambridge CB3 0HE, United Kingdom
[2]Department of Physics, Slovak Technical University, Ilkovicova 3, SK-812 19 Bratislava, Slovak Republic
[3]Department of Chemistry, Imperial College of Science, Technology, and Medicine, South Kensington SW7 2AY, United Kingdom

This paper deals with the determination of transition state structures using large scale ab-initio plane wave calculations. We firstly comment on classical gradient based methods which were previously proposed in the literature, and underline the difficulties in using them within the framework of plane wave density functional theory calculations. Consequently, in order to surmount these difficulties we propose a concerted constrained minimisation method, and describe its implementation. As an application of the method, we report the determination of the transition state for the conversion of methanol into dimethyl ether inside chabazite.

1. Introduction

In recent years, new algorithms have been proposed for locating transition state (TS) structures within large scale ab-initio plane wave techniques. These techniques, sometimes referred to as Car-Parrinello (CP) methods (1,2), are based on the density functional theory (DFT) approach (3,4) to calculate the total energy of solids. These calculations are based on the Born Oppenheimer approximation and approximate the system of interest as an infinite periodic structure. Because they effectively deal only with one-electron equations, such techniques can be applied to systems with unit cells containing hundreds of atoms, hence their description as large scale ab-initio techniques.

2. Finding transition state structures within large scale ab-initio methods

The main problem in determining transition structures within this framework arises from the fact that reliable second derivatives of the total energy with respect to the position of nuclei are only available at a high computational cost. Only first derivatives

[4]Permanent address: LEPES-CNRS, BP 166, 38042 Grenoble Cedex 9, France.

(gradients, or forces) can be obtained cheaply computationally if one uses a plane wave basis set to expand the electronic wave functions.

TS structures correspond to stationary points on a Potential Energy Surface (PES) where the Hessian matrix (H) has one and only one negative eigenvalue while all the others are positive. The classical way to locate stationary points on a PES is to use quasi-Newtonian techniques (5) based on local second order expansion of the total energy. Consequently, such techniques cannot be implemented efficiently within CP codes, and in stead we have to use techniques which only require the calculation of the forces on atoms to locate TS structures.

Recently, two main types of algorithms which only require the calculation of gradients of the energy have been proposed in order to locate saddle points within large scale DFT techniques. The first of these algorithms is the so called "ridge method" (6) in which one walks down the ridge separating reactants and products valleys towards its minimum. Formally, though such a method should be generally applicable, in practice it appears that it is extremely hard to walk down the ridge once it has been approximately located at some point. Indeed, when far away from the saddle point, features of the PES near the ridge when calculated within DFT can appear to be extremely sharp. In many practical cases, though we were able to determine two different structures on either side of the ridge (which only differed in the Cartesian coordinates of the nuclei by less than 0.01 Å), the calculated gradients were still so different that they did not contain any valuable information about the direction of the descending ridge (see figure 1). This behaviour would not cause any difficulty if it was cheap to verify that the structures are on either side of the ridge. However, the only reliable way to check that the two respective structures do lie on either side of the ridge is to check that they actually transform into reactants and products when relaxed to equilibrium. Such a relaxation process when repeated many times as one iteratively approaches the ridge, will require extensive computation, and make such method extremely computationally intensive.

The second class of algorithm which has been proposed is based on adaptive chain methods linking both the reactant and product structures on the PES(7). These algorithms are based on the fact that on the Mean Energy Path (MEP) linking the two minima (reactant and products), the gradient is continuously parallel to this very same path. In practice, one should first determine an approximate starting path, choose an appropriate sampling of points along it and then at each chosen point perform a projection of the gradient perpendicular to the local direction of the approximate path. A first step of minimisation is then performed using these projected gradients in order to find a second (and better) approximate path. This process is then iterated up to convergence (see figure 2). Since the interpolation function must contain the information about the PES in between two different points where the calculation is actually performed, such a method requires a good interpolation of the approximate MEP to be found at each iteration. However, if this is not the case, the minimisation process can be subject to inconsistency. In particular, it is clear that if the postulated curve locally follows a steepest slope, the local gradients will be parallel to the postulated curve. If the interpolation coefficients are determined so that nearest points connect (which is usually the case when interpolating a series of points), then the calculation will appear to be locally converged when the curve follows the local steepest slope. It follows that such a method can be applied in cases where the interpolation function has the same structure as the actual PES, i.e. if the interpolation function contains long distance information about the PES. However, because the PES in large scale DFT calculations results from self consistent complex calculations, finding such a function seems to be out of reach, and we have found that the method is rather likely to lead to incorrect results.

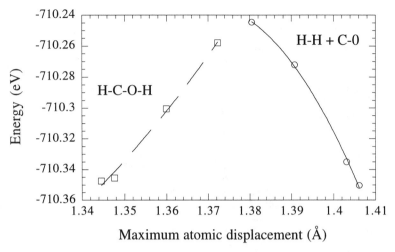

Figure 1: Sharp shape of the PES close to the ridge separating the reactant and product valleys for the adsorption of dihydrogen onto carbon monoxide. The left side corresponds to the product valley, and the right side correspond to the reactant valley.

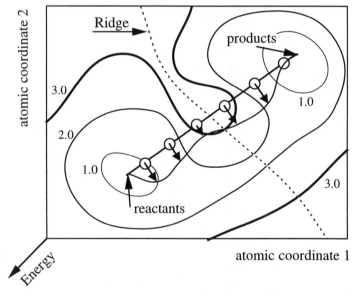

Figure 2: Schematic representation of the different steps for the adaptive chain method. This is a two dimensional projection of the potential energy surface. The coordinates projected in the plane correspond to two atomic coordinates. The energy dimension is perpendicular to this plane. Three different energy contours, denoted by the values 1.0, 2.0, and 3.0 are represented. In the lower left corner is the reactant valley, and in the right upper corner is the product valley. The ridge separating both valleys is shown by a dashed line. In this example, the first transformation curve is the straight line linking the structures of the reactants and products. The second curve is constructed using a projection of the gradient perpendicular to the local direction of the straight line.

3. Constrained minimisation to locate transition structures for large scale ab-initio techniques

Both methods described above are based on the idea that finding a saddle point can be done by walking on the potential energy surface. In order to find a minimum (stable structure) using gradients only, one does not need to strictly control the chosen minimisation path. All the different paths for which the gradients are iteratively minimised lead more or less rapidly to the minimum. On the contrary, if one wants to walk to a saddle point using gradient only techniques, at each step the path has to be controlled carefully. More importantly, errors at each step add up and consequently make the algorithm unstable. This is why we believe that finding a saddle point using gradient only techniques cannot be performed efficiently by explicitly exploring the PES. We believe that an implicit exploration of the PES is more likely to allow the determination of transition structures.

Any implicit exploration of the PES implies a conceptual redefinition of the TS search problem. As far as transition structures are concerned, one should not consider them as saddle points only but should consider the underlying chemical process. A transition structure lies somewhere at the highest energy point along the mean energy transformation path between reactant and products. It usually corresponds to the formation of a new bond and the consequent breaking of a pre-existing bond in the reactant structure.

Once some prior chemical knowledge on the reaction is obtained, it is often fairly easy to impose the reactant structure to continuously transform into the product structure by forcing a new bond to form or equally by forcing an existing bond to break. As an example, let us consider the case of two structures A-H + B which transforms into A + H-B by destruction of the covalent bond linking the hydrogen atom H to fragment A in order to form a new covalent bond with fragment B (H-B). In this process, the underlying transformation mechanism is driven by the increasing distance between fragment A and atom H, or equally by the decreasing distance between atom H and fragment B. One consequently realises that in such a case, finding an MEP reduces to finding the minimum energy structure for any fixed distance between H and fragment A or between H and fragment B.

Obviously, such a method would not apply in the general case where many atoms move collectively unless the enforced displacement of just one atom triggers the whole transformation. Dealing with many bonds forming and breaking is complicated using such an algorithm, and is not recommended in the first instance. However, we have found that this algorithm is robust and efficient on a wide variety of cases such as the example described in section 4. We will now discuss the technical details of the implementation of such a method within the DFT-CP framework.

Following the procedure described for constrained molecular dynamics (8), particularly as the minimisation process is based on a gradient minimisation, minimising the total energy of a given system containing a constrained bond distance is best performed by modifying the forces directly. The constrained minimisation we suggest implementing uses one undetermined Lagrange multiplier to represent the magnitude of the additional fictitious force ($\mathbf{g}$) directed along the constrained bond. This Lagrange multiplier is subsequently chosen so as to keep the interatomic distance constant between two chosen atoms a and b. The forces ($\mathbf{f}^a$, $\mathbf{f}^b$) calculated for the two constrained atoms should be modified by a force $\mathbf{g}$, such as:

$$f^a_{mod} = f^a + g$$

$$f^b_{mod} = f^b - g \qquad (1)$$

$$g = \lambda r_{ab}$$

Because of the simplicity of the constraint, the Lagrange multiplier λ (one multiplier only) can be calculated analytically during the calculation, and does not require an iterative technique. The implementation appears to be straightforward. Determining λ involves solving a second degree polynomial equation in λ which has two roots. The lowest absolute value root (and consequently the lowest value of the modified force g) is the one which should be chosen because it leads to a smaller modification of the minimisation surface when compared to the actual PES.

However when implementing this technique, much attention should be paid to the minimisation algorithm itself. Because a constraint is included, the PES which is explored during the minimisation of the total energy changes at every step. In this case, the use of a conjugate gradient scheme appears to be quite inefficient, and a steepest descent scheme is much more appropriate. However, close to the minimum, the use of conjugate gradient steps is found to speed up convergence. Moreover in many cases when implemented within a CP type scheme, since we are dealing with a minimisation leading to structures far from the minima in the valleys, it is sometimes found that the electronic system did not relax to the Born-Oppenheimer surface. This can be tested by introducing some noise in the converged plane wave coefficients, and reconverging in order to check that the minimum had actually been reached.

Since this method is based on minimisation techniques, one difficulty is to find an adequate starting configuration for the search. For instance, if trying to find a point on the MEP which is located on the reactant side, one cannot converge properly starting from a configuration which is on the products side, and will obtain an energy after minimisation which is too high. However, by generating starting configurations as linear combinations of the two different structures on either side of the ridge, such inappropriate starting configuration can often be spotted because the calculated MEP consequently present a jump in energy.

In many cases, a good way to be sure that the starting configuration is on the right side of the ridge is to perform a concerted approach to the TS structure from both sides of the ridge. Starting from the two stable structures, one can construct two new structures closer to the saddle point on both sides of the ridge. If the two new structures have approximately the same energy, this ensures that they are situated on either side of the ridge, and should have an energy which is below that of the saddle point. They consequently provide good starting points for the construction of new trial configurations in order to reach the transition structure. In the next section, we shall present an illustration of the method .

4. Transition state determination for the conversion of methanol into dimethyl ether inside chabazite

The calculation of the many body PES for the reaction is obtained from the calculation of the electronic ground state of the system within density functional theory in the Generalised Gradient Approximation of Perdew and Wang (9). The system of interest is enclosed in a supercell to which periodic boundary conditions are applied. Norm conserving pseudopotentials are used to represent the nuclei and core electrons, with the valence electrons treated explicitly, the wave functions being expanded in terms of plane waves. To reduce the size of the basis set required, the potential for oxygen and carbon have been optimised using Qc filter tuning (10), allowing a plane wave cut-off of 620 eV. Brillouin zone sampling was carried out only at the Γ point. By monitoring the convergence of energy differences with respect to plane wave cut-off and Brillouin zone sampling, we estimate the errors due to these to be approximatively less than 0.03 eV. The computational cost of the non-local pseudopotential implementation has been minimised by carrying out the projection in real space (11). The accuracy of the pseudo potential has been checked previously (12). The calculations were carried out using the CETEP code (13) running on a massively parallel computer, the Hitachi SR2201 located at the University of Cambridge High Performance Computing Facility.

The reaction we are concerned with takes place inside a zeolite cage, namely chabazite. Zeolites are microporous silicate structures which can absorb certain organic molecules (14). When some of the silicon atoms are substituted by aluminium, the four-fold coordinated aluminium atom has a negative charge which is then balanced by a near by cation. If this cation is a proton, a Bronsted acid is formed. Some substituted zeolites (for instance, chabazite) are known to act as catalysts for organic reactions such as the one we are concerned with here. In the present calculation, we have been using the Al-substituted chabazite structure containing a proton which had been optimised previously (12). The zeolite structure for these calculations contains 37 atoms in the unit cell (11 Si, 24 O, 1 Al and 1 H).

We have been investigating the main step in the methanol to dimethyl ether conversion reaction following a direct associative mechanism as suggested by (15). However, supercell calculations which take into account the full zeolite structure clearly show that at a loading of two methanols per acid site the framework proton is transferred to one of the methanol molecules (16). Thus the starting state for the reaction is a methanol molecule and methoxonium ion. The reaction is formally described inside chabazite by the following equation:

$$H_3C\text{-}OH \quad + \quad H_3C\text{-}OH_2^+ \quad \text{->} \quad H_3C\text{-}OH_2\text{-}CH_3^+ \quad + \quad H_2O \qquad (2)$$

We have considered two stable configurations. The first one corresponds to two methanol molecules inside the cage. One of the molecules is chemisorbed onto the chabazite framework close to the Al atom and forms a methoxonium ion which was shown to be stable in this configuration in (12). The second methanol molecule is positioned at the opposite side of the cage in order to minimise electronic repulsion with the methoxonium ion (figure 3). The second structure corresponds to a protonated DME which is positioned far away from the aluminium atom which corresponds to one of the stable structures on the product side of the reaction (figure 4). In this structure, a water molecule is also absorbed onto the zeolite framework close to the position of the aluminium atom.

The calculation of the reaction pathway has been carried out using the constrained bond method described above. Figure 5 shows the variation of the total energy as a function of the constrained bond distance. In this case, the constrained distance was chosen to be between the carbon atom in the methoxonium ion and the oxygen atom of the second methanol molecule. The calculated energy barrier for the conversion of two methanol molecules into dimethyl ether is found to be 0.7 eV. This value is about 0.2 eV lower than a previously published one which involved a direct interaction between the zeolite and the CH_3 radical of the transition state (14). Here, the reaction takes place across the zeolite channel and the CH_3 radical in the transition state is at the centre of the channel as shown on figure 6. The transition state can be seen to be composed of three distinct fragments: H_2O, CH_3, and CH_3OH. However, there is partial bonding between these fragments with charge density distributed over these

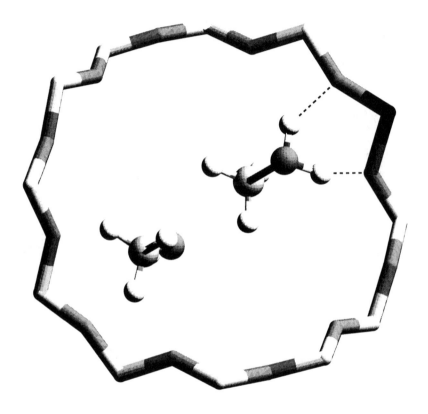

Figure 3: Equilibrium structure for the two methanol molecules inside a chabazite unit cell containing a single Bronsted acid site. One methanol molecule is reduced and transformed into a methoxonium ion. The second methanol molecule is positioned away from the first one. The cylindrical ring schematically represents the chabazite framework. The black cylinder indicates the position of the aluminium atom.

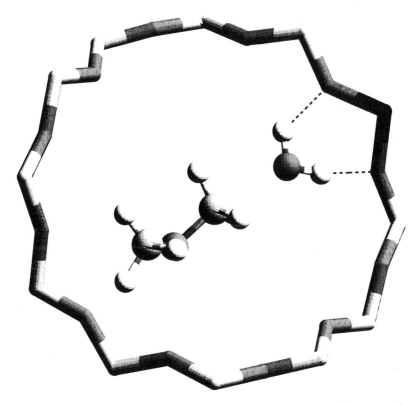

Figure 4: Equilibrium structure for protonated dimethyl ether and water molecule inside a chabazite unit cell. The water molecule is adsorbed on the zeolite framework close to the aluminium atom.

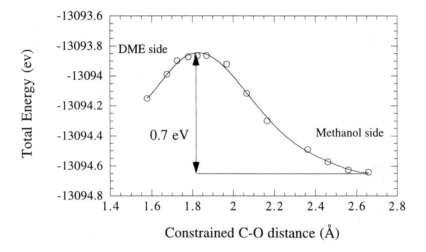

Figure 5: Energy of the different minimum energy structures on the MEP as a function of the constrained C-O distance (Å). The circles correspond to calculated points. The line is a sixth order polynomial fit to the points.

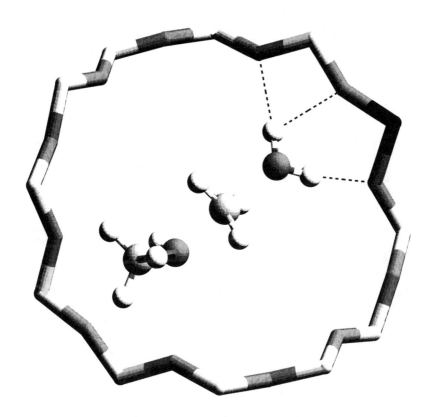

Figure 6: Transition state structure calculated along the reaction path linking the structures presented in figures 3 and 4. The water fragment is absorbed onto the chabazite framework. The methanol fragment is at the opposite end of the unit cell from the water molecule. The CH$_3$ fragment is positioned in the middle of the channel.

356

bonds and so it is not possible to assign an exact integral charge to any particular fragment.

Comparing this result with the results in (14), one can confirm that the formation of DME should indeed be of the S_N2 type. Moreover, this result brings new insight on the nature of the interaction between the CH_3 fragment in the transition state and the zeolite. Comparing our results to (14), one can conclude that the smaller this interaction, the lower the activation barrier. It consequently appears that the main role of the zeolite is to bring an extra proton which allows the backside S_N2 reaction to take place, but not to stabilise the CH_3 fragment of the transition state.

5. Conclusion

Determining transition state structures has always been a challenging task. Finding transition states in the absence of second derivatives of the total energy is an even greater problem. We have investigated some of the major algorithms proposed in the literature and concluded that because of the sharp features of many PES, either the ridge method or the adaptive chain method cannot be generally used. We suggested that a robust way to find the transition state is to perform an implicit exploration of the PES using a constrained bond method together with a concerted approach to the saddle point from either side of the ridge. This procedure has been applied successfully to a number of different reactions but it is not easily generalised to reactions in which the transition state is reached by the collective motion of large numbers of atoms.

The constrained bond method has been applied to determine one transition state for the conversion of methanol into DME inside chabazite. We have found a low energy barrier of 0.7 eV when the reaction takes place at the centre of the zeolite channel, through backside S_N2 mechanism. We have shown that the smaller the interaction between the CH_3 fragment of the transition structure and the zeolite framework, the lower the activation energy for the reaction.

References

(1) Car R. and Parrinello M., Phys. Rev. Lett. **1985**, 55, 2471

(2) Payne M.C., Teter M.P., Allan D.C., and Joannopoulos J.D., Rev. Mod. Phys. **1992**, 64, 1046

(3) Hohenberg P., Kohn W., Phys. Rev. **1964**, 136, B864

(4) Kohn W., and Sham, Phys. Rev. **1965**, 146, A1133

(5) Schlegel H.B., Adv. Chem. Phys. **1987**, 67, 249

(6) Ionova I.V., and Carter E.A., J. Chem. Phys. **1994**, 10, 6562

(7) Ulitsky A., and Elber R., J. Chem. Phys. **1990**, 92, 1510

(8) Allen M.P., and Tildesley D.J., *Computer Simulation of Liquids*, Oxford University Press: Oxford, 1994; p. 92

(9) Perdew J.P., In Electronic Structure of Solids '91; Zeische P. and Eschring H., Ed.; Akademie Verlag: Berlin, 1991

(10) Lee M.-H.Thesis, University of Cambridge, UK, 1995

(11) King-Smith R.D., Payne M.C., and Lin J.S., Phys. Rev. B **1991**, 44, 13063

(12) Shah R., Gale J.D., and Payne M.C., J. Phys. Chem. **1996**, 100, 11688; Shah R., Gale J.D., and Payne M.C., J. Phys. Chem. **1997**, 101, 4787

(13) Clarke L.J., Stich I., Payne M.C., Comp. Phys. Comm. **1992**, 72, 14

(14) Sauer J., Chem. Rev. **1989**, 89, 199

(15) Blaszkowski S.R., and van Santen, R.A., J. Am. Chem. Soc. **1996**, 118, 5152

(16) Stich, I., Gale, J.D., Terakura, K. and Payne, M.C., Chem. Phys. Lett. **1998**, 283, 402.

Chapter 28

Acidic Catalysis by Zeolites: Ab Initio Modeling of Transition Structures

Joachim Sauer, Marek Sierka, and Frank Haase

Humboldt-Universität, Institut für Chemie, Arbeitsgruppe Quantenchemie, Jägerstraße 10-11, D-10117 Berlin, Germany

Two different methods for treating transition states in large catalytic systems are presented: (i) a combined quantum mechanics - empirical valence bond (QM-EVB) method for localizing transition structures in embedded clusters and (ii) the constraint Car-Parrinello molecular dynamics for a fully periodic ab initio treatment. The power of the methods is demonstrated for two different elementary reactions in zeolite catalysts, proton jump between different proton localization sites and C-C formation between two adsorbed methanol molecules yieldig ethanol and water.

Zeolites are microporous materials which are extensively used in the chemical industry as heterogeneous catalysts. They contain Brønsted acidic OH groups, $SiO(H)Al(OSi)_3$, which represent the active sites and are capable of activating chemical bonds. While the different zeolites have this type of active site in common, they differ by the structure of the (alumo)silicate framework and, hence, by the shape and size of the micropores.

Theoretical studies of the catalytic activity involve different stages of increasing complexity.

(1) The Brønsted acidity is characterized by calculating *energies of deprotonation*. This has been done for different framework structures *(1, 2)*. This is a direct, but hypothetical process and does not involve a transition structure.

(2) The catalytic activity may be related to the *proton mobility* within the cavity of the *unloaded* zeolite. Introducing an Al atom into the three-dimensional cornersharing net of SiO_4 tetrahedra creates a negative charge which is compensated by the acidic proton. The proton can be attached to any of the four oxygen atoms of the tetrahedron. Depending on the particular framework considered, the different proton positions have different energies. It is known that, e.g., in H-faujasite protons occupy preferentially the O3 and O1 sites. In this study, we are interested in the transition structure and the energy barrier for proton jumps between framework oxygen atoms belonging to the same AlO_4 tetrahedron.

(3) The proton jump from the Brønsted site to an adsorbed molecule which is activated for further reaction by protonation. We have shown for methanol by different methods that for a loading of one molecule per Brønsted site the protonated methanol in the zeolite cavity is a low lying transition structure or a local minimum about 10-20 kJ/mol above the neutral methanol hydrogen-bonded to the zeolite surface *(3, 4)*.

(4) The ultimate interest is in elementary steps of the catalytic conversion of the feedstock molecules activated by protonation. We are particularly interested in methanol which is the feedstock of the two industrially important processes, methanol to olefin (MTO) and methanol to gasoline (MTG). For the methanol conversion different mechanisms have been proposed. They are based on transition structures and intermediates localized by density functional calculations which represent the catalysts by a small cluster model *(5)*. A particularly interesting step in this mechanism is the formation of the first C-C bond which we study here.

Localizing transition structures is routine now for gas phase reactions, but there are substantial difficulties for reactions involving solid catalysts. A realistic approach which aims at reactivity differences between different zeolites with different framework structures must include periodic boundary conditions and take the structure of the micropores into account. This means that the number of degrees of freedom is very large which makes the localization of transition structures very demanding. We present two strategies to overcome such difficulties. The first is the use of a combined quantum mechanics - interatomic potential approach (QM-Pot; hybrid or embedded cluster method) for getting information on the potential energy surface. To allow bond breaking and making in the interatomic potential part we adopt Warshel's empirical valence bond (EVB) method *(6)*. Transition structures are localized by the a modification of the trust region method *(7-9)*. For so many degrees of freedom (up to 435 in this study) this can be completed only because an initial transition structure and an initial Hessian are calculated with the EVB potential functions only. The final transition structure search is made using combined QM-EVB gradients which are also used to update the Hessian. We demonstrate the success of the method by localizing transition structures for a well defined elementary step, proton jumps between different framework oxygen atoms in two different zeolite structures (stage 2 above). Significantly different barriers are predicted for different crystallographic sites in a given zeolite and for zeolites with different framework structure.

The second strategy aims at exploring larger regions of a potential energy surface with a rich structure of local minima and saddle points. It is the method of constraint molecular dynamics *(10)* which we use in combination with a full ab initio calculation of the zeolite applying periodic boundary conditions (Car-Parrinello MD) *(11)*. Such simulations are computationally very demanding and presently limited to catalysts with modest unit cell sizes. We study the C-C bond formation between two methanol molecules in chabazite (stage 4 above).

Combined QM-EVB Studies of Proton Jumps in Zeolites

Methods. The embedding scheme used *(1, 12)* partitions the entire system (S) into two parts: the inner part (I) containing site in question, and the outer part (O). Saturating the dangling bonds of the inner part with link atoms (hydrogen atoms) yields the cluster (C). The energy of the total system is obtained approximately from the QM energy of the cluster and the difference between the energies of the total periodic system and the cluster, calculated by the interatomic potential functions. Hence, when calculating the forces on an atom α of the I region all three energies make contributions:

$$F_{\alpha \in I} = F_{QM}(C) + F_{Pot}(S) - F_{Pot}(C) \tag{1}$$

For an atom of the O region, the force is obtained from the interatomic potential function alone. A link atom is not moved according to the force acting on it. It is instead kept fixed on the bond which it terminates *(12)*.

Our subtraction scheme (equation 1) requires an analytical interatomic potential function both for the outer and the inner region. For treating reactions including transition structure (TS) searches this becomes a serious problem. Interatomic potential functions generally are not capable of describing bond breaking and bond making processes. The empirical valence bond (EVB) approach of Warshel *(6)* is a solution which we adopt. The single minimum potential energy surfaces (PES) of the reactant and product states are coupled within an empirical 2x2 valence bond problem to yield the adiabatic PES which describes both minima connected by the transition structure. The crucial part of the EVB model is the exchange element V_{12} which we calculate following Chang and Miller *(13)*:

$$[V_{12}(\mathbf{r})]^2 = A \exp (\mathbf{B}^T \Delta \mathbf{r} + \Delta \mathbf{r}^T \, \mathbf{C} \, \Delta \mathbf{r}) \tag{2}$$

where $\Delta \mathbf{r} = \mathbf{r} - \mathbf{r}_0$, $\mathbf{r}$ is the current structure, $\mathbf{r}_0$ is a reference structure in terms of internal coordinates, and A, **B** and **C** are a constant scalar, vector and matrix, respectively, which are derived using quantum mechanically calculated energies, gradients and Hessian for the TS. Differently from the original proposal we define V_{12} in terms of internal coordinates of the atoms with the largest displacement along the reaction path only, and not of all the atoms. This allows us to derive values of A, **B**, **C** and $\mathbf{r}_0$ using quantum mechanical calculations on small cluster models for the TS. These values are assumed to be transferable and used to describe the reaction in the periodic lattice. Our modification of the method relies on the observation that the reaction coordinate involves only negligible motions of atoms sufficiently distant from the reaction site.

We use a modification of the trust region method *(7-9)* with BFGS *(7)* and GMSP *(14)* update methods of the Hessian for localizing minima and transition structures, respectively. First, the optimum cell parameters for the periodic zeolite lattice are determined using potential functions alone (constant pressure optimization). Then a constant volume optimization of all degrees of freedom is performed within the combined QM-EVB scheme. The initial Hessian matrix calculated at the EVB level is updated in every iteration. We stress that localizing the TS is achieved without explicit calculation of the Hessian for the QM part, which in many cases would be computationally extremely expensive. Convergence in the gradient norm of $0.0001 \, E_h \, a_0^{-1}$ is usually achieved within 12 cycles for the minimum and within 20 cycles for the TS search (H-faujasite, 435 degrees of freedom). For structures with a smaller QM region (3 or 4 tetrahedra) the exact combined QM-EVB Hessian matrices are evaluated at the end of the search to proof that a stationary point of correct order is found.

For the QM part we apply the density functional (DFT) method and adopt the B3-LYP *(15)* functional. The basis set is triple-zeta on oxygen and double-zeta on all other atoms. Polarization functions are added to all atoms. The interatomic potential functions have the functional form of a shell-model ion-pair potential which has the advantage of including the mutual polarization of the inner and outer regions. They are parametrized on B3LYP calculations using the same basis set as for the QM cluster *(16)*.

The present implementation*(17)* of the QM-EVB scheme uses the GULP program*(18)* for evaluating the interatomic potential function contributions and the TURBOMOLE *(19, 20)* program for the quantum mechanical calculations.

Problem and Models. Experimetally, the jump rate and the barrier are not easily accessible. The zeolite has to be completely free of any adsorbed molecule and it is not easy to completely dehydrate and deammoniate the samples. NMR experiments have been performed to estimate the activation energy for proton jumps in zeolites, but the results are scanty and inconsistent *(21-24)*. For example for faujasite type zeolites the barriers of 20-40 kJ/mol (Si/Al = 1.2 and 2.6) *(21)* and 61 kJ/mol (Si/Al = 3) *(24)* have been reported. Early quantum mechanical calculations by Sauer et al. *(25)* used finite cluster models and arrived at an estimate of 52 ± 10 kJ/mol for the proton jump barrier. This applies to an "average" zeolitic Brønsted site, but cannot explain differences between different zeolites.

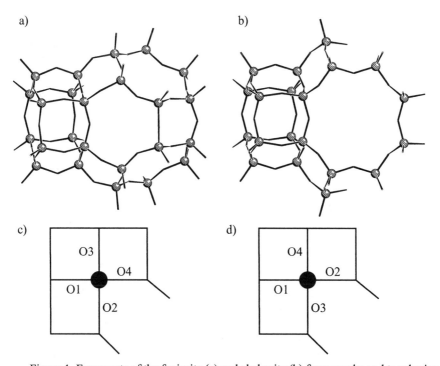

Figure 1. Fragments of the faujasite (a) and chabasite (b) frameworks and topologies around an Al site (c and d, respectively) which are identical for both frameworks, except the numbering.

The combined QM-EVB method presented above properly accounts for the structure of the periodic lattice and the long range crystal interactions at modest computational expense. Specifically, we tackle the problem of differences between the proton jump barriers between different crystallographic positions within the H-faujasite lattice and between two different zeolite frameworks, H-faujasite and H-chabasite.

In the unit cells of faujasite, Figure 1a, and chabasite, Figure 1b, all the tetrahedral atoms are crystallographically equivalent but there are four inequivalent oxygen positions, labelled O1 - O4. Both frameworks consists of four- and six-membered silicate rings and include the double six-membered ring (hexagonal prism) as secondary building unit. The framework topology around the Al atom is similar for both frameworks (Figure 1c). We perform QM-EVB calculations using different sizes of the QM region (cluster model). The first consists of three TO_4 tetrahedra (3T model, Figure 2a). For H-faujasite we define a large QM region consist of 3 four-membered silicate rings ($4T^3$ model, Figure 2b). For comparison, we also examine the finite cluster model used by Sauer et. al. *(25)* (Figure 2c).

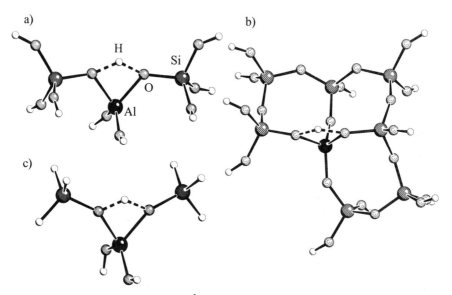

Figure 2. Embedded 3T and $4T^3$ cluster models (a and b, respectively) and free space cluster model (c).

Results and Discussion. Table I shows bond angles and distances which characterize the proton jump transition structure. For 3T cluster models embedded into the two framework structures the exact combined QM-EVB Hessian marices have been calculated and we confirmed that they have one negative eigenvalue (imaginary frequencies of about 1300 cm^{-1}). For the larger models the Hessian initialized by the EVB method, but updated using gradients from the combined QM-EVB method, always retained one negative eigenvalue during the whole optimization process. In all the structures the proton, the Al atom and the two bridging oxygen atoms form a plane within 5 degree. The proton is located between the two oxygen atoms with O-H distances 120.4 - 127.7 pm. The formation of TS involves large distortion of the O-Al-O tetrahedral angle by about 30 degrees. This distortion is larger for the free cluster model due to the lack of framework constraints. The structures obtained for the O1-O4 jump in H-faujasite using the 3T and 4T³ cluster models are virtually identical.

Table I. Bond Distances (pm) and Angles (Degree) of Proton Jump Transition Structures.

zeolite	jump sites	QM cluster	r(O-H)	r(O'-H)	r(Al-H)	O-Al-O'
free cluster			122.5	122.5	194.9	74.8
H-chabasite	O1-H-O2	3T	123.3	126.3	194.3	76.9
H-faujasite	O1-H-O4	3T	127.7	125.3	194.4	78.0
	O1-H-O4	4T³	127.0	125.1	194.2	77.6
	O1-H-O2	4T³	126.7	120.4	187.2	77.7
	O1-H-O3	4T³	123.6	121.6	185.7	77.0
	O3-H-O4	4T³	123.4	121.2	184.0	77.7
	O3-H-O2	4T³	123.7	124.5	194.3	76.3

Table II shows the reaction energies and activation barriers calculated for the proton jumps. The total QM-EVB energies can be decomposed into the quantum mechanical contribution, $\Delta E_{QM//QM\text{-}EVB}^{(\#)}$, and the long range contribution $\Delta E_{LR//QM\text{-}EVB}^{(\#)}$ both evaluated for the structure obtained from QM-EVB optimizations. For almost the same cluster size the quantum mechanical contribution of the embedded cluster is larger than the barrier for the free cluster which is due to the constraint imposed by the framework on the structure adjustment for the transition structure. Both clusters consists of three tetrahedra, but differ in the termination, OH for the embedded cluster and H for the free space cluster. Moreover, the long range interaction also makes a positive and significant contribution to the energy barrier. Since both effects neglected in the free space clusters work in the same direction, the results for embedded cluster calculations are substantially larger than the free cluster result. This is different from deprotonation and ammonia adsorption energies for which the two effects act in opposite direction (1, 2).

Increasing the quantum mechanical part from the 3T to the 4T³ model reduces the long range contribution to 1/3 while the total barrier changes from 107 to 101 kJ/mol only. This is the expected behaviour as in the limit of a very large model the long range correction should vanish. For H-chabasite we could afford fully periodic single point calculations using the Dsolid code (26). The QM-EVB and the periodic QM results agree within 1 kJ/mol.

Table II: Reaction Energies, ΔE, and Activation Barriers, ΔE#, as well as Quantum Mechanical and Long Range Contributions to the Barrier, QM//QM-EVB and LR//QM-EVB, for Proton Jumps in Zeolites, kJ/mol.

zeolite	positions	QM cluster	ΔE	ΔE#	QM	LR
free cluster			0.0	42.0	42.0	-
H-chabasite	O1-H-O2	3T	21.7	75.7	51.2	24.5
H-faujasite	O1-H-O4	3T	22.2	107.0	66.0	41.0
	O1-H-O4	4T^3	13.8	100.9	87.0	13.9
	O1-H-O2	4T^3	19.6	98.2	77.8	20.4
	O1-H-O3	4T^3	5.8	105.6	89.5	16.1
	O3-H-O4	4T^3	8.0	109.6	94.0	15.6
	O3-H-O2	4T^3	13.8	66.9	59.9	7.0

The calculations reveal that proton jump barriers show significant differences between different zeolites, i.e. between H-chabasite and H-faujasite, and there are also large differences between the different sites and paths in a given zeolite. In H-faujasite the lowest barrier found is 67 and the highest one 110 kJ/mol (4T^3 model, Table 2). It has been suggested *(24)* that the proton jump barrier is proportional to the acidity differences between the two oxygen sites involved in the jump. Our results do not support this assumption. For example, for the O1-O4 and O2-O3 jumps in H-faujasite the energy barriers differs by more than 30 kJ/mol while the relative energies of the two minima are the same (13.8 kJ/mol). These relative energies correspond directly to differences in the deprotonation energies, the usual measure of acidity.

Constraint Car-Parrinello Molecular Dynamics for C-C Bond Formation between Two Methanol Molecules in H-Chabasite

Simulations. Constant temperature MD simulations at 500 K are performed. First, unconstraint simulations are made for the initial state of two adsorbed methanol molecules and the product state, adsorbed ethanol and water molecules. Because the likely reaction path is known to involve a rather high activation barrier the method of the 'Blue Moon ensemble' is adopted *(27)*. It requires the choice of a holonomic constraint which is introduced into the Car-Parrinello Lagrangian.

The C-C distance is the natural choice of a reaction coordinate for C-C bond formation. By performing several MD simulations for different values of the constraint the reaction coordinate is scanned from the average C-C distance in the reactant to the average C-C distance in the product state. Hence, this method does not localize a single transition structure, but samples the phase space in its vicinity. Free energy differences are then computed by integrating the statistically averaged constrained forces along the chosen reaction coordinate. The electronic problem is solved by density functional theory employing the Perdew-Burke-Ernzerhof functional. We use Vanderbilt ultrasoft pseudopotentials and expand the Kohn-Sham orbitals in a plane wave basis set including all plane waves with kinetic energies up to 25 Ry.

Results. We have carried out constraint MD simulations for six values of the C-C distance between 150.8 pm, the equilibrium distance of the ethanol molecule, and 359.8 pm, the C-C distance in the adsorbed methanol dimer. The following observations are made:
(1) The zeolitic proton is immediately transferred to the methanol dimer and virtually no fluctuations are seen which return the proton.
(2) At a C-C distance near the transition state (210 - 230 pm) the C-O bond of the protonated methanol molecule becomes significantly elongated.
(3) The transition state structures comprise a carbonium ion and a weakly interacting water molecule. Configurations are seen in which the proton of the carbonium ion is shared between the two carbon atoms similar to a non-classical bridged ethyl cation.

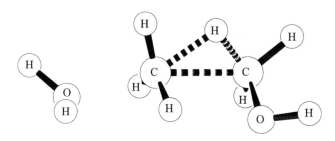

Figure 3. Constraint MD simulation of two methanol molecules in H-chabasite. Snapshot showing a configuration close to the transition state.

(4) The water molecule is split off from the transition state complex.
(5) Eventually the proton of the carbonium ion is transfered back to the zeolite framework restoring the acid site. Later on the ethanol molecule forms a hydrogen-bonded surface complex with its OH group pointing to the zeolitic proton.

Figure 4 shows the statistically averaged constrained forces along the reaction coordinate. Integration from the initial state to the transition state yields a free energy of activation of about 200 kJ/mol. To estimate the error connected with the relatively small number of integration points we made an additional MD simulation at r(CC)=4.1 bohr close to the transition state. The free energy of activation increased by 20 kJ/mol. Hence, our final estimate of the free energy barrier is 220 ± 20 kJ/mol. We also computed an energy profile of the reaction by constrained optimization (zero Kelvin) of representative configurations taken from the 500K trajectories. According to this profile the activation energy is 180 kJ/mol.

Blaszkowski et al. *(5)* calculated activation energies for the same reaction, but replaced the periodic zeolite structure by a small cluster model, $HOHAl(OH)_2OH$. It comprises just the acidic proton and the AlO_4-tetrahedron terminated by hydrogen atoms. The activation energy calculated was 310 kJ/mol, significantly larger than our value. This is not surprising because the transition structure is an ionic complex which in our periodic scheme can be stabilized by the zeolitic framework. However, the transition structure found in the cluster calculations is similar to that found in our CPMD simulations.

366

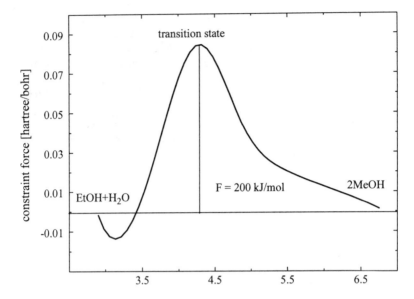

Figure 4. Statistically averaged constrained forces as a function of the C-C distance (cubic spline of 6 points).

Acknowledgement

This work has been supported by the "Max-Planck-Gesellschaft" and the "Fonds der Chemischen Industrie". Computer time on a CRAY-T3E has been provided by the "Höchstleistungsrechenzentrum Stuttgart (HLRS)".

Literature Cited

1. Eichler, U.; Brändle, M.; Sauer, J. *J. Phys. Chem. B* **1997**, *101*, 10035.
2. Brändle, M.; Sauer, J. *J. Am. Chem. Soc.* **1998**, *120*, 1556.
3. Haase, F.; Sauer, J. *J. Am. Chem. Soc.* **1995**, *117*, 3780.
4. Haase, F.; Sauer, J.; Hutter, J. *Chem. Phys. Lett.* **1997**, *266*, 397.
5. Blaszkowski, S. R.; van Santen, R. A. *J. Am. Chem. Soc.* **1997**, *119*, 5020.
6. Aqvist, J.; Warshel, A. *Chem. Rev.* **1993**, *93*, 2523.
7. Fletcher, R. *Practical Methods of Optimization;* John Wiley&Sons: New York, NY, 1987.
8. Helgaker, T. *Chem. Phys. Lett.* **1991**, *182*, 503.
9. Culot, P.; Dive, G.; Nguyen, V. H.; Ghuysen, J. M. *Theoret. Chim. Acta* **1992**, *82*, 189.
10. Ryckaert, J. P.; Cicotti, G. *J. Chem. Phys.* **1983**, *78*, 7368.
11. Galli, G.; Parrinello, M. In *Computer Simulations in Material Science;* Meyer, M.; Pontikis, V., Eds.; Kluwer: Dordrecht, 1991; pp 283-304.

12. Eichler, U.; Kölmel, C.; Sauer, J. *J. Comput. Chem.* **1997**, *18*, 463.
13. Chang, Y.-T.; Miller, W. H. *J. Phys. Chem.* **1990**, *94*, 5884.
14. Bofill, J. M. *Chem. Phys. Lett.* **1996**, *260*, 359.
15. Becke, A. D. *J. Chem. Phys.* **1993**, *98*, 5648.
16. Sierka, M.; Sauer, J. *Faraday Discuss.* **1997**, *106*, 41.
17. Sierka, M. *QMPOT*; Humboldt-University: Berlin, 1998.
18. Gale, J. D. *J. Chem. Soc., Faraday Trans.* **1997**, *93*, 629.
19. Ahlrichs, R.; Bär, M.; Häser, M.; Horn, H.; Kölmel, C. *Chem. Phys. Lett.* **1989**, *162*, 165.
20. Treutler, O.; Ahlrichs, R. *J. Chem. Phys.* **1995**, *102*, 346.
21. Freude, D.; Oehme, W.; Schmiedel, H.; Staudte, B. *J. Catal.* **1974**, *32*, 137.
22. Freude, D. *Chem. Phys. Lett.* **1995**, *235*, 69.
23. Baba, T.; Inoue, Y.; Shoji, H.; Uematsu, T.; Ono, Y. *Microporous Mater.* **1995**, *3*, 647.
24. Sarv, P.; Tuherm, T.; Lippmaa, E.; Keskinen, K.; Root, A. *J. Phys. Chem.* **1995**, *99*, 13763.
25. Sauer, J.; Kölmel, C. M.; Hill, J.-R.; Ahlrichs, R. *Chem. Phys. Lett.* **1989**, *164*, 193.
26. *DSOLID*; Molecular Simulations Inc.: San Diego, 1994.
27. Carter, E. A.; Ciccotti, G.; Hynes, J. T.; Kapral, R. *Chem. Phys. Lett.* **1989**, *156*, 472.
28. Munson, E. J.; Kheir, A. A.; Lazo, N. D.; Haw, J. F. *J. Phys. Chem.* **1992**, *96*, 7740.

ENZYMATIC REACTIONS

Chapter 29

Molecular Dynamics Simulations of Substrate Dephosphorylation by Low Molecular Weight Protein Tyrosine Phosphatase

Karin Kolmodin, Tomas Hansson, Jonas Danielsson, and Johan Åqvist

Department of Molecular Biology, Uppsala University, Biomedical Center, Box 590, S-751 24 Uppsala, Sweden

Dephosphorylation of phosphotyrosine by the low molecular weight protein tyrosine phosphatase proceeds via nucleophilic substitution at the phosphorous atom yielding a covalent enzyme-substrate intermediate that is subsequently hydrolyzed. The reactive nucleophile is the thiolate anion of Cys12. Here we calculate the free energy profiles of putative reaction mechanisms by molecular dynamics and free energy perturbation simulations, utilizing the empirical valence bond method to describe the reaction potential surface. Binding calculations addressing the protonation state of the enzyme-substrate complex are also performed. The calculations give a consistent picture of the catalytic mechanism that is compatible with experimental data.

Phosphorylation and dephosphorylation of tyrosine residues in proteins are important regulatory mechanisms involved in cellular processes such as cell growth, proliferation and differentiation (*1-7*). Protein tyrosine kinases are responsible for the phosphorylation of tyrosine residues, resulting in either activation or deactivation of the substrate protein. The kinases are counteracted by protein tyrosine phosphatases (PTPases) which hydrolyze the tyrosylphosphates. Considerable progress has been made in recent years towards elucidating the catalytic machinery of PTPases and several enzymological studies as well as crystal structures have been reported (*8-17*). Despite these advances there are, however, a number of fundamental questions regarding the catalytic reaction mechanism and pathway that remain unanswered. In this work we employ the empirical valence bond (EVB) method (*18-20*) together with molecular dynamics (MD) simulations and free energy perturbation (FEP) to investigate the mechanism of the low molecular weight (low M_r) PTPases.

The low M_r protein tyrosine phosphatases are small cytosolic enzymes (18,000 Da) without sequence homology to other proteins in the protein tyrosine phosphatase family, such as PTP1B and *Yersinia* PTPase. However, they possess the active site signature motif C-(X)$_5$-R-(S/T) common to all PTPases, where the eight amino acid residues comprise the phosphate binding loop or P-loop. The fact that kinetic properties such as formation of an E-S intermediate, pH-rate profiles etc. are similar for different types of PTPases (*21-24*) indicates that they employ a common mechanism for catalysis. The physiological substrates of the low M_r PTPases are yet unknown, but they readily hydrolyze both phosphotyrosyl proteins and peptides as well as small aryl phosphates, and are themselves regulated by tyrosine phosphorylation at position Y131 and Y132 (*25, 26*).

The catalytic reaction has been shown to proceed via a double displacement mechanism (*27*) involving a phophoenzyme intermediate where the phosphate group becomes covalently bound to the cysteine residue in the active site motif (Cys12 in low M_r PTPase) (*28-33*). The formation of this intermediate is accomplished by a substitution reaction where Cys12 attacks the phosphorous atom and the leaving group is protonated by an appropriately positioned acid (Asp129 in low M_r PTPase) (*34, 35*). This aspartate residue is thought to subsequently activate a water molecule which hydrolyzes the phosphorylated cysteine in the next step (Figure 1). It is still unclear whether the catalytic cysteine is in its thiol or anionic form in the free enzyme. Zhang *et al.* (*36*) measured the pK_a value of Cys12 by alkylation with iodoactetate and iodoacetamide. These results suggest a pK_a of 6.75-7.52 depending on the reagent used, whereas Evans *et al.* (*24*) estimate the pK_a of the same residue to be below 4.0 from pH-rate profiles.

Earlier calculations by us on the low M_r PTPase have shown that the enzyme environment significantly stabilizes proton transfer from the cysteine residue to one of the substrate (phenylphosphate) oxygens thereby enabling activation of the nucleophile for attack on the phosphate group. It is thus clear that the pK_a of Cys12 is lowered relative to the normal value of cysteine in solution (pK_a~8.3). However, if Cys12 is already ionized in the free enzyme it seems likely that the substrate then would bind as the monoanion. Otherwise, a total negative charge of three on the reacting groups is expected to lead to strong electrostatic repulsion between the substrate and the nucleophile, thereby displacing the substrate from its optimal binding position. Crystal structures of the Cys→Ser mutant in *Yersinia* PTPase in complex with SO_4^{2-} (*8*) as well as the corresponding mutant in PTP1B in complex with small peptides containing phosphorylated tyrosine (*12, 13*) also show more or less identical binding conformations compared to the native low M_r PTPase in complex with SO_4^{2-}. Since the serine residue in the two mutants is undoubtedly protonated this indicates that these different complexes all have one proton and bear the same overall charge of -2 on the relevant groups.

In the present study we investigate the energetics of the first half of the catalytic reaction in the low M_r PTPase, i.e. formation of the phosphoenzyme intermediate, using the EVB approach. We address the activation of the reactive cysteine, the nucleophilic substitution at the phosphorous atom and the degree of concertedness of this substitution with leaving group protonation by Asp129. Calculations on the mutation of Asp129 to alanine are also reported and suggest an

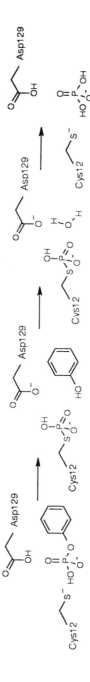

Figure 1. The reaction mechanism catalyzed by low M_r PTPase.

alternative mechanism for this mutant. The protonation state of the reaction complex is examined by binding energy calculations. The results of these different simulations give a detailed picture of the mechanism of catalysis that is consistent with experimental data.

Computational Models and Methods

The reaction potential energy surface is represented by the EVB model that has been described in detail elsewhere (*18-20*). The valence bond structures used in the present calculations are shown in Figure 2 and the reaction is thus modeled in terms of conversions between these different states. For example, nucleophilic activation is described by the process $(\Phi_1 \rightarrow \Phi_2)$ and the nucleophilic substitution is represented, e.g., by $(\Phi_2 \rightarrow \Phi_3 \rightarrow \Phi_4)$. Since the phosphate oxygens in the enzyme are not equivalent, due to restricted rotation as seen in Figure 3, it is necessary to consider three separate VB structures for each state with a singly protonated phosphate group. This leads to a total of 14 different VB states all of which have been considered here. In addition, we also examine some cases with a total charge of -3 on the reacting fragments, but these are not depicted in Figure 2 since it was found that substrate binding does not seem possible in this case (see below). The EVB hamiltonian was calibrated against relevant solution reactions utilizing experimental energetics data as well as semi-empirical (AM1/SM2 and PM3/SM3) and *ab initio* (HF/6-31G*) calculations of geometries. As described elsewhere (*18, 20*) this involves determining gas-phase energy difference $\Delta\alpha_{ij}$ as well as off-diagonal matrix elements H_{ij} between pairs of VB states so that the EVB potential surface reproduces experimental reaction free energies and barrier heights of relevant reference reactions in solution. This calibration procedure thus involves simulations of uncatalyzed reaction steps with the reacting fragments in water and fitting the above parameters so that calculated and observed free energies coincide.

In the case of proton transfer steps, such as $(\Phi_1 \rightarrow \Phi_2)$, $(\Phi_4 \rightarrow \Phi_5)$ and $(\Phi_4 \rightarrow \Phi_6)$, the pK_a difference between donor and acceptor together with linear free energy relationships (LFERs) for proton transfer states were used as described in (*37, 38*) for calibration of the relevant EVB parameters. Calibration of nucleophilic displacement steps, such as $\Phi_2 \rightarrow \Phi_4$, Φ_5, Φ_6 (via Φ_3) was done using data from Kirby and Younas (*39*) on hydrolysis with phenol leaving groups, from Åkerfeldt (*40*) on hydrolysis of phosphorothioic acids, from Borne and Williams (*41*) on equilibrium constant dependence on leaving group pK_a and from Guthrie's thermodynamic data on phosphoric acid derivatives (*42*). Details of this rather extensive thermodynamic analysis will be published elsewhere. However, the high energy transient intermediate structure Φ_3 deserves a comment here as its properties are not directly accessible to experiment. The geometries of various pentacoordinated species with different combinations of axial groups were optimized with the AM1/SM2 and PM3/SM3 hamiltonians (*43*). For the case with RS– and PhO– as axial groups both AM1/SM2 and PM3/SM3 locate a similar minimum. We used the geometry of Φ_3 from the former calculations which gives both axial ligand distances of ~2.4 Å. For more nucleophilic axial ligands, (e.g. $CH_3O–$) this type of minimum has smaller distances to P and a slightly associative character, i.e. the total axial bond order to P increases somewhat, while with less nucleophilic ligands the minimum is slightly dissociative.

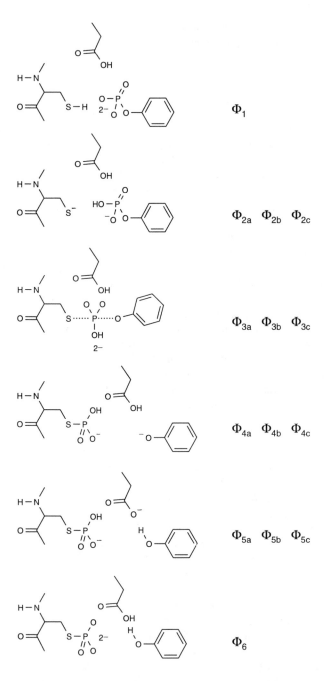

Figure 2. Valence bond states used in the EVB calculations.

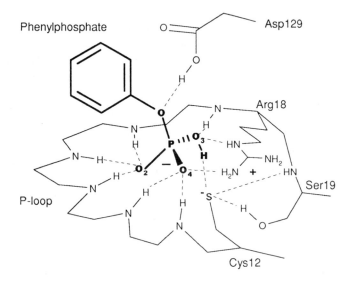

Figure 3. Schematic view of the active site in low M_r PTPase. The VB state shown is Φ_2, where the proton in positioned on the substrate phosphate group.

be regarded as an intermediate between the associative (pentacovalent) and dissociative (metaphosphate) extremes, but with slightly dissociative character. The actual free energy of this state (Φ_3) is not of major importance for our calibrations since it only serves as a reference for the flanking barriers. However, its energy has been estimated with hydroxyl ligands in (42) to be 22 kcal/mol above Φ_2 (with the same ligands). AM1/SM2 calculations yield 24 kcal/mol with –OH ligands in reasonable agreement with this estimate, while 12.5 and 12.9 kcal/mol is obtained for the symmetric cases with RS– and PhO– ligands, respectively. These latter values were thus used as our estimates for this state.

The force field parameters for the different VB states were taken as far as possible from the GROMOS potential (44), which was also used to model the rest of the system. However, bonds within the reaction fragments were represented by Morse potentials using standard bond lengths and dissociation energies. Charges for the non-standard moieties involving S–P bonding were also derived from AM1/SM2 calculations and merged with those of the standard GROMOS fragments to maintain compatibility with these.

The protein coordinates used in the MD simulations are those of bovine liver low M_r PTPase in complex with sulfate ion published by Su *et al.* (9). The phenylphosphate substrate was modeled into the crystal structure as described in (38). The phosphorous atom was positioned approximately where the sulfate atom is found in the crystal structure, letting the phenyl ring perfectly fit in the narrow hydrophobic binding slot. In addition to Cys12, seven residues close to the reaction center were considered to be charged: Arg18, Arg53, Asp48, Asp56, Arg58, His72, Asp92, whereas His66 and Asp129 (the general acid) were uncharged. Other charged groups distant from the reaction center were replaced by neutral dipolar groups. The simulation protocol was essentially as described in (38), but with somewhat longer trajectories, 47-83 FEP λ-points and 5 ps simulation for each value of λ. All MD/FEP/EVB calculations were carried out using the program Q (45). The reaction center was surrounded by a 16 Å sphere of SPC water in the solution (calibration) simulations and by a sphere of the same size containing both protein and water in the enzyme simulations. Protein atoms outside of this sphere were restrained to their crystallographic coordinates and interacted only via bonds across the boundary. The water surface was subjected to radial and polarization surface restraints according to a new model (Marelius, J.; Åqvist, J., in preparation) similar to those of (46) and (47). The MD trajectories were run at a constant temperature of 300 K using a time step of 1 fs. A non-bonded cut-off radius of 10 Å was used together with the LRF method (48) for longer range electrostatics.

Results

The resulting free energy profile from the EVB/FEP/MD calculations is summarized in Figure 4. It can be seen that the enzyme exerts a substantial catalytic effect on the reaction and generally stabilizes the high-energy structures by about 10 kcal/mol or more compared to the uncatalyzed reaction. With the proton on the most favorable phosphate oxygen, the calculated reaction barrier is 14-15 kcal/mol which is in excellent agreement with the reported turnover rate of around 10-100 s^{-1} (34, 35). It

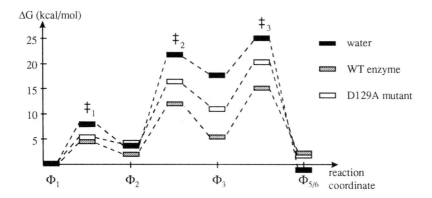

Figure 4. Free energy profile of the reaction mechanism in aqueous solution based on experimental data, and the corresponding reaction in WT and D129A low M_r PTPase obtained from MD/FEP/EVB calculations.

should, however, be noted that hydrolysis of the phosphoenzyme is proposed to be the rate determining step of the reaction (33, 49).

In addition, we find that the protein environment facilitates proton transfer from Cys12 to the phosphate group of the substrate, thus ensuring availability of the nucleophilic anion for the substitution reaction. The small difference in free energy between Φ_1 and Φ_2 (1.5 kcal/mol) indicate that the pK_a of the cysteine is close to that of the substrate, i.e. it is lowered by the enzymatic environment. It has previously been shown that a hydrogen bond from Ser19 is important for lowering the pK_a of Cys12 (24, 38).

The simulated reaction profiles of the three possible proton transfers ($\Phi_1 \rightarrow \Phi_{2a,b,c}$) show that there is no significant discrimination of the acceptor. Proton transfer is feasible from the cysteine to any of the three oxygens, with a slight preference for the two which are hydrogen bonded to Arg18 (O_3 and O_4 in the crystal structure). The position of the proton does not have any major effect on the catalysis of the approach to the transition state ($\Phi_{2a,b,c} \rightarrow \Phi_{3a,b,c}$). On the other hand, it appears that stabilization of the phosphoenzyme intermediate resulting from leaving group departure is more sensitive to the nature of phosphate protonation. When the proton is bound to oxygen O_3, which accepts a hydrogen bond from N_ε of Arg18, it can be engaged in hydrogen bonding to the negatively charged Asp129. When the proton is bound to O_2 the distance to Asp129 becomes too large to allow such hydrogen bonding. For the third case with the proton bound to O_4 we observe a stabilization of the phosphoenzyme intermediate that is somewhere in between the other two cases.

Simulations of the P–O bond cleavage and leaving group departure clearly indicate that bond cleavage at the bridging oxygen has to be concerted with protonation of the leaving group in order to depress a charge separation in the active site. This is also in agreement with interpretations of experimental data. The bond cleavage was first simulated along the stepwise pathway, $\Phi_3 \rightarrow \Phi_4$ followed by $\Phi_4 \rightarrow \Phi_5$, which is predicted to be energetically unfavorable in the enzyme, yielding a barrier of ~22 kcal/mol. A developing negative charge on the leaving group oxygen apparently cannot be stabilized by the proton environment and, since the binding cavity is very narrow, solvating water molecules are excluded from the active site. The concerted pathway, ($\Phi_3 \rightarrow \Phi_5$) is strongly facilitated by the enzyme and the resulting negative charge on Asp129 is, unlike the phenolate ion, accessible to solvent. The apparent need for concerted leaving group protonation may indicate that the nucleophilic attack in the second half of the overall reaction (Figure 1), where the phosphoenzyme is hydrolyzed by a water molecule, also has to be concerted with proton abstraction by Asp129 from the water nucleophile. This issue is, however, left for future work.

An alternative mechanism for the protonation of the leaving group has also been investigated. Here the phosphate group itself was used as the acid for leaving group protonation ($\Phi_{3a,b,c} \rightarrow \Phi_6$) resulting in a double negative charge on the cysteinyl phosphate. It appears that also this reaction is significantly catalyzed by the enzyme and it is therefore possible that such a reaction mechanism may be utilized in mutants lacking the general acid/base Asp129. In order to examine this issue we calculated the free energy profile for substrate dephosphorylation also for the D129A mutant assuming leaving group protonation by the phosphate monoanion. The resulting free energy profile for the D129A mutant (Figure 4) shows that it is considerably less efficient than the wild type enzyme but still retains catalytic power. The difference in activation energy between WT and D129A at the rate limiting transition state for substrate dephosphorylation is approximately 5 kcal/mol, which corresponds to a rate decrease by a factor of 4000 for the mutant. This appears to be consistent with experimental data which show that the D129A mutant is about 2000-3000 fold slower than the wild type (*34, 35*). The calculated activation energies of both the wild type enzyme and the D129A mutant are thus consistent with kinetic rate experiments, but as noted above, the dephosphorylation of the protein may be the rate limiting step for the wild type enzyme (*33, 49*).

The protonation state of the reaction complex is one of the main issues in elucidating the catalytic mechanism of PTPases. The rates of reaction of substrates such as phenyl phosphate in solution are usually much faster for the monoanion than for the dianion (*39*). It is also noteworthy that PTPases generally display optimal activity in the acidic pH range. Furthermore, as mentioned above the crystal structures of various inhibitor complexes with WT and catalytic Cys→Ser mutated enzymes generally show van der Waals contact between Cys/Ser and the ligand oxygens. Since some of these complexes, such as the Cys→Ser mutants with various ligands (*8, 12, 13*) and WT low M_r PTPase in complex with HEPES (*17*), clearly have a total charge of -2 on the nucleophile-substrate moiety, it seems reasonable to expect that the corresponding complex of the active enzyme has the same protonation state. The most straightforward way to examine this issue is to try to evaluate the difference in substrate affinity for different protonation states. We performed FEP calculations where the phenylphosphate substrate was transformed from monoanion to dianion in aqueous solution and in the solvated protein with Cys12 in its anionic form. The calculated difference in binding free energy was ~15 kcal/mol for the monoanion to dianion perturbation, indicating that there is no affinity for the substrate dianion with Cys12 ionized. It can also be seen from the MD structures that the doubly charged substrate is significantly displaced from its initial binding site due to electrostatic repulsion with the nucleophile. With Cys12 ionized and the substrate in its monoanionic form, on the other hand, the structural agreement between MD and the crystal complexes are excellent. We also performed calculations of binding free energies using the linear response approach described in (*51, 52*), which confirmed the hypothesis that a doubly negative substrate does not have affinity for low M_r PTPase with Cys12 in its anionic form.

Discussion

In this work we have used MD/FEP/EVB simulations to investigate details of the substrate dephosphorylation mechanism in low M_r PTPase. The reaction mechanism with leaving group departure concerted with protonation by Asp129 is found to be fully compatible with experimentally observed reaction rates. It appears that the enzyme is designed to stabilize negative charge more or less precisely in the plane of the so called P-loop where the equatorial phosphate oxygens are positioned during nucleophilic substitution. For a stepwise mechanism where the leaving group departs in its anionic form, and thus moves out of the 'focus' of the P-loop, the calculations predict very unfavourable energetics due to insufficient solvation of the phenol anion. On the other hand, the concerted protonation by Asp129 gives significant catalysis and the resulting negative charge on the aspartate becomes more solvent accessible.

Simulations of the D129A mutant reaction also give highly unfavourable energetics with phenolate anion on the effective leaving group and suggest that there would be no enzymatic activity of D129A if this were the reaction pathway. However, our calculations predict that for this mutant a mechanism where the protonated phosphate group itself is involved in concerted protonation may be employed. The calculated increase in activation energy (relative to Φ_1) for this mechanism is 5 kcal/mol or about a factor 4000 slower catalysis. Hence, it appears that this type of mechanism can explain why the D129A mutant still retains considerable activity.

We have found that with a singly protonated Cys12-phosphate complex the calculated energetics is consistent with experimental data. However, it has been proposed that both the cysteine and the substrate are fully ionized in the reactive complex (23, 53), which would give a total charge of -3 rather than -2 on the reactants. Although such a situation would seem very unfavorable in view of the resulting electrostatic repulsion (the corresponding reaction is not observed at all in solution (39)), one cannot exclude the possibility that the enzyme might stabilize the -3 complex significantly. To address this problem we carried out calculations of the different binding affinity of phenylphosphate mono- and di-anion for the thiolate form of the enzyme. The results, however, clearly predict that the dianion does not bind when Cys12 is ionized and the cysteine-phosphate distance increases significantly with respect to the typical distances seen in various enzyme-ligand complexes. The fact that pH-rate profiles for k_{cat} show a descending slope of -1 on the basic side of the pH-optimum with an apparent pK_a that is substrate dependent and agrees exactly with the pK_a of the substrate phosphate group (21, 54), also suggests that the ES complex is singly protonated. In this context it is also interesting to note the recently reported crystal structure by Zhang et al. (17) of the low M_r PTPase in complex with vanadate. This structure shows a typical trigonal bipyramidal conformation with the vanadate covalently linked to Cys12. Since vanadate at the given concentration exists as $H_2VO_4^-$ in solution between pH 4 and 8.3 (55) it is most likely that the observed complex with low M_r PTPase actually corresponds to $(Cys12)S-VO_4H_2^{2-}$. This structure would thus correspond closely to our high-energy intermediate (Φ_3). Furthermore, the reported binding of inorganic phosphate is three times stronger at pH 5 than at pH 7.5 (17) which agrees well with the relative monoanion concentration (33%) at the higher pH value.

Regarding the actual pK_a of Cys12 the situation is somewhat unclear. While reaction experiments with iodoacetate have indicated that its pK_a is as high as 7.5 (36), pH-rate profiles show a typical ionization around pH 4-5 that is usually attributed to the catalytic cysteine. At any rate it is clear that its pK_a is lowered in the free enzyme from the normal value of ~8.3, which can be explained by the efficient stabilization of negative charge in the active site. Exactly how the two ionizations of Cys12 and the substrate phosphate groups are shifted upon binding is less clear. However, the characteristic k_{cat} vs. pH rate profile with optimum around 5-6 flanked by an acidic limb with slope +1 and a basic limb with slope -1 at least appears consistent with two possible ionizations of the Cys-phosphate moiety.

As noted above, the characteristic geometry of the active site with backbone amide groups and the Arg18 side chain focused towards the center of the P-loop appears ideally adapted for stabilizing negative charge in the plane where the equatorial oxygens are positioned in the transition state(s). This situation resembles that encountered, e.g. in the Ras-RasGAP and transducin α structures, where GTP is proposed to be hydrolyzed by analogous mechanisms *(56-58)*.

In summary, we have shown that the enzymatic environment catalyzes the reaction pathway where a singly protonated reaction complex of total charge -2 forms a phosphoenzyme intermediate, by almost eight orders of magnitude. The resulting reaction barrier is compatible with experiments and the calculations indicate that the present model of the reaction pathway is a good representation of the actual catalytic process in the low M_r PTPase.

Acknowledgments

Support from the EC Biotechnology Program DGXII and the NFR is gratefully acknowledged.

Literature Cited

1. Hunter, T. *Cell*, **1989**, *58*, 1013-1016.
2. Ramponi, G.; Manao, G.; Camici, G.; Cappugi, G.; Ruggiero, M.; Bottaro, D. P. *FEBS letters*, **1989**, *250*, 469-473.
3. Fischer, E. H.; Charbonneau, H.; Tonks, N. K. *Science*, **1991**, *253*, 401-406.
4. Ruggiero, M.; Pazzagli, C.; Rigacci, S.; Magnelli, L.; Raugei, G.; Berti, A.; Chiarugi, V. P.; Pierce, J. H.; Camici, G.; Ramponi, G. *FEBS letters*, **1993**, *326*, 294-298.
5. Mondesert, O.; Moreno, S.; Russel, P. *J. Biol. Chem.* **1994**, *269*, 27996-27999.
6. Barford, D.; Jia, Z.; Tonks, N. K. *Nature Struct. Biol.* **1995**, *2*, 1043-1053.
7. Stone, R. L; Dixon, J. E. *J. Biol. Chem.* **1994**, *269*, 31323-31326.
8. Stuckey, J. A.; Schubert, H.L.; Fauman, E. B.; Zhang, Z.-Y.; Dixon, J. E.; Saper, M. A. *Nature*, **1994**, *370*, 571-575.
9. Su, X.-D.; Taddei, N.; Stefani, M.; Ramponi, G.; Nordlund, P. *Nature*, **1994**, *370*, 575-578.
10. Logan, T. M.; Zhou, M. M.; Nettesheim, D. G.; Meadows, R. P.; Van Etten, R. L.; Fersik, S. W. *Biochemistry*, **1994**, *33*, 11087-11096.
11. Zhang, M.; Van Etten, R. L.; Stauffacher, C. V. *Biochemistry*, **1994**, *33*, 11097-11105.
12. Barford, D.; Flint, A. J.; Tonks, N. K. *Science*, **1994**, *263*, 1397-1404.
13. Jia, Z.; Barford, D.; Flint, A. J., Tonks, N. K. *Science*, **1995**, *268*, 1754-1758.
14. Yuvaniyama, J.; Denu, J. M.; Dixon, J. E.; Saper, M. A. *Science*, **1996**, *272*, 1328-1331.
15. Bilwes, A. M.; den Hertog, J; Hunter, T.; Noel, J. P. *Nature*, **1996**, *382*, 555-559.
16. Fauman, E. B.; Yuvaniyama, C.; Schubert, H. L.; Stuckey, J. A.; Saper, M. A. *J. Biol. Chem.* **1996**, *271*, 18780-18788.

382

17. Zhang, M.; Zhou, M.; Van Etten, R.L.; Stauffacher, C.V. *Biochemistry,* **1997,** *36,*15-23.
18. Warshel, A. *Computer Modeling of Chemical Reactions in Enzymes and Solutions*; Wiley, New York, NY, 1991.
19. Warshel, A.; Sussman, F.; Hwang, J.-K. *J. Mol. Biol.* **1988,** *201,*139.
20. Åqvist, J.; Warshel, A. *Chem. Rev.* **1993,** *93,* 2523-2544.
21. Zhang, Z.-Y.; Malachowski, W. P.; Van Etten, R. L.; Dixon, J. E. *J. Biol. Chem.* **1994,** *269,* 8140-8145.
22. Zhang, Z.-Y. *J. Biol. Chem.* **1995,** *270,* 11199-11204.
23. Denu, J. M.; Zhou, G.; Guo, Y.; Dixon, J. E. *Biochemistry,* **1995,** *34,* 3396-3403.
24. Evans, B.; Tishmack, P. A.; Pokalsky, C.; Zhang, M.; Van Etten, R. L. *Biochemistry,* **1996,** *35,* 13609-13617.
25. Rigacci, S.; Degl'Inocenti, D.; Bucciantini, M.; Cirri, P.; Berti, A.; Ramponi, G. *J. Biol. Chem.* **1996,** *271,* 1278-1281.
26. Tailor, P.; Gilman, J.; Williams, S.; Couture, C.; Mustelin, T. *J. Biol. Chem.* **1997,** *272,* 5371-5374.
27. Saini, M. S.; Buchwald, S. L.; Van Etten, R. L.; Knowles, J. R. *J. Biol. Chem.* **1981,** *256,* 10453-10455.
28. Chiarugi, P.; Marzocchini, R.; Raugei, G.; Pazzagli, C.; Berti, A.; Camici, G.; Manao, G.; Cappugi, G.; Ramponi, G. *FEBS letters,* **1992,** *310,* 9-12.
29. Cirri, P.; Chiarugi, P.; Camici, G.; Manao, G.; Raugei, G.; Cappugi, G.; Ramponi, G. *Eur. J. Biochem.* **1993,** *214,* 647-657.
30. Guan, K.-L.; Dixon, J. E. *Proteins: Struct. Funct. Genet.* **1991,** *9,* 99-107.
31. Zhang, Z.-Y.; Dixon, J. E. *Advan. Enzymol.* **1994,** *68,* 1-36.
32. Zhang, Z.-Y.; Wang, Y.; Dixon, J. E. *Proc. Natl. Acad. Sci. USA,* **1994,** *91,* 1624-1627.
33. Davis, J. P.; Zhou, M.-M.; Van Etten, R. L. *J. Biol. Chem.* **1994,** *269,* 8734-8740.
34. Zhang, Z.; Harms, E.; Van Etten, R. L. *J. Biol. Chem.* **1994,** *269,* 25947-25950.
35. Taddei, N.; Chiarugi, P.; Cirri, P.; Fiaschi, T.; Stefani, M.; Camici, G.; Raugei, G.; Ramponi, G. *FEBS letters,* **1994,** *350,* 328-332.
36. Zhang, Z.-Y.; Davis, J. P.; Van Etten, R. L. *Biochemistry,* **1992,** *31,* 1701-1711.
37. Åqvist, J.; Fothergill, M. *J. Biol. Chem.* **1996,** *271,* 10010-10016.
38. Hansson, T.; Nordlund, P.; Åqvist, J. *J. Mol. Biol.* **1996,** *265,* 118-127.
39. Kirby, A. J.; Younas, M. *J. Chem. Soc.* **1970,** 1165-1172.
40. Åkerfeldt, S. *Acta Chem. Scand.* **1963,** *17,* 319-328.
41. Bourne, N.; Williams, A. *J. Org. Chem.* **1984,** *49,* 1200-1204.
42. Guthrie, J. P. *J. Am. Chem. Soc.* **1977,** *99,* 3991-4001.
43. Cramer, C. J.; Truhlar, D. G. *Science,* **1992,** *256,* 213-217.
44. van Gunsteren, W. F.; Berendsen, H. J. C. *Groningen Molecular Simulation (GROMOS) Library Manual,* Biomos B. V., Nijenborgh 16, Groningen, The Netherlands, 1987.
45. Åqvist, J. *Q,* Version 2.1; Uppsala University, 1996.
46. King, G.; Warshel, A. *J. Chem. Phys.* **1989,** *91,*3647-3661.
47. Essex, J.W.; Jorgensen, W. L. *J. Comput. Chem.* **1995,** *16,* 951-972.

48. Lee, F. S.; Warshel, A. *J. Chem. Phys.* **1992**, *97*, 3100-3107.
49. Zhang, Z.-Y.; Van Etten, R. L. *Biochemistry*, **1991**, *30*, 8954-8959.
50. Zhang, Z.-Y.; Van Etten, R. L. *J. Biol. Chem.* **1991**, *266*, 1516-1525.
51. Åqvist, J.; Medina, C.; Samuelsson, J.-E. *Protein Eng.* **1994**, *7*, 385-391.
52. Marelius, J.; Graffner-Nordberg, M.; Hansson, T.; Hallberg, A.; Åqvist, J. *J. Comput.-Aided Mol. Design,* **1998**, *12*, 000-000.
53. Hengge, A. C.; Zhao, Y.; Wu, L.; Zhang, Z.-Y. *Biochemistry*, **1997**, *36*, 7928-7936.
54. Zhang, Z.-Y.; Van Etten, R. L. *Arch. Biochem. Biophys.* **1990**, *282*, 39-49.
55. Pope, M. T.; Dale, B. W. *Q. Rev. Chem Soc.* **1968**, *22*, 527-548.
56. Scheffzel, K.; Ahmadian, M. R.; Kabsch, W.; Weismüller, L.; Lautwein, A.; Schmitz, F.; Wittinghofer, A. *Science*, **1997**, *277*, 333-338.
57. Sondek, J.; Lambright, D. G.; Noel, J. P.; Hamm, H. E.; Sigler, P. B. *Nature*, **1994**, *372*, 276-279.
58. Langen, R.; Schweins, T.; Warshel, A. *Biochemistry*, **1992**, *31*, 8691-8696.

Chapter 30

Transition States in the Reaction Catalyzed by Malate Dehydrogenase

Mark A. Cunningham and Paul A. Bash

Department of Molecular Pharmacology and Biological Chemistry, Northwestern University Medical School, Chicago, IL 60611

A QM/MM hybrid method was used to investigate the reaction pathway in malate dehydrogenase (MDH). A minimum energy surface was computed and a putative reaction pathway was established in which the proton transfer from malate to His-177 was found to occur prior to the hydride transfer to NAD^+. Transition states along the minimum energy pathway were also identified.

Enzymes are critical components in the biochemical machinery employed by living organisms, serving as efficient, precisely regulated and extraordinarily specific catalysts. Understanding the mechanisms by which enzymes achieve their remarkable catalytic abilities has been a long-standing goal of biochemists. Despite the tremendous progress toward this goal, there exist a number of issues which remain difficult to address with current experimental methodology. This is due to the fact that the critical catalytic steps associated with molecular recognition and capture of the substrate and the ensuing chemical events—making and breaking bonds—take place on exceedingly short time scales. The transition states on the reaction pathway, which by their very nature exist only briefly, represent saddle points on the free energy surface. The energy barriers posed by these transition states determine the reaction rate but experimental determination of the transition state structures remains problematic. Crystal structures are insightful and some systems have been crystallized with so-called transition state analogs but interpretation of the results is not always straightforward. This situation has motivated our efforts to develop numerical tools that can be used to address some of these critical issues of molecular recognition and catalytic processes in enzymes.

Given that the Schrödinger equation provides an appropriate framework for describing the quantum mechanical behavior of atoms and molecules, one might hope that the study of enzyme-catalyzed reactions might prove to be a reasonably straightforward exercise. Unfortunately, there are significant restraints imposed by both the sophistication of the algorithms which produce approximate solutions of the

Schrödinger equation and the computing power available, even in supercomputing environments. High-level *ab initio* quantum mechanical calculations based on Hartree-Fock or density functional methods are capable of generating enthalpies of formation, for example, which agree with experimental measurements to within 4–8 kJ/mol (1) but such calculations are limited in practice to systems involving perhaps a dozen heavy atoms. Computations of entire enzyme-substrate complexes composed of thousands of atoms are clearly outside the realm of feasibility for these methods. As an alternative, one might consider using one of the semi-empirical quantum mechanical methods (2) and recent improvements in the computational efficiency of these algorithms suggests that simulation of thousand atom systems is feasible (3–4). One method that is capable of providing a dynamical simulation of large systems like the enzyme-substrate complex is the molecular mechanics model (5). In this method, the atoms are treated classically, with chemical bonds represented by force constants that define the bond lengths and angles between adjacent bonds. Atoms which are not bound interact electrostatically and via a Lennard-Jones potential. The charge distributions and parameters which define the model are determined by calibrating to the known structures and infrared spectra of small molecules in the gas phase and measured thermodynamic properties in the condensed phase. The molecular mechanics model provides a rough approximation to solutions of the Schrödinger equation but, in practice, provides realistic simulations of enzyme systems.

One issue that can be addressed with molecular mechanics methods is the formation of the Michaelis complex, which can be considered to be the point in the reaction sequence when the substrate has been captured and oriented by the enzyme. Hydrogen bonds stabilizing the complex have formed but no covalent bonds have been reorganized. Figure 1 is a depiction of the Michaelis complex of malate and MDH and was obtained from a dynamical simulation using a molecular mechanics hamiltonian (6). The structure of the Michaelis complex for malate and MDH cannot be determined experimentally; the reaction proceeds too quickly but a crystal structure of citrate bound in the active site of MDH has been obtained (7). The structure displayed in Figure 1 was obtained by substituting a malate substrate into a conformation analogous to the one occupied by citrate in the experimentally-defined crystal structure and then allowing the protein and malate substrate to equilibrate into a minimum-energy configuration. As a measure of the ability of the molecular mechanics method to realistically reproduce the actual protein environment, we can compute the differences in position between atoms in the x-ray crystal structure and those determined by the numerical model. The root mean squared difference summed over α-carbon atoms is 0.35 Å; for all atoms, the figure rises to 0.89 Å. These values are quite reasonable and we can conclude that the numerically-derived structure of the Michaelis complex illustrated in Figure 1 is a realistic representation of the actual Michaelis complex of the MDH:malate:NAD$^+$ system.

QM/MM Method

Unfortunately, the molecular mechanics model cannot realistically describe the subsequent processes of breaking and forming chemical bonds. A more accurate description of the inherently quantum phenomena associated with bond formation is

386

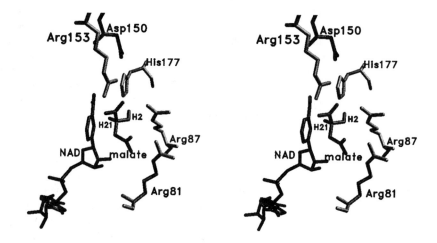

Figure 1. Stereo view of the active site of MDH, with the substrate malate and cofactor NAD⁺.

necessary. Rather than resort to a cluster-type approach to address the issue of electronic structure during the catalysis process, we note first that the molecular mechanics method does provide a realistic description of the dynamics of the protein as a whole. If we further postulate that the quantum chemical activity is confined to a small region near the active site, then it is possible to construct a hybrid model in which only a relatively few atoms are described quantum mechanically and the remainder are treated with molecular mechanics (8–10). This approach has the advantage that the effects of the enzyme environment on the reaction are included explicitly, unlike the cluster models in which enzyme environmental effects are more crudely approximated. Unfortunately, even with the reduction in the size of the problem to be treated quantum mechanically from thousands of atoms to tens of atoms, it is not possible to incorporate any of the high-level *ab initio* quantum methods and retain any hope of running dynamical simulations.

While this is not as satisfying a situation as one might hope, it is still possible to generate reasonable simulations with the semi-empirical models. In Table 1, we list a small sampling of values from the so-called G2 test set which compare computed enthalpies of formation with experimental measurements (1). The G2 method is a composite theory based on the Hartree-Fock 6-311G(d,p) basis set and several basis extensions. Electron correlation is incorporated by means of Møller-Plesset perturbation theory and quadratic configuration interaction. At the present time, it represents the state-of-the-art in high-level *ab initio* quantum methods. The rms variance for the G2 method over the entire 148-member test suite was 5.1 kJ/mol, with a maximum deviation of 34 kJ/mol. The rms variance of the density functional (DFT) method B3LYP was 10.2 kJ/mol, with a maximum deviation of 84 kJ/mol. We haven't computed AM1 enthalpies for all the members of the G2 test set but a quick inspection of Table 1 indicates that the variance for the AM1 method will be significantly larger than that of either of the two *ab initio* methods. On the other hand, we are not interested in reproducing the entire slate of chemical reactions for all known elements. Rather, we are interested in a specific few interactions which may occur in a particular enzyme-substrate system. Because the AM1 model is defined by a set of empirically-determined parameters, we might anticipate that suitable adjustments to those parameters will produce a model hamiltonian which can be quite accurate for a limited set of interactions.

Table I. Errors in estimation of enthalpies of formation.

Molecule	ΔH^f_0 (kJ/mol)	|G2 – Expt.|	|DFT – Expt.|	|AM1 – Expt.|
Methane	−74.8±0.4	2.9	6.7	37.6
Ethane	−84.0±0.4	2.1	2.5	10.9
Benzene	82.3±0.8	16.3	18.8	9.2
Formic acid	−378.3±0.4	8.4	3.8	28.8
Acetic acid	−432.2±1.7	6.3	10.9	1.7
Methylamine	−23.0±0.4	0.0	13.4	7.9
Trimethylamine	−23.8±0.8	5.9	0.8	16.7
Acetaldehyde	−165.9±0.4	5.4	1.3	7.9
Pyridine	140.4±0.8	9.2	0.8	10.9

Calibration of the QM model hamiltonian. We must first specify the important interactions made between the enzyme and substrate. Figure 2 depicts the active site of the MDH:malate:NAD$^+$ complex, with the key atoms labelled. We note that the crystal structure and other biochemical data (11–12) indicate that a proton is transferred from the O2 oxygen atom of the malate substrate to the NE2 nitrogen atom in the imidazole ring of the His-177 residue in MDH and a hydride ion is transferred from the C2 carbon atom in malate to the C4N carbon atom in the nicotinamide ring of the NAD$^+$ cofactor. In addition to requiring that the quantum hamiltonian provide a reasonable description of the structures of the key elements, we want to ensure that both the proton- and hydride-transfer reactions are accurately represented. To do so, we need to include analogous reactions in small molecule systems, for which experimental data are available and for which high-level *ab initio* quantum calculations can be performed.

Focusing for the moment on the proton transfer reaction, we note that the proton is derived from a hydroxyl group on the malate substrate. We used methanol as an analog for the proton donor and an imidazole ring as an acceptor analog. To fit the parameters of the AM1 model hamiltonian, we chose as target data the experimental enthalpies of formation of the reactants, methanol and imidazole, and products, methoxide and imidazolium. We also include experimental dipole moments for methanol and imidazole, theoretical dipole moments for methoxide and imidazolium (from HF/6-31G(d) calculations) and structural information obtained from high-level (MP2/6-31G(d)) *ab initio* quantum calculations: bond lengths, angles and dihedral angles. The nonlinear optimization problem associated with this step has been outlined by Dewar and Thiel (13).

Some results of the hamiltonian parameter fit are listed in Table II. The optimized geometries have an rms difference from the target bonds of 0.011 Å, the computed angles differ by 1.06 degrees and dihedrals differ from their target values by 0.19 degrees. The AM1-SSP enthalpies agree with the target values to within 3 kJ/mol; dipole moments agree to within 0.3 Debye. The enthalpy of reaction for the proton transfer from methanol to imidazole, $\Delta\Delta H_f^0 = \Delta H_f^0(\text{products}) - \Delta H_f^0(\text{reactants})$, is found experimentally to be 656.7 kJ/mol; the AM1-SSP value is 661.3 kJ/mol and standard AM1 method produces 681.5 kJ/mol. The optimized AM1-SSP value agrees to within 5 kJ/mol, which is the same level of accuracy obtained with very computationally intensive G2 calculations in the G2 test set. The optimized AM1-SSP model hamiltonian may not fare well if tested in a broad range of problems, at least for the case of the proton transfer reaction between malate and His-177, we are reasonably confident that the AM1-SSP model will produce accurate results.

Table II Parameter fitting for the proton transfer reaction.

	ΔH_0^f (kJ/mol)			$|\mu|$ (D)		
	Target[a]	AM1	AM1-SSP	Target[b]	AM1	AM1-SSP
Methanol	–201.9	–234.2	–201.2	1.70	1.62	1.97
Methoxide	–138.8	–161.0	–136.2	2.16	1.38	2.09
Imidazole	146.3	212.2	145.5	3.80	3.60	3.69
Imidazolium	739.9	820.6	741.7	1.74	1.63	1.76

[a]Values are from Ref. 14. [b]Values are from Ref. 15.

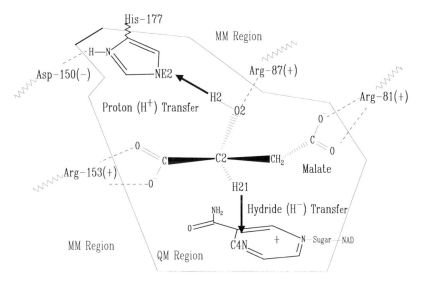

Figure 2. Schematic drawing of the active site of the MDH:malate:NAD$^+$ complex.

QM/MM interactions. The next issue we must address concerns the interactions between the atoms described with a quantum mechanical potential, in what is termed the QM partition of the model, and those described classically through molecular mechanics, or in the MM partition. The hybrid model includes both electrostatic and van der Waals interactions between atoms in the two partitions. Explicitly, the model incorporates the following terms:

- QM electron–MM partial charge electrostatic potential. Atoms in the MM partition have no electrons; their effective charges are positioned at the atom center.
- QM nucleus–MM partial charge electrostatic potential.
- QM/MM van der Waals potential, which models the electronic repulsion and dispersion properties that are missing because the MM atoms have no electrons.

In the case where an atom in the QM partition is bonded to an atom in the MM partition, as would be the case if we chose to draw the QM boundary between the alpha-carbon of a protein residue and the beta-carbon of its side chain, the model incorporates fictitious ``link'' atoms which serve to terminate the QM electron density along the bond. The link atoms have no interactions with atoms in the MM partition but do contribute to the energy and forces felt by atoms in the QM partition. There are no adjustable parameters for the link atoms. Neither of the QM/MM electrostatic potential terms contain any free parameters but the van der Waals interactions must be calibrated to realistically represent the forces on atoms in the QM partition due to atoms in the MM partition.

There are no experimental data to guide this parameterization effort. Instead, we rely on high-level *ab initio* quantum calculations of a water molecule interacting with the small molecule analogs. Hartree-Fock calculations using the 6-31G(d) basis set were performed, optimizing the individual structures of water and the analog molecules. A series of HF/6-31G(d) optimizations were then performed with a single degree of freedom: the distance between the water molecule and a specific target atom in the analog molecule. The 6-31G(d) basis set has been shown to generate reasonable structures but the interaction energies for neutral molecules are systematically underestimated (16). We have used a scale factor of 1.16 to compensate for this bias on all of the interactions between neutral reactants (17).

The set of calculations performed to establish the *ab initio* target values was then repeated with the hybrid model. Geometries of the small model analog molecules were optimized with the AM1-SSP quantum model hamiltonian. The water molecules were treated with molecular mechanics and we used the TIP3P model of Jorgensen, et al (18). The van der Waals parameters for atoms in the small molecule analogs were then adjusted to reproduce both the interaction energies and the optimal intermolecular distances obtained from the Hartree-Fock calculations. We have plotted the interaction energies obtained from the AM1-SSP quantum calculations against those obtained from the Hartree-Fock method in Figure 3. Overall, a 0.7 kJ/mol rms deviation was observed for interaction energies.

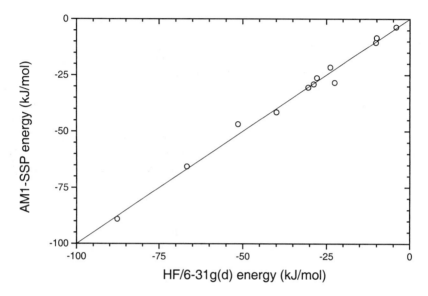

Figure 3.   Calibration of QM/MM interactions.   Hybrid method interaction energies are shown plotted against HF/6-31G(d) values.   The solid line is drawn to guide the eye.

Proton transfer in solution

The pieces are now in place to perform simulations with the calibrated hamiltonian parameters. Before investigating the reaction mechanism in the context of the enzyme, however, we can touch base with experiment one last time by examining the proton transfer reaction between methanol and imidazole in solution. The free energy change ΔG in solution can be obtained from the experimental pK_a values of the reactants by means of the following relation:

$$\Delta G = -2.3RT[pK_a(\text{imidazole}) - pK_a(\text{methanol})]. \tag{1}$$

The pK_a value of methanol (19) is 15.5 and of imidazole (20) is 6.05, yielding an experimental value for ΔG of 53.6 kJ/mol. We computed the free energy change in solution utilizing a free energy perturbation method (21).

In this formalism, the system is characterized by a hamiltonian $H(\mathbf{p},\mathbf{q},\lambda)$ which is a function of the coordinates $\mathbf{q}$ and conjugate momenta $\mathbf{p}$ and a multidimensional coupling parameter λ. The parameter λ serves to define a pathway between two states A and B. We define a sequence of discrete states represented by values λ_i, $i = 1,...,N$, which transform the state A into the state B via a series of suitably small steps. The free energy difference between two adjacent states is given by the following relation:

$$\Delta G(\lambda_i \rightarrow \lambda_{i+1}) = -RT \ln\langle\exp\{[H(\mathbf{p},\mathbf{q},\lambda_{i+1}) - H(\mathbf{p},\mathbf{q},\lambda_i)]/RT\}\rangle, \tag{2}$$

where the term enclosed in the angle brackets $\langle\rangle$ represents an ensemble average. The total free energy change from state A to state B is just the sum over all the intermediate steps, as given below:

$$\Delta G(A \rightarrow B) = \Sigma_i \, \Delta G(\lambda_i \rightarrow \lambda_{i+1}). \tag{3}$$

In computing the free energies, we use the ergodic hypothesis and assume that a time-averaged sampling over the structures as they evolve dynamically is equivalent to the actual ensemble average over all possible configurations. An estimate of the computational error which arises do to our discrete method employed in Equations 2 and 3 can be obtained by computing the free energy change for the inverse reaction, proceeding from state B to state A. That is, at each state λ_i, we compute the free energy change for both the forward $\lambda_i \rightarrow \lambda_{i+1}$ and backward $\lambda_i \rightarrow \lambda_{i-1}$ directions.

To simulate the proton transfer reaction between methanol and imidazole, immersed the solute molecules in an 18-Å radius ball of TIP3P water. This produced a model consisting of 2388 atoms: the 15 atoms of methanol and imidazole which are to be treated quantum mechanically, and 791 water molecules in the MM partition. A deformable, stochastic boundary condition (22–23) was enforced on atoms in the region from 16 to 18 Å. The initial configuration of methanol and imidazole was transformed into methoxide and imidazolium by moving the proton from the O2 oxygen atom in methanol to the NE nitrogen atom in imidazole in a series of 0.05 Å steps. The distances between the proton and the two heavy atoms were constrained at each step, along the path shown in Figure 4. Because the free energy is a thermodynamic state variable, the difference in free energy between states A and B is

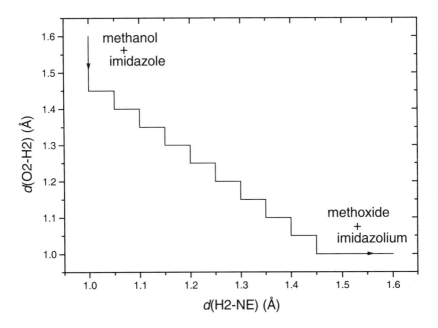

Figure 4. Path taken during proton transfer reaction. The distances between the proton and the heavy atoms were constrained at each intermediate step.

path-independent. Each intermediate state was equilibrated for 20 ps and data collected for 10 ps using 1 fs molecular dynamics time steps. The computed free energy changes were 51.0 kJ/mol in the forward direction and 49.8 kJ/mol in the backward direction, which compare quite favorably with the experimental value of 53.6 kJ/mol. The free energy profile for the forward path is depicted in Figure 5. The rather jagged patch in the center of the path is an artifact of the path we chose and not physically significant. Only the end point values are meaningful.

We note that the solvating water molecules have a significant influence on the reaction. In the gas phase, the experimental free energy change was 656.7 kJ/mol. In solution, there is a dramatic energy stabilization of the charges on the product species. It will be interesting to compare this result with the equivalent reaction in the enzyme environment.

Transferability of the QM/MM parameters

One last concern we should address before examining the reaction mechanism in the enzyme is the transferability of the QM/MM parameters. These parameters were established by the microsolvation procedure described above, defining the van der Waals parameters according to interactions with water molecules. The proton transfer reaction between methanol and imidazole in solution was well-described by the QM/MM method but in the protein, the interactions with residues in the active site will not always be with oxygen atoms. In particular, nitrogen atoms in the guanidinium groups of active-site arginine residues will play a key role in stabilizing the malate substrate. As a final check on the model hamiltonian, we have examined some of the key interactions between atoms in the QM partition and side chains in the protein. We have performed an extensive study of these interactions in small model systems (5) and found that, without further refinement, the parameters obtained through the microsolvation process can adequately describe the QM/MM interactions. In Figure 6, we illustrate one example of these studies, in which a methyl-guanidinium (representing an arginine residue) interacts with an acetate ion (representing one of the carboxylate groups of malate). The methyl-guanidinium was placed in the MM partition and the acetate in the QM partition. An analysis of the entire system was also performed with a density functional model where, again, the reactant structures were optimized independently and then were translated with one degree of freedom to obtain the interaction energy as a function of distance. Figure 6 is representative of the agreement we found; energies were reproduced within 20 kJ/mol and minimum-interaction distances to within 0.3 Å.

Reaction mechanism in the enzyme

With the calibration of the semi-empirical hamiltonian and van der Waals interaction parameters, we are now prepared to run simulations in the enzyme environment. We constructed a model of the enzyme by considering all of the amino acid residues within an 18 Å radius of the C2 carbon atom of malate, the NAD^+ cofactor and 39 water molecules deduced from the crystal structure. Another 105 water molecules were added by superimposing a 20 Å ball of TIP3P water and then removing all TIP3P molecules within 3.1 Å of nonhydrogen protein atoms, substrate, cofactor or crystal

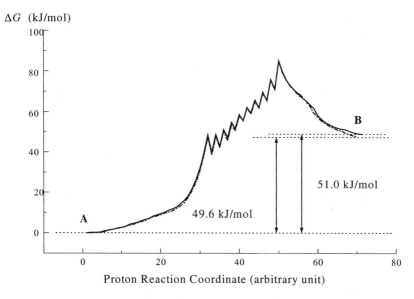

Figure 5. Free energy profile of the proton transfer between methanol and imidazole. This curve represents the forward difference path. The backward difference is similar.

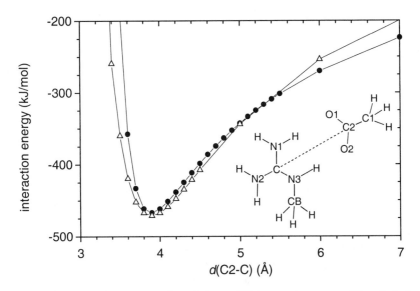

Figure 6. Interaction energy of methyl-guanidinium and acetate. Closed circles (•) are the QM/MM energies and open triangles (Δ) are the results of DFT calculations.

water molecules. Finally, another 27 water molecules were added from a resolvation procedure like that just described but after 40 ps of molecular dynamics calculations with all atoms fixed except water molecules. Using solely a molecular mechanics description for all atoms, the system was heated from 0 K to 300 K over an interval of 20 ps with atom velocities assigned from a Gaussian distribution every 2 ps in 30 K increments. The system was then equilibrated for another 80 ps, followed by 40 ps of data collection to define the Michaelis complex illustrated in Figure 1. At this point, the QM/MM calculations were initiated and 20 ps of equilibration were performed, followed by 20 ps of data collection.

One concern about the QM/MM method is whether the dynamical behavior of atoms in the QM partition is accurately depicted. We can compute the root mean square deviation between structures and find that the difference between the Michaelis complex defined by the MM simulations and the equilibrated structure defined by the QM/MM calculations is 0.16 Å for alpha-carbon atoms and is 0.39 Å for all atoms. We conclude that, when properly calibrated, the QM/MM method provides a realistic dynamical model of the enzyme-substrate system.

To explore the reaction mechanism of the MDH:malate:NAD$^+$ system, we could employ the free energy perturbation method we utilized in the proton transfer study between methanol and imidazole. A somewhat less computationally-intensive alternative approach is to examine the minimum energy surface of the reaction. We started with the QM/MM equilibrated structure defined above and annealed it from 300 K to 0 K over 20 ps while constraining the H2 proton and H21 hydride to be equidistant (1.3 Å) from the donor and acceptor atoms. We then energy minimized the resulting configuration for 5000 steps, maintaining the constraints on the proton and hydride positions. The distances between the O2 oxygen atom of the malate substrate and the H2 proton; between the H2 proton and the NE2 nitrogen atom of His-177 of MDH; between the C2 carbon atom of the malate substrate and the H21 hydride ion; and between the H21 hydride ion and the C4N carbon atom of NAD+ were varied on a four-dimensional grid with 0.2 Å spacing. An energy minimization of 1000 steps was performed at each grid point, resulting in 675 separate minimum energy values in the four-dimensional space.

We produced a minimum energy surface in the following manner. For each of two degrees of freedom: the distances between the proton and the hydride ion and the malate substrate, the minimum energy value was sought from the possible values of the other two degrees of freedom. A plot of this reaction surface is depicted in Figure 7. What is striking about the minimum energy surface is the large barrier facing an initial hydride transfer. The minimum energy pathway clearly indicates that the proton transfer occurs first, followed by the hydride transfer. This is somewhat surprising due to the fact that the intermediate state after the proton transfer has taken place will have a net charge on the substrate of −3e. In the gas phase, we might have expected the hydride reaction to proceed first, to minimize the charge separation in the intermediate state. As we saw in the study of the proton transfer in solution, however, solvation effects can be quite large. The environment of the enzyme clearly provides some stabilizing effects on the charged intermediate state (5). We can see how important these solvation effects are by the following analysis. For each of the minimum energy states defined by the surface in Figure 7, we performed a single-point energy calculation in which all the charges on MM atoms were set to zero, thereby removing

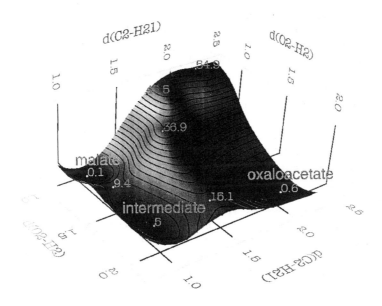

Figure 7. Minimum energy surface of the enzyme catalyzed reaction. Reprinted by permission of *Biochemistry*. Copyright 1997 American Chemical Society.

the principle solvation effects from the calculated energies. We plot these data in Figure 8 and note that there is now a large barrier opposing the initial proton transfer. Without the solvation effects due to the enzyme, the hydride transfer would occur first. Consequently, we can see directly in the MDH:malate:NAD$^+$ system that the catalytic properties of the enzyme are not solely due to the proper orientation of the substrate and proximity of the reacting element. Solvation effects due to other residues present in the active site can have significant impact on the reaction mechanism.

Conclusions

There are two transition states along the minimum energy pathway depicted in Figure 7, that represent the barriers presented to the proton and hydride transfers. We find a barrier of approximately 39 kJ/mol (9.4 kcal/mol) for the proton-transfer transition state and a barrier of 63 kJ/mol (15.1 kcal/mol) for the hydride-transfer transition state. This is in with the experimental observation that the hydride-transfer reaction is the rate-limiting chemical step in the reaction pathway (11–12).

reaction pathway We have demonstrated that realistic simulations of enzyme-substrate systems are possible with currently available computing resources. The keys elements in our method are the use of a quantum mechanical description of atoms in the active site of the enzyme, calibration of the semi-empirical quantum hamiltonian, calibration of the interactions between atoms in the QM partition and those in the MM partition and, of course, good experimental data on the crystal structure of the enzyme. Without the quantum description of atoms in the active site, we would be unable to treat the important bond formation events that define the catalytic process. Unfortunately, limitations of present algorithms and computing resources require that we use a semi-empirical quantum method but we have demonstrated that it is possible to calibrate the method against experimental data and produce results which rival those of the best *ab initio* quantum approaches, albeit in a limited set of circumstances. Furthermore, by calibrating the interactions of atoms in the QM partition with those in the MM partition, it is possible to produce realistic dynamics calculations. In essence, one can turn on the quantum description of the atoms without affecting the dynamics of the system as a whole. In this way, we can directly compute the effects of the protein matrix on the reaction mechanism, without resorting to any *ad hoc* schemes for estimating the influence of active-site residues. We note that recent experiments by John Burgner's group at Purdue are consistent with our proposed mechanism, lending some credence to our belief that the simulations produce a realistic description of the enzyme-substrate system.

Finally, for the case of the MDH:malate:NAD$^+$ system, it appears as though the enzyme produces an environment much like the solvating environment of aqueous solution, providing for the stabilization of charged intermediate states. Additionally, following the lock-and-key hypothesis, the active site residues serve to orient the malate substrate into a configuration which is optimal for the subsequent chemical events to take place. We are now in the process of extending this effort to consider other aspects of the MDH system, such as substrate specificity and the changes in activity brought about by modification of important subgroups. These are questions of prime importance to the complete understanding of enzyme systems and are questions

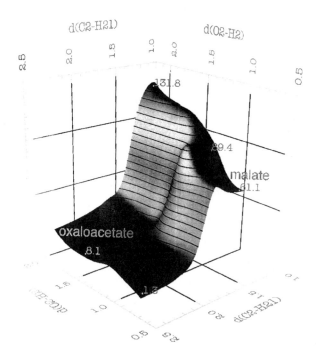

Figure 8. Minimum energy ``gas phase'' reaction surface. Note that this figure is reoriented with respect to Figure 7. Reprinted by permission of *Biochemistry*. Copyright 1997 American Chemical Society.

which can be addressed by numerical simulation, working in concert with careful experimentation.

References

1. Curtiss, L. A.; Raghavachari, K.; Redfern, P. C.; Pople, J. A. *J. Chem. Phys.* **1997,** *106,* 1063–1079.
2. Dewar, M. J. S.; Zoebisch, E. G.; Healy, E. F.; Stewart, J. J. P. *J. Am. Chem. Soc.* **1985,** *107,* 3092–3099.
3. Dixon, S.L.; Merz, K.M. Jr. *J. Chem. Phys.* **1996,** *104,* 6643-6649.
4. Stewart, J.J.P. *Int. J. Quantum Chem.* **1996,** *58,* 133-146.
5. Brooks, C. L.; Karplus, M.; Pettit, B. M. *Proteins: A Theoretical Perspective of Dynamics, Structure and Thermodynamics;* Advances in Chemical Physics, LXXI, John Wiley and Sons, New York, 1988.
6. Cunningham, M. A.; Ho, L. L.; Nguyen, D. T.; Gillilan, R. E.; Bash, P. A. *Biochemistry* **1997,** *36,* 4800-4816.
7. Hall, M. D.; Banaszak, L. J. *J. Mol. Biol.* **1993,** *232,* 213–222.
8. Warshel, A.; Levitt, M. *J. Mol. Biol.* **1976,** *103,* 227–249.
9. Singh, U. C.; Kollman, P. A. *J. Comp. Chem.* **1986,** *7,* 718.
10. Field, M. J.; Bash, P. A.; Karplus, M. *J. Comp. Chem.* **1990,** *11,* 700–733.
11. Parker, D. M.; Lodola, A.; Holbrook, J. J. *Biochem. J.* **1978,** *173,* 959–967.
12. Lodola, A.; Shore, J. D.; Parker, D. M.; Holbrook, J. J. *Biochem. J.* **1978,** *175,* 987–998.
13. Dewar, M. J. S.; Thiel, W. *J. Am. Chem. Soc.* **1977,** *99,* 4899–4907.
14. Lias, S. G.; Bartmess, J. E.; Liebman, J. F.; Holmes, J. L.; Levin, R. D.; Mallard, W. G. *J. Phys. Chem. Ref. Data* **1988,** *17,* Suppl. 3.
15. McClellan, A. L. *Tables of Experimental Dipole Moments,* Rahara Enterprises, El Cerrito, CA, Volume 2.
16. Hehre, W. J.; Radom, L.; Scheyer, P.; Pople, J. A. *Ab Initio Molecular Orbital Theory,* Wiley, New York, 1988.
17. MacKerell, A. D., Jr.; Karplus, M. *J. Phys. Chem.* **1991,** *95,* 10559–10560.
18. Jorgensen, W. L.; Chandrasekar, J.; Madura, J.; Impey, R. W.; Klein, M. L. *J. Chem. Phys.* **1983,** *79,* 926.
19. Maskill, H. *The Physical Basis of Organic Chemistry,* Oxford University Press, New York, 1989.
20. Dawson, R. M. C.; Elliot, D. C.; Elliot, W. H.; Jones, K. M. *Data for Biochemical Research,* Third Edition, Oxford University Press, New York, 1986.
21. Bash, P. A.; Ho, L. L.; MacKerell, A. D., Jr.; Levine, D.; Hallstrom, P. *Proc. Natl. Acad. Sci. USA* **1996,** *93,* 3698–3703.
22. Brooks, C. L.; Karplus, M. *J. Mol. Biol.* **1989,** *208,* 159–181.
23. van Gunsteren, W. F.; Berendsen, H. J. C. *Mol. Phys.* **1977,** *34,* 1311–1327.

Chapter 31

Modelling of Transition States in Condensed Phase Reactivity Studies

Neil A. Burton, Martin J. Harrison, Ian H. Hillier, Nicholas R. Jones, Duangkamol Tantanak, and Mark A. Vincent

Department of Chemistry, University of Manchester, Manchester M13 9PL, United Kingdom

In this Chapter we describe the calculation of transition states for condensed phase reactions involving catalysis by an enzyme and by a titanosilicate. Transition structures for hydride transfer between NADPH and FAD in the enzyme glutathione reductase have been determined using a QM/MM potential, for two forms of the enzyme, *e. coli* and human. Both the transition state structures and energies, and the associated kinetic isotope effects are found to differ for the two enzymes. A cluster has been used to model the transition state for alkene epoxidation by hydrogen peroxide, catalysed by titanosilicates, predicting attack of the oxygen atom of the titanium(IV)–hydroperoxide intermediate that is closer to the metal centre.

The majority of chemical processes and all biochemical ones occur in the condensed, rather than the gas phase. Methods for modelling the potential energy surfaces associated with gas phase reactions are well established and widely available in quantum chemistry packages. Here, for small and medium size systems results of a "chemical" accuracy may be obtained using widely available distributed computational resources. There is now increasing effort to achieve this degree of realism for condensed phase systems, particularly to understand reactivity and catalysis and hence to aid molecular design across a range of important chemical areas. It is to solve these problems that the traditional methods of both computational chemistry and solid state physics are now being focused. In the latter area, periodic calculations based upon a plane wave pseudopotential (PWP) approach, incorporating the work of Car and Parrinello to permit dynamical simulations to be performed, have been used to study solid state structure and reactivity. Such methods become extremely computationally demanding for large unit cells, and indeed may be inappropriate for systems lacking natural periodicity. However, as long as sufficiently large unit cells are employed, this approach may even be effectively used to study the electronic structure of isolated molecules.

The extension of traditional gas phase electronic structure calculations to model the condensed phase is based upon an embedding or hybrid approach. For most systems to be modelled, there is usually a chemical centre, where electronic effects, ranging from electron polarisation to bond breaking and forming occurs. These electronic effects are modulated by an environment, which itself takes no direct part in the chemical process, but of course, can respond both structurally and electronically to the primary process being modelled.

In the area of biochemistry, understanding the relationship between enzyme structure and its catalytic effect has been aided by the use of such hybrid calculations *(1–4)*. Here the active site of the enzyme, including the substrate and important catalytic residues are treated quantum mechanically, whilst the remainder of the enzyme is modelled at a lower level, employing one of the force fields commonly used in macromolecular modelling.

The Hybrid Model

In the traditional hybrid model the system is divided into two regions, one described at an appropriate level of quantum mechanics (QM) and the other using a molecular mechanical (MM) force field. The effective Hamiltonian for the system is written as the sum of the Hamiltonians for the QM and MM regions and the interaction between them;

$$\hat{H}_{eff} = \hat{H}_{QM} + \hat{H}_{MM} + \hat{H}_{QM/MM} \tag{1}$$

the corresponding total energy $\left(E_{tot}\right)$ can be similarly written as

$$E_{tot} = E_{QM} + E_{MM} + E_{QM/MM}. \tag{2}$$

The interaction term is given by

$$\hat{H}_{QM/MM} = -\sum_{i,s} \frac{q_s}{R_{is}} + \sum_{m,s} \frac{Z_m q_s}{R_{ms}} + \sum_{m,s} \left[\frac{A_{ms}}{R_{ms}^{12}} - \frac{B_{ms}}{R_{ms}^{6}} \right] \tag{3}$$

where m and s label the QM and MM atoms respectively, and i is the index of the electrons of the QM system. The first term of the right hand side of equation *(3)* accounts for the effect of the formal charges of the MM fragment (q_s) on the QM part and may readily be incorporated into standard QM codes. The final two terms are the remaining Coulombic and Lennard–Jones interaction terms between the QM and MM fragments. We utilise the code GAUSSIAN *(5)* to enable both semi–empirical and *ab initio* QM calculations to be carried out within a variety of hybrid schemes. The Gaussian link structure allows evolving electronic structure methods to be coupled to a range of descriptions of the MM region, thus enabling all the standard facilities within both the QM and MM programs to be efficiently utilised. For example for the macromolecular modelling work we use the code AMBER *(6)* and associated force fields to carry out the MM calculations. Such a hybrid QM(Gaussian)/MM(Amber) implementation allows the geometry of the QM fragment to be optimised in the field of the MM portion using the redundant internal coordinate algorithm, and conversely the geometry of the MM portion can be optimised in the field of the QM fragment. Although such optimisations of the two portions are not directly coupled at present, an iterative optimisation procedure does allow for the effective optimisation of the combined QM/MM system. The characterisation of transition geometries can now be routinely carried out by calculation of the second derivatives of the QM/MM energy. The van der Waals (vdw) contributions to the second derivatives are calculated by finite difference of the vdw gradients whereas the electrostatic terms can be calculated either by finite difference or analytically using standard methods. In the optimisations the position of several atoms may be frozen to speed convergence by increasing the relative mass of these atoms in the cartesian to internal coordinate conversions. A similar procedure is followed in the calculation of the vibrational frequencies enabling characterisation of the constrained stationary state.

When there are covalent linkages between the QM and MM portions their treatment is somewhat arbitrary. Hydrogen termination is used to satisfy the valence of the QM portion, placed at a fixed distance from the appropriate QM atom. No interactions between these hydrogen termination atoms and the MM portion is included and to avoid unrealistic interactions, the charges of the MM junction atoms attached to the QM system are set to zero, with the charges on the remaining MM atoms scaled to conserve total charge.

Such hybrid methods are also being used to model solid state catalysis. We have shown that catalysis by Bronsted acid sites in zeolites may be modified by the inclusion of the electrostatic field due to the infinite lattice *(7)*.

In this paper we exemplify our current work by presenting preliminary predictions of transition states both for a reaction catalysed by an enzyme, and by a zeolite based metal oxide system.

Hydride Transfer and Kinetic Isotope Effects in Glutathione Reductase.
Redox reactions involving hydride transfer are implicated in a number of important biological reactions, involving enzymes such as dehydrogenases and reductases *(8)*. These reactions have traditionally been studied theoretically using a range of quantum mechanical methods, applied to considerably simplified models of the actual biological system *(9–11)*. Both the degree of realism of the model system and the level of sophistication of the quantum mechanical treatment have progressively advanced, with a recent study involving the use of density functional theory (DFT)

methods *(10)*. Such calculations are often directed towards interpretation of the measured kinetic isotope effects (KIE) associated with the hydride transfer. However, to date, an isolated gas phase molecular system has always been used, with no explicit consideration of the role of the enzyme on the energetics, or structure, associated with the hydride transfer.

An experimentally much studied system is the flavoenzyme, glutathione reductase (GR), which catalyses the reduction of glutathione disulfide (GSSG) to the biological antioxidant glutathione (GSH) *(12)*. The initial, rate determining step of the reaction, is the transfer of a hydride ion from the coenzyme NADPH (Nicotinamide Adenine Dinucleotide Phosphate) to the permanently bound cofactor FAD (Flavin Adenine Dinucleotide), to produce the reduced enzyme as a stable intermediate *(13)*

$$E + NADPH + H^+ \rightarrow EH_2 + NADP^+.$$

The first stage of this process is

$$NADPH + FAD \rightarrow NADP^+ + FADH^-$$

the transition state structure for this reaction being shown in Figure 1.

We here describe hybrid QM/MM calculations of the hydride ion transfer in the GR enzyme, this being the first prediction of the transition states and KIEs for this reaction, explicitly including the structure of the enzyme.

We have studied both the human and *e.coli* forms of the enzyme to test suggestions, based upon KIE data *(14)*, that there are differences in the corresponding transition states *(8)*. Structures for these enzymes *(15)* were obtained from the Brookhaven database *(16)*. These were built with AMBER4.0, solvated with 256 TIP3P *(17)* water molecules and the structures optimised using the AMBER force field. The QM region was chosen to contain the nicotinamide moiety of NADP and the flavin rings of FAD, in addition to a few extra-ring atoms, as shown in Figure 2. The valencies of the QM fragment were satisfied by the addition of hydrogen atoms to the QM link atoms. The potential energy surface for hydride transfer was explored by energy minimisation of structures in which the distance between the FAD nitrogen (N_1) and the transferred hydrogen atom (H_t) was decreased in steps of 0.1Å. We use the AM1 QM Hamiltonian *(18)* and optimise the geometry of both the QM and MM moieties for each such structure. Further refinement of the transition state geometry from this procedure leads to a properly characterised transition state structure having a single imaginary frequency. The primary KIEs were calculated for these structures within the rigid rotor, harmonic oscillator approximation (Bigeleisen equation *(19)*) with a Wigner tunnelling correction *(20)*.

The calculated barrier height and structural parameters of the transition state are shown in Table I.

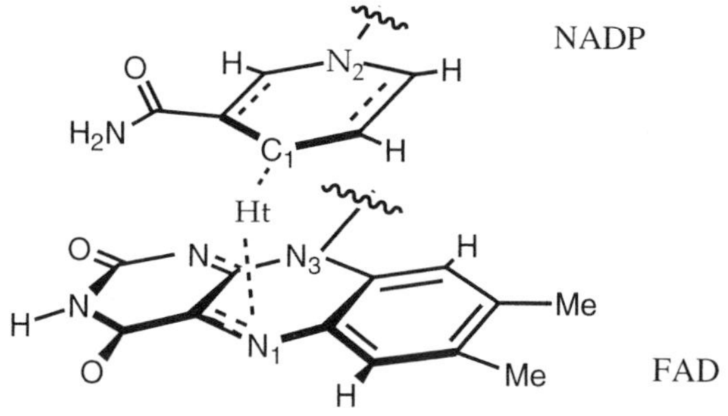

Figure 1. Transition state geometry showing atom numbering used in models of the hydride transfer.

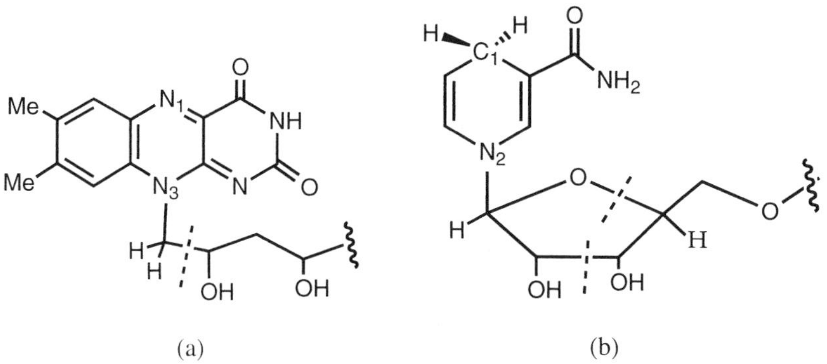

(a)

(b)

Figure 2. The QM FAD and NADPH moieties. The dashed lines indicate the QM–MM junction. (a) Flavin Adenine Dinucleotide (FAD) (isoalloxazine end). (b) Nicotinamide Adenine Dinucleotide (NADPH) (nicotinamide end).

Table I. Calculated Barrier Height and Structural Parameters of Transition State

	Barrier (kcal. mol^{-1})	N_3–C–C–N_1 dihedral (°)	N_2–C–C–C_1 dihedral (°)	N_1–H_t (Å)
E.coli	26.8	5.8	1.6	1.33
Human	20.3	4.3	3.1	1.50
No MM[a]	29.5	–	–	–
No Glu184[a]	25.4	–	–	–
No MM	37.0	5.6	3.2	1.20

[a] Evaluated for human enzyme stationary structures.

Inspection of all the structures shows marked ring pucker both in the isoalloxazine and the nicotinamide rings, with the transferring atoms N_1 and C_1 moving out of the plane of the ring by approximately 5°. Such structural changes have been noted in *ab initio* model studies on hydride transfer between two nicotinamide rings *(21)*. In contrast to the previous calculations on model systems *(9–11)*, our calculations are able to identify the important role of the enzyme on transition state structure and energetics. Thus, the human enzyme yields a lower barrier and earlier transition state than the *e.coli* enzyme (Table I), which is reflected in the smaller KIE calculated for the human enzyme (Table II).

Table II. Imaginary Frequencies (cm^{-1}) of Transition States and KIEs (k_H/k_D)

	$v(H)$	$v(D)$	KIE Semi-classical	KIE with correction[a]
Human	1373i	1109i	2.9	3.7
E.Coli	1820i	1420i	3.1	4.4
No MM	1699i	1332i	3.0	4.2 (4.2)[b]

[a] Wigner tunnel correction *(20)*.

[b] Result from *ab initio* model studies *(10)* in parenthesis.

Although the measured KIEs *(14)* for the two enzymes studied here are considerably smaller than the values we calculate (due to the composite nature of the rate constants) a value of 3.99 is found for the spinach enzyme *(14)* (for which no structure is available).

The role of the enzyme in lowering the barrier can be readily studied by removal of the formal atomic charges on selected residues followed by re–evaluation of the energy of the stationary structures. Thus, removal of the charges on the nearby Glutamate 184 residue in the human form, which binds the amide group of the NADP, (no Glu 184, Table I), shows that this residue contributes about 50% of the total barrier lowering effect of this enzyme, as judged by comparison with the calculation in which all MM charges have been removed. Location of the transition state in the total absence of the MM environment, leads to a considerably larger barrier than for either of the enzymes studied, and a correspondingly later transition state (Table I).

Although the present calculations use a semi–empirical Hamiltonian, they show that the actual enzyme structure is reflected in both the barrier to the reaction and the KIE, and provide a way forward for more realistic modelling of enzyme catalysed reactions.

The Mechanism of Alkene Epoxidation by Hydrogen Peroxide Catalysed by Titanosilicates. There is considerable current interest in transition states of epoxidation of alkenes, particularly in view of the use of such reactions to produce stereoselective products. Epoxidation using a variety of organic reagents such as performic acid, dioxirane and oxaziridine has been modelled using high level *ab initio* methods, with emphasis on the synchronous or asynchronous nature of the oxygen transfer *(22)*. Alkene epoxidation by hydrogen peroxide can be catalysed by titanium containing zeolites such as TS–1, TS–2 and Ti–MCM–41, which involve framework Ti(IV) species *(23)*. The active species is generally considered to be a hydroperoxide species accommodated in the titanium coordination sphere, and possibly hydrogen bonded to a water or alcohol molecule (Figure 3a) *(24)*.

We have carried out preliminary cluster calculations in order to identify the transition state for this epoxidation reaction. On–going studies are to include solid state effects *via* an embedded cluster approach. However, the bare cluster calculations are a necessary first step in understanding the catalytic role of the Ti(IV) active site. We have identified a low energy transition state which involves transfer of the oxygen atom (O(1)) that is directly bonded to the titanium atom, rather than the terminal oxygen atom (O(2)) (Figure 3b). The calculated transition state has a number of interesting structural features when compared both to the reactant structure, and to the corresponding transition state involving an organic peroxy species. The reactant structure (Figure 3a) has the peroxy group bound sideways, similar to the arrangement in related metal complexes *(25)*. In the transition state there is a lengthening of the Ti–O(1) length and a more substantial reduction in the Ti–O(2) length, corresponding to the development of Ti–O(2) single bond. This structure suggests an early transition state reminiscent of the synchronous structure found for the epoxidation of ethene by performic acid. Not only are the C–C and C–O lengths very similar in both transition states, but our calculated barrier (11.9 kcal mol^{-1}, at the B3LYP/3–21G* level) is close to the value for the organic peroxidation (14.1 kcal mol^{-1}, B3LYP/6–31G*) *(22)*. However, in the latter reaction, there is attack by the terminal oxygen atom of the performic acid, with proton migration to the third oxygen atom of the acid. We have located the corresponding transition state for the titanosilicate catalysed reaction (Figure 3c). Here it is the methanol molecule that assists in the deprotonation of the peroxy

408

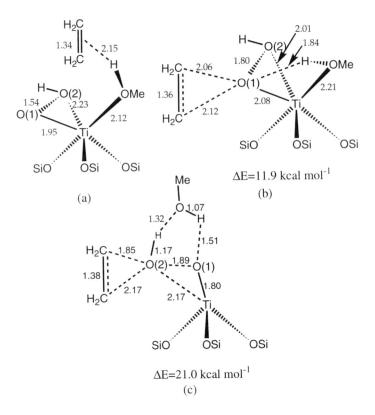

Figure 3. (a) Hydroperoxy intermediate, and (b), (c) transition structures for attack of each hydroperoxide oxygen atom on ethene. ΔE is the calculated barrier. (Silicon atoms are terminated by hydrogen atoms).

species. However, the barrier (21 kcal mol^{-1}) is considerably higher, and the transition state structure suggests an asynchronous mechanism.

Summary and Concluding Remarks

We have shown how electronic structure calculations can be used to gain insight into complicated reactions occurring in the condensed phase. For the enzyme catalysed reaction involving hydride ion transfer, the enzyme has an important role in lowering the barrier to the reaction. In the case of the epoxidation reaction, promoted by the titanosilicate, similarities and differences compared to the reaction with organic reagents have been identified. Here the role of more distant atoms of the catalyst, not yet included in the cluster calculation, remain to be identified.

Acknowledgments

We thank EPSRC for support of this research.

Literature Cited

1. Aqvist, J.; Warshel, A. *Chem. Rev.* **1993**, *93*, 2523.
2. Singh, U. C.; Kollman, P. A. *J. Comp. Chem.* **1986**, *7*, 718.
3. Field, M. J.; Bash, P. A.; Karplus, M. *J. Comp. Chem.* **1990**, *11*, 700.
4. Harrison, M. J.; Burton, N. A.; Hillier, I. H. *J. Am. Chem. Soc.* **1997**, *119*, 12285.
5. Frisch, M. J.; Trucks, G. W.; Schlegel, H. B.; Gill, P. M. W.; Johnson, B. G.; Robb, M. A.; Cheeseman, J. R.; Keith, T. A.; Petersson, G. A.; Montgomery, J. A.; Raghavachari, K.; Al–Laham, M. A.; Zakrzewski, V. G.; Ortiz, J. V.; Foresman, J. B.; Cioslowski, J.; Stefanov, B. B.; Nanayakkara, A.; Challacombe, M.; Peng, C. Y.; Ayala, P. Y.; Chen, W.; Wong, M. W.; Andrés, J. L.; Replogle, E. S.; Gomperts, R.; Martin, R. L.; Fox, D. J.; Binkley, J. S.; Defrees, D. J.; Baker, J.; Stewart, J. J. P.; Head–Gordon, M.; Gonzalez, C.; Pople, J. A. *GAUSSIAN 94*, (Gaussian Inc. Pittsburgh PA, 1995).
6. Pearlman, D. A.; Case, D. A.; Caldwell, J. C.; Seibel, G. L.; Singh, U. C.; Weiner, P.; Kollman, P.A. AMBER 4.0, University of California, San Francisco, 1992.
7. Sherwood, P.; de Vries, A. H.; Collins, S. J.; Greatbanks, S. P.; Burton, N. A.; Vincent, M. A.; Hillier, I. H. *Faraday Discuss.* **1997**, *106*, 79.
8. Muller, F. *Chemistry and Biochemistry of Flavoenzymes: Volume II*, CRC Press, 1992.
9. Sustmann, R.; Sicking, W.; Schultz, G. E. *Angew. Chem. Int. Ed. Eng.* **1989**, *28*, 1023.
10. Andres, J.; Moliner, V.; Safont, V. S.; Domingo, L. R.; Picher, M. T. *J. Org. Chem.* **1996**, *61*, 7777.

410

11. Mestres, J.; Duran, M.; Bertran, J. *Bioorg. Chem.* **1996**, *24*, 69.
12. Voet, D.; Voet, J. G. *Biochemistry*; Wiley: New York (NY), 1990.
13. Blankenhorn, G. *Eur. J. Biochem.* **1976**, *67*, 67.
14. Vanoni, M. A.; Wong, K. K.; Ballou, D. P.; Blanchard, J. S. *Biochem.* **1990**, *29*, 5790; Sweet, W. L.; Blanchard, J. S. *Biochem.* **1991**, *30*, 8702.
15. Karplus, P. A.; Schulz, G. A. *J. Mol. Biol.* **1987**, *195*, 701; *J. Mol. Biol.* **1989**, *210*, 163; Mittl, P. R. E.; Schulz, G. E. *Protein Science* **1994**, *3*, 799.
16. Brookhaven National Laboratories, Associated Universities Inc. (E.C.1.6.4.2 , 1GET and 1GRB).
17. Jorgensen, W. L.; Chandrasekhar, J.; Madura, J.; Impey, R. W.; Klein, M. L. *J. Chem. Phys.* **1983**, *79*, 926.
18. Dewar, M. J. S.; Thiel, W. *J. Amer. Chem. Soc.* **1977**, *99*, 4499.
19. Bigeleisen, J.; Wolfsberg, M. *Adv. Chem. Phys.* **1958**, *1*, 15; Bigeleisen, J. *J. Chem. Phys.* **1949**, *17*, 675.
20. Wigner, E. *Z. Phys. Chem.* B **1933**, *19*, 203.
21. Wu, Y. D.; Lai, D. K. W.; Houk, K. N. *J. Amer. Chem. Soc.* **1995**, *117*, 4100.
22. Houk, K. N.; Liu, J.; DeMello, N. C.; Condroski, K. R. *J. Am. Chem. Soc.* **1997**, *119*, 10147.
23. Murugavel, R.; Roesky, H. W. *Angew. Chem. Int. Ed. Engl.* **1997**, *36*, 477.
24. Khouw, C. B.; Dartt, C. B.; Labinger, J. A.; Davis, M. E. *J. Catal.* **1994**, *149*, 195.
25. Boche, G.; Möbus, K.; Harms, K.; Marsch, M. *J. Am. Chem. Soc.* **1996**, *118*, 2770.

Chapter 32

Transition States for *N*-Acetylneuraminic Acid Glycosyltransfer: Catalysis via a Transition State Hydrogen Bond

Benjamin A. Horenstein

Department of Chemistry, University of Florida, Gainesville, FL 32611

A computational analysis of glycosyltransfer reactions of α–carboxylate substituted and α-H substituted pyranosyl oxocarbenium ions has been undertaken to model the glycosyltransfer of N-acetylneuraminic acid (NeuAc). Ion-molecule complexes, transition states, and products have been identified for reaction of each oxocarbenium ion with water. The carboxylate group adjacent to the anomeric center of NeuAc may function as a catalyst during glycosidic bond formation or cleavage reactions. Both acid-base catalysis and electrostatic catalysis are possible functions that the carboxylate group could play. As such, glycosyltransfer of N-acetyl neuraminic acid may be considered as a "mini" model for glycosylases. The calculated transition state barrier for capture of an α-carboxylate pyranosyl oxocarbenium ion by water is 1.9 kcal/mol higher than the barrier for capture of an α-H substituted pyranosyl oxocarbenium ion, pointing to stabilization of the oxocarbenium ion by the carboxylate group. The transition state for capture of the α-carboxylate pyranosyl oxocarbenium ion features a hydrogen bond between the attacking water and carboxylate group, with an estimated energy of ~8 kcal/mol. This result leads to the observation that while the carboxylate group may stabilize the oxocarbenium ion, it also facilitates its capture. The timing of proton transfer with respect to glycosidic bond cleavage/formation is also discussed and compared to the available experimental data. Some possible implications for glycosylase catalysis are presented.

N-acetyl neuraminic acid (NeuAc, Figure 1, structure A, R = H) possesses a highly acidic carboxyl residue with a pKa < 3 adjacent to the anomeric carbon.(*1*) Because glycosidicaly bound NeuAc is involved in a number of different cell-surface glycoprotein and glycolipid binding determinants,(*2*) an understanding of the mechanisms of the enzymes that maintain NeuAc glycosides would facilitate design of inhibitors and allow modulation of cell-surface glycosylation patterns. The key enzymes involved in maintenance of NeuAc glycosides are sialyltransferases, neuraminidases, and trans-sialidases. Sialyltransferases transfer N-acetyl neuraminic acid to terminal saccharide units on glycoproteins and glycolipids, and are the key biosynthetic enzyme for creation of this linkage in bacteria and higher eukaryotes, including humans.(3) Neuraminidases are hydrolases, which cleave glycosidicaly bound NeuAc. The best known neuraminidase is that from influenza virus, currently a target of drug design.(4,5) Trans-sialidases are found in the trypanosome *Trypanosoma cruzi*, a human parasite responsible for Chagas' disease.(6) This unusual enzyme transfers sialic acid from host cells to the parasite, possibly facilitating its entry into host cells or providing immunomasking to the trypanosome. From the point of view of inhibitor and drug design, there is considerable interest in catalytic mechanisms for glycosyltransfer of N-acetylneuraminic acid. The mechanistic aspects of NeuAc glycosyl transfer are of interest in their own right, because of the presence of the carboxylate group adjacent to the anomeric center. A brief discussion of glycosyltransfer in terms of acetal hydrolysis highlights why this is the case.

The formation and hydrolysis of glycosidic bonds to saccharide residues may be generally represented as an interconversion of an acetal **2a** and hemiacetal **2c**. (Figure 2) Acid-catalyzed acetal hydrolysis can proceed via oxocarbenium ion intermediates **2b**, which are stabilized relative to 2° carbocations due to participation of lone pair electrons on the oxygen atom adjacent to the acetal carbon. Mechanistic studies of oxocarbenium ions has led to the conclusion that if the "R" group for **2b** is an electron releasing aryl substituent, the lifetime of **2b** can be enhanced in aqueous solution due to extensive charge delocalization into the aromatic ring via π-interactions with the carbocationic center.(7) When "R" is alkyl, as in glucose, the barrier for capture by solvent in these systems is extremely small, and the lifetime of the oxocarbenium ion is on the order of 10^{-12} s.(8) In the presence of anions, alkyl oxocarbenium ions are thought not to exist as intermediates, instead, encounter complexes of the nascent oxocarbenium ion and anion collapse directly to product.(9)

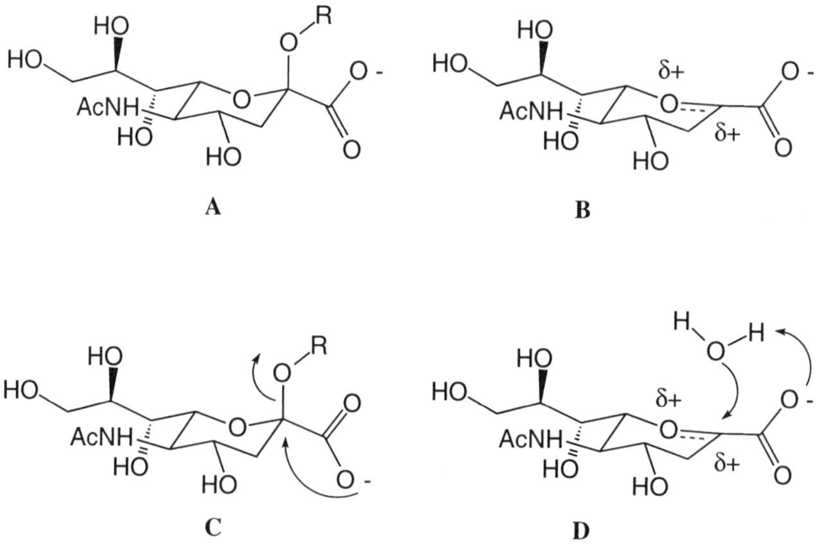

Figure 1. N-acetyl Neuraminic acid (NeuAc) and possible mechanistic roles for its α-carboxylate group.

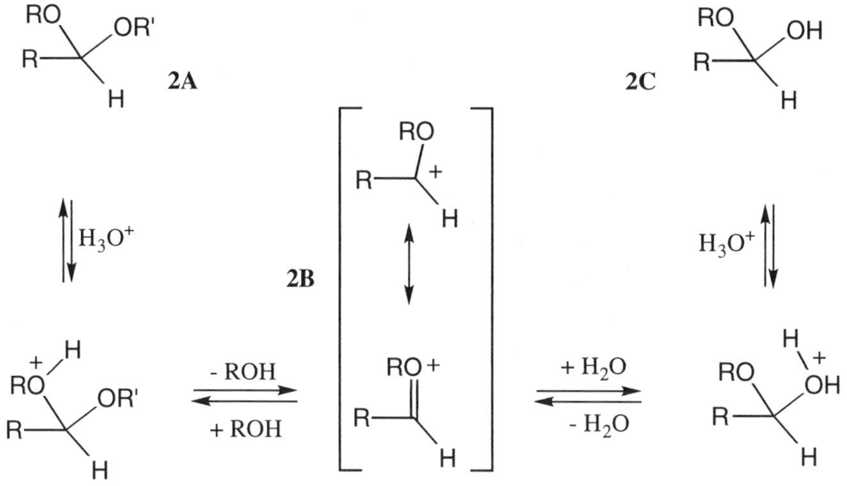

Figure 2. A general scheme for specific acid-catalyzed acetal hydrolysis.

Also, alkyl oxocarbenium ions are so reactive that they are trapped by water without the observation of general base catalysis.(7) Reconsider the scheme presented (Figure 2) if the acetal substrate had a carboxylate rather than a hydrogen at the acetal carbon, as would be the case for glycosyltransfer of N-acetylneuraminic acid. As shown in Figure 1, the carboxylate could electrostaticaly stabilize an oxocarbenium ion like transition state and/or oxocarbenium ion intermediate (**B**), act as an intramolecular nucleophile (**C**), and/or mediate general acid-base catalysis (**D**). The carboxylate therefore mimics the chemical reactivities proposed for carboxyl amino acid side chains found in glycosylase active sites.(10,11)

Experimental and computational approaches have been applied to NeuAc glycosyltransfer. Sinnott and colleagues have provided evidence that solvolysis of aryl glycosides of N-acetylneuraminic acid can proceed with nucleophilic participation of the carboxylate group in the transition state,(12) but other studies argue that nucleophilic participation is not universal. Both kinetic isotope effect studies and solvent trapping studies for acid-catalyzed solvolysis of the cytidine monophosphate glycoside of NeuAc (CMP-NeuAc) at pH 5 were inconsistent with a nucleophilic role for the carboxylate group, and further suggested that the carboxylate group stabilized a discrete oxocarbenium ion intermediate in solution.(13) Further support for this suggestion was found by azide trapping studies during solvolysis of CMP-NeuAc solvolysis, which have indicated that an oxocarbenium ion intermediate with a lifetime of ~ 0.1 ns is formed after rate-limiting scission of the glycosidic bond.(14) Ab-initio calculations at the B3LYP/6-31G(d) level of theory were used to estimate that an α-carboxylate substituted oxocarbenium ion would be ~17 kcal/mol more stable than an α-H substituted oxocarbenium ion in an aqueous environment.(15) The lifetimes of α-H oxocarbenium ions such as those derived from 2-deoxyglucose and glucose in water are on the order of 10^{-11} - 10^{-12} s, (8,9,16) and a lifetime of ~ 1 second would be predicted for an α-carboxylate oxocarbenium ion if the carboxylate-derived stabilization were applied only to the oxocarbenium ion and not at all to the transition state for its capture by solvent. The short 0.1 ns lifetime for the NeuAc oxocarbenium ion demonstrates that the activation barrier for capture of the oxocarbenium ion must also be lowered in concert with the stabilization of the oxocarbenium ion itself, though the longer lifetime of the NeuAc oxocarbenium ion relative to the glucosyl oxocarbenium ion shows that the activation barrier for capture of the former is still larger. Azide trapping experiments conducted in H_2O and D_2O

allowed calculation of an apparent solvent isotope effect of 1.1 for trapping the oxocarbenium ion by water.(*14*) The simplest interpretation of this result is that a proton is not in flight in the transition state, arguing against classical general base catalysis in the transition state. Using density functional theory, (B3LYP/6-31G(d)) a transition state structure was identified for capture of an α-COO substituted oxocarbenium ion by water.(*14*) The structure featured a hydrogen bond between the carboxylate and attacking water, but proton transfer was not a part of the reaction coordinate. This result suggested that hydrogen bonding might serve to stabilize the shift in charge from the oxocarbenium ion to the attacking water that occurs during solvent capture of the NeuAc oxocarbenium ion.

The goals of the work presented here were to estimate the relative barriers for water-capture of an α-carboxylate substituted oxocarbenium ion as a model for the NeuAc oxocarbenium ion, versus the same reaction for an H-substituted pyranosyl oxocarbenium ion, and explore the role that the carboxylate group plays in this process.

Model and Methods

The basic model chemistry consisted of either an α-H or α-COO$^-$ substituted pyranosyl oxocarbenium ion microsolvated by two water molecules. Stationary points were sought which corresponded to the oxocarbenium ion/water complex, the transition state for collapse of the complex, and the protonated product hemiacetal/hemiketals, as shown in structures **3-6** and **8-9** of Figure 3. Except when noted otherwise, the geometry optimizations used a self-consistent reaction field (SCRF) solvation model to account for solvent polarity effects.(*17,18*) Frequency calculations were performed on all stationary points for zero-point energy corrections and for calculation of isotope effects. RHF frequencies were scaled by 0.9. The mass-weighted cartesian displacements corresponding to the single imaginary frequency of transition states were examined to confirm that the transition state mode corresponded to the expected nuclear displacements. Single point energy evaluations employed the same theory and basis as the geometry optimization but solvent polarity effects were modeled with the polarized continuum model of Tomasi, as implemented in Gaussian94(*19*) using the SCIPCM model (self consistent isodensity polarized continuum model).(*20*) Kinetic and equilibrium isotope effects were calculated with QUIVER(*21*) which uses the Bigeleisen-Mayer formulation for calculation of the isotope effect. (*22*) Input to the program consists of the optimized geometry and force constants obtained from *ab-initio* calculations.

The computational challenge was to identify stationary points for termolecular systems consisting of as many as 11 heavy atoms. The potential energy surfaces (PES's) in the regions of both minima and transition states were extremely flat, and required calculation of force constants at every point in the optimizations for convergence. Transition states corresponding to **7**(Figure 3) which has only one water molecule, were identified at RHF/6-31G(d), RHF/6-31G(d,p), and B3LYP/6-31G(d) (*23,24*) levels of theory; these results are included in the discussion for comparison. The RHF/6-31G(d,p) level of theory was employed for all other cases where the oxocarbenium ion/water complex consisted of two water molecules. The modeled chemistry involved charged species, and the charge distributions change over the reaction coordinate. It was anticipated that some accounting of solute-solvent electrostatic interaction would be required to successfully model the chemistry, and as will be presented below, this was found to be the case. The second water molecule serves to microsolvate the nucleophilic water, and thus provides a step toward a more realistic model. The SCRF and SCIPCM models primarily serve to account for the effect of solvent (water) polarity on the calculated energy, but they can not account for solute-solvent interactions in an explicit way. Different approaches to solvation modeling have been reviewed.(*25*)

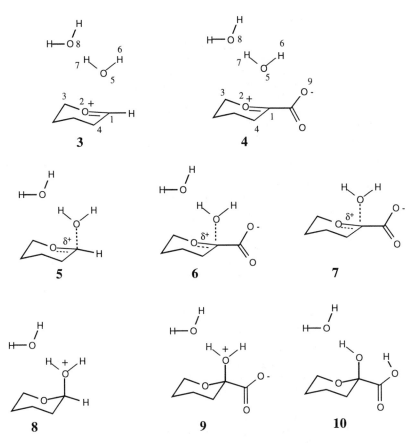

Figure 3. Model structures corresponding to ion-molecule complexes **3-4**, transition states **5-7**, and products **8-10**, for reaction of α–H or α–COO-substituted oxocarbenium ions with water. The atom numbering scheme for structures **3-4** is used for the description of geometric parameters presented in Tables I and II.

Hydrogen bonding is an important structural feature of all of the stationary points identified, and it is important to consider the level of theory employed in this work in light of some of some recent results for similar hydrogen bonded systems. Correlation effects appear to be important to obtain the most accurate geometry and energy, but Hartree-Fock level theory with reasonably large basis sets give results which show many of the same trends observed for calculations using density functional theory or second order perturbation theory.(26,27) McAllister's recent work (26) on the formate-formic acid strong hydrogen bond has shown that HF level calculations underestimate the strength of the hydrogen bond by ~ 5 kcal/mol relative to MP2 or DFT calculations at an internuclear O-O distance of 2.5 Å. The calculations also revealed that the energy is quite sensitive to the bond angle defining the hydrogen bond (O-H-O) with substantial penalties for deviation from linearity. In this case however the HF calculations underestimate the *penalty* for deviation from linearity by ~ 2.3-2.5 kcal/mol, so for hydrogen bonds that are far from linearity, use of HF level theory underestimated the bond energy by ~2.5 kcal/mol overall with respect to the correlated methods.

Optimized structures and energetics

From experimental studies it is estimated that the α-carboxylate substituted neuraminyl oxocarbenium ion has a lifetime in water of ca. 10 - 25 times larger than oxocarbenium ions derived from 2-deoxy pyranosides.(*8,13,14,16*) This indicates that the relative activation barrier for capture of the neuraminyl oxocarbenium ion is approximately 1.4 -2.0 kcal/mol higher. In order to model this chemistry, structures for, and the relative energetics of, oxocarbenium ion/water ion-molecule complexes and the transition states for their collapse were sought. In initial work, density functional theory (DFT) was successfully employed to locate a transition state (Figure 3, structure **7**) corresponding to capture of the α-carboxylate oxocarbenium ion by a single water, not having a second "microsolvating" water. Table I presents a comparison of some key geometric parameters as a function of theory, basis set, and solvation model. It is clear that apart from the distance between the anomeric carbon and attacking water oxygen, the structures are quite similar. Notably, all models show similar geometry for the pyranosyl ring and similar bond lengths for the hydrogen bond between the attacking water and carboxylate group. Also, optimized geometric parameters are only moderately sensitive to whether the calculation was for the gas phase or if the SCRF solvation model was used. In all cases the position of the hydrogen in the water-carboxylate hydrogen bond is asymmetric, residing on the water oxygen. Work to extend application of DFT to the other microsolvated models presented in Figure 3 is in progress. Table II presents some key geometric parameters obtained at the RHF/6-31G(d,p) level corresponding to the microsolvated structures shown in Figure 3. The two ion-molecule complexes **3** and **4** have distances d(1-5) of 2.385 and 2.569 Å between the nucleophilic water oxygen and the oxocarbenium ion carbon. This distance is too far for significant covalent bonding, but is sufficiently close that a significant dipole-charge interaction can occur. The greater distance for **4** may be a reflection of carboxylate stabilization, i.e. less ion-dipole interaction is required to stabilize complex **4**. It was anticipated that the oxocarbenium ions would be flat about the $C-O^+=C-C$ dihedral but this was not the case, with calculated dihedrals θ(3-2-1-4) of 12.7° and 14.7°, respectively. This is probably because the models do not have solvating water on both faces of the oxocarbenium ion. The second "solvating" water in both complexes appears to be involved in a conventional hydrogen bond to the nucleophilic water, with O-O internuclear distances d(5-8) of 2.845 and 2.908 Å. The endocyclic C-O bond d(1-2) is very short for each complex, having substantial π-character. For the α-carboxylate substituted oxocarbenium ion, the nucleophilic water is involved in a hydrogen bond to the carboxylate group, with an O-O internuclear distance d(5-9) of 2.907 Å and the hydrogen atom being asymmetrically disposed, residing squarely on the water oxygen. In proceeding to the transition state, each complex features significant pyrimidalization at the anomeric carbon, with C-O internuclear distances d(1-5) of a conventional hydrogen bond to the nucleophilic water, with O-O internuclear distances d(5-8) of 2.845 and 2.908 Å.

The endocyclic C-O bond d(1-2) is very short for each complex, having substantial π-character. For the α-carboxylate substituted oxocarbenium ion, the nucleophilic water is involved in a hydrogen bond to the carboxylate group, with an O-O internuclear distance d(5-9) of 2.907 Å and the hydrogen atom being asymmetrically disposed, residing squarely on the water oxygen. In proceeding to the transition state, each complex features significant pyrimidalization at the anomeric carbon, with C-O internuclear distances d(1-5) of 1.851 and 1.909 Å for transition structures **5** and **6**, respectively. It is interesting that although progress to the transition state is accompanied by large changes in the C-O internuclear distance between the anomeric carbon and attacking water, very little change is found for the endocyclic C-O bond. This bond still has substantial π-character, and therefore these transition state structures are oxocarbenium ion-like. In transition state **6** the α-carboxylate group is again hydrogen bonded to the attacking water, but the internuclear O-O distance is 2.540 Å. This short internuclear distance is indicative of a strong hydrogen bond, but the angle O-H-O is acute at 130.5 ° which weakens the strength of the bond. The energetics of the hydrogen bond will be discussed below. Inspection of the geometric parameters in Tables I and II reveals that the addition of the second microsolvating water to the model results in an earlier transition state for **6** than for **7**, with the forming glycosidic bond ~ 0.11 Å longer in **6**. This shows that the transition state structure is sensitive to explicit solvation, and emphasizes the importance of this feature of the model.

Table I. Selected geometric parameters^a for transition state 7.

computational model	d(1-5)	d(1-2)	d(1-4)	d(5-6)	d(5-7)	d(5-9)	θ(3-2-1-4)
RHF/6-31G(d)	1.807	1.300	1.505	1.012	0.950	2.409	36.6
RHF/6-31G(d,p)	1.843	1.296	1.503	1.010	0.946	2.400	35.6
RHF/6-31G(d,p) SCRF^b	1.800	1.298	1.507	0.992	0.947	2.418	37.5
B3LYP/6-31G(d)	2.069	1.302	1.505	1.035	0.972	2.496	27.9
B3LYP/6-31G(d) SCRF^b	2.037	1.300	1.506	1.029	0.972	2.495	29.0

^{a}Distances d are in angstroms, dihedral angle θ is in degrees. ^{b}The SCRF solvation model employed a dielectric of 78, and a cavity radius as determined in Gaussian94.

For the α–H pyranosyl reaction, the ZPE-corrected barrier for **3→5** is 1.3 kcal/mol, while for the α–COO- oxocarbenium ion capture, **4→6**, the barrier is 3.2 kcal/mol. These results are in excellent agreement with experiments that showed that the α-carboxylate substituted NeuAc oxocarbenium ion is longer lived than the glucosyl oxocarbenium ion in aqueous solution. The difference in barrier heights of 1.9 kcal/mol would correspond to a difference in rate of 23 at 37 °C; the experimentally estimated difference is 10-25. The importance of including a computational solvation model bears mention. The gas phase ZPE-corrected energy barriers for **3→5** and **4→6** are calculated to be 5.1 and 3.1 kcal/mol, respectively. This represents a reversal in the predicted relative reactivity for **3** and **4**, and is not in agreement with experimental results.

Table II. Selected geometric parameters^a and energy for 3-6,8

	d(1-5)	d(1-2)	d(1-4)	d(5-6)	d(5-7)	d(5-9)	θ^b	d(5-8)	Energy^c
3	2.385	1.240	1.483	0.945	0.954	na^d	12.7	2.845	-421.18218
4	2.569	1.250	1.491	0.946	0.950	2.907	14.7	2.908	-608.32710
5	1.851	1.280	1.498	0.949	0.974	na	31.6	2.674	-421.18007
6	1.909	1.289	1.503	0.960	0.957	2.540	34.2	2.820	-608.32197
8	1.555	1.331	1.514	0.952	1.007	na	43.6	2.524	-421.18219

^{a}Distances "d" are in angstroms, dihedral angle θ is in degrees. ^{b} "θ" designates the dihedral angle about atoms 3-2-1-4. ^{c}Energy values are in Hartrees and are zero-point corrected. ^{d}"na" = not applicable.

Hydrogen bonding and transition state stabilization

An α-carboxylate substituted oxocarbenium ion should enjoy an estimated 17 kcal/mol stabilization relative to the α-H substituted oxocarbenium ion.(15) During bond formation, it was anticipated that a major component of the barrier would be a penalty for loss of electrostatic stabilization at the charged anomeric carbon. However, at least two factors could mitigate against this. First, as bond formation proceeds, charge is *transferred* to the attacking water; this is evident for conversion of the ion molecule complex 3 to the transition state 5, in which Mulliken charge analysis revealed an increase of +0.135 charge units at the attacking water. For the α-carboxylate oxocarbenium ion, the attacking water develops +0.042 units of charge in the process 4→6; the balance is dispersed into the carboxylate group. To the extent that the carboxylate group electrostaticaly stabilizes this shift in charge, the barrier would be reduced. Secondly, the carboxylate can hydrogen bond to the attacking water which will also stabilize the transition state. While hydrogen bonds are considered to have a substantial electrostatic component,(27) this portion of transition state stabilization would apply to the charge transferred to the water only, and should not have an impact on charge at the oxocarbenium ion itself.

Two approaches were taken to assess the contribution of the hydrogen bond to transition state stabilization as illustrated in Figure 4. Figure 4 panel A, presents a "top" view of transition state 6 before and after rotation of the attacking water dimer by ~ 60° so as to break the hydrogen bond to the carboxylate group. Actually, in this conformation the two hydrogens of the attacking water form a weak bifurcated bond to the carboxylate oxygen atom. To obtain the ΔE, a partial optimization was performed in which the new torsion and the glycosidic C-O bond were frozen, and a new single point electronic energy was determined using the SCIPCM method with a dielectric of 78. Relative to the transition state, this conformation was ca. 8.7 kcal/mol higher in energy. The second approach represented in Figure 4 panel B involved determination of single point energies as the carboxylate group was rotated to planarity with the oxocarbenium ion atoms, and thus breaking the hydrogen bond to the attacking water. The difference in energy obtained was 9.2 kcal/mol, which reflects loss of the hydrogen bond *and* the penalty for placing the carboxylate group coplanar with the ring oxocarbenium ion atoms.(15) This latter quantity was estimated to be 2.0 kcal/mol by rotating the carboxylate group to planarity in the absence of the two water molecules. The difference between these two quantities, 7.2 kcal/mol, is the estimate for loss of the hydrogen bond. With either of the two above approaches, zero point energy corrections could not be applied because when either the water ensemble or carboxylate were rotated, the resulting conformations were presumably not stationary points. The two approaches provide estimates that are in good agreement, suggesting that the transition state hydrogen bond is worth 7.2-8.7 kcal/mol, a value indicative of a strong hydrogen bond which is consistent with the short O-O internuclear distance of 2.54 Å. In transition state 6, the O5-H6-O9 bond angle is quite small at 130.5° so the full potential bond strength is not realized. Counterbalancing factors effecting the strength of the hydrogen bond are likely to operative. Later transition states will have more acute O-H-O bond angles, weakening the hydrogen bond. Later transition states should also raise the acidity of the hydrogen bonded proton of the attacking water, providing a better pK_a match between donor water and acceptor carboxylate, strengthening the hydrogen bond.

The estimate that the hydrogen bond in transition state 6 is worth 7.2-8.7 kcal/mol suggests that in the absence of this hydrogen bond, the overall barrier to α-COO oxocarbenium ion attack by water would be 9.1-10.6 kcal/mol higher than for attack on an α-H substituted oxocarbenium ion. This provides a basis for explaining why the oxocarbenium ion derived from NeuAc is not significantly more long lived than the glucosyl oxocarbenium ion: the ability of the carboxylate to interact with attacking water in the transition state provides a considerable amount of catalysis via the hydrogen bond. This type of catalysis has been discussed for the base-catalyzed addition of hydroxylic solvents to carbenium ions.(28)

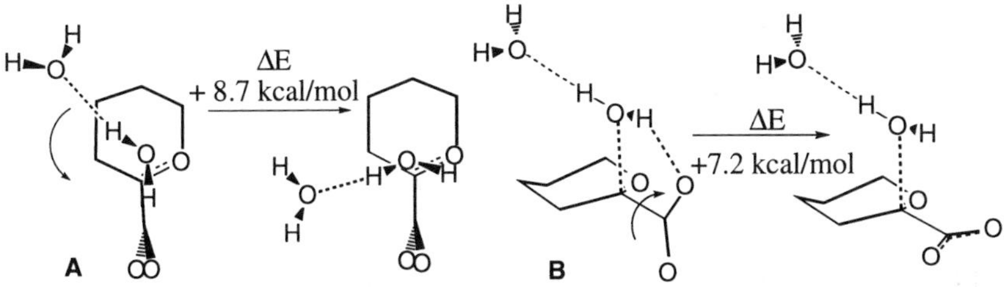

Figure 4. Two estimates of water/carboxylate interaction energy. The starting structure for these calculations was transition state **6**; the ΔE values are not ZPE-corrected.

The relative timing of proton transfer and heavy atom motion.

During capture of the α-H substituted oxocarbenium ion **3**, the attacking water is enjoying a hydrogen bond to the second "solvating" water, but no proton transfer has yet occurred at transition state **5**. For the resulting product **8**, the new anomeric hydroxyl is protonated and this species is a "stable" intermediate ~ 1.3 kcal/mol lower in energy than transition state **5**. It was reported recently that protonated axial 1-hydroxy pyran (analogous to **8** without the microsolvating water), is not a stable geometry at the MP2/6-31G(d) level of theory.(*29*) Instead, this species underwent spontaneous glycosidic (C-O) bond cleavage to afford the oxocarbenium ion-water complex. In the present work, this behavior was reproduced at the RHF/6-31G(d,p) level, but it is significant that when the second microsolvating water was included the protonated hydroxypyran **8** was stabilized to exist as an intermediate. It will be important to see if this observation holds with correlated methods. Gas phase semi-empirical calculations predict that protonated acetals can exist as intermediates.(*30*) The reason that the intermediacy of **8** is important can be considered in terms of the microscopic reverse for its formation, i.e. **8** → **5** → **3**. This process corresponds to specific acid catalyzed hydrolysis of an acetal. If **8** were not an intermediate, this would imply that the specific acid catalyzed process involves concerted protonation and glycosidic bond cleavage, which would not be in agreement with experimental solvent deuterium isotope effects.(*31*)

The situation for the α-carboxylate substituted oxocarbenium ion is similar with regard to transition state proton transfer. It is clear that transfer of H6 from the attacking water to the carboxylate is not significant in the transition state which argues against concerted general base catalysis. Unlike the protonated hemiacetal **8**, the protonated product **9** was not an intermediate at the RHF/6-31G(d,p) level. Partial geometry optimizations of **9** were performed with the OH bond (O5-H6, Figure 3) frozen, followed by a full optimization. Proton transfer to the carboxylate group to afford **10** is exothermic by at least 25 kcal/mol, with no apparent barrier. Since **9** could not be identified as an intermediate, it is surmised that proton transfer to the carboxylate occurs asynchronously in the transition state, in other words proton transfer lags far behind heavy atom motion corresponding to formation of the glycosidic bond. Current work is aimed at assessing the effect of electron correlation on the stability of **9**. The results for cutoff model **11** (Figure 5) suggest that **9** will not be an intermediate when correlation effects are included, because a geometric minimum placing the bridging hydrogen on the "glycosidic" oxygen of **11** could not be identified. It is possible that additional solvation at the carboxylate group would lower its basicity enough to permit species such as **9** or **11** to exist as intermediates, and this is under investigation.

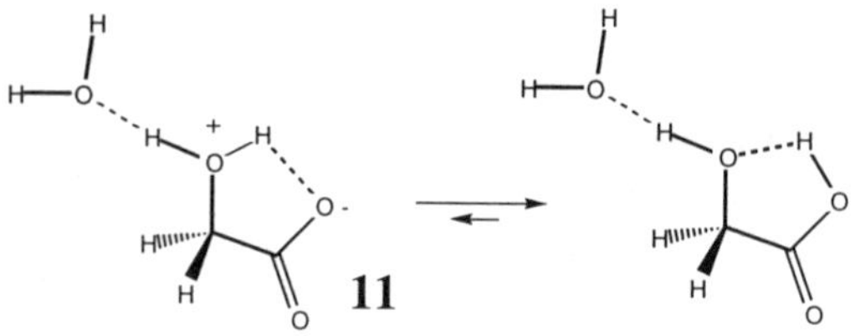

420

Figure 5. A calculation at B3LYP/6-31+G(d)/SCRF theory to test the intermediacy of **11**.

Solvent deuterium isotope effects.

Solvent isotope effects are a useful approach to obtain information on the timing of proton transfer in acetal hydrolysis.(*31*) Specific acid catalysis generally represents a rapid pre-equilibrium protonation step (Figure 2) with associated inverse solvent isotope effects where k_{H_2O}/k_{D_2O} is typically 0.3-0.4. General acid catalysis will typically have a normal isotope effect with $k_{H_2O}/k_{D_2O} > 1$ because the observed isotope effect can be dominated by a primary contribution from a proton in flight at the transition state.

It was of interest to determine if the model and level of theory employed in this work could make good predictions of solvent isotope effects for conversion of the ion-molecule complex **3** to the protonated hemiacetal **8**. Note that the equilibrium isotope effect for conversion of **3** to **8** can loosely be considered as the microscopic reverse of the specific acid catalyzed hydrolysis of an acetal, and so the isotope effect will be normal in this case, not inverse. The equilibrium solvent isotope effect for conversion of **3** to **8** would be expected to be ~ 2.1 based on fractionation factor analysis.(*32*) A value of 1.17 was calculated using the structures and frequencies obtained for **3** and **8**, using QUIVER. While the correct trend was observed, substantial error is present. A likely source of error arising from the model may be that the attacking water is incompletely solvated, having only one hydrogen bonding partner. Support for this idea was found when the contribution to the isotope effect for the hydrogen bonded hydrogen of the attacking water (atom # 7, Figure 3) was calculated. Fractionation factor analysis predicts a contribution of 1.45 and a value of 1.35 was calculated with QUIVER, reasonably good agreement. On the other hand, the "free" non hydrogen bonded hydrogen (atom # 6, Figure 3) of the attacking water should also have the same contribution, but calculation indicated an inverse value of 0.87, suggesting that this site is the culprit. Better modeling of the solvent isotope effect is anticipated when models with additional solvation are employed. Turning now to the α-carboxylate system, we reported that the experimental solvent deuterium isotope effect for capture of the N-acetyl neuraminyl oxocarbenium ion was $k_{H_2O}/k_{D_2O} = 1.1$.(*14*) Calculation of this kinetic isotope effect using the frequency data for models **4** and **6**, afforded a value of 1.03, in reasonably good agreement with the experimental results. This essentially unity kinetic isotope effect shows that the net vibrational environment of the deuteriums on water remains unchanged between the ion-molecule complex **4** and the transition state **6**. The structural implication is that proton transfer has not proceeded far if at all, and heavy atom motion is early at the transition state, since otherwise a normal secondary isotope effect approaching 2.1 would arise if C-O bond formation were significant.(*32*)

A relationship to glycosylase catalysis.

An old and unresolved question in the area of glycosyltransfer centers on the nature of intermediates which may arise during the hydrolysis and synthesis of the glycosidic bond.(*10,11*) The Phillips hypothesis(*33*) concerning the mechanism of the retaining glycosylase hen egg white lysozyme (HEWL) included the proposition that after cleavage of the glycosidic bond, an enzyme bound oxocarbenium ion was formed which ion-paired with the carboxylate of active site residue Asp 52 (Figure 6, upper branch). In this model, catalysis is dominated by active site electrostatics. This is strongly supported by the theoretical analysis of Warshel and coworkers, in which the calculated rate acceleration of HEWL relative to the solution hydrolysis was in excellent agreement with experiment.(*34,35*) Alternatively, a nucleophilic role for Asp52 (Figure 6, lower branch) has also been proposed.(*36*) The high reactivity found for alkyl and glycosyl oxocarbenium ions in solution would argue that such an ion pair would collapse rapidly to the acylal, or not form at all, such that the acylal would be reached in a concerted fashion from the bound reactant. Indeed, kinetically competent acylals have been isolated for other retaining glycosylases via use of fluorinated sugars with good leaving groups.(*37*) One may raise the point that fluorination may tip the balance in favor of the covalent path, since fluorinated oxocarbenium ions should be especially unstable. On the other hand, a contemporary crystallographic study of HEW lysozyme has lead to the suggestion that Asp 52 is too distant to form an acylal covalent intermediate without significant strain,(*38*) a point in favor of the ion-pair mechanism. New support for the chemical *feasibility* of glycosyl oxocarbenium ion /active site carboxylate ion pairs has been presented in this work. Both experimental and computational studies show that the NeuAc oxocarbenium ion exists as an intermediate in solution, despite the presence of the potentially nucleophilic carboxylate residue.

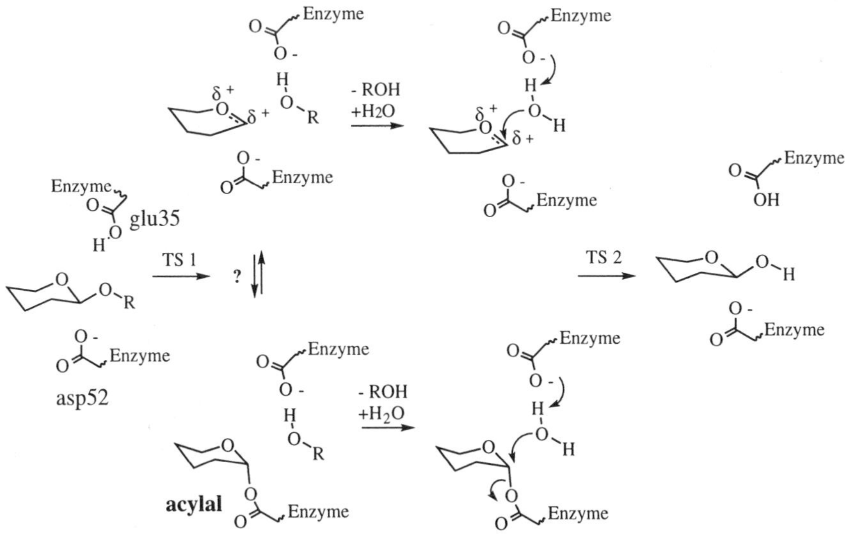

Figure 6. Lysozyme mechanism with respect to active-site Asp and Glu carboxyls. The substituents on the pyranosyl rings have been omitted for clarity.

Further, our results indicate that the NeuAc oxocarbenium ion self-catalyzes its capture by water via a strong transition state hydrogen bond. Extended to retaining glycosylases, the conjugate base of the general acid group (e.g. Glu35 in HEWL) may catalyze water addition to an enzyme bound glycosyl oxocarbenium ion in a similar way.

These speculative suggestions bear the possibility of computational analysis in the future. At least for the present an explicit *ab-initio* treatment of the enzyme is well out of reach. It is hoped however that carefully constructed model systems may allow comparative analyses of the upper and lower branches of the mechanistic

422

scheme set out in Figure 6. In particular, it would be of interest to characterize the potential energy surfaces for each branch and use this data to calculate the kinetic isotope effects, which could be compared to experiment. Thus far only α-H and ^{18}O leaving group kinetic isotope effect data have been reported for lysozyme, but not primary ^{14}C isotope effects at the anomeric carbon.(*39,40*) Carbon isotope effects at the anomeric carbon can be diagnostic for discrimination between the two paths shown in Figure 6.(*41,42*)

Conclusions.

Microsolvated transition states for capture of α–H and α–COO$^-$ substituted oxocarbenium ions by water have been identified at the RHF/6-31G(d,p) level of theory using an aqueous solvation model. The difference in barrier heights for capture of the α-H and α-COO- substituted oxocarbenium ions is calculated to be 1.9 kcal/mol, in good agreement with the estimated experimental values of 1.4 and 2.0 kcal/mol. While the electrostatics of placing a carboxylate group proximate to the oxocarbenium ion are highly favorable for its stabilization as an intermediate, the ability of the α-carboxylate group to hydrogen bond to an attacking water provides significant transition state stabilization which is worth an estimated 7.2-8.7 kcal/mol. Proton transfer to the carboxylate is highly asynchronous, occurring sometime after significant formation of the glycosidic bond. One implication is that the microscopic reverse, protonation of the glycosidic oxygen by an α-carboxyl group will be asynchronous and *early* with respect to glycosidic bond cleavage. Future studies of this system may then be useful to model general acid catalyzed glycoside hydrolysis.

Acknowledgments

The National Science Foundation is thanked for support of this research under CAREER award MCB-9501866. The Northeast Regional Data Center at the University of Florida is thanked for computer resources allocated under the Research Computing Initiative. Professor Martin Saunders is thanked for providing a copy of QUIVER.

References

1. Schauer, R. *Adv. Carbohydr. Chem. Biochem.* **1982**, *40* 131.
2. Varki, A. *Glycobiology* **1993**, *3*, 97.
3. Varki, A. *J. Biol. Chem.* **1993**, *268*, 16155.
4. Taylor, G. *Curr. Opin. Struc. Biol.* **1996**, *6*, 830.
5. Parr, I. & Horenstein, B.A. *J. Org. Chem.* **1997**, *62*, 7489.
6. Schenkman, S.; Eichinger, D.; Pereira, M.E.A.; and Nussenzweig, V. *Ann. Rev. Microbiol.* **1994**, *48*, 499.
7. Richard J. P. *Tetrahedron*, **1995**, *51*, 1535.
8. Amyes, T.L. and Jencks, W.P. *J. Am. Chem. Soc.* **1989**, *111*, 7888.
9. Banait, N.S. and Jencks, W.P. *J. Am. Chem. Soc.* **1991**, *113*, 7951.
10. Sinnott, M.L. *Chem. Rev.* **1990**, *90*, 1171.
11. Withers, S.G. *Pure App. Chem.* **1995**, *67*, 1673.
12. Ashwell, M.; Guo, X.; Sinnott, M.L. *J. Am. Chem. Soc.* **1992**, *114*, 10158.
13. Horenstein, B.A. and Bruner, M. *J. Am. Chem. Soc.* **1996**, *118*, 10371.
14. Horenstein, B.A. and Bruner, M. *J. Am. Chem. Soc.* **1998**, *120*, 1357.
15. Horenstein, B.A. *J. Am. Chem. Soc.* **1997**, *119*, 1101.
16. Huang, X.C, Surry, C., Hiebert. T., Bennet, A.J. *J. Am. Chem. Soc.* **1995**, *117*, 10614.
17. Onsager, L. *J. Am. Chem. Soc.* **1936**, *58*, 1486.
18. Wong, M.W.; Frisch, M.J.; Wiberg, K.B. *J. Am. Chem. Soc.* **1991**, *113* , 4776.
19. Gaussian 94, revision C.3, Frisch, M.J.; Trucks, G.W.; Schlegel, H.B.; Gill,

P.M.W.; Johnson, B.G.; Robb, M.A.; Cheeseman, J.R.; Keith, T.; Petersson, G.A.; Montgomery, J.A.; Raghavachari, K.; Al-Laham, M.A.; Zakrzewski, V.G.; Ortiz, J.V.; Foresman, J.B.; Cioslowski, J.; Stefanov, B.B.; Nanayakkara, A.; Challacombe, M.; Peng, C.Y.; Ayala, P.Y.; Chen, W.; Wong, M.W.; Andres, J.L.; Replogle, E.S.; Gomperts, R.; Martin, R.L.; Fox, D.J.; Binkley, J.S.; Defrees, D.J.; Baker, J.; Stewart, J.P.; Head-Gordon, M.; Gonzalez, C.; and Pople, J.A. Gaussian, Inc., Pittsburgh PA, 1995.

20. Miertus, S.; Scrocco, E.; Tomasi, J. *J. Chem. Phys.* **1981**, *55*, 117.
21. Saunders, M.; Laidig, K.E.; Wolfsberg, M. *J. Am. Chem. Soc.* **1989**, *111*, 8989.
22. Bigeleisen, J. and Mayer, M.G. *J. Chem. Phys.* **1947**, *15*, 261.
23. Becke, A.D. *J. Chem. Phys.* **1993**, *98*, 5648.
24. Lee, C.; Yang, W.; Parr, R.G. *Phys. Rev. B* **1988**, *37*, 785.
25. Warshel, A. and Chu, Z.T. in *Structure and Reactivity in Aqueous Solution, Characterization of Chemical and Biological Systems*; Cramer, C.J. and Truhlar, D.G Eds.; ACS Symp. Ser. 568; American Chemical Society: Washington D.C. 1994; pp 71-94.
26. Smallwood, C.J. and McAllister, M.A. *J. Am. Chem. Soc.* **1997**, *119*, 11277.
27. Scheiner, S. *Hydrogen Bonding, A Theoretical Perspective*; Oxford University Press: New York, 1997.
28. Richard, J.P. and Jencks, W.P. *J. Am. Chem. Soc.* **1984**, *106*, 1396.
29. Smith, B.J. *J. Am. Chem. Soc.* **1997** *119*, 2699.
30. Deslongchamps, P.; Dory, Y.L.; Li, S. *Can. J. Chem.* **1994**, *72*, 2021.
31. Cordes, E.H. and Bull, H.G. *Chem. Rev.* **1974** ,*74*, 581.
32. Schowen, K.B.J. in *Transition States of Biochemical Processes*, Gandour, R.D. and Schowen, R.L. Eds., Plenum Press: New York, 1978 pp. 225-283.
33. Phillips, D.C. *Proc. Royal. Acad. Sci.* **1967**, *57*, 484.
34. Warshel, A. *Proc. Natl. Acad. Sci. USA*, **1978**, *75*, 5250.
35. Warshel, A. and Weiss, R.M. *J. Am. Chem. Soc.* **1980**, *102*, 6218.
36. Koshland, D.E. *Biol. Rev.*, **1953**, *28*, 416.
37. Withers, S. G.and Street, I. P. *J. Am. Chem. Soc.* **1988**, *110*, 8551.
38. Strynadka, N.C.J.and James, M.N.G. *J. Mol. Biol.* **1991**, *220*, 401.
39. Dahlquist, F.W.; Rand-Meir, T.; Raftery, M.A. *Biochemistry*, **1969**, *8*, 4214.
40. Rosenberg, S. and Kirsch, J.F. *Biochemistry*, **1981**, 3196.
41. Schramm, V.L. in *Enzyme Mechanism from Isotope Effects*, ed. P.F. Cook, CRC Press: Boca Raton, 1991 pp 367-388.
42. Huang, X.; Tanaka, K.S.E.; Bennet, A.J. *J. Am. Chem. Soc.* **1997**, *119*, 11147.

Chapter 33

Molecular Dynamics and Quantum Chemical Study of Endonuclease V Catalytic Mechanism

M. Krauss[1], N. Luo[2], R. Nirmala[2], and R. Osman[2]

[1]Center for Advanced Research Biotechnology, National Institute of Science and Technology, Rockville, MD 20850
[2]Department of Physiology and Biophysics, Mt. Sinai School of Medicine, New York, NY 10029

Endonuclease V initiates repair of damaged DNA, that contains the thymine dimer, by cleavage of the glycosidic bond through the attack of an amine nucleophile. The transition state for this process is described using a series of model calculations that focus on the electronic characteristics that assist in stabilization of the transition state in the enzyme active site. The inherent geometrical and electronic features of the transition state are obtained in an *in vacuo* calculation which is then compared to situations where H-bond stabilization of the developing charge is included. The model of the endo V active site includes representations of the Glu-23 and Arg-26 residues. The guanidinium side chain of the arginine residue does not transfer a proton to the thymidine carbonyl in the developing anionic base even when optimizations are intitiated with a proton equidistant between the N of arginine and the O of the base. In the transition state structure, charge separation in the glycosidic bond does not significantly delocalize into the base or sugar, so the stabilization energy due to H-bonding is not large. The activation energy of the glycosidic cleavage catalyzed by a neutral amine is calculated to be about 30 kcal/mol.

1. Introduction

Environmental factors, such as high energy radiation, alkylating agents, and UV light, produce a spectrum of damaged DNA which may have severe biological consequences. DNA repair is therefore an essential component for the survival of a biological system. An important category of DNA repair is the multistep process that consists first of base excision followed by the disruption of the strand, whether by a β-elimination or the hydrolysis of the phosphodiester bond. The glycosylases that function in this category can be classified into monofunctional and multifunctional enzymes (1,2). Monofunctional glycosylases remove the damaged base and leave an abasic site (AP) (3). The AP site is then processed by another class of enzymes, the AP-endonucleases. In the multifunctional class of enzymes, the bacteriophage T4 endonuclease V (endo V) removes pyrimidine (thymine) dimers in a combined glycosylase/AP lyase activity. An imino enzyme-DNA intermediate results from the glycosylase step. This covalent attachment facilitates the catalytic β-elimination of the phosphodiester bond at the abasic site (4,5).

Two different nucleophiles have been observed attacking the C1' atom in the sugar, a hydroxyl anion and an amine (2). An example of hydroxyl anion attack is the mono-functional glycosylase, uracil-DNA glycosylase (UDG), which removes uracil from DNA (6-9). Water in the UDG active site may be activated by an Asp yielding the hydroxyl anion nucleophile. An example for the amine, is the bifunctional glycosylase/apyrimidinic (AP) lyase, T4 endonuclease V (endoV), which catalyzes the cleavage of the N-glycosyl bond and the disruption of the phosphodiester bond at the resultant apyrimidinic site. The availability of the crystal structures of the UDG (6-9) and endo V (10,11) enzymes in relevant DNA complexes presents an opportunity to address the question of kinetic selectivity on a fundamental molecular level. As in other enzymes, kinetic selectivity of DNA repair enzymes, as distinguished from a static selectivity of damage recognition, depends on the ability of the enzyme to lower the transition state for the rate determining processes. There have been a number of suggestions that, while the two enzymes differ in the choice of nucleophile, both enhance the hydrolysis of the glycosidic bond by an activating ionic hydrogen bond to the pyrimidine which either prepares the nucleotide for a nucleophilic attack on the sugar or stabilizes the resulting separation of charge during the reaction (2). Activating proton interactions can fundamentally alter or strongly polarize electronic structure and provide an element to the microscopic mechanism that an enzyme is uniquely constructed to deliver. However, while in UDG the reaction terminates at that point, in endoV the apyrimidinic site remains attached to the enzyme to facilitate the lyase step. From kinetic isotope effects, transition state structures have been deduced for hydrolysis of the glycosidic bond in nucleosides (12-14). The transition state is deduced to proceed with little participation of the nucleophile but these studies have considered purine nucleosides that may be initially protonated. This again raises the question of strong hydrogen bonding to the pyrimidine base in endoV or even a protonation of the base. Protonation of the thymidine carbonyl has been suggested for endo V (15). Breaking the glycosidic bond leads to an oxocarbonium cation electronic structure in the sugar and ultimately to the suggestion that substitution of the sugar by a pyrrolidine residue would act as a transition state analogue (16). The electronic character of the transition state in the glycosylase enzymes is required to determine the validity of these analogues. Mutant studies in endoV have established that the terminal amine and a Glu-23 are essential for catalysis but their precise roles in the two steps of the mechanism is not determined (5,15,17-19). The microscopic mechanism has not been determined either for other examples of multifunctional enzymes such as endonuclease III and formamidopyrimidine glycosylase (20). It is interesting to note, however, that prior protonation of a guanine is suggested to activate the glycosidic bond.

This note will focus on the glycosylase reaction in endonuclease V but will also examine the inherent characterisitcs of the transition state for the cleavage of the glycosidic bond by the two nucleophiles. The details of the microscopic mechanism for both classes are still not clear regarding the reaction path such as the characteristics of the transition state, the importance of activation of the glycosidic bond prior to or concurrent with nucleophilic attack, and the role played by the Glu-23 in both the glycosylase and lyase steps in endo V.

The ab initio quantum chemical determination of the transition states for the in vacuo reactions is straightforward. The *in vacuo* behavior of the reaction path is very different between the attack of the hydroxyl anion and the amine. It is well known that in the S_N2 reaction initiated by an anion *in vacuo*, the ionic hydrogen or multi-polar bond at long range is sufficiently strong to drive the reaction. In water, the solvation energy of the small hydroxyl anion is greater than that of the transition state that develops leading usually to a substantial activation energy. For the amine attack, there is no ionic interaction and the in vacuo activation energy is already very substantial because of charge separation along the reaction coordinate. Thus, a preliminary analysis of the amine reaction does not require the careful attention to the solvation of the nucleophile even to obtain qualitatively relevant results. At this time the inherent differences between the developing transition state for the hydroxyl and amine attacks will be noted but the hydroxyl reaction will not be followed into the enzyme environment. However, this will be done for the glycosidic cleavage reaction in endonuclease V. Models will be used to analyze the stabilizing influence of the cationic hydrogen bonds to the pyrimidine base but a minimal model of the reaction in endonuclease V will be presented at this time. A molecular dynamics simulation starting from the crystal structure of a mutant enzyme-substrate complex is used to reconstruct the wild type enzyme and define a quantum motif for a native enzyme model. Quantum chemical analysis of the reactive behavior at the active site will be investigated by incorporating effective fragment potentials (EFP) to represent those protein residues that are not directly involved in the chemistry but affect the reaction path through their electrostatic interactions or hydrogen bonding. Subsequent to cleavage of the glycosidic bond, a number of proton transfers is required for base product release as well as opening of the sugar ring. The nature of the residues involved in these steps is not clarifed by the crystal structures. Although this preliminary calculation will stop at the cleavage of the glycosidic bond, the final transition state structure should be relevant to subsequent behavior. The transition state structures will be compared to the experimental transition states that have been deduced from isotopic variations of the rates for various nucleosides.

Reaction Path Calculation: The Glycosylase Step Quantum Motif

In the multifunctional enzymes the role of the nucleophile is served by an amine, which is an integral part of the protein. In endoV the N-terminal amine plays the role of the nucleophile that attacks the C1' and forms and imino intermediate. The protonation state of the terminal amine is unclear. A self-consistent pK_a calculation (21) of endo V and the endo V-DNA complex determines that the pK_a of the amine terminus changes from 8.4 to 7.2 upon DNA binding. Thus, in the complex the nucleophile is about 50% neutral. In the present study we have assumed that the attacking amine terminus is neutral. In the enzyme-DNA structure derived from the MD simulation, Arg-26 is H-bonded to the O_2 carbonyl of thymine and Glu-23 is positioned in close proximity to the N-terminus. Mutations have shown Glu-23 to be essential for glycosylase activity . One interpretation emphasizes the stabilization of the positive charge of the imino intermediate while the other suggests that the negative charge of Glu-23 enhances the nucleophilicity of the amino terminus through a proton transfer mechanism. The importance of stabilization is difficult to reconcile with recent binding experiments demonstrating a positively charged pyrrolidine-based inhibitor does not discriminate between the wild type enzyme and the E23Q mutant (15).

Whether the proton transfer mechanism extends to the protonation of the base and activation of the glycosidic bond is also an important consideration in the quantum calculation of the active site reaction path. Only double stranded DNA is a substrate for endo V. The adenine complementary to the 5'-thymine of the dimer flips out of the stacking arrangement and is inserted into a pocket inside the protein. Arg-26 makes a specific hydrogen bond to the O2 position of the 5' thymine. This is unlikely by itself to be the protonating residue that activates the glycosidic bond, analogous to the His-286 in UDG (6), because the calculated pK_a of this arginine is 11.5. In addition, in vacuo ion-pairs involving protonated arginine have been observed suggesting the difficulty in transferring the proton (22). Although initial protonation has been suggested in the glycosylase step (15), the crystal structure does not obviously reveal the donating residue. Protonation could be concurrent from the terminal amine nucleophile but this amine is found to be on the opposite side of the 5'-thymine with respect to the sugar.

In order to obtain the structure of the enzyme environment around the thymine dimer, the enzyme mutant was restored to its native form by replacing Gln-23 with Glu-23. The relaxation of the structure of the complex with classical mechanics after the native protein is restored to Glu-23 is an imperative first step. The molecular dynamics (MD) simulation starts from the x-ray structure but transforms Gln-23 back to Glu-23. The complex is embedded in a periodic box of water with a total of 22650 atoms in the system. To equilibrate the water and ions, the system was heated to 600K while keeping the protein-DNA complex in a frozen conformation. A 200ps MD simulation was run at 600K, the system was minimized, reheated to 300K and the constraints on the solute were gradually relaxed over a 100ps time interval. At the end of this process a 700ps trajectory was run on the entire system. AMBER 4.1 was used for all MD calculations (23). The development of the active site H-bonding was examined both through snapshots along the trajectory and a statistical proximity analysis of a given donor or acceptor. For the purpose of constructing the minimal active site model, we note that Oϵ2 of Glu-23 interacts with the O4' of the sugar through a water about half the time. The water exchanges rapidly within the simulation time. Only one water is used in the active site but a network of waters may be involved. Oϵ2 also comes close to the nearby terminal amine. Oϵ1 of Glu-23 interacts with a water, that does not exchange over the simulation time, and with the backbone amide of Arg-3.

The minimal qauntum motif constructed from these simulations include, thymidine with a reduced C5-C6 bond to model the dimer and surrounded by abbreviated models for Arg-26, Glu-23, and the Thr-2 terminal amine. Three waters have been included in the model: two H-bonding to Arg-26 and one to Glu-23. This structure is depicted in fig.1a. The phosphates were found to be shielded by arginine and waters and will

A

B

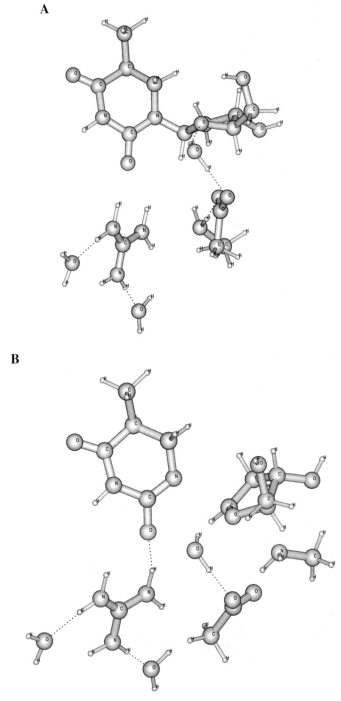

1. Conformations of endonuclease V active site model for: a) reactant, b) 'transition state', c) product.

C

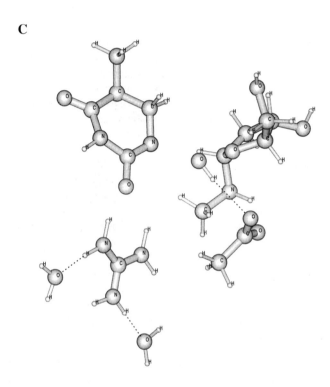

be neglected at this time since they do not play an essential role in the cleavage of the glycosidic bond.

Quantum Chemistry Methods

Optimization of the reactants and the cleavage transition state are all obtained at the RHF level using the GAMESS code (24) with effective core potentials and their concomitant double-zeta level orbitals (25). Since we are interested here in the glycosidic cleavage reaction, only the reactants and transition states are reported here. In vacuo transition states are obtained for both the attack of the hydroxyl anion, ammonia, and methyl amine. Exploration of the effect of hydrogen bonding on the base uses effective fragment potential (EFP) models of protonated and neutral methyl amine bound to one or both carbonyl oxygens on the pyrimidine base as seen in fig. 2. The EFP method is a critical feature of the theoretical model. It is based on the separation of the of the chemical system into two components, a quantum region (QR) and an EFP spectator region (SR). The total Hamiltonian for such a system is defined as the sum of the QR and SR hamiltonians plus an interaction term, V_{QR-SR},

$$H' = H_{QR} + H_{SR} + V_{QR-SR}$$
(1)

The QR is treated using traditional ab initio methods while the SR is replaced by effective potentials which simulate the quantum interaction between the QR and SR regions. The EFP allows a realistic treatment of the enzymatic reaction in the catalytic active site by representing the protein interactions in the quantum hamiltonian in a computationally tractable method (26,27). The EFP are implemented in the latest versions of the GAMESS code. The EFP accurately represent the electrostatic, polarization, exchange repulsion, and charge transfer effects for non-bonded interactions in the quantum hamiltonian. At the present time the all-electron chemically reacting region can be optimized in the field of the frozen and constraining EFP or protein environment. Gradient optimization of the quantum region within a fixed EFP environment is implemented in the code. The EFP environment represents the static effective dielectric in a very fine-grained manner. The non-bonded EFP have been shown to be accurate relative to all-electron calculations for the determination of the rotational barrier for an amide (28) and for the optimization of the excited states of formamide solvated by water (29). When the substrate binds the solvent is partially excluded and the reacting region is isolated from the solvent and essentially interior. The MD simulation shows that both the Glu-23 and Arg-26 residues are partially exposed in the enzyme (67% and 48%, respectively) and buried after the substrate binds (100% and 97%, respectively).

The electrostatic and polarization components of the EFP are generated by the GAMESS code producing distributed moments through the octupole for the electrostatic EFP (30) and through the dipole for the polarization EFP (31). The points are located at atom and bond mid-points and have been found to accurately represent the fields at hydrogen-bonded distances. The exchange repulsion (ER) and charge transfer (CT) terms are distinct interactions and can be well estimated by the restricted variational space (RVS) methodology (32). The exchange repulsion and charge transfer have similar dependence on overlap between the interacting moieties which allow them to be modeled by the same function. An RVS analysis is used to

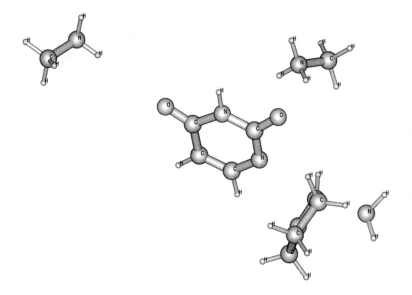

2. Model of H-bonding to 5-methyl, 6-hydro thymidine carbonyls using lysine EFP representation.

determine these terms which can be compared well to the difference between the total interaction energy and the sum of the electrostatic and polarizability terms calculated with the EFP. Utility programs are used to fit the exchange repulsion and charge transfer as well as the charge penetration correction to the electrostatic interaction. The repulsive potentials for the pair interactions likely to arise in protein calculations have been generated and are used in the EFP for the Lys, Arg, and Glu.

Analysis of Nucleophilic Transition States

Nucleophilic attack of a hydoxyl anion was considered in two cases, the neutral and protonated nucleoside. The phosphate was deleted since its charge would prevent the reaction without the protein shielding. The 3' phosphate was first clipped at the oxygen bound to the sugar leaving a 3'OH. The protonated nucleoside behaves differently from the neutral and this is evident in the electronic structure of the reactant molecules. Gradient optimization of the neutral and the protonated uracil nucleoside shows that upon protonation, the glycosidic bond length is increased by 0.04Å and the bond order decreases by 20%. Nucleophilic attack of OH⁻ on the C1' in the cation nucleoside occurs without a barrier yielding two neutral species with the base in a higher energy tautomer where the proton is located on O2. .

For *in vacuo* nucleophilic attack of an hydroxyl anion on the neutral nucleoside, the sugar 3'OH was reduced to a hydrogen for simplicity. As seen in fig.3a, at the TS the system separates into three charged components with the hydroxyl oxygen, O_h, 2.26Å from C1' and in turn C1' is separated from N1 on uracil by 2.02Å with an angle of 168 from O-C1'-N. This is an S_N2 transition state which is informative on the electronic character of the sugar and the base as well as their mutual geometric arrangement. The mode behavior of the principle negative eigenvalue shows the concerted motion of the leaving group and nucleophile with respect to the C1' in the nucleoside characteristic of an S_N2 reaction. The sugar cation has many of the characteristics that have been found in analyzing isotopic rate data (12-14). The calculated C1'-O4' bond distance is 1.36Å which is shorter than the 1.43Å calculated in the nucleoside reactant but larger than the 1.25Å calculated for the isolated sugar cation. In the isolated cation, the bond order is 1.5 but only 1.1 in the TS. The C1'-O4' distance has been deduced to be 1.30Å for a TS in a nucleoside hydrolase (12). Stabilization of the TS with aspartate residues has been suggested by either hydrogen-bonding to the ribosyl hydroxyls or interaction with the oxocarbonium cation. However, the atomic charge populations deduced from the charge density find that although the ether oxygen is less negative in the TS, it is still appreciably negatively charged with a Mulliken population of -0.38. Any stabilization by an asparatate residue would have to be through a water or metal cation intermediate. The cationic charge is localized on the C1' which is coupled to the localized negative charge of N1 on the uracil.

The bond orders reflect the geometries. A small bond order of 0.32 is found for the bond intiated between O and C1' while the cleaved glycosidic bond is reduced to 0.09. The bond order for C1'-O4' is only slightly increased to 1.02 from the 0.94 in the isolated nucleoside. The charge localization on N1 is reflected in the bond orders of 1.87 and 1.86, respectively, for the C2O and C4O carbonyl bonds of uracil. There is little charge transferred to the carbonyl bonds and the bond orders of N1-C2 and N1-C6 alter only slightly from the isolated nucleoside.

Using ammonia as the nucleophile in a simple in vacuo model of the endoV glycosylase step determines a transition state initially with the ribose characterized as an oxocarbonium cation with the C1'-O4' distance of 1.31Å. The separation and geometry between nucleophile, sugar, and base are somewhat different from the hydroxyl anion with the ammonia adopting a tetrahedral conformation consonant with its developing cationic character. This leads to a smaller angle of 134 for N-C1'-N. The N-C1' distance is 2.23Å similar to that in the anion but the C1'-N distance to the base of 2.77Å is longer. However, a second negative eigenvalue of the Hessian persists with the uracil anion slowly rotating so the carbonyl oxygen can form a long-range electrostatic interaction with the hydrogen on the developing 'ammonium cation'. This interaction is reduced with a methyl amine nucleophile as seen in fig.3b but the overall geometry of the transition state does not alter much. Another very small negative eigenvalue persists because of the weak interaction between the hydrogens on the methyl group with the carbonyl. Both the N-C1' and C1'-N distances increase to 2.32Å and 2.79Å, respectively. Even adding two water EFP to screen the O2 does not sufficiently shield this interaction even though the methyl group is kept about 0.5Å farther away and the N-C1' and C1'-N distances decrease slightly. The TS geometries and bond orders are summarized in Tables 1 and 2, respectively, for the amine nucleophile. This weak interaction results in a large activation energy of 46 and 39 kcal/mol, respectively for ammonia and methyl amine. Even the slight polarization in methyl amine has an appreciable effect in this calculation. However, these large enthalpic activation energies very much exceed the probable activation energy of about 20 kcal/mol estimated from the observed rate (33).

The large activation energy reflects the small interaction between the neutral amine nucleophile and the sugar. While the bond order for the glycosidic bond is decreased below 0.05 signifying almost complete breakage, the N-C1' bond order is 0.19 in vacuo and 0.28 when the base is H-bonded by a protonated lysine EFP on C2O and a neutral lysine EFP on C4O. Stabilization of the charge developing on the base is often noted but the localization of the charge at N1 warns us that the ionicity of the carbonyl bonds may not change sufficiently to alter the activation energy appreciably.

A cationic arginine residue H-bonds to the C2O of the thymine in the active site of endoV. In the next step of modelling the possibility of H-bond stabilization of the cleavage with the ammonia nucleophile, a cationic H-bond interacts with the C2O carbonyl of uracil using a protonated lysine EFP as described in fig. 2. As can be seen in Table 1 the change from the in vacuo transition state is modest. The nucleophile approaches C1' a little closer and the C2O bond has stretched due to the interaction with the ionic H-bond. Although the interaction has not substantially altered the localization of the charge at N1 in the base, it does yield a modest reduction in the activation energy of about 8 kcal/mol. This is due to the non-linear H-bond that resulted after optimization. The proton donates into the maximum of the oxygen electron density which would be in a ring of about 30° away from linear. Nonetheless, the activation energy is still very high. Adding a neutral lysine EFP to C4O also yields a reduction in the activation energy of 8 kcal/mol. The neutral lysine interaction does not contribute as the distance between N-C4 increases to 5.49Å during optimization while the N-C2 value goes to 2.72Å compared to 2.80Å when there is

A

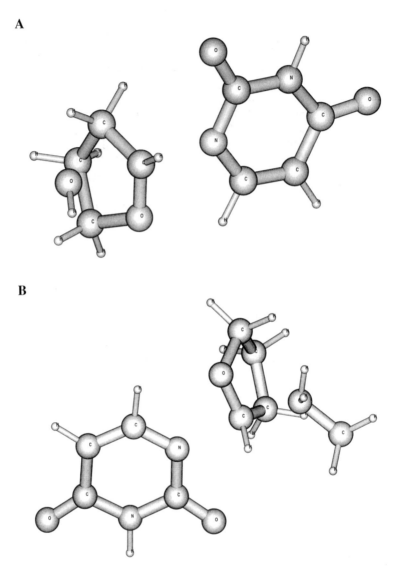

B

3. Conformations of *in vacuo* transition states for the nucleophiles: a) hydroxyl anion, b) methyl amine.

Table 1. Transition State Geometries.

	N-C1'	C1'-N1	C1'-O4	N1-C2	N1-C6	C2-O	C4-O	H-C2O	H-C4O
1.	2.234	2.769	1.308	1.360	1.374	1.279	1.249		
2.	2.317	2.792	1.306	1.360	1.373	1.278	1.248		
3.	2.016	2.689	1.326	1.338	1.390	1.295	1.240	1.811	
4.	2.085	2.652	1.319	1.342	1.386	1.293	1.243	1.966	3.253
5.	2.935	2.551	1.290	1.315	1.468	1.290	1.242	1.978	

1. in vacuo ammonia, all geometries in Å.
2. in vacuo methyl amine
3. lysp EFP interacting with C2O
4. lysp, lys EFP interacting with C2O and C4O, respectively
5. model of endo V active space with arg and glu efp

Table 2. Bond Orders in Bonds Formed or Altered by Glycosidic Cleavage

	N-C1'	C1'-N1	C1'-O4'	N1-C2	N1-C6	C2-O	C4-O
1.	0.68	0.94	1.05	0.98	1.91	1.94	
2.	0.19	<0.05	1.20	1.34	1.19	1.66	1.89
3.	0.13	0.20	0.94	1.18	1.07	1.85	1.89
4.	0.28	<0.05	1.15	1.39	1.11	1.60	1.91
5.	<0.05	<0.05	1.27	1.64	0.91	1.57	1.93

1. *in vacuo* nucleoside
2. *in vacuo* ammonia nucleophile
3. *in vacuo* methyl amine nucleophile
4. lysp EFP on C2O carbonyl
5. endo V model with 5,6 hydro, methyl thymidine

only the cationic H-bond on C2O. Only the cationic H-bond contributes to the reduction in the activation energy.

The preliminary model for the endo V active site includes the essential Arg and Glu and three waters determined in an MD simulation. The EFP for these residue models are fixed in the relative positions found in the MD snapshot at 700 psec with respect to the nucleotide elements to which they bind. The Glu EFP is fixed relative to the sugar to which it H-binds while the arg EFP is H-bound to the thymine carbonyl C2O. This allows relative motion when the glycosidic bond is broken. For this calculation there is a number of negative eigenvalues reflecting the substantial interaction between the Glu and the amine nucleophile. Although the RMS gradient fell to below 0.001 when the distance between the amine nitrogen and C1' was about 2.6Å, it gradually increased as the attraction of the Glu to the amine increases this distance. The calculation was stopped at a point where the RMS gradient again fell below 0.001 and remained for at least three iterations. The resulting geometry is shown in fig.1b and described in Tables 1 and 2. The amine nucleophile is now found to be almost completely unbound from the sugar. The weak interaction is now overwhelmed by the presence of the Glu which has an electrostatic attraction to the amine that culminates in this geometry in an H-bond between an amine hydrogen and the Oϵ1 of the Glu. This lead to the product in an appropriate position for the Glu to remove a proton and yield the Schiff base intermediate. The partially saturated thymine behaves very similar to the uracil in other respects with regard to the development of a partial oxocarbonium cation and the localization of the charge on the N1 position.

Conclusion

The in vacuo transition state conformations are in reasonable agreement with stationary states obtained by modeling in elements of the protein environment. Several features are worth emphasizing for the case of the amine nucleophile which is the primary concern of this paper. First, the neutral amine interaction with the sugar is weak and is partially disrupted by other, competitive interactions. This is particularly true when the arginine binding to the thymine and the presence of Glu-23 is modeled into the calculation. Second, the glycosidic bond is essentially broken in the TS. Since the bond is broken but not replaced by a comparable bond elsewhere, the activation energy for attack of the neutral nucleophile is considerably larger than the expected value for the experimental reaction. This suggests that a proton transfer step may be required to lower the barrier. Stabilization by ionic hydrogen bonds to the base have often been proposed but only half the necessary reduction of the barrier is calculated in the model. Third, the separation of charge in the cleavage of the glycosidic bond is localized both in the sugar and base. The localization of charge in the sugar yields the short C1'-O4' bond that is characteristic of the oxocarbonium cation described by the analysis of isotopic data in a number of studies. Additional substantial change in the barrier could result from protonation of the base or direct H-bonding to the separating charge. Examination of the MD active site does not suggest that either of these possibilities is likely. The H-bond to C2O is with an arginine which is very unlikely to transfer a proton even in vacuum but is found to be solvated partially in the active site. An optimization was performed in the transition state conformation where the proton in the H-bond was placed equidistant between the arginine nitrogen and the thymidine oxygen of the anionic base which found that it preferred the arginine. There is no other likely residue to donate a proton to the base. The calculation shows a decrease of the terminal amine pK_a upon substrate binding. We have initially examined the behavior of the neutral system since there is no direct path to move the amine proton to the base. However, the final conformation of the product in the active site (fig.1c) finds the Glu-23 within H-bonding distance of the the amine. In addition to stabilizing the developing chargeon the amine, this conformation is also appropriate to initiate proton transfer from the amine, even perhaps in the course of the nucleophilic attack. This possibility is now under study. In addition, the derivation of a potential energy surface for the initial step in the glycosylase step will be used to conduct a molecular dynamics simulation to estimate the free energy barrier for this reaction.

Acknowledgement: We thank S.Worthington and H.Gilson for their assistance in the construction of the effective fragment potentials. The figures were prepared using Molden3.2 which was obtained from G.Schaftenaar, CAOS/CAMM Center Nijmegen, The Netherlands 1996. This work was supported in part by PHS grant CA 63317 (R.Osman).

References

1. Lloyd,R.S.; Van Houten,B., in "DNA Repair Mechanisms: Impact on Human Disease and Cancer", Voss,J.M.H., ed., R.G.Landes Co., Austin,TX, **1995**; pp25-66.

2. Krokan,H.E.; Standal,R.; Slupphaug,G. Biochem.J. **1997**,325,1.

3. Sun,B.; Latham,K.A.; Dodson,M.L.; Lloyd,R.S., J.Biol.Chem. **1995**,19501-19508.

4. Manoharan,M.; Mazumder,A.; Ranson,S.C.; Gerlt,J.A.; Bolton,P.H., J.Am.Chem.Soc. **1988**,110,2690-2691.

5. Dodson,M.L.; Schrock,R.D.III; Lloyd,R.S., Biochemistry **1993**,32,8284.

6. Savva,R.; McAuley-Hecht,K.; Brown,T.; Pearl,L., Nature **1995**,373,487.

7. Mol,C.D.; Arvai,A.S.; Slupphaug,G.; Kavil,B.; Alseth,I.; Krokan,H.E.; Tainer,J.A, Cell **1995**,80,869.

8. Mol,C.D.; Arvai,A.S.; Sanderson,R.J.; Slupphaug,G.; Kavil,B.; Krokan,H.E.; Mosbaugh,D.W.; Tainer,J.A., Cell **1995**,82,701.

9. Slupphaug,G.; Mol,C.D.; Kavil,B.; Arvai,A.S.; Krokan,H.E.; Tainer,J.A., Nature **1996**,384,87.

10. Morikawa,K.; Ariyoshi,M.; Vassylyev,D.G.; Matsumoto,O.; Katayanagi,K.; Ohtsuka,E., J.Mol.Biol. **1995**,249,360.

11. Vassylyev,D.G.; Kashiwagi,T.; Mikami,Y.; Ariyoshi,M.; Iwai,S.; Ohtsuka,E.; Morikawa,K., Cell **1995**,83,773.

12. Horenstein,B.A.; Parkin,D.W.; Estupinan,B.; Schramm,V.L., Biochemistry **1991**,30,10788.

13. Horenstein,B.A.; Schramm,V.L. Biochemistry **1993**,32,7089.

14. Parkin,D.W.; Schramm,V.L. Biochemistry **1995**,34,13961.

15. Schrock,R.D.,III; Lloyd,R.S., J.Biol.Chem. **1991**,266,17631.

16. McCullough,A.K.; Scharer,O.; Verdine,G.L.; Lloyd,R.S., J.Biol.Chem.**1996**,271,32147.

438

17. Doi,T.; Recktenwald,A.; Karaki,Y.; Kikuchi,M.; Morikawa,K.; Ikehara,M.; Inaoka,T.; Hori,N.; Ohtsuka,E., Proc.Natl.Acad.Sci.U.S.A. **1992**,89,9420.

18. Manuel,R.C.; Latham,K.A..; Dodson,M.L.; Lloyd,R.S., J.Biol.Chem. **1995**,270,2652.

19. Iwai,S.; Maeda,M.; Shirai,M.; Shimada,Y.; Osafune,T.; Murata,T.; Ohtsuka,E., Biochemistry **1995**,34,4601.

20. Mazumder,A.; Gerlt,J.A.; Absalon,M.J.; Stubbe,J.; Cunningham,R.P.; Withka,J.; Bolton,P.H., Biochemistry **1991**,30,1119-112.

21. Mehler,E.L. J.Phys.Chem. **1996**,100,16006.

22. Price,W.D.; Jockusch,R.A.; Williams,E.R. J.Am.Chem.Soc.**1997**,119,11988.

23. Pearlman,D.A.; Case,D.A.; Caldwell,J.C.; Ross,W.S.; Cheatham III,T.E.; Ferguson,D.M.; Seibel,G.L.; Chandra Singh,U.; Weiner,P.; Kollman,P.A., AMBER 4.1, University of California, San Francisco, CA,**1995**.

24. Schmidt,M.W.; Baldridge,K.K.; Boatz,J.A.; Elbert,S.T.; Gordon,M.S.; Jensen,J.H.; Kokeski,S.; Matsunaga,N.; Nguyen,K.A.; Su,S.; Windus,T.L.; Dupuis,M.; Montgomery,J.A. General atomic and molecular electronic structure system, GAMESS, J.Comput.Chem. **1993**,14,1347.

25. Stevens,W.J.; Basch,H.; Krauss,M. J.Chem.Phys. **1984**,81,6026.

26. Jensen,J.H.; Day,P.N.; Gordon.M.S.; Basch,H.; Cohen,D.; Garmer,D.R.; Krauss,M.; Stevens,W.J., "An effective fragment method for modeling intermolecular hydrogen-bonding effects on quantum mechanical calculations", in *Modeling the Hydrogen Bond*, ed. Douglas A.Smith, ACS Symposium Series 569, **1994**, p.139.

27. Day,P.N.; Jensen,J.H.; Gordon,M.S.; Webb,S.P.; Stevens,W.J.; Krauss,M.; Garmer,D.R.; Basch,H.; Cohen,D., J.Chem.Phys. **1996**,105,1968.

28. Chen,W.; Gordon,M. J.Chem.Phys.**1996**,105,11081.

29. Krauss,M.; Webb,S.P., J.Chem.Phys. **1997**,107,5771.

30. Stone,A.J.; Alderton,M., Mol.Phys. **1985**,56,1047-1064.

31. Garmer,D.R.; Stevens,W.J., J.Phys.Chem. **1989**,93,8263-8270.

32. Stevens,W.J.; Fink,W.H. , Chem.Phys.Lett. **1987**,139,15.

33. Nyaga,S.G.; Dodson,M.L.; Lloyd,R.S. Biochemistry **1997**,36,4080.

Chapter 34

Charge Transfer Interactions in Biology: A New View of the Protein–Water Interface

Gautham Nadig[1], Laura C. Van Zant[2], Steve L. Dixon[2], and Kenneth M. Merz, Jr.[1]

[1]152, Davey Laboratory, Department of Chemistry, Pennsylvania State University, University Park, PA 16802
[2]Computational Chemistry Laboratory, Terrapin Technologies, Inc., 750 Gateway Boulevard, South San Francisco, CA 94080

In this paper we employ linear-scaling quantum mechanical methodologies to carry out the first fully quantum mechanical calculation on a protein/water system (~5,000 atoms total). These calculations demonstrate for the first time that the superposition of a number of small charge transfer interactions at the protein/water interface results in a substantial transfer of charge from the protein surface to the surrounding solvent. Furthermore, we show that the charge transfer interaction is a significant contributor to the overall interaction energy in hydrogen bonding complexes - even more so than the closely related polarization interaction. Finally, we discuss the theoretical and experimental ramifications of the charge transfer interaction for biomolecules in aqueous solution.

The exact nature of electrostatic interactions in biological systems has been of intense interest for a number of years.(1-4) The fundamental aspects of the electrostatic interaction between two molecules can be apportioned as shown in Figure 1. In the classic electrostatic interaction there is an interaction between the occupied molecular orbitals (MOs) which does not result in the mixing or exchange of electrons between the two molecular species. This interaction typically dominates the total interaction between two molecules as they come together to form a complex (see discussion below). Thus, it is not surprising that significant effort has focused on developing reliable point charge representations of biomolecules(5,6) that represent the classic electrostatic interaction that can then be used in both explicit solvent(2) and implicit solvent models.(1) The exchange repulsion interaction represents the interaction between the occupied MOs of two molecules and represents the repulsive interactions arising from the exchange and delocalization of electrons between these two systems. This is typically a large interaction and in closed-shell complexes (e.g., the water dimer) is usually unfavorable. This interaction is generally treated using Lennard-Jones type terms in classical force fields and in most force fields it is the balance between the electrostatic and exchange repulsion (plus dispersion) terms that governs interactions between two molecules (e.g., hydrogen bonding). In recent years the polarization interaction has become of much greater interest, because of the realization that in order to obtain an accurate representation for condensed phase systems like liquid water, *etc.* it is critical to account for polarization which can be 10-20% of the total interaction energy.(*7*) The polarization effect can be visualized as the intramolecular reorganization of electrons through the mixing of occupied and unoccupied molecular orbitals (see Figure 1). In terms of an atomic point charge representation the polarization interaction manifests itself through a redistribution of the electrons within a pair of molecules as they form a complex while at the same time maintaining the net charge on the molecules in question.

Related to the polarization interaction is the charge transfer (CT) interaction. However, instead of an intramolecular redistribution of electrons an intermolecular mixing occurs between the occupied and virtual orbitals of the two systems that are in close contact (see Figure 1). This interaction has two net effects: (1) it stabilizes (or destabilizes) complex formation - this affects the total interaction energy and (2) it results in the net partial transfer of charge (*i.e.*, electrons) from one complexing molecule to the other. Little work has focused on this interaction in protein systems, but recently it has been suggested that it plays a modest role in explaining so-called cation-π interactions.(*3,8,9*) This is surprising because it is known that the magnitude of the CT interaction energy was in many cases twice that of the polarization energy.(*10,11*) It also has been generally assumed that the net result of the CT interaction (*i.e.*, a net transfer of charge) in most cases is very small (a few hundredths of an electron) and that its overall effect on the charge distribution of a molecule is small. This, indeed, turns out to be the case for two small molecules interacting with one another (*e.g.*, the water dimer).

Another aspect which has made it difficult to determine the importance of the CT interaction in a macromolecular system is the need to carry out fully quantum mechanical calculations on systems containing thousands of atoms. Using standard strategies it is only possible to carry out semiempirical and *ab initio* calculations on systems up to ~1000 and ~100 atoms, respectively. However, in recent years solutions to solving the coulomb bottleneck in *ab initio* calculations(*12,13*) and on the matrix diagonalization problem in semiempirical theory(*14,15*) now makes it possible to study large molecular systems using quantum mechanical methodologies. Thus, the calculations presented herein represent the first time that fully quantum mechanical studies on a small protein embedded in explicit solvent have been performed.

The major cold shock protein of *E.Coli*, CspA , a small hydrophilic protein with 69 amino acid residues was used as a model system in our study. We chose to carry out calculations on this system due to its relatively small size and its near neutral charge under physiological conditions. The crystal structure at 2.45 Å resolution was used as the initial model of CspA.(*16*) Since under normal physiological conditions proteins exhibit rich conformational dynamics which play an essential role in protein function(*17,18*), we decided to study the dynamics of CspA by performing molecular dynamics simulations on a protein/water system using the AMBER(*5*) force field with the TIP3P(*19*) water model and the SANDER(*20*) MD module.(*21*) From a 500ps sampling phase we extracted 100 coordinate sets that we then carried out single point semiempirical (PM3(*22*)) SCF calculations using the DivCon program.(*23*) Each single point PM3 calculation required ~8 hours of computer time on a SGI Origin 200 workstation. Thus, these are still expensive calculations, but in the absence of our linear-scaling quantum mechanical code(*23*) these calculations would of been impossible to carry out in a reasonable amount of time. We studied CspA in two charged states, one in which the system has a unit negative charge and the other has a net neutral charge.(*24*) An *in vacuo* simulation of CspA was also performed to serve as a control. Figure 2 depicts the total charge on the protein over the period of the simulation. The charges presented are Coulson charges(*25*) and are not electrostatic potential fit(*26*) charges. ESP charges are better at reproducing the multipolar (*e.g.*, dipole, quadrupole, *etc.*) characteristics of a molecule than are Coulson charges, but it

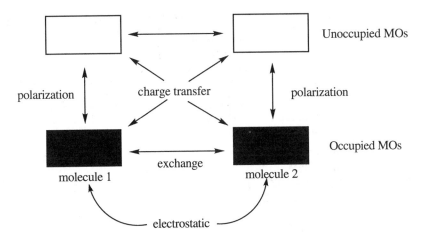

Figure 1. Molecular orbital interactions between two molecules upon complexation that give rise to the electrostatic, exchange repulsion, polarization and charge transfer interactions.

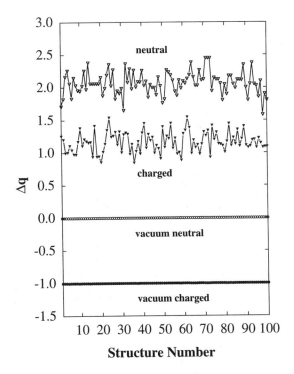

Figure 2. Variation of the net charge on the CspA protein in the vacuum and in aqueous solution. In all cases on going from vacuum to the aqueous phase results in an increase of the charge on the protein by ~2 units of charge in the positive direction.

is impossible to use the ESP fitting procedure on systems as large as those studied herein.(26) As expected the net charge in vacuo was integral in all cases (*i.e.*, -1 or 0). When we calculate the protein charge in the presence of the water molecules contained in the system we found that there was a significant amount of charge transferred from the protein surface to the surrounding solvent. The average total charge on the protein was 1.167 with a 13 % fluctuation for the negatively charged CspA system, and 2.07 with an 8 % fluctuation for the neutral case. Thus, while the net charge of the system consisting of the protein and water was conserved, we found that the protein transfers roughly two (2) units of charge to the solvent regardless of the initial charge state of the protein.

Table I. The Average Charge ($<\Delta q>$) Transferred to Water by Charged/Polar Residues of CspA.

Residue	$<\Delta q>$
Ser 1	+0.35
Lys 3	+0.05
Lys 9	+0.07
Asn 12	+0.01
Asp 14	-0.21
Lys 15	+0.05
Asp 23	-0.22
Asp 24	-0.19
Lys 27	+0.07
Asp 28	-0.21
Gln 37	+0.06
Asn 38	+0.02
Asp 39	-0.18
Lys 42	+0.06
Asp 45	-0.21
Glu 46	-0.22
Gln 48	+0.03
Glu 55	-0.22
Lys 59	+0.06
Asn 65	-0.02
Leu 69	-0.47

a) A "+" sign indicates that the residue is accepting charge (*i.e.*, electrons) from its surroundings, while a "-" sign indicates that this residue is donating charge (*i.e.*, electrons) to its surroundings.

In order to determine if specific groups of atoms or if all atoms contribute equally to the observed charge transfer we analyzed the contribution of each amino acid residue to the overall CT. We were able to track down several residues that were the main contributors to the observed transfer of charge (see Table I). We found, albeit not surprisingly, that the atoms involved are the polar and charged residues, Lys, Asp and Glu. Thus, for example the Lys residues each accept ~0.05e of charge, while the Glu/Asp residues each transfer about 0.2e of charge to the surrounding solvent environment. Ser 1 is on the N-terminus of CspA and is positively charged so it accepts a large (0.35e) amount of charge, while Leu 69, which is the C-terminus, is negatively charged and donates a large amount of charge (0.47e) to the surrounding environment. The remaining residues within CspA transfer or accept ~0.05e or less.

Thus, in terms of the transfer of charge the Asp/Glu and the C-terminal protein are the most important overall (*i.e.*, the carboxylate group), while positively charged groups like Lys tend to slightly counterbalance the net transfer of charge. We note that CspA does not have an Arg residue so we are unable to comment on how much charge it might transfer to solvent.

The observation that charge is transferred between the protein/water interface is interesting, but it does not address what is the magnitude of the charge transfer interaction energy. In order to estimate the strength of this interaction we have made use of the Morokuma decomposition method(*10*) as implemented by GAMESS(*27*) and the results of these calculations on a series of hydrogen bonded complexes are presented in Table II. These systems were chosen since they are representative of the types of interactions present at the protein/water interface. The calculations could only be carried out at the 6-31G level of theory due to convergence problems at higher levels of theory. Thus, while the absolute energies may be too high, we expect that the relative energy ordering for the various interaction energies will not change significantly. Furthermore, it has been shown that the Morokuma methodology can be unstable when very large basis sets (*i.e.*, much larger than the 6-31G basis set) are used in calculations of this type.(*11*) The electrostatic portion of the interaction in all cases is the leading contributor to formation of a stable hydrogen bonded complex. The exchange interaction is destabilizing, while the polarization interaction is ~10% of the total interaction energy in all cases as has been suggested previously.(*7*) The CT interaction accounts for ~20% of the total interaction energy in all cases. Thus, the CT interaction is more important than the polarization interaction by about ~10% of the total interaction energy.

Table II. Decomposition of the Total Interaction Energy (ΔE_{tot}) into Polarization (ΔE_{pol}), Charge Transfer (ΔE_{CT}), Exchange Repulsion (ΔE_{ex}) and Electrostatic (ΔE_{el}) contributions.[a]

System	ΔE_{pol}	ΔE_{CT}	ΔE_{ex}	ΔE_{el}	ΔE_{totals}
Methyl formamide-water	-1.36	-2.84	7.40	-14.11	-11.45
Acetate anion-water	-2.53	-5.18	16.34	-32.53	-25.61
Methyl amine-water	-0.46	-0.76	2.29	-3.81	-2.85
Methyl ammonium-water	-3.50	-4.71	14.72	-31.26	-24.63
Water -Water	-0.80	-1.70	6.31	-11.13	-7.49

a) All energies in kcal/mol.

To insure that the observed transfer of charge is not dependent on the use of a semiempirical Hamiltonian we used a higher level of theory (*e.g.*, the 6-311G** *ab initio* basis set) to calculate the amount of charge transferred for the selection of molecules given in Table II. Importantly, we observe the transfer of charge in both the *ab initio* and semiempirical cases (see Table III). Most critically, we find that the PM3 Coulson derived charges consistently underestimate the amount of charge transferred suggesting that the Coulson charges obtained in our PM3 calculations on the CspA/water system result in an underestimation of the amount of charge transferred. The configurations used to calculate the Coulson charges for the CspA/water system were obtained from an MD simulation using the AMBER force field model. Thus, the configurations used in the PM3 calculations were not derived using forces calculated from the PM3 Hamiltonian. The question then arises what effect differences in PM3 and AMBER derived geometries might have on the observed CT. Short of carrying out a fully quantum mechanical PM3 MD simulation on the protein/water system the best way to address this is to determine what the observed net transfer of charge is when the quantum mechanical calculations are carried out at the optimized geometry obtained using the AMBER force field or PM3. These are presented in Table IV. From these calculations we observe that the transfer of charge on average is not extremely sensitive to whether the AMBER or PM3 geometry is used for the determination of the Coulson charges. This is also consistent with the observation that the fluctuation in the net transfer of charge is relatively small over one hundred separate configurations (see Figure 2). Thus, we conclude that neither the method employed nor the geometry used has a large enough effect on our calculated results for the CspA/water system to invalidate our observation of significant transfer of charge at the protein/water interface.

Table III. Calculated Charge Transfer to a Water Molecule as Determined Using *ab initio* and the Semiempirical PM3 Hamiltonian.

System[a]	6-311G**[b]	PM3[c]	PM3[d]
Acetate anion-water	-0.067	-0.062	-0.035
Methyl amide-water	-0.085	-0.049	-0.013
Methyl amine-water	-0.023	-0.058	+0.001
Methyl ammonium-water	+0.006	-0.033	+0.031
Water-Water	-0.006	-0.031	+0.006

a) The HF/6-311G** optimized geometry was used in all cases.
b) Calculated using electrostatic potential (ESP) fitting methods.
c) Calculated using ESP fitting methods.
d) Coulson charges.

Table IV. Calculated Charge Transfer to a Water Molecule as Determined Using the Semiempirical PM3 Hamiltonian.

System	PM3[a]	PM3[b]
Acetate anion-water	-0.072	-0.055
Methyl amide-water	-0.026	-0.028
Methyl amine-water	+0.001	+0.003
Methyl ammonium-water	+0.030	+0.006
Water-Water	-0.020	-0.015

a) Coulson charges determined at the PM3 optimized geometry.
b) Coulson charges determined at the AMBER optimized geometry.

In order to understand the ramifications of our observations we need to consider both the CT interaction (a component of the total interaction energy) and the result of this interaction which is the actual transfer of charge. In force field methodologies the CT interaction energy is embedded into the model itself through a suitable selection of the atomic point charges used in the determination of the electrostatic component of the force field. This is similar to what is done in the case of the polarization interaction unless one adds in the polarization effect explicitly.(28) Much effort has been focused on inclusion of polarization interactions in classical models(28-30) and given that the CT interaction is as important as the latter interaction it is clear that more effort should be directed towards the development of models that incorporate CT interactions.

The end result of a CT interaction is the actual transfer of charge.(*31*) This results in a significant alteration of the charge distribution within a molecule and, hence, effects the electrostatic interactions within a molecule. In force field methods the charge distribution is typically generated such that neutral units are assembled (*e.g.*, by residue or by functional unit). However, our results indicate that this may not be an accurate representation. We have focused on the transfer of charge between a protein surface and the surrounding solvent, but clearly CT interactions within a protein, for example, also result in the net redistribution of charge. The latter effect is not typically accounted for in classical force field models. Our observations also have ramifications for continuum electrostatic models like Poisson-Boltzmann methodologies.(*1*) In these methods the atomic point charge model used ensures that the formal charge on a protein (or other biomolecule) is retained, while our results suggest that this is not an entirely accurate description. These methods do not explicitly incorporate CT, but rather (as in force field methods) it is effectively included via the approximations inherent in the method itself.

Charge transfer is observed in many experimental systems (*e.g.*, photochemical generation of CT complexes(*32*), *etc.*), but these are typically smaller systems and not large biomolecules. Charge transfer salts like tetrathiofulvalene-tetracyanoquinodimethanide (TTF-TCNQ) have been experimentally observed to transfer 0.6e of charge from TTF to TCNQ.(*33*) Computationally, CT effects have been observed, for example, to be very important in metal complexes in that specific water molecules are required to accurately represent the experimental absorption spectra of a metal ligand complex.(*34*) However, due to the highly polarizable nature of the previous two examples charge transfer was not unexpected, while it is more of a surprise at the water/protein interface. Experimental verification for our results could be done using crystallography, but very high resolution structures are needed to get good quality charge distributions(*35,36*) and protein structures are not typically determined at very high resolution. Small molecule systems are and it should be possible to observe CT in crystals containing small molecules and water. For example, a recent article demonstrating the accurate experimental determination of charge density provides a methodology that could determine the amount of charge transfer in small peptide crystals.(*37*) Indeed, the DL-proline·H$_2$O system they looked at might of revealed charge transfer effects, but the authors refined the experimental data with the constraint that no charge transfer between the water molecule and the amino acid was possible. Clearly, there are experimental systems that undergo CT and, more importantly, there are ways to assess if CT is present in proteins.

In summary, we have carried out the first fully quantum mechanical calculations on an explicit protein/water system (~5000 atoms total) using the semiempirical PM3 Hamiltonian. From the calculations presented herein we have arrived at two significant conclusions: (1) We have found that charge transfer interactions account for ~20% of the total interaction energy involved in hydrogen bonded complexes and this is ~10% higher than the effect polarization has on calculated interaction energies. While the latter observation supports previous conclusions(*10,11*) we more importantly (2) observe that CT interactions result in the net transfer of charge from the surface of a protein to the surrounding solvent and that this transfer of charge can be quite substantial. Given these observations we conclude that CT interactions as well as the transfer of charge could have a significant impact on both our experimental and theoretical understanding of biomolecules in aqueous solution.

Acknowledgments

This work has been supported by the DOE (DE-FGO2-96ER62270) and the NIH (GM44974). Generous support from the Pittsburgh Supercomputer Center and the

446

Cornell Theory Center through a MetaCenter grant is also acknowledged. Fruitful discussions with Greg Farber and Steve Scheiner are also acknowledged.

Literature Cited

(1) Honig, B.; Nicholls, A. *Science (Wash.)* **1995**, *268*, 1144-1149.
(2) Warshel, A.; Åqvist, J. *Annu. Rev. Biophys. Biphys. Chem.* **1991**, *20*, 267-298.
(3) Dougherty, D. A. *Science (Wash.)* **1996**, *271*, 163-167.
(4) Burley, S. K.; Petsko, G. A. *Adv. Protein Chem.* **1988**, *39*, 125-189.
(5) Cornell, W. D.; Cieplak, P.; Bayly, C. I.; Gould, I. R.; Merz, K. M., Jr.; Ferguson, D. M.; Spellmeyer, D. C.; Fox, T.; Caldwell, J. W.; Kollman, P. A. *J. Am. Chem. Soc.* **1995**, *117*, 5179-5197.
(6) Storer, J. W.; Giesen, D. J.; Cramer, C. J.; Truhlar, D. G. *J. Comput.-Aided Mol. Design* **1995**, *9*, 87-110.
(7) Gao, J.; Xia, X. *Science* **1992**, *258*, 631-635.
(8) Kumpf, R. A.; Dougherty, D. A. *Science (Wash.)* **1993**, *261*, 1708-1710.
(9) Caldwell, J. W.; Kollman, P. A. *J. Am. Chem. Soc.* **1995**, *117*, 4177-4178.
(10) Kitaura, K.; Morokuma, K. *Int. J. Quant. Chem.* **1976**, *10*, 325-340.
(11) Cybulski, S. M.; Scheiner, S. *Chem. Phys. Lett.* **1990**, *166*, 57-64.
(12) White, C. A.; Johnson, B. G.; Gill, P. M. W.; Head-Gordon, M. *Chem. Phys. Lett.* **1994**, *230*, 8-16.
(13) Strain, M. C.; Scuseria, G. E.; Frisch, M. J. *Science (Wash.)* **1996**, *271*, 51-53.
(14) Dixon, S. L.; Merz, K. M., Jr. *J. Chem. Phys.* **1996**, *104*, 6643-6649.
(15) Lee, T.-S.; York, D. M.; Yang, W. *J. Chem. Phys.* **1996**, *105*, 2744-2750.
(16) Schindelin, H.; Jiang, W.; Inouye, M.; Heinemann, U. *Proc. Nat. Acad. Sci.* **1994**, *91*, 5119.
(17) Brooks, C. L., III; Karplus, M.; Pettitt, B. M. *Proteins: A Theoretical Perspective of Dynamics, Structure, and Thermodynamics*; John Wiley and Sons: New York, 1988; Vol. LXXI.
(18) McCammon, J. A.; Harvey, S. C. *Dynamics of Proteins and Nucleic Acids*; Cambridge University Press: New York, 1987.
(19) Jorgensen, W. L.; Chandrasekhar, J.; Madura, J.; Impey, R. W.; Klein, M. L. *J. Chem. Phys.* **1983**, *79*, 926.
(20) Pearlman, D. A.; Case, D. A.; Caldwell, J. C.; Seibel, G. L.; Singh, U. C.; Weiner, P.; Kollman, P. A. ; University of California, San Francisco, 1991.
(21) The coordinates for CspA were obtained from the Protein Data Bank (pdb entry pdb1mjc.ent) and the protein was placed at the center of a pre-equilibrated box of TIP3P Monte Carlo equilibrated water molecules. The box was truncated so as to include a 14 Å. thick solvent layer from the protein surface. The AMBER all atom force field with parm94.dat parameter set was used to model the protein. A uniform dielectric constant of ε=1 was used. Molecular dynamics simulations in the NTP ensemble consisted of a initial 100 steps of energy minimization using the steepest descent algorithm, followed by 575 picoseconds of dynamics with a 1 femtosecond timestep, in which the initial 75 ps was considered the equilibration period and the remaining 500 ps was the production run. All through out the

simulations, we used a 15 Å non-bonded cut -off radius and a 25 time step update frequency of the non bonded pairlist. Coordinates were saved every 5 ps for future analysis.

(22)   Stewart, J. J. P. *J. Comp. Chem.* **1989**, *10*, 209-220.

(23)   Dixon, S. L.; Merz, K. M., Jr. *J. Chem. Phys.* **1997**, *107*, 879-893.

(24)   The unit negative charge was neutralized by protonating a surface Histidine residue and not through the addition of a counterion.

(25)   Pople, J. A.; Beveridge, D. L. *Approximate Molecular Orbital Theory*; McGraw-Hill: New York, 1970.

(26)   Besler, B. H.; Merz, K. M. J.; Kollman, P. A. *J. Comput. Chem.* **1990**, *11*, 431-439.

(27)   Frisch, M. J.; Baldridge, K. K.; Boatz, J. A.; Elbert, J. A.; Elbert, S. T.; Gordon, M. S.; Jensen, J. H.; Koseki, S.; Matsunaga, N.; Nyugen, K. A.; Su, S. J.; Windus, T. L.; Dupuis, M.; Montgomery, J. A. *J. Comput. Chem.* **1993**, *14*, 1347.

(28)   Allen, M. P.; Tildesley, D. J. *Computer Simulation of Liquids*; Clarendon Press: Oxford, 1987.

(29)   Dang, L. X.; Rice, J. E.; Caldwell, J.; Kollman, P. A. *J. Am. Chem. Soc.* **1991**, *113*, 2481.

(30)   Sprik, M.; Klein, M. L. *J. Chem. Phys.* **1988**, *89*, 7556-7560.

(31)   This is not always true. In some cases the amount of charge transferred between molecules 1 and 2 (see Figure 1) are equivalent yielding no net transfer of charge.

(32)   Turro, N. J. *Modern Molecular Photochemistry*; The Benjamin/Cummings Publishing Company, Inc.: Menlo Park, California, 1978.

(33)   Coppens, P. *X-Ray Charge Densities and Chemical Bonding*; Oxford University Press: Oxford, 1997.

(34)   Stavrev, K. K.; Zerner, M. C.; Meyer, T. J. *J. Am. Chem. Soc.* **1995**, *117*, 8684-8685.

(35)   Pearlman, D. A.; Kim, S.-H. *J. Mol. Biol.* **1990**, *211*, 171-187.

(36)   Pearlman, D. A.; Kim, S.-H. *Biopolymers* **1985**, *24*, 327-357.

(37)   Koritsanszky, T.; Flaig, R.; Zobel, D.; Krane, H.-G.; Morgenroth, W.; Luger, P. *Science (Wash.)* **1998**, *279*, 356-358.

Chapter 35

Modeling the Citrate Synthase Reaction: QM/MM and Small Model Calculations

Adrian J. Mulholland[1] and W. Graham Richards[2]

[1]School of Chemistry, University of Bristol, Bristol BS8 1TS, United Kingdom
[2]New Chemistry Laboratory, Oxford University, Oxford OX1 3QT, United Kingdom

In this chapter, we review calculations on the mechanism of the enzyme citrate synthase. Transition state and stable intermediate structures have been optimized for small models of the reaction, at semiempirical and *ab initio* levels. The reaction in the enzyme has been studied with combined quantum mechanical/molecular mechanical (QM/MM) methods. The first step of the reaction (deprotonation of acetyl-CoA by Asp-375), and the resulting nucleophilic intermediate, have been examined in detail. The results indicate that the enolate of acetyl-CoA is the likely intermediate, and that it is stabilized by normal hydrogen bonds from His-274 and a water molecule. The results do not support the proposal that a 'low-barrier' hydrogen bond stabilizes this intermediate in citrate synthase.

Achieving a deeper understanding of enzyme catalytic processes is a problem of great practical and fundamental significance. Computer simulations can make an important contribution by providing a description of enzyme mechanism at the molecular level (*1-3*). Calculations can be used to study unstable species, to evaluate possible alternative mechanisms, and to calculate energetic contributions to catalysis. These are central considerations in an enzyme-catalyzed reaction, which are difficult to address by experiment alone. A first step in the modeling process is the investigation of possible reaction intermediates and transition states (TSs), to identify basic features of the potential energy surface governing reaction. One approach is to perform 'supermolecule' calculations on clusters of small molecular fragments representing functional groups of the enzyme and substrate (*2,4*) by standard quantum chemical techniques. TS and stable complex structures can then be optimized. By necessity, only a small portion of the enzyme can be treated (although recent advances allow single point semiempirical calculations on small proteins (*5*)).

© 1999 American Chemical Society

The surrounding protein and solvent is likely to have a significant effect on the reaction and should be represented. Secondly, from a practical point of view, it is often difficult to perform geometry optimizations in such calculations, because it can be difficult to apply constraints which realistically mimic the covalent and non-bonded interactions at the active site but at the same time are flexible enough to allow optimization. Optimization without constraints can lead to the fragments drifting in ways which would be impossible in the confines of the active site, producing complexes which are not relevant to the reaction in the enzyme. One way to overcome these problems is to use combined quantum mechanical/molecular mechanical (QM/MM) methods, which have been the focus of much recent research (6-10). The essence of the QM/MM approach is that a small region is treated quantum mechanically, and is coupled to a simpler molecular mechanics description of the remainder of the system (11). This allows the reaction in the enzyme to be treated while including the effects of the protein environment (1), and TS structures can be optimized (12). QM/MM calculations have given useful insight into a number of enzyme reactions (13-18). Most applications to date have been at the semiempirical molecular orbital (MO) level of QM treatment because of the demands of higher level calculations. Semiempirical methods offer the advantage of being highly computationally efficient, which allows molecular dynamics simulations to be performed (19,20), and the free energy profile to be calculated (21). They are, however, also subject to errors in some cases, making it important to test their reliability for a given application (22).

Citrate Synthase

Citrate synthase catalyzes the first step in the citric acid cycle, namely the formation of citrate from acetyl-CoA and oxaloacetate (23,24). Rate-limiting (23,25) deprotonation of acetyl-CoA forms a nucleophilic intermediate (26-28), which subsequently attacks the carbonyl carbon of oxaloacetate. The carbonyl group of oxaloacetate is polarized at the active site (29,30), which may assist the condensation. Citryl-CoA is thought to be formed as an intermediate by this step (31). Hydrolysis of the thioester bond then allows release of the products. Crystallographic and mutagenesis experiments have identified Asp-375 as the catalytic base (Figure 1) for deprotonation of acetyl-CoA (32,33); His-274 also interacts with this substrate (34) and plays a catalytic role (35) (numbering for pig citrate synthase). This mechanism appears to be conserved across all (S-)citrate synthases (36). It has been uncertain whether the nucleophilic intermediate is the enolate or enol of acetyl-CoA (24), or alternatively an 'enolic' form, sharing a proton with His-274 in a 'low-barrier' hydrogen bond (37,38). This is an important question, relating directly to the mechanisms used by this and other enzymes to stabilize reaction intermediates and attain rapid reaction rates (17). It has been suggested that the enolate is too unstable to be consistent with the observed rate of reaction, unless it is significantly stabilized by the enzyme (39), leading to the proposal that concerted acid-base catalysis to form the enol was more probable (32,33). However, the lack of an effective general acid at the active site (His-274 is neutral) means that the enol form of acetyl-CoA appears to be less stable (17).

Citrate synthase is not dependent on metal ions for catalysis (23), and so the stabilization of any intermediate must be due to interactions with the protein itself. Similar problems of instability of reaction intermediates exist for a variety of enzymes, and it has been suggested that the necessary stabilization in general is provided by a hydrogen bond between a charged intermediate and a neutral general acid at the active site (His-274 in citrate synthase). The hydrogen bond was proposed to take on the special character of a low-barrier hydrogen bond in the intermediate complex, that is to say the hydrogen bonded proton would become effectively shared between the bonded partners due to a small or non-existent barrier to proton transfer. According to this proposal (37,38), such a bond is expected to be of considerably higher energy than a conventional hydrogen bond, and therefore to stabilize the high energy intermediate. Low-barrier hydrogen bonds have been put forward as a mechanism for stabilizing intermediates in many enzymes (37,38,40), but this proposal has been controversial (41-46).

The metabolic importance of citrate synthase has led to the characterization of the enzyme from a number of organisms by a wide variety of biochemical techniques (23,30,32,36). The availability of these data make the enzyme an attractive system to study by simulation techniques. In this chapter we review calculations on citrate synthase, including recent *ab initio* QM/MM studies (47), and show how they have contributed to our understanding of catalysis by this important enzyme.

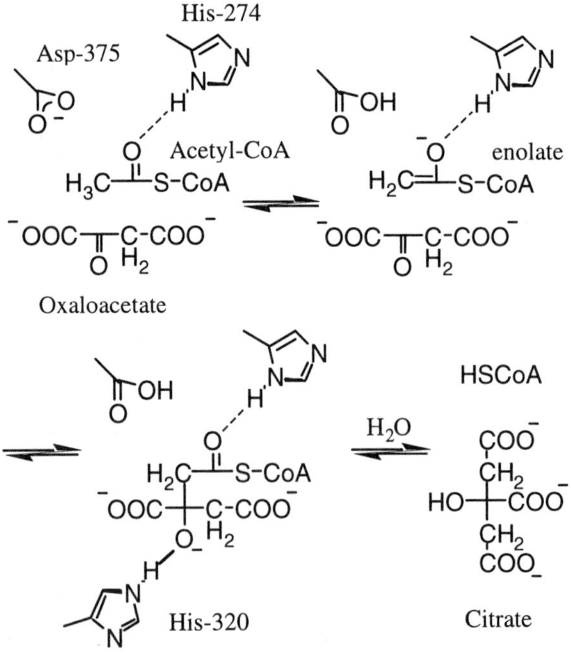

Figure 1. The reaction mechanism of citrate synthase. Acetyl-CoA is deprotonated by Asp-375, giving the enolate. The enolate of acetyl-CoA is stabilized by hydrogen bonds from His-274 (and a water molecule), and is the nucleophile for attack on the second substrate, oxaloacetate.

Calculations on Small Models of the Citrate Synthase Reaction

We have performed calculations on the first stage of the citrate synthase reaction, on the second substrate, oxaloacetate, and on a simple model of the condensation reaction. The focus of these calculations was to model the nucleophilic intermediate produced by the initial (rate-limiting) step, to examine whether the enolate, enol or enolic form of acetyl-CoA is the likely intermediate, and how it is stabilized by the enzyme.

Acetyl-CoA Enolization. We have studied small models of the first step of the citrate synthase reaction (*4*), to test the accuracy of semiempirical methods for this step, and to provide models of the TS and stable complexes for use in the QM/MM calculations. The AM1 Hamiltonian (*48*) was chosen for the semiempirical calculations (see below), using sulfur parameters (*49*) which have been found to perform well (*50*). Before beginning a study of a reaction employing semiempirical molecular orbital methods, it is important to test their accuracy for the system of interest by comparison with high-level *ab initio* and/or experimental results. The sidechain of His-274 was represented by (5-)methylimidazole, as in the protein it is believed to be neutral, bearing a proton on ND1 (*17,24*). Asp-375 was represented by acetate. AM1 gives a deprotonation enthalpy of 347.7 kcal/mol for 5-methyl-imidazole (using $\Delta H_f(H^+)=367.2$ kcal/mol), compared to an experimental value of 352.9 kcal/mol for the more stable 4-methyl isomer (*51*). For imidazole, the AM1 deprotonation enthalpy (347.6 kcal/mol) is also too low compared to experiment (352.5 kcal/mol (*51*)), although close to the MP2/6-31+G(*d*)//6-31+G(*d*) *ab initio* value (*52*). AM1 overestimates the proton affinity of acetate (354.8 kcal/mol, AM1; versus 352.0, MP3/6-311++G(*d,p*)//6-31G(*d*) (*53*); 349.0, experiment (*51*)).

Acetyl-CoA was represented by the thioester methylthioacetate (CH_3SCOCH_3) in the model calculations. Thioesters are more acidic than oxygen esters, which is believed to be a factor in their selection in biochemical reactions (*54*). AM1 gives too small a deprotonation enthalpy for methylthioacetate (360.2 kcal/mol) compared to MP3/6-31+G(3*df*,2*p*)//MP2/6-31+G(*d*) ($\Delta H=366.3$ kcal/mol, including RHF/6-31+G(*d*) vibrational and scaled zero-point energy corrections), although it is close to the MP2/6-31+G(*d*) result ($\Delta H=361.4$ kcal/mol, (*55*)). The AM1 deprotonation enthalpy of the enol of methylthioacetate is 346.8 kcal/mol, 1.8 kcal/mol higher than the MP3/6-31+G(3*df*,2*p*)//MP2/6-31+G(*d*) value ($\Delta H=345.0$ kcal/mol). AM1 performs better than PM3 for these reactions. The AM1 tautomerization energy for conversion of the keto form of methylthioacetate (13.4 kcal/mol) to the enol is considerably lower than *ab initio* findings ($\Delta H=21.3$ kcal/mol ($\Delta E=21.4$ kcal/mol), MP3/6-31+G(3*df*,2*p*)//MP2/6-31+G(*d*)). AM1 and *ab initio* electrostatic potential fitted charges are similar for all three forms of the thioester, indicating that the electronic distribution of the thioester is treated well by the semiempirical method (*55*). Importantly, AM1 shows the same increase of negative charge on the carbonyl oxygen and the sulfur atom due to deprotonation of methylthioacetate. The increase of negative charge on the sulfur is larger, indicating that charge acceptance by this atom may stabilize the thioester enolate (*55*).

Optimization of Transition State Structures. The TS for proton abstraction from methylthioacetate (representing acetyl-CoA) by acetate (representing Asp-375) has been optimized at the AM1 level, and subsequently at the RHF/6-31+G(*d*) level (*4*). The initial geometries were taken from a representative high-resolution crystal structure (*34*). Semiempirical TSs for this model, (and for proton transfer in a model of the hydrogen bond between His-274 and acetyl-CoA, and both reactions in a larger model, see below) were located using the SADDLE method (*56*) in the MOPAC 6.0 program (*57*). These approximate TSs were refined by gradient norm minimization using NLLSQ or SIGMA, and finally elimination of unwanted additional imaginary frequencies by minimizing along these modes in dynamic reaction coordinate (DRC) calculations. Normal mode analysis then ascertained that the TS geometries were first order saddle points (i.e. points with one imaginary frequency). We found the SADDLE procedure useful for location of approximate TSs for proton transfer while preventing unrealistic 'drifting' of the components of the complexes in ways which would not be possible in the active site. In this technique, the superimposed geometries of the reactants and products are taken as starting points and gradually moved towards each other in turn along a connecting vector. At each point one geometry is optimized while constrained to remain the same distance from the other structure, and the procedure is repeated until the structures approach to within a specified distance of one another, and should serve as an approximate TS. For each model, the products of a reaction step were built initially using the same heavy atom positions as for the reactant species by the addition of hydrogen atoms in the appropriate positions. This approach is suitable for proton transfer reactions in which large conformational changes do not accompany reaction. The AM1 TSs were good starting points for *ab initio* TS optimizations for models A and B. *Ab initio* force constants were used to begin higher level geometry optimizations, first at the RHF/6-31G(*d*) level, and then RHF/6-31+G(*d*). The geometries of the reactant and product complexes for each step produced by AM1 intrinsic reaction coordinate (IRC) calculations were also optimized at semiempirical and *ab initio* levels.

The products of the proton abstraction (the thioester enolate and acetic acid) are calculated to be highly unstable compared to the reactant complex at all the *ab initio* levels. In fact, when correlation energy is included at the MP2 level with either basis set, or when zero-point corrections are included, the 'TS' structure on the potential energy surface lies lower in energy than the product complex. The 'TS' is 14.8 kcal/mol higher than the reactant complex, and the product complex 17.0 kcal/mol higher, at the MP2/6-31+G(*d*)//6-31+G(*d*) level (*4*). This emphasizes the requirement for stabilization of the enolate if it is to be an intermediate in the citrate synthase reaction. A similar situation was found for the hydrogen bonded complex of 5-methylimidazole and the enolate of methylthioacetate ($^-CH_2COSCH_3$), a model of the hydrogen bonded intermediate. In this case, TSs for proton transfer from the imidazole ring to the enolate oxygen were optimized at the RHF level with both basis sets, but again proved to be lower in energy than the 'product' complex of the enol and methylimidazolate when correlation or zero-point corrections were applied. It is highly unfavorable to transfer a proton from methylimidazole to the thioester enolate. The only stable complex at the *ab initio* levels tested was that of the enolate accepting a normal hydrogen bond.

Clearly the enolate must be stabilized if it is to be a reaction intermediate, and calculations on the reaction should include those groups responsible for stabilizing it, in particular the sidechain of His-274. It was not possible to perform *ab initio* TS searches for a model including all three components, because of the computational demands of force constant calculation for this comparatively large system. TS optimization was possible with AM1 in a model including the sidechains of Asp-375 and His-274 (represented as in the bimolecular complexes by acetate and 5-methylimidazole) and methylthioacetate to represent acetyl-CoA (*4*). TSs were found by the methods described above for both proton abstraction by acetate from methylthioacetate, and for proton transfer to the resulting thioester enolate oxygen from methylimidazole.

The influence of the hydrogen bond from methylimidazole on the structure of the TS for the first reaction step (abstraction of a proton from 'acetyl-CoA' by 'Asp-375' to form the thioester enolate) can be seen by comparing the AM1 TS structures with and without the hydrogen bond donor present (Figure 2). The hydrogen bond stabilizes the enolate product more than the keto form, as it is stronger with the negatively charged enolate. In line with expectations from the Hammond postulate, the hydrogen bonding group makes the TS earlier, i.e. more like the reactants (acetate and methylthioacetate). The separation of the heavy atoms which exchange the proton (O of acetate and C of methylthioacetate) is little changed (increased from 2.61Å to 2.63Å on inclusion of methylimidazole), but the C···H and O···H distances are similar to one another for the TS including the hydrogen bond donor (C···H= 1.32Å, O···H=1.34Å) unlike the TS including only acetate and methylthioacetate (C···H=1.37Å, O···H=1.26Å). The hydrogen bond from methylimidazole is calculated by AM1 to stabilize the enolate by 5.3 kcal/mol relative to the keto(substrate) form of the thioester. The barrier to the reaction is reduced by only 0.2 kcal/mol, from 11.9 kcal/mol to 11.7 kcal/mol. These effects are likely to be underestimated, because AM1 gives erroneously low energies for strong hydrogen bonds (*50*); the AM1 energy of the hydrogen bond between methylimidazole and the thioester enolate (15.9 kcal/mol) is too low (-ΔH = 25.1 kcal/mol, MP2/6-31+G(*d*)/6-31+G(*d*) (*4*)). The difference between the AM1 hydrogen bond energy in the bimolecular complex and the AM1 stabilization calculated for the same interaction in the reacting system is due to interactions with acetate/acetic acid in the reacting system. This shows that it is important to consider the reacting system as a whole to evaluate the contribution of the hydrogen bond, which cannot be found from hydrogen bond energies a for bimolecular complex alone. The relative energies of the stable complexes calculated by AM1 for the three component system are reasonable, because of a cancellation of errors: the underestimation of the increase in hydrogen bond energy is similar to the overestimation of the stability of the (acetic acid and thioester enolate) products (*4*).

The AM1 energy change for the transfer of a proton from methylimidazole to the enolate oxygen in the bimolecular model (8.4 kcal/mol) is close to *ab initio* findings (6.9 kcal/mol, MP2/6-31+G(d)//6-31+G(*d*), (*4*)). The AM1 barrier for this process is too high, but again the indications are that this method can give reasonable results for the relative energies of the keto, enolate and enol forms of acetyl-CoA in models of the citrate synthase reaction. For the largest model, AM1 gives the keto form as most stable, followed by the enolate, with the enol highest in energy. PM3 was found to give some hydrogen bond energies better, but described the TS for

454

conversion of the enolate to the enol poorly. Most importantly, PM3 gives an incorrect charge distribution for the imidazole ring in the keto, enolate and TS complexes compared to *ab initio* charges (*4*), due to an error in the PM3 nitrogen parameters (*58*). It is vital that the interactions between the QM and MM systems are treated accurately, requiring that the charge distribution of the QM system is reasonable. This was not the case for the PM3 treatment of the methylimidazole/acetyl-CoA (keto or enolate) system, and so AM1 was chosen for QM/MM calculations on the reaction. The incorrect treatment of the charge distribution of N-H groups may limit the usefulness of PM3 for QM/MM calculations on enzyme reactions.

Oxaloacetate and the condensation step of the reaction. Citrate synthase binds oxaloacetate in a compact 'citrate-like' conformation (*59*). In particular, the α-keto-carboxylate moiety is observed bound approximately planar, interacting with arginine and other residues at the active site (*33*). The planar conformation is of higher energy than the gas-phase minimum, which has the carboxylate perpendicular to the adjacent carbonyl group. The AM1 barrier for rotation of the carboxylate (2.4 kcal/mol) is very close to the MP2/6-31+G(*d*) result (2.3 kcal/mol), with the maximum in both cases at an O-C-C-O dihedral angle of approximately 0° (*59*). It is notable that *ab initio* calculations without diffuse functions failed to give the correct geometry and gave an erroneously low rotational barrier. This is a reminder of the importance of diffuse functions in *ab initio* calculations for some properties of anionic systems.

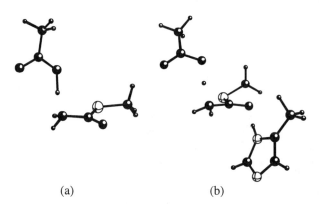

(a)   (b)

Figure 2. AM1 TS structures for proton abstraction by acetate (representing Asp-375) from methylthioacetate (representing acetyl-CoA) in (a) a model containing only these two molecules; and (b) a model also containing 5-methylimidazole, representing His-274. The hydrogen bond from the imidazole ring stabilizes the thioester enolate product of the reaction, resulting in an earlier TS in (b) (see text). (Adapted from ref. 4).

It proved difficult to optimize a TS structure for models of the condensation step of the reaction (i.e. the nucleophilic attack of the enolate or enol of acetyl-CoA on the carbonyl carbon of oxaloacetate). For example, oxaloacetate and the enolate are both negatively charged, and so repel one another. A TS structure was optimized using acetone as to represent the carbonyl group which is the target of the nucleophilic attack, and a simplified thioester in which the methyl group of methylthioacetate bonded to sulfur was replaced by a hydrogen atom. This model is not ideal, but TS optimization including the methyl group encountered difficulties due to rotation of the methyl group. The TS structure was first optimized by semiempirical methods (AM1, PM3 and MNDO). Again, the AM1 TS was a good starting point for *ab initio* optimizations, at the RHF/6-31+G(*d*) and MP2/6-31+G(*d*) levels (*28*). The AM1 barrier to the reverse reaction (decomposition of the addition product) is 3.5 kcal/mol, comparable to the MP2/6-31+G(*d*) barrier (4.8 kcal/mol, (*28*)). The barrier to the addition reaction from the fully optimized reactant complex is dominated by the need to overcome the favorable complexation energy of the ion-acetone complex, which is not relevant to the enzyme reaction. Compared to the separated reactants, the TS is 8.9 kcal/mol higher in energy (AM1). An equivalent model of the addition of the enol to acetone was also optimized at the AM1 level, and the barrier to addition was considerably larger (33.9 kcal/mol). As expected, the neutral enol form is a much weaker nucleophile than the charged enolate. For the reaction in the enzyme to proceed efficiently, it appears that the intermediate must have at least a high degree of enolate character. If the enol of acetyl-CoA were formed instead, it would react only very slowly and would slow the overall rate of production of citrate. These results are for a model system, and a full investigation will require the reaction in the enzyme to be studied. Given good models of the TS and product for the addition process, this can be accomplished by QM/MM techniques. The restraints on the reacting groups due to the protein can be included by empirical force field terms. Also, the QM region would be affected by the atomic partial charges of the MM surroundings, and so the polarization of the oxaloacetate carbonyl when bound to the enzyme (which may assist the condensation step (*30*)) can be represented.

QM/MM Calculations on Acetyl-CoA Enolization in Citrate Synthase

QM/MM calculations on the enolization of citrate synthase have been carried out at the AM1 level of QM treatment to examine the effect of the enzyme on the energetics of the reaction (*1,17*). The structure of chicken citrate synthase with acetyl-CoA and *R*-malate bound (*34*) was used, with the necessary small modifications to convert *R*-malate to oxaloacetate. Two sizes of system were studied, including all residues with one or more atoms within a specified radius of the terminal methyl carbon of acetyl-CoA: radii of 17Å and 20Å were employed. Each of these systems included all the residues identified as important for binding or catalysis. Atoms within 14Å of the center of the simulation zone were not restrained, while those further from the center were restrained or constrained to their crystallographically determined positions. It was found that His-274 and His-320 are probably neutral, singly protonated on ND1 and NE2 respectively, whereas the protonation state of His-320 (which binds oxaloacetate) was uncertain. In the 17Å system this residue was treated as positively charged, and in the 20Å system, calculations were performed with His-320 either positively charged or neutral. Waters observed in the crystal structure were included, and tests indicated that addition of further water at the active site was not required. Details of the calculations and the parameters used have been published previously (*17*). The simulation package CHARMM (*11,60*) was used throughout.

The sidechains of Asp-375, His-274 and the thioester portion of acetyl-CoA were treated quantum mechanically by AM1, with all other atoms treated by MM. Initially the energy of the citrate synthase-substrate complex was minimized fully. The conversion of the keto (substrate) form of acetyl-CoA to the enolate by proton abstraction by Asp-375, and of the resulting enolate to the enol by donation of a proton from His-274 were studied by minimization along the respective reaction pathways. The TSs for each step were modeled approximately by minimization with the ratio of the proton-heavy atom distance to the heavy atom separation restrained to its value in the exact (gas-phase) TS described above. The resulting enolate and enol forms were minimized to the same tolerance as for the substrate complex, without restraints on the active site. The calculations were repeated in the reverse direction; that is, from the resulting enol to the enolate and then the keto form of acetyl-CoA, and essentially the same results were found, indicating that the calculated energies were not affected by systematic drift.

According to the AM1/CHARMM results, the enolate form is of considerably lower energy than the enol within citrate synthase. The energies were calculated to be affected significantly by many groups in the enzyme, but the overall pattern remained the same in all the calculations: the substrate (keto) form was calculated to be most stable, followed by the enolate, with the enol highest in energy. In the larger (20Å) system, the enolate form was calculated to lie 2.7 kcal/mol higher than the substrate, and the enol complex 14.5 kcal/mol higher than the enolate, with His-320 treated as positively charged. The charge of His-320 significantly affected the calculated reaction energies (with His-320 neutral, the energy of the enolate complex is 15.7 kcal/mol above the substrate complex, and the enol 8.2 kcal/mol above the enol) but did not change the conclusion that the difference in energy between the enolate and enol is large. Large barriers were found for both reaction steps, although these are given less accurately by AM1 than the relative energies of the different forms. Analysis of the contributions of the surrounding MM groups to the reaction energetics showed that many groups affected the calculated energies, particularly charged groups close to the active site. Oxaloacetate significantly destabilizes the enolate form relative to the substrate because of its proximity and high negative charge. Other notable contributors were found to include a water molecule (Wat-585) which appears to be conserved in structures of citrate synthase complexes, and stabilizes the enolate form relative to the substrate by hydrogen bonding. A serine sidechain (Ser-244) was found to donate a hydrogen bond to the imidazole of His-274 and to stabilize the enol form (*17*). Sequence alignments (*36*) indicate that Ser-244 is a conserved residue. The stabilization provided by this interaction was calculated to be insufficient to make the enol form of comparable energy to the enolate, and so its role appears to be to position His-274 to bind substrate, and to ensure its neutrality. It may also assist in strengthening the hydrogen bond between the enolate of acetyl-CoA and His-274 by cooperative (three-body) interactions.

457

Ab Initio QM/MM Calculations. The semiempirical QM/MM calculations provided no evidence of equalization of the pK_a of the enol of acetyl-CoA and His-274 at the active site of citrate synthase (i.e. the energy difference between the enolate and enol forms was calculated to be large). Approximate equalization of the pK_as of the intermediate and His-274 would be expected according to the proposal that a low-barrier hydrogen bond is responsible for stabilizing the intermediate. The results therefore do not support this hypothesis for citrate synthase. It is of course important to test these conclusions by more sophisticated calculations, and this has recently been achieved by QM/MM calculations on this reaction step at the *ab initio* level (*47*). The CHARMM package (*60*) has recently been interfaced (*9*) with the quantum chemistry program GAMESS (*61*) to allow *ab initio* QM/MM calculations. This implementation has been tested in detail (*9*).

The *ab initio* QM/MM calculations on citrate synthase have built on the earlier semiempirical studies. The starting points for the *ab initio* calculation were the structures of the keto, enolate and enol forms of acetyl-CoA produced by AM1/CHARMM minimization, in the 17Å radius simulation system described above. The QM system and the QM/MM partitioning were slightly different, with different treatments tested (*47*). It is vital that the structures are well minimized for meaningful comparisons of the calculated energies. This is ensured by extensive prior QM/MM minimization at the AM1 level, which allows the MM region to attain a well-minimized structure suitable for the complex under consideration. *Ab initio* QM/MM calculations are very demanding of computer time, and so it is not possible to carry out extensive minimization at the higher level. In the present case, *ab initio* QM/MM minimizations were limited to of the order of 100 steps. This is sufficient to optimize the geometry of the QM system properly, as shown by comparison to the geometries of fully optimized bimolecular complexes (*4*). It is also sufficient to allow the MM region to undergo the necessary small relaxation to accommodate the fairly small electronic and structural changes at the active site on going to the higher level treatment. The changes in the MM region during the reaction show the importance of allowing the structure of the protein to move in such calculations. To ensure further that the structures were properly relaxed, they were minimized first at the RHF/3-21G(*d*) and then the RHF/6-31G(*d*) QM/MM level. As with the earlier calculations, the QM system bears an overall net negative charge, and so it is important to consider effects of including diffuse functions. Model calculations showed that the structures optimized with and without diffuse functions are very similar (*4*), so that in this case a good geometrical description can be achieved by optimization without them. The energetic consequences were tested by single point calculations at the RHF/6-31+G(*d*) level. It was not possible to carry out QM/MM calculations at the correlated level but the effects of electron correlation on the QM system have been tested by single point MP2/6-31G(*d*) and MP2/6-31+G(*d*) calculations on the QM/MM optimized geometries.

The conclusions of the *ab initio* QM/MM calculations are the same as those of the earlier semiempirical calculations in that energy difference between the enolate and enol forms is calculated to be large, and not markedly reduced by the enzyme. Electron correlation significantly affects the reaction energetics, but does not change this finding. The energies along the reaction pathway calculated at the RHF/6-31G(*d*) level are somewhat different from the earlier AM1 QM/MM results, which is partly due to small differences in the QM region, and partly to the different level of QM treatment. Qualitatively the results are similar. At all levels the enolate is calculated to be significantly lower in energy than the enol, and therefore the enolate is the more likely intermediate in the citrate synthase reaction. The calculations do not indicate that the effective acidities of the intermediate and His-274 are similar at the active site, and so do not support the proposal that a low-barrier hydrogen bond is involved in stabilizing the intermediate. A model of the enolic intermediate is also calculated to be of higher energy than the enolate. His-274 and the conserved water molecule, Wat-585, are both calculated to stabilize the enolate relative to the substrate. It appears that the best description of the nucleophilic intermediate in citrate synthase is the enolate of acetyl-CoA, stabilized by conventional hydrogen bonds from His-274 and a water molecule.

Discussion

QM/MM calculations are an effective tool to study enzyme reactions. They allow the structures of reaction intermediates and TS models to be optimized, including the effects of the protein environment which are essential in realistic calculations. These calculations can be carried out at the semiempirical or *ab initio* levels of QM treatment. For the foreseeable future, *ab initio* QM/MM calculations are likely to remain computationally intensive, and may be of most use for testing and calibrating more approximate treatments such as semiempirical QM/MM or empirical valence bond (*3*) methods. Calculations on small models of the type outlined here can also be very useful in this regard. They can provide energies and fully optimized structures of TSs and intermediates for detailed comparisons. One very attractive approach to modeling enzyme reactions is the use of semiempirical methods tailored to reproduce high level results for a particular reaction. The parameters which are integral to the semiempirical treatment can be optimized to fit high level *ab initio* or experimental results, which has proved useful for QM/MM calculations (*7*), and for simulations of organic reactions (*62,63*). Concerns have been expressed that it may be difficult simultaneously to fit well several properties of higher level potential surfaces in this way, however (*64*). The inadequacies of Hartree-Fock level treatments for many reactions are well known, in particular with small basis sets, but correlated *ab initio* QM/MM calculations are likely to be impractical in many cases. Properly tested semiempirical QM/MM treatments may well be preferable to low-level *ab initio* QM/MM calculations, in particular as they allow more extensive investigations of potential energy surfaces and of reaction dynamics. The MM parameters used to represent van der Waals interactions between the QM and MM system should be optimized to reproduce interaction energies and geometries calculated at the *ab initio* level (*6,7,11*). Given such flexibility and computational efficiency, combined with testing against *ab initio* QM and QM/MM results, the prospect of semiempirical QM/MM simulations providing further useful insight into enzyme reactions is bright. Another important consideration is that the effects of solvent dielectric screening should be included in a full treatment. This can be done through continuum (*5*) or more explicit solvation models, or simple charge scaling schemes to reproduce more sophisticated calculations.

The finding that catalysis in citrate synthase does not involve a low-barrier hydrogen bond agrees with calculations on triosephosphate isomerase, in which the analogous catalytic His-95 has been found to have no tendency to donate a proton to the enediolate intermediate, and instead to stabilize it by a normal hydrogen bond (*42*). Experimental results show that while unusually short hydrogen bonds can be formed in enzymes, including citrate synthase, they do not lead to exceptionally large binding energies (*41,65*). The strongest, shortest hydrogen bonds are found for charged interactions between groups with approximately equal pK_as (*66,67*), but there is no special stabilization associated with disappearance of the barrier to proton transfer (*46*), or at ΔpK_a=0 (*45*). Theoretical analysis indicates that low-barrier hydrogen bonds do not offer an energetic advantage for enzyme catalysis (*43*). Low-barrier hydrogen bonds may form in some proteins (*40*), but it is not necessary for hydrogen bonds to be of this type to stabilize high energy intermediates significantly, as shown by the highly favorable (gas phase) interaction energy of the thioester enolate/5-methylimidazole complex (26.6 kcal/mol, MP2/6-31+G(*d*)//6-31+G(*d*), (*4*)) which has a conventional hydrogen bond geometry, and represents the hydrogen bonded intermediate in citrate synthase.

Conclusions

Semiempirical and *ab initio* QM/MM calculations, and calculations on small models, indicate that the enolate of acetyl-CoA is the likely nucleophilic intermediate in citrate synthase, stabilized by hydrogen bonds from His-274 and a water molecule. The enolate character of the intermediate is likely to be important for the second step of the reaction (nucleophilic attack on the carbonyl carbon of oxaloacetate) to proceed efficiently. The results provide no support for the hypothesis that the hydrogen bond between His-274 and the intermediate is of the low-barrier type, and show that normal hydrogen bonds can provide significant stabilization of the enolate intermediate.

Acknowledgments. A.J.M. is a Wellcome Trust International Prize Travelling Research Fellow (grant ref. no. 041229). We thank Dr. Guy Grant for helpful discussions.

Literature Cited

1. Mulholland, A. J.; Karplus, M. *Biochem. Soc. Trans.* **1996**, *24*, 247.
2. Mulholland, A. J.; Grant, G. H.; Richards, W. G. *Protein Engng.* **1993**, *6*, 133.
3. Åqvist, J.; Warshel, A. *Chem. Rev.* **1993**, *93*, 2523.
4. Mulholland, A. J.; Richards, W. G. *J. Phys. Chem. A*, in press.
5. York, D. M.; Lee, T.-S.; Yang, W. *J. Am. Chem. Soc.* **1996**, *118*, 10940.
6. Gao, J.; Freindorf, M. *J. Phys. Chem. A* **1997**, *101*, 3182.
7. Bash, P. A.; Ho, L. L.; MacKerell, A. D., Jr.; Levine, D.; Hallstrom, P. *Proc. Natl. Acad. Sci. USA* **1996**, *93*, 3698.
8. Eurenius, K. P.; Chatfield, D. C.; Brooks, B. R.; Hodoscek, M. *Int. J. Quantum Chem.* **1996**, *60*, 1189.
9. Lyne, P. D.; Hodoscek, M.; Karplus, M., in preparation.
10. Stanton, R. V.; Hartsough, D. S.; Merz, K. M. *J. Comp. Chem.* **1995**, *16*, 113.
11. Field, M. J.; Bash, P. A.; Karplus, M. *J. Comp. Chem.* **1990**, *11*, 700.

12. Moliner, V.; Turner, A. J.; Williams, I. H. *Chem. Commun.* **1997**, *14*, 1271.
13. Lyne, P.; Mulholland, A. J.; Richards, W. G. *J. Am. Chem. Soc.* **1995**, *117*, 11345.
14. Cunningham, M. A.; Ho, L. L.; Nguyen, D. T.; Gillilan, R. E.; Bash, P. A. *Biochemistry* **1997**, *36*, 4800.
15. Ranganathan, S.; Gready, J. E. *J. Phys. Chem. B* **1997**, *101*, 5614.
16. Harrison, M. J.; Burton, N. A.; Hillier, I. H. *J. Am. Chem. Soc.* **1997**, *119*, 12285.
17. Mulholland, A. J.; Richards, W. G. *Proteins: Structure, Function and Genetics* **1997**, *27*, 9.
18. Chatfield, D. C.; Eurenius, K. P.; Brooks, B. R. *J. Mol. Struct. (Theochem)* **1998**, *423*, 79.
19. Liu, H.; Müller-Plathe, F.; van Gunsteren, W. F. *J. Mol. Biol.* **1996**, *261*, 454.
20. Hartsough, D.; Merz, K. M., Jr. *J. Phys. Chem.* **1995**, *99*, 11266.
21. Bash, P. A.; Field, M. J.; Karplus, M. *J. Am. Chem. Soc.* **1987**, *109*, 8092.
22. Mulholland, A. J.; Richards, W. G. *Int. J. Quantum Chem.* **1994**, *51*, 161.
23. Srere, P. In *Advances in Enzymology*; Meister, A, Ed.; Wiley: New York, 1975, Vol. 43; pp 57-102.
24. Remington, S. J. *Curr. Opinion Struct. Biol.* **1992**, *2*, 730.
25. Lenz, H.; Buckel, W.; Wunderwald, P.; Biederman, G.; Buschmeier, V.; Eggerer, H.; Cornforth, J. W.; Redmond, J.; Mallaby, R. *Eur. J. Biochem.* **1971**, *24*, 207.
26. Wlassics, I. D.; Anderson, V. E. *Biochemistry* **1989**, *28*, 1627.
27. Clark, J. D.; O'Keefe, S. J.; Knowles, J. R. *Biochemistry* **1988**, *27*, 5961.
28. Mulholland, A. J.; Richards, W. G. *J. Mol. Struct. (Theochem)*, in press.
29. Kurz, L. C.; Drysdale, G. R. *Biochemistry* **1987**, *26*, 2623.
30. Kurz, L. C.; Shah, S.; Frieden, C.; Nakra, T.; Stein, R. E.; Drysdale, G. R.; Evans, C. T.; Srere, P. A. *Biochemistry* **1995**, *34*, 13278.
31. Löhlein-Werhahn, G.; Goepfert, P.; Kollmann-Koch, A.; Eggerer, H. *Biol. Chem. Hoppe-Seyler* **1988**, *369*, 417.
32. Alter, G. M.; Casazza, J. P.; Zhi, W.; Nemeth, P.; Srere, P. A.; Evans, C. T. *Biochemistry* **1990**, *29*, 7557.
33. Karpusas, M.; Branchaud, B.; Remington, S. J. *Biochemistry* **1990**, *29*, 2213.
34. Karpusas, M.; Holland, D.; Remington, S. J. *Biochemistry* **1991**, *30*, 6024.
35. Evans, C. T.; Kurz, L. C.; Remington, S. J.; Srere, P. A. *Biochemistry* **1996**, *35*, 10661.
36. Russell, R. J. M.; Hough, D. W.; Danson, M. J.; Taylor, G. L. *Structure* **1994**, *2*, 1157.
37. Gerlt, J. A.; Gassman, P. G. *J. Am. Chem. Soc.* **1993**, *115*, 11552.
38. Cleland, W. W.; Kreevoy, M. M. *Science* **1994**, *264*, 1887.
39. Thibblin, A.; Jencks, W. P. *J. Am. Chem. Soc.* **1979**, *101*, 4963.
40. Gerlt, J. A.; Kreevoy, M. M.; Cleland, W. W.; Frey, P. A. *Chemistry and Biology* **1997**, *4*, 259.
41. Usher, K. C.; Remington, S. J.; Martin, D. P.; Drueckhammer, D. G. *Biochemistry* **1994,** *33*, 7753.
42. Alagona, G.; Ghio, C.; Kollman, P. A. *J. Am. Chem. Soc.* **1995**, *117*, 9855.

43. Warshel, A.; Papazyan, A. *Proc. Natl. Acad. Sci. USA* **1996**, *93*, 13665.
44. Guthrie, J. P. *Chemistry and Biology* **1996**, *3*, 163.
45. Shan, S.; Loh, S.; Herschlag, D. *Science* **1996**, *272*, 97.
46. Scheiner, S.; Kar, T. *J. Am. Chem. Soc.* **1995**, *117*, 6970.
47. Mulholland, A. J.; Lyne, P. D.; Karplus, M., in preparation.
48. Dewar, M. J. S.; Zoebisch, E. G.; Healy, E. F.; Stewart, J. J. P. *J. Am. Chem. Soc.* **1985**, *107*, 3902.
49. Dewar, M. J. S.; Yuan, Y.-C. *Inorg. Chem.* **1990**, *29*, 3881.
50. Zheng, Y.-J.; Merz, K. M., Jr. *J. Comp. Chem.* **1992**, *13*, 1151.
51. Meot-Ner, M. *J. Am. Chem. Soc.* **1988**, *110*, 3071.
52. Bash, P. A.; Field, M. J.; Davenport, R. C.; Petsko, G. A.; Ringe, D.; Karplus, M. *Biochemistry* **1991**, *30*, 5826.
53. Wiberg, K. B.; Breneman, C. M.; LePage, T. J. *J. Am. Chem. Soc.* **1990**, *112*, 61.
54. Solomons, T. W. G. In *Organic Chemistry*; Wiley: New York, 1992; pp 581-583.
55. Mulholland, A. J.; Richards, W. G., in preparation
56. Dewar, M. J. S.; Healy, E. F.; Stewart, J. J. P. *J. Chem. Soc. Faraday Trans. 2* **1984**, *80*, 227.
57. Stewart, J. J. P. *J. Computer-Aided Mol. Design* **1990**, *4*, 1.
58. Rzepa, H. S.; Yi, M. *J. Chem. Soc., Perkin Trans. 2* **1990**, 943.
59. Mulholland, A. J.; Richards, W. G. *J. Mol. Struct. (Theochem)*, in press.
60. Brooks, B. R.; Bruccoleri, R. E.; Olafson, B. D.; States, D. J.; Swaminathan, S.; Karplus, M. *J. Comp. Chem.* **1983**, *4*, 187.
61. Schmidt, M. W.; Baldridge, K. K.; Boatz, J. A.; Elbert, S. T.; Gordon, M. S.; Jensen, J. H.; Koseki, S.; Matsunaga, N.; Nguyen, K. A.; Su, S.; Windus, T. L.; Dupuis, M.; Montgomery, J. A., Jr. *J. Comp. Chem.* **1993**, *14*, 1347.
62. Gonzalez-Lafont, A.; Truong, T. N.; Truhlar, D. G. *J. Phys. Chem.* **1991**, *95*, 4618.
63. Rossi, I.; Truhlar, D. G. *Chem. Phys. Lett.* **1995**, *233*, 231.
64. Peslherbe, G. H.; Hase, W. L. *J. Chem. Phys.* **1996**, *104*, 7882.
65. Wang, Z.; Luecke, H.; Yao, N.; Quiocho, F. *Nature Structural Biology* **1997**, *4*, 519.
66. Majerz, I.; Malarski, Z.; Sobczyk, L. *Chem. Phys. Lett.* **1997**, *274*, 361.
67. Meot-Ner, M.; Sieck, L. W. *J. Am. Chem. Soc.* **1986**, *108*, 7525.

Chapter 36

Isotope Effects on the ATCase-Catalyzed Reaction

J. Pawlak[1], M. H. O'Leary[2], and P. Paneth[1]

[1]Institute of Applied Radiation Chemistry, Technical University of Łódź (Lodz), Żeromskiego 116, 90-924 Łódź, Poland
[2]Dean's Office, School of Natural Sciences and Mathematics, California State University, Sacramento, CA 95819–6123

QM/MM calculations were performed on the reaction between carbamyl phosphate and aspartate catalyzed by aspartate transcarbamoylase (ATCase). The putative tetrahedral intermediate and transition states for two chemical steps were characterized. Carbon isotope effects were calculated theoretically and compared with the experimental data.

Modern computational chemistry offers a pletora of methods to model chemical and biochemical processes. However, due to the complexity of real systems and limitations in the computing power of presently available computers, compromises must be made regarding the applied theory level and/or the size of the model. Thus, one of the fundamental questions that has to be answered in all theoretical studies concerns reliability of the method used. It is not surprising that calculations performed at different levels of theoretical scrutiny frequently lead to different properties of the modeled systems, such as geometry, charge distribution, mechanism etc. Deciding which results are correct and which are wrong is not necessarily an easy task. Neither is finding a good criterion upon which to ground such decisions. Theoreticians tend to favor giving the highest theoretical level available at the moment the status of a reference against which results of lower level calculations are tested. Inevitably the reference changes with the development of computational methodologies. Experimentalists rather look for agreement between calculated and experimental properties such as geometries and energies of modeled systems. This approach is however possible only for stable reactants. For short-lived intermediates, intermediates on the enzyme surface (which are not amenable for direct measurements), or transition states, such comparison obviously is not possible.

In recent years hybrid QM/MM methods emerged, which simultaneously take advantage of the precision of quantum methods and the speed of molecular mechanics, permitting us to tackle reactivity in enzymatic and condense phases processes (*1-4*). Having experimental background we are inclined to use measurable quantities to verify calculated results. Isotope effects are thought to be extremely sensitive probes of transition state properties and thus should provide a very good test of theoretical calculations. Our previous experience with modeling isotope effects indicated that semiempirical methods alone are not suitable for calculations of isotope effects (*5,6*). However, we did not obtain much improvement in quantification of isotope effects by either ab initio or DFT methods, especially for the primary kinetic isotope effects (*6*). At the same time we are biased toward low level calculations which would allow improved analysis of the experimental findings, rather than making calculations of isotope effects a quest of its own. Thus, the QM/MM technology seemed to be a very attractive alternative.

Herein we report hybrid a quantum mechanical – molecular mechanics (QM/MM) approach to the mechanism of aspartate transcarbamylase (ATCase) reaction. ATCase catalyzes the carbamylation of aspartate by carbamyl phosphate in the first committed step of pyrimidine biosynthesis in *E. coli* (*7*):

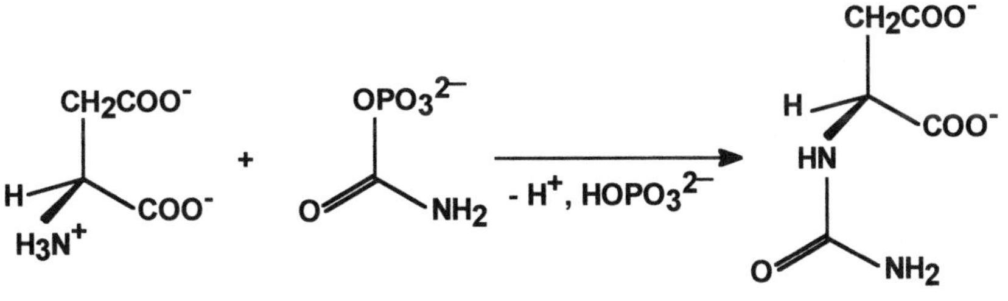

For our purpose it is an ideal model reaction because:
1. experimental isotope effects have been reported (*8*) for
   - different atoms (hydrogen, carbon and nitrogen),
   - equilibrium and kinetic conditions,
   - wild type and mutated enzymes,
   - holoenzyme and the catalytic subunit. Further,
2. numerous x-ray structures of the enzyme are available (*9*) providing a good starting point for modeling the active site of the enzyme.

Computational Details

Models. The model of reactants in the active site pocket was built based on the initial coordinates (PDB structure) of aspartate transcarbamylase provided by Professor Evan Kantrowitz. These corresponded to the catalytic trimer "frozen" in the active form through N-(phosphonoacetyl)-L-aspartate (PALA), an analogue of the putative intermediate, bound in the active site. The model of the active site of ATCase included aminoacids and water molecules within a 6 Å radius of PALA. From this structure 6 aminoacids and 7 water molecules which were not in the immediate vicinity of the substrates were removed. N-terminal ends of aminoacids were capped by hydrogen atoms to form -NH_2, and C-terminal ends were capped by amino groups to form -$CONH_2$. PALA was replaced by reagents. The net charge of +1 was assigned to the model of the wild type and 0 for the histidine-134 changed into alanine (H134A) enzymes on the basis of expected protonation states of aminoacids and reactants at physiological pH.

Models of reactants in aqueous solution were obtained by placing each of the reactants in a large box water molecules with distribution obtained from the TIP3P model (*10*) and removing all water molecules outside a 6 Å radius around a reactant. Models obtained in this way contained 67 and 69 explicit water molecules for aspartate and carbamyl phosphate, respectively.

Hybrid QM/MM calculations. Two methods were used for calculations of reactants in the enzyme active site. In both cases the reactants were treated quantum mechanically while the model of the enzyme pocket was treated at the molecular mechanics level.

In the first method the geometry of reagents was optimized in a rigid structure of the active center. Mixed mode, as implemented in HyperChem (*11*) was used for geometry optimization. The classical region was included as electrostatic potentials, which perturbed the properties of the quantum mechanical region. The interactions of the charges in the classical region with the quantum mechanical region were treated by including them in the one-electron core Hamiltonian for the quantum region. The quantum atoms were those of the reactants treated at the semiempirical level by the PM3 Hamiltonian (*12*). The remaining part of the system was treated classically. It included 23 aminoacids, which are either within the active site or are essential for the catalysis and 11 molecules of water. In a single point calculation of the whole system (233 heavy atoms and 257 hydrogens) the charge distribution was calculated semiempirically. These charges were then included in the semiempirical Hamiltonian at the geometrical location of atoms of the aminoacids and neighboring water molecules. Thus the semiempirical calculations of the reacting molecules were performed within the electrostatic field of the atoms comprising the active site of the enzyme.

In the second method geometry of a part of the active site molecules, treated at the molecular mechanics level, was also optimized. An arbitrary selection was made that atoms comprising the active site residues and hydrogens of water molecules which are contained in a sphere with 3.8 Å radius from reactants were allowed to change positions while positions the outer sphere atoms were left constrained at the crystallographic values. The calculations were carry out in turns at the quantum level and at the molecular mechanics level. The quantum calculations proceeded exactly in the same manner as described above for the first („rigid") method, the only difference being that part of the active site atoms changed position between steps. The molecular mechanics calculations were carried out for atoms in the sphere layer described by the 3.8 Å sphere minus the inner part – reactants. During this step reactant (or transition state) atoms were defined in terms of molecular mechanics force field but their positions were constrained (as well as positions of atoms in the outer sphere from 3.8

to 6 Å). The number of atoms whose positions were optimized (approximately 90) changed dynamically between steps.

Calculations of geometries in aqueous solution were performed in the same manner as those described above for the first method but instead of atoms of the active site the water molecules were used in the molecular mechanics part.

The optimization of geometries of reactants and intermediates, carried out using the RHF PM3 Hamiltonian and the Polak-Ribiere (conjugate gradient) algorithm for reactants and intermediates, was terminated when the RMS gradient reached 0.01 kcal(molÅ)$^{-1}$. Structures of transition states were obtained using the Eigenvector Following method. In these cases the RMS gradients of about 0.3 kcal(molÅ)$^{-1}$ were obtained. Force field calculations were subsequently performed to ensure the optimized geometries correspond to either reactants (no imaginary frequencies) or transition states (exactly one imaginary frequency).

The molecular mechanics optimization of active site atoms was performed using the ESFF force field as implemented in the Discover 95.0/3.0 program (*13*) and was terminated when the gradient reached 0.1 kcal(molÅ)$^{-1}$. The calculations using the flexible method were carried out until no changes in energies were observed in both molecular mechanics and quantum calculations. These are illustrated in Figure 1.

Isotope effects calculations. The force field calculations were repeated for all isotopic species and the resulting isotopic frequencies, ν, were used in calculations of isotope effects according to the Bigeleisen complete equation (*14*):

$$^{H}k = \frac{\nu_L^{\neq}}{\nu_H^{\neq}} \times \prod_i^{3n^P-6} \frac{u_{iL}^R \cdot \sinh(u_{iH}^R/2)}{u_{iH}^R \cdot \sinh(u_{iL}^R/2)} \times \prod_i^{3n^{\neq}-7} \frac{u_{iH}^{\neq} \cdot \sinh(u_{iL}^{\neq}/2)}{u_{iL}^{\neq} \cdot \sinh(u_{iH}^{\neq}/2)} \tag{1}$$

where $u=h\nu/k_BT$, h and k_B are Planck and Boltzman constants, respectively, and T is absolute temperature. Northrop notation (*15*) is adopted for the isotope effects, with leading superscript denoting heavy isotope and subscripts identifying individual reactions in a reaction scheme. Isotope effects were calculated using the Isoeff program (*16*).

Our model calculations show that the complete equation is more robust than the primary equation which is based on frequencies together with principal moments of inertia and molecular masses. The latter is more sensitive to errors in moments of inertia and frequencies. Small frequencies can introduce sizable errors due to rounding of calculated values. We have also tested the influence of considerable residual gradients on the final isotope effects since we were unable to converge transition states to the RMS gradients below 0.3 kcal(molÅ)$^{-1}$. However, we have found that although the absolute values of the frequencies were off by nearly 20 cm^{-1} no errors in isotope effects were observed. This is because isotope effect calculations are reasonably insensitive to the absolute value of a frequency and depend mainly on the isotopic difference.

Results and Discussion

We restrain our discussion to carbon isotope effects although nitrogen isotope effects should prove even more sensitive probe. However, we were unable to achieve convergence using the AM1 Hamiltonian (*17*) the first QM/MM method. The PM3 calculations are known to have problems with nitrogen, especially when this atom is involved in hydrogen bonding(*5,18,19*). These problem are nicely illustrated by the calculations of the overall equilibrium isotope effect of the ATCase reaction. Since these were calculations of the equilibrium isotope effects they involved only optimization of stable molecules; aspartate, carbamyl phosphate, and N-carbamylaspartate. In Table 1 we compare results from QM-PM3/TIP3P and QM-AM1/TIP3P obtained using the first hybrid method.

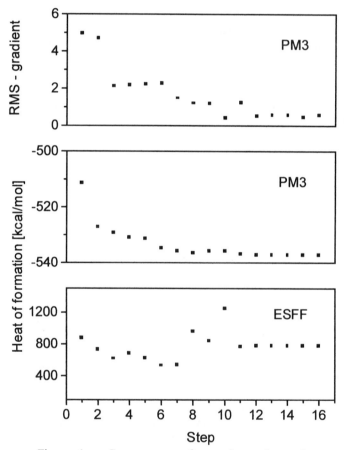

Figure 1. Convergence of energies and starting semiempirical gradient during optimization of both quantum and molecular mechanics regions.

Table 1. Overall equilibrium isotope effects on ATCase reaction

method isotope effect	PM3/TIP3P	AM1/TIP3P	experimental
^{13}C	1.0110	1.0094	-
^{15}N	1.0113	1.0001	0.9990 ± 0.0002

As can be seen the carbon equilibrium isotope effects obtained using both method are both about 1%. The nitrogen equilibrium isotope effects, on the other hand, are significantly different. The PM3 result indicates large positive isotope effect while the value obtained by the AM1 method is practically unity (no isotope effect). The comparison of these results with the experimental value of the overall nitrogen equilibrium isotope effect of 0.999 clearly indicates that the result obtained using the QM-PM3/TIP3P scheme is erroneous.

Typically equations describing apparent kinetic isotope effect as a function of commitments and intrinsic isotope effects, i.e. isotope effects on individual steps, are developed assuming no isotopic fractionation upon binding of reactants by the protein. We have showed previously both experimentally and theoretically that heavy-atom binding isotope effects can be substantial (18,20). These isotope effects were thought to be important only for atoms directly involved in binding, e.g., oxygen, hydrogen and nitrogen. Our present results indicate that they may be also important for carbon. The comparison of carbamyl phosphate in aqueous solution and bound in the active site indicates substantial stiffening of the net bonding to carbon, as illustrated by the bond lengths given in Table 2.

Table 2. Bond lengths to carbon in carbamyl phosphate

environment bond	water [Å]	enzyme [Å]
C=O	1.244	1.227
C-O(P)	1.293	1.322
C-N	1.457	1.418

As can be seen both the C=O and C-N bonds are substantially shortened upon going from aqueous solution to the active site pocket. On the contrary, the bond between the carbon of carbamyl phosphate and bridging oxygen is elongated, which may reflect enzyme's ability to prepare the phosphate moiety for the departure. Furthermore, the carbonyl bond is strongly polarized due to a hydrogen bond between the oxygen and a hydrogen of one of the water molecules present in the vicinity, which should result in restraining bending motion of the carbonyl group. These changes manifest themselves in inverse binding carbon isotope effect equal to 0.996 for both wild type and mutated enzyme. At first glance the 0.4 % inverse carbon isotope effect seems not realistic. However, we have observed earlier inverse carbon kinetic isotope effects of the similar magnitude (0.995 – 0.9975) on the PEP carboxylic carbon in the PEPC catalyzed reaction (21). Recently Cook reported inverse carbon isotope effects in some decarboxylases (Cook, P.F. University of Oklahoma, personal communication, 1998).

Energetics patterns obtained for wild type and H134A mutated enzymes are similar although different in absolute values by about 90 kcal/mol. More useful for comparison is a pathway with reactants bound in the active site superimposed as shown in Figure 2. Only properties of isomeric structures are shown. Calculated heats of formation of the ternary enzyme – reactants complex with the protonated aspartate are much larger, -489.7 and -452.8 kcal/mol for wild type and H134A enzymes, respectively. The energy difference between complexes comprising protonated and deprotonated forms of aspartate are very different for these enzymes; 104.5 kcal/mol for the wild type enzyme and only 49.4 kcal/mol for the H134A mutated enzyme. The fate of the proton transferred to the protein should be determined in order to correct the absolute values. This big difference between the two protonation states with environment is worth noticing.

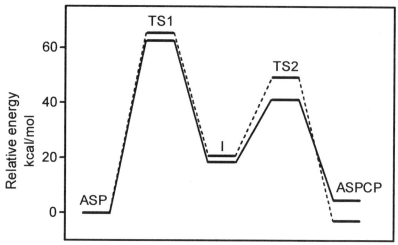

Figure 2. Energetic pathways for wild type (solid line) and H134A mutated (dashed line) ATCase.

Two important differences in the energetic pathways of the wild type enzyme and its H134A mutation should be noted. Firstly, the irreversible step, the dephosphorylation, proceeds with larger activation energy for the mutant while the activation energies of the reverse decomposition of the tetrahedral intermediate back to the first transition state are very close. This means that the commitment for this step ($b=k_9/k_8$ in the original paper, see scheme below for the description of rate constants) should be smaller for the mutant than for the wild type enzyme. This conclusion is opposite to the one reached based on the assumption of equal intrinsic isotope effects made in the analysis of experimental isotope effects. Secondly, the dephosphorylation step is much more exothermic in the case of mutated enzyme; activation energies of the enzyme-bound products going back to the second transition state are about 36 and 51 kcal/mol for wild type and H134A enzymes, respectively. For the wild type enzyme this activation energy is less than for the reaction of intermediate partitioning back to the first transition state (about 44 kcal/mol). This means that in the case of reaction catalyzed by the wild type enzyme the subsequent step – release of products from the enzyme – may also contribute to the commitments.

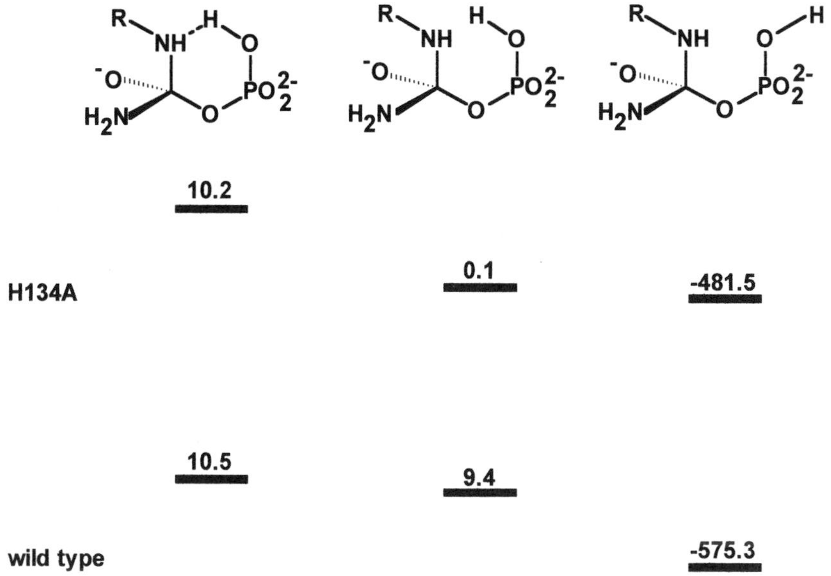

10.2

H134A

0.1

-481.5

10.5

9.4

wild type

-575.3

Figure 3. Optimized structures of the tetrahedral intermediate I. $R = -CHCO_2^-(CH_2CO_2^-)$. For the first two structures the energy difference from the third one is given. For the third structure heats of formation in kcal/mol are given.

An important feature of the mechanism which emerged from our calculations is synchronicity of the C-N bond making between the aspartate and carbamyl phosphate and N$\cdots$H$\cdots$O intramolecular proton transfer between nitrogen and phosphate moiety. Both these changes occur within the same transition state (TS1 in Figure 2). We were unsuccessful in finding any alternative stepwise pathway.

We have identified a few structures of the putative tetrahedral intermediate. These are schematically presented in Figure 3. In the first structure the transferred proton is still in hydrogen bonding contact with the nitrogen atom. In the second the N$\cdots$H distance changes to 3.3 Å causing this hydrogen bond to break. Finally, the newly formed H-O bond is rotated by 140 degrees away from the nitrogen. For both enzymes the most stable is this third form of intermediate (Figure 3) and therefore this was used in calculations of isotope effects.

In order to discuss calculated isotope effects it is necessary to comment on the relationship between calculated stationary points and the reaction scheme used for the analysis of experimental results. Presented below is the scheme which comprises the major steps of the ATCase catalysis:

$$\text{E·CP} + \text{ASPH}^+ \underset{k_2}{\overset{k_1}{\rightleftharpoons}} \text{E·CP·ASPH}^+ \underset{k_4}{\overset{k_3}{\rightleftharpoons}} [\text{E·CP·ASPH}^+]^* \underset{k_6}{\overset{k_5}{\rightleftharpoons}}$$

$$\text{EH·CP·ASP} \underset{k_8}{\overset{k_7}{\rightleftharpoons}} \text{EH·I} \overset{k_9}{\longrightarrow} \text{EH} + \text{ASPCP} + \text{HPO}_4^{2-}$$

The first step is binding of the protonated aspartate, followed by the conformational change in the protein, and subsequent deprotonation of the aspartate, with proton being lost to the enzyme. Conversion of reactants to the tetrahedral intermediate is characterized by the forward rate constant labeled k_7. Dephosphorylation of this intermediate leads in the next step to products. It should be noted that this final step includes both formation of products in the active site and their release from the enzyme. We have pointed out on the basis of energetic profiles that this step may be different for the wild type and H134A mutated forms of the enzyme. Our active site model is based on the crystallographic data of the reactive (R) conformation of the enzyme. Thus no conclusion about the conformational change can be drawn from our results and the corresponding step (with rate constants k_3 and k_4) is transparent for our discussion.

Calculated isotope effects are given in Table 3. The first column lists equilibrium carbon isotope effects on binding already discussed above. The next two columns give kinetic carbon isotope effects on going from either protonated ($^{13}k_{3-7}$) or deprotonated ($^{13}k_7$) aspartate to the first transition state. Since neither deprotonation, nor conformational changes are expected to introduce substantial isotopic fractionation these two isotope effects are expected to be close, which is the case for both enzyme forms. The last two isotope effects are connected with forward ($^{13}k_9$) and backward ($^{13}k_8$) decomposition of the intermediate I. It was assumed in the discussion of the experimental results that these isotope effects should be equal (8b). Our results indicate, however, that these effects are considerably different. This is because the reaction coordinate for going from the intermediate to the first transition state is dominated by the proton transfer and not by the C-N bond formation. Thus the calculated isotope effect is small. The isotope effect on the forward reaction, on the other hand, has a considerable isotope effect, corresponding to early transition state of the C-O bond breaking process.

Table 3. Calculated carbon isotope effects for individual steps of the ATCase reaction.

isotope effect	$^{13}K_{binding}$	$^{13}k_{3-7}$	$^{13}k_7$	$^{13}k_8$	$^{13}k_9$
enzyme					
Wild type	0.9964	1.0344	1.0346	1.0075	1.0185
H134A	0.9963	1.0389	1.0407	1.0099	1.0155

Comparison of our theoretical values to the experimentally measured is possible for the carbon kinetic isotope effect on the H134A enzyme. Modifications of the equation (5) from the reference (8b) which include isotope effect on binding, and numerical values for commitments (a=c=0, and b=0.62) lead to the following expression:

$$^{13}k_{app} = {}^{13}K_{binding} \; {}^{13}k_7 \; \frac{{}^{13}k_9 / {}^{13}k_8 + 0.62}{1.62} \tag{2}$$

Substitution of calculated individual isotope effects leads to ^{13}k equal to 1.0404 which matches nicely the experimental value of 1.0413 ± 0.0011 for the H134A enzyme. Such comparison is, unfortunately, not possible for the native enzyme since the observed value depends on the concentration of aspartate. Comparison of the two rows in Table 3 supports our earlier conclusion (Pawlak, J.; O'Leary, M.H.; Paneth, P. unpublished data) that the intrinsic isotope effects for mutated enzymes might be quite different from isotope effects on the corresponding steps in wild type enzymes.

Only initial results from the second method (flexible active site) are available. We have optimized reactants and the first transition state in the active site pocket. Structures of the transition state obtained by means of these two methods are very similar – proton transfer is slightly more advanced in the structure obtained using flexible model of the active site. Structures of the reactants, on the other hand, are substantially different. The role of histidine-134 which emerges from these calculations nicely fits conclusions based on experimental data; in the enzyme – reactants complex it is in hydrogen bonding contact with one of the oxygens of phosphate moiety confirming its role in binding carbamyl phosphate. In the transition state it is rotated (Figure 4) so that the hydrogen bonding between histidine proton and carbonyl oxygen is established. Thus this residue proves important for catalysis by polarizing the carbonyl bond.

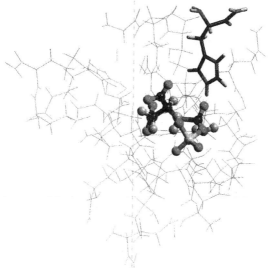

Figure 4. Active site model of ATCase with bounded transition state (rendered as sticks and balls). H134 is highlighted. For color, see the color insert.

472

Acknowledgements

Financial support from the State Committee for Scientific Research (KBN), Poland (grant 6P04A05810), and the Fogarty-NIH (grant LWL/62-130-11101), as well as computer time allocation grants from the Poznań Supercomputer & Networking Center, Poland and the Interdisciplinary Center for Modeling, Warsaw, Poland are gratefully acknowledged.

Literature Cited

1. Warshel, A.; Levitt, M. *J. Mol. Biol.* **1976**, 103, 227.
2. Gao, J. *Rev. Comput. Chem.* **1996**, 7, 119.
3. Bakowies, D.; Thiel, W. *J. Phys. Chem. B* **1996**, 100, 10580.
4. Merz, K.; this volume.
5. Gawlita, E.; Szylhabel-Godala, A.; Paneth, P. *J. Phys. Org. Chem.* **1996**, 9, 41.
6. Czyryca, P.; Paneth, P. *J. Org. Chem.* **1997**, 62, 7305
7. Alewell, N.M. *Ann. Rev. Biophys. Biophys. Chem.* **1989**, 18, 71.
8. (a) Parmentier, L.E.; O'Leary, M.H.; Schachman, H.K.; Cleland, W.W. *Biochemistry* **1992**, 31, 6570. (b) Parmentier, L.E.; Weiss, P.M.; O'Leary, M.H.; Schachman, H.K.; Cleland, W.W. *Biochemistry* **1992**, 31, 6577. (c) Waldrop, G.L.; Turnbull, J.L.; Parmentier, L.E.; O'Leary, M.H.; Cleland, W.W.; Schachman, H.K. *Biochemistry* **1992**, 31, 6585. (d) Waldrop, G.L.; Turnbull, J.L.; Parmentier, L.E.; Lee, S.; O'Leary, M.H.; Cleland, W.W.; Schachman, H.K. *Biochemistry* **1992**, 31, 6592. (e) Parmentier, L.E.; O'Leary, M.H.; Schachman, H.K.; Cleland, W.W. *Biochemistry* **1992**, 31, 6598.
9. (a) Kantrowitz, E.R.; Lipscomb, W.N. *Science* **1988**, 241, 669. (b) Krause, K.L.; Volz, K.W.; Lipscomb, W.N. *J. Mol. Biol.* **1987**, 193, 527.
10. Jorgensen, W.L.; Chandrasekhas, J.; Madura, J.D.; Impey, R.W.; Klein, M.L. *J. Chem. Phys.* **1983**, 79, 926.
11. Hyperchem v. 5.01, HyperCube, Inc. FL.
12. (a) Stewart, J.J.P. *J. Comp. Chem.* **1989**, 10, 209. (b) Stewart, J.J.P. *J. Comp. Chem.* **1989**, 10, 221.
13. InsightII User Guide, October 1995, San Diego: Biosym/MSI 1995.
14. Melander, L. *Isotope Effects on Reaction Rates*, Ronald Press Co., New York, **1960**.
15. Northrop, D.B. *Methods. Enzymol.* **1982**, 87, 607.
16. ISOEFF ver. 6ha and ver. 7 (for Windows) available upon request from P. Paneth (paneth@ck-sg.p.lodz.pl). MOPAC/AMPAC, AMSOL, SIBIQ, GAMESS, GAUSSIAN and HYPERCHEM formats are currently supported.
17. Dewar, M.J.S.; Zoebisch, E.G.; Healy, E.F.; Stewart, J.J.P. *J. Am. Chem. Soc.* **1985**, 107, 3902.
18. Gawlita, E.; Anderson, V.E.; Paneth, P. *Eur. Biophys. J.* **1994**, 23, 353.
19. Schröder, S.; Daggett, V.; Kollman, P. *J. Am. Chem. Soc.* **1991**, 113, 8922.
20. Gawlita, E.; Anderson, V.E.; Paneth, P. *Biochemistry*, **1995**, 34, 6050.

Chapter 37

Enzymatic Transition State Structures Constrained by Experimental Kinetic Isotope Effects: Experimental Measurement of Transition State Variability

Paul J. Berti and Vern L. Schramm

Department of Biochemistry, Albert Einstein College of Medicine, 1300 Morris Park Avenue, Bronx, NY 10461–1602

An individual kinetic isotope effect (KIE) provides bond vibrational and geometric information for the differences in atomic environment between an atom in a reactant and its transition state. The magnitude of KIEs measured at every atom surrounding the sites of bond-making and bond-breaking in a reaction places tight experimental constraints on calculations to determine the geometry of the transition state structure. Interpretation of enzymatic KIEs into the transition state structure first requires quantitation of the extent to which non-chemical enzymatic steps suppress the chemical isotope effects. Experimental methods are now available to solve this problem for most enzymes. The experimental transition state structure that matches the family of intrinsic KIEs is located by a systematic search through reaction space of leaving-group and attacking-group bond orders. The variation of transition state structures through reaction space is estimated from ab initio optimizations of small model compounds. This approach has been used with several N-ribohydrolases and transferases. The molecular electrostatic potentials at the van der Waals surfaces for these transition state structures have been used to design transition state inhibitors. Novel transition state inhibitors with affinities up to 10^6-fold greater than those for the substrates have been obtained using these methods. Comparison of solution and enzymatic transition state structures provides fundamental information on how the enzyme stabilizes the transition states. The uncatalyzed and enzymatic hydrolysis and thiolysis of nicotinamide from NAD^+ are used as examples of the practice of enzymatic transition state analysis.

The transition state structures for enzymatic reactions are explored within the kinetic isotope effect (KIE) formalism of Bigeleisen and Wolfsberg, which holds that intrinsic KIEs report on the differences in environment for an atom between the reactant and transition state of a reaction (1). The isotope effect partition function contains contributions from three sources: the change in molecular mass and moment of inertia, the vibrational zero-point energy differences and the isotopic effect on excited state vibrational levels. The zero-point energy differences between the isotopically labeled atoms in the initial and transition states is usually the dominant term for KIE measurements, and is what we consider when describing isotope effects qualitatively. The vibrational contributions to isotope effects arise from the quantum treatment of harmonic oscillators; equilibrium isotope effects or KIEs are insensitive to isotopic substitutions in classical treatments. Although the Bigeleisen/Wolfsberg treatment of the relationship between KIEs and transition state structure requires several assumptions and approximations, these terms appear in both the numerator and denominator of the equations, and tend to cancel each other out. Excellent discussions of these principles and their applications to enzymatic reactions can be found in the articles of Sunhel and Schowen (2), Huskey (3), and in the discussion of Sims and Lewis on predicting transition state structure from isotope effects (4). Experimental KIEs provide constraints to which the model of the computed enzyme-stabilized transition state must comply. The procedure requires:

a) the synthesis of several reactant NAD$^+$ molecules, each with an individual isotopic label which is expected to be perturbed as the reactant is converted to the transition state

b) the measurement of several primary and secondary KIEs to high accuracy

c) conversion, if necessary, of experimental KIEs to intrinsic isotope effects

d) construction of test transition states throughout reaction space, calculation of KIEs for all test transition state structures, and selection of the transition state structure for which the calculated KIEs match the experimental ones

e) calculation of the wave function and comparison of the molecular electrostatic potential at the van der Waals surfaces to characterize the change between reactant and transition state.

The wave function contains the complete description of the state change from reactant to transition state. The information can be used to describe geometric and energetic changes at the transition state, within the limits of the accuracy of the transition state model. Transition state information can be used to design analogues of the transition state as inhibitors, or to explore the effect of different catalysts, reactants, or chemical environments on the nature of the transition state. This chapter will summarize the experimental and computational approaches used to determine enzymatic transition state structures. The chemical example used to illustrate the procedure is the hydrolysis and thiolysis of the N-ribosidic bond of nicotinamide adenine dinucleotide (NAD$^+$). Comparison of the transition states will be made for the uncatalyzed reaction and the same reaction catalyzed by diphtheria toxin and the thiolysis reactions catalyzed by pertussis toxin (5-10).

Determination of Intrinsic Isotope Effects

Synthesis of a family of specifically isotopically labeled substrates is a necessary prelude to generate the primary data set of experimental KIEs. In some cases, sensitive mass spectrometric methods can be used to measure natural abundance KIEs but these methods are not as generally applicable as the competitive method with radioisotopic labeling. Methods for KIE determination have been reviewed, and with some recent developments for simultaneous determinations of natural abundance KIEs by NMR analysis, are well represented in the literature (11-15). The factors which cause experimentally observed enzymatic KIEs to be less than intrinsic are enzyme-mediated non-chemical steps including substrate binding and product release, which are called commitment factors. An analysis of commitment factors and some methods to correct experimental KIEs to intrinsic ones have been developed by Northrop (16). The expression for the determination of intrinsic KIE for any enzymatic reaction has the form:

$$KIE_{exptl} = \frac{KIE_{intrinsic} + C_f + C_r \cdot K_{eq}}{1 + C_f + C_r} \qquad (1)$$

where KIE_{exptl} and $KIE_{intrinsic}$ are the experimental and intrinsic KIEs, C_f and C_r are forward and reverse commitment factors and K_{eq} is the equilibrium isotope effect for the isotope of interest. K_{eq} can be determined experimentally, or calculated with acceptable accuracy from the vibrational frequencies of the substrates and products, for example using the QUIVER program of Saunders et al. (17). C_f and C_r are experimentally determined by the substrate trapping experiments developed by Rose (18). C_f measures the relative rates of substrate dissociation from the enzyme•substrate complex *versus* the rate of forward reaction to yield products. If the forward chemical step is faster than the rate of substrate dissociation (*i.e.* if C_f is large), then the irreversible catalytic step will be substrate binding. Since KIEs measured by the competitive method yield KIEs in k_{cat}/K_M, that is, up to the first irreversible step, the measured KIEs will report on the rate of substrate association, and give no information on the enzyme's chemical mechanism. C_r is the analogous figure for the reverse reaction. The presence of large C_f or C_r values is sometimes referred to as 'kinetic complexity'. A variety of methods to quantitate or avoid commitment factors has been developed, including multiple KIEs and KIE measurements for pre-steady-state sub-stoichiometric turnovers of enzyme-substrate complexes (19-21). These techniques have made it possible to measure intrinsic isotope effects for most enzymatic reactions, and therefore provide access to experimental transition state information.

Transition State Structures for Uncatalyzed and Enzymatic Hydrolysis of NAD[+]

KIEs of NAD[+] Hydrolysis. The isotopically labeled NAD[+] molecules synthesized for transition state analysis were produced by a combination of enzymatic and chemical methods to incorporate the labels indicated in Figure 1 and Table 1 (5,22). A KIE is the ratio of rate constants (*e.g.* $^1k/^3k$) for substrates containing different isotopes at a defined position. In a typical experiment to determine a ^3H KIE, the ratio of $^1H/^3H$ in the products of a reaction taken partly to completion is determined and compared with the ratio in the original substrate. Since ^1H is not directly measurable by liquid scintillation counting, a ^{14}C label at a remote (isotopically insensitive) position is used as a reporter for the reaction rate of ^1H. The ratio of $^{14}C/^3H$ (= $^1H/^3H$) is determined in the products and compared to the original ratio in the substrate to give the experimental KIE. NAD[+] substrates incorporate an isotopic label in the sensitive position near the scissile bond and a remote label to report on the reaction rate without label. In some cases, stable isotope labels are quantitated directly by mass spectrometry. The magnitudes of the kinetic isotope effects are small, but the accuracy of the measurements provides confidence that KIEs of less than 1 part per 100 can be quantitated. Determining KIEs for as many atoms as practical surrounding the reactive bond helps to constrain the model of the transition state in the subsequent computational steps.

Intrinsic Isotope Effects. Uncatalyzed solvolysis usually exhibits intrinsic isotope effects. Comparison of KIEs from uncatalyzed and enzyme-catalyzed reactions provides a transition state benchmark for the enzymatic transition state (Table 1). NAD[+] hydrolysis is highly exergonic, so the chemical step is essentially irreversible, and only the forward commitment of substrate binding could reduce the intrinsic KIE. Substrate trapping experiments (18) provide a direct experimental measure of commitment factors and are summarized in Figure 2. NAD[+] commitment experiments with diphtheria toxin demonstrated that the forward commitment factor,

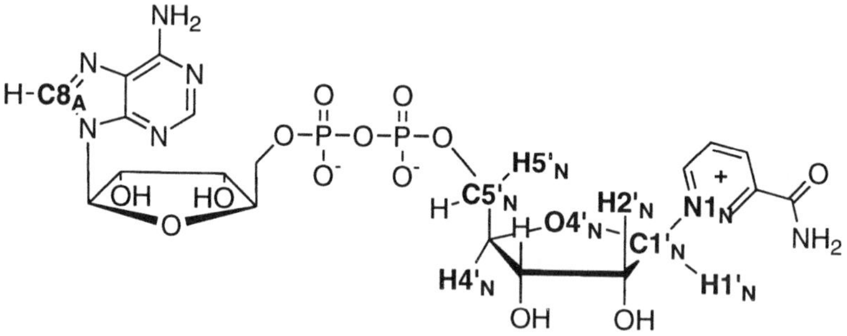

Figure 1. The structure of NAD$^+$ with atomic labels.

Table 1. Experimental and calculated KIEs of NAD$^+$ hydrolysis and thiolysis. Calculated KIEs are for the experimental transition state structures (see text). KIEs determined at 373 K for solvolysis, 310 K for enzymatic reactions.

Label of interest	KIE type	Solvolysis Experimental	Calc'd	Diphtheria toxin Experimental	Calc'd	Pertussis toxin Experimental	Calc'd
1-^{15}N	primary	1.020 ± 0.007	1.020	1.030 ± 0.004	1.029	1.023 ± 0.003	1.022
1'-^{14}C	primary	1.016 ± 0.002	1.019	1.034 ± 0.004	1.034	1.049 ± 0.003	1.047
1-^{15}N, 1'-^{14}C	double primary	1.034 ± 0.002	-	1.062 ± 0.010	-	1.069 ± 0.003	-
1'-^{3}H	α-secondary	1.194 ± 0.005	1.198	1.200 ± 0.005	1.186	1.199 ± 0.009	1.056
2'-^{3}H	β-secondary	1.114 ± 0.004	1.114	1.142 ± 0.005	1.141	1.105 ± 0.004	1.110
4'-^{3}H	γ-secondary	0.997 ± 0.001	-	0.990 ± 0.002	-	0.991 ± 0.002	-
5'-^{3}H	δ-secondary	1.000 ± 0.003	-	1.032 ± 0.004	-	1.020 ± 0.004	-
4'-^{18}O	α-secondary	0.988 ± 0.007	0.986	0.988 ± 0.003	0.985	-	0.990

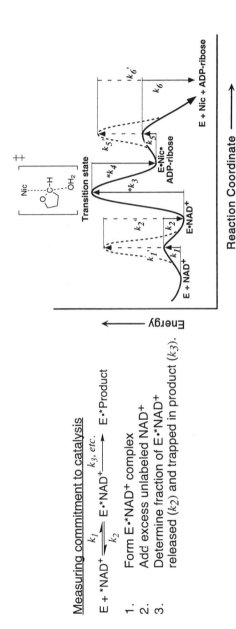

Figure 2. The kinetic steps of substrate binding, release and catalysis which define substrate commitment. Binding isotope effects are assumed to be small relative to KIEs, thus reactant NAD+ molecules which bind to the enzyme and do not equilibrate with both free NAD+ and the transition state contribute only binding isotope effects to the experimental measurement.

C_f, is 0.003, a negligible value (6). Similar experiments with pertussis toxin gave C_f < 0.02 under selected conditions, also a negligible value for KIE measurement (8-10). These results establish that intrinsic isotope effects are observed for solution NAD^+ hydrolysis and for that catalyzed by diphtheria and pertussis toxins. The KIEs from these reactions can be used to directly compare the transition state structures.

Isotopically labeled atoms which are in less constrained environments at the transition state than in reactant give normal KIEs (heavy isotope reacts more slowly) while those in more constrained environments give inverse KIEs (heavy isotope reacts more rapidly). Inspection of the KIEs in Table 1 indicates several isotope effects which differ significantly between the solvolytic reaction and the enzymatic ones. Since the KIEs arise from transition state environments, the experimental data provide direct evidence that the transition states for uncatalyzed and enzymatic hydrolysis differ significantly.

Optimized Atomic Models of the Transition State. A structure interpolation approach has been developed to permit the simultaneous adjustment of all bond lengths and angles of a transition state model in a manner consistent with ab initio predictions for the atomic rearrangement (5,6). This method creates many transition state structures throughout reaction space for which KIEs are then calculated.

Test transition state structures are generated by interpolation between two reference structures. For the hydrolysis of NAD^+, the reference structures were the reactant, NAD^+, and {oxocarbenium ion + nicotinamide}, which would be the intermediates in a classical $D_N + A_N$ mechanism. The structure of the ribosyl moiety of the transition state was assumed to vary between the two reference structures only as a function of its oxocarbenium ion character, OC. By definition, OC = 0 in the reactant NAD^+, and OC = 1 in the oxocarbenium ion. OC is a function only of the bond orders to the leaving group ($n_{LG,TS}$) and nucleophile ($n_{Nu,TS}$) in the transition state model, and to the leaving group in the reactant ($n_{LG,reactant}$):

$$OC = 1 - (n_{Nu,TS} + n_{LG,TS}) / n_{LG,reactant} \qquad (2)$$

Each internal coordinate (bond length, bond angle, or torsional angle) of the ribose ring was a combination of the first and second reference structures. For example, for a bond length (r) in the ribosyl ring in the test transition state (r_{TS}):

$$r_{TS} = r_{reactant} + OC^{1.8} \bullet (r_{oxocarbenium} - r_{reactant}) \qquad (3)$$

Ab initio model calculations showed that, due to transition state imbalance effects, rearrangement of the ribosyl ring was a non-linear function of OC (23,24). The C1'-O4' bond order, which is diagnostic of the ribosyl oxocarbenium ion character, varied approximately as a function of $OC^{1.8}$, rather than OC in equation 3. This is close to the value of the exponent, OC^2, that would have been predicted by Kresge (discussed in 23). The structure of the nicotinamide ring was a function of the leaving group bond order only; the N1-C1'-H1' bond angle was adjusted to reflect the relative bond orders between C1' and the leaving group and nucleophile (5). Many test transition state structures throughout reaction space were generated and the KIEs calculated for each one.

Predicted KIEs for the Models of the Transition State. To calculate KIEs, the structures of reactant NAD^+ and the test transition state structure are defined, along with the force constants governing the molecular internal coordinates, that is, bond stretches, bond angle bends, and bond torsions. The reactant-transition state partition

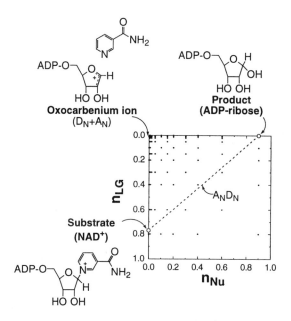

Figure 3. Reaction space diagram for NAD^+ hydrolysis. Reaction space is illustrated by plotting the leaving group bond order (n_{LG}) on the ordinate and nucleophile bond order on the abscissa (n_{Nu}). The reaction proceeds from the **Substrate** in the lower left corner to **Product** in the upper right. A classical D_N+A_N (S_N1) reaction mechanism involves formation of an **Oxocarbenium ion** intermediate, with complete loss of the leaving group, before the nucleophile bond order starts to increase. A concerted, synchronous A_ND_N (S_N2) mechanism follows the dotted diagonal line, where the increase in n_{Nu} at each point matches exactly the loss of n_{LG}.

function of Bigeleisen and Wolfsberg (*1*) is incorporated into the program BEBOVIB (*25*) and is used to predict the KIE for each isotopic label in each test transition state structure. A contour plot is generated for each isotopic label showing the area of reaction space where the calculated KIEs match the experimental ones. There is a unique point in reaction space where the contours for all the isotopic labels converge; this is the experimental transition state structure.

The ab initio optimization of the oxocarbenium ion reference structures in the structure interpolation method are for the structure in vacuo. Calculations using a continuum dielectric model for water solvation (*26*) indicated negligible changes in the optimized structures and negligible differences in calculated KIEs (<0.001 in all cases).

Transition State for NAD+ Solvolysis. Matching the experimental and predicted KIEs for the transition state of NAD+ solvolysis at pH 4.0 (Table 1) indicated a nicotinamide leaving group bond of 2.6 Å (0.02 Pauling bond order) and the attacking water nucleophile at 3.0 Å (0.005 Pauling bond order). The experimental solvolysis transition state structure is an important benchmark for solving the enzymatic transition states. First, the quality of the match of all the calculated versus experimental KIEs help validate the method of structure interpolation that was used to generate test transition state structures. This is both from the point of view of the algorithm, based on the use of oxocarbenium ion character to define the ribosyl ring structure, and of the accuracy of the two reference structures. Second, the non-enzymatic transition state is not subject to distorting interactions that may be imposed by the enzyme during catalysis. Thus, the solvolytic reaction presents an "undistorted" transition state structure. In contrast, there is clear evidence in the diphtheria (*6*) and pertussis (*10*) toxin-catalyzed reactions of KIEs arising at atoms 4'-^{3}H and 5'-^{3}H specifically from enzyme interactions that are not present the solution chemistry. These conclusions for the enzymatic reactions are possible only by direct comparison of the KIEs for solvolysis and enzymatic reactions.

Transition State for NAD+ Hydrolysis by Diphtheria Toxin. For enzymatic reactions, the family of experimental KIEs rarely demonstrate an exact match to those predicted by purely computational means. Hydrolysis of NAD+ by diphtheria toxin gives a family of KIEs which are readily matched to a transition state structure with a Pauling bond order of 0.02 (bond length = 2.6 Å) to the leaving nicotinamide and a bond order to the attacking water nucleophile of 0.03 (2.5 Å). Since both the leaving group and attacking nucleophile are participating at the transition state, the reaction is an A_ND_N (S_N2) mechanism. It differs from canonical nucleophilic displacements by the advanced degree of leaving group bond extension and the low bond order to the attacking nucleophile. The participation of the water nucleophile is significant, since calculations in which the nucleophile is absent caused a significant deviation between predicted and observed KIEs. The transition state structure for pertussis toxin-catalyzed NAD+ hydrolysis has also been determined and is very similar to that for diphtheria toxin (*6,8*).

There are small differences between the experimental and calculated α–secondary KIEs, 1'-^{3}H and 4'-^{18}O, but much larger differences for the remote 4'-^{3}H and 5'-^{3}H KIEs (Table 1). The 4'-^{3}H and 5'-^{3}H KIEs for the solvolytic reaction are near unity, so the much larger experimental KIEs for the enzymatic reaction show that geometric and/or electrostatic distortion is applied by the enzyme as the reactant NAD+ is converted to the enzyme-bound transition state. Transition state stabilization by hydrogen bonds, ionic and hydrophobic interactions prevent the transition state from adopting the minimum energy structure observed in the uncatalyzed reaction.

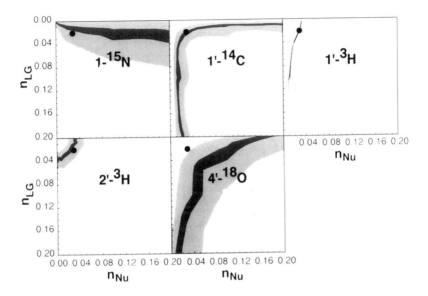

Figure 4. Match of calculated *versus* experimental KIEs. For each isotopic label, the shaded area represents the match of the calculated with the experimental KIE as a function of n_{LG} and n_{Nu}. The light shading represents the 95 % confidence interval of each measured KIE, with dark shading representing the exact measured KIE (approximately ± 0.001). The experimental transition state structure is indicated by "•". Note that only the top-left (dissociative) corner of reaction space is shown.

Transition State for Pertussis Toxin. Pertussis toxin ADP-ribosylates protein $G_{i\alpha1}$ *in vivo*, using a cysteine residue as the ADP-ribose acceptor, and with the thiolate of the cysteine as the nucleophile. KIEs have been measured for the ADP-ribosylation reaction (Table 1) and the transition state structure determined. The forward commitment to catalysis, C_f, was measured and found to be negligible. Reverse commitment (C_r) occurs at low temperature and pH. Intrinsic chemical KIEs were expressed at 30 °C and pH 8. Solvent deuterium isotope effects are inverse, indicating that the cysteine thiol must be ionized to the thiolate anion prior to transition state formation. The structure interpolation method yielded an experimental transition state structure that matched well with the experimental KIEs. The transition state had Pauling bond order to the nicotinamide leaving group of 0.11 (bond length = 2.1 Å) and 0.09 (2.5 Å) to the incoming sulfur nucleophile.

Transition State Features Identified by KIEs. The transition state structures that must be adopted to match the experimental KIEs for the NAD+ solvolytic and enzymatic hydrolysis reactions contain features which could not have been anticipated by chemical intuition. In both the uncatalyzed and enzymatic reactions, the values of the experimental KIEs required that the total bond order to C1' at the reaction center be increased at the transition state relative to the reactant (*5,6*). The summation of C1' bond order increases from 3.62 in reactant NAD+ to 3.78 for the diphtheria toxin-stabilized hydrolysis transition state, and to 3.92 in the transition state for uncatalyzed hydrolysis. For the total bond order to increase at the reaction center in a mechanism which has low orders to both the leaving group and the attacking nucleophile is counterintuitive. The explanation for the increased bond order can be seen in oxocarbenium ions optimized using a hybrid density functional theory minimization with the 6-31+G** basis set (*5,27,28*). In the oxocarbenium ion of the transition state, C1' is sp²-hybridized and the empty p-orbital to the leaving group forms a π–bond with O4' which also has increased sp² character. This increases the O4'-C1' bond order. Hyperconjugation of the C2'-H2' bond with the C1' p-orbital also contributes to the π-bonding. Inductive effects from the positive charge of the transition state also cause increased bond order to C1'. These interactions are illustrated in Figure 5.

A second unexpected feature of the enzyme-stabilized transition state is the remote isotope effects at H4' and H5' (Table 1) which are near unity in the uncatalyzed reaction. The two atomic interactions considered to explain these KIEs were geometric distortion in the C4'-C5'-O5' bond or a change in the ionization of the neighboring phosphate group as a consequence of interaction with the enzyme.

The results from ab initio calculations for the distortional effect on the remote 5'-³H KIE are shown in Figure 6. As the bond angle is varied, the bending force constants are altered. The experimentally observed KIE corresponds to a modest bond angle distortion of +/- 15° from the reactant bond angle. A change in the protonation of the phosphate group also influences the electron distribution of this group. A small molecule model, ethyl phosphate in neutral and monoanion forms, was optimized and the magnitude of the ³H-KIE predicted with QUIVER (*17*) to be 1.049. Thus, a change in ionization state is also sufficient to account for the observed isotope effect. One or a combination of these effects is the likely explanation for the experimental observation.

The ADP-ribosylation reaction of pertussis toxin provides an opportunity to explore the effect of strong nucleophiles on transition state structure. From chemical intuition, one would expect that the proximity of a powerful nucleophile, the thiolate anion, would cause increased nucleophilicity at the transition state. It is satisfying to

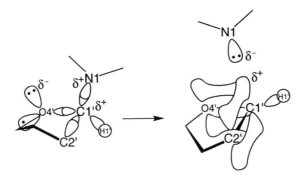

Figure 5. Rearrangement of molecular orbitals between reactant NAD$^+$ and an oxocarbenium ion. (Reproduced with permission from ref. 5. Copyright 1997 American Chemical Society.)

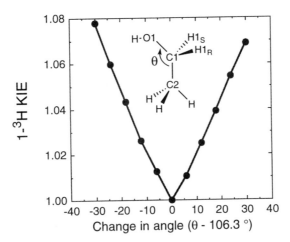

Figure 6. Distortional KIEs as modeled with EtOH. The bond angle, $\theta = \angle$C2-C1-O1 is analogous to the $\angle$C4'-C5'-O5' bond angle of NAD$^+$, and the 1-^{3}H KIE analogous to the 5'-^{3}H KIE. (Reproduced with permission from ref. 6. Copyright 1997 American Chemical Society.)

see the substantial increase in the $1'$-^{14}C KIE which increases from 1.6% in the uncatalyzed solvolysis to 3.4% in the enzymatic hydrolysis where increased nucleophilic participation occurs from H_2O, to the KIE of 4.9% with the attacking thiolate anion. In fully developed, symmetrical S_N2 displacements of the type observed in S-adenosylmethionine synthetase (29) the ^{14}C isotope effect is at its Streitswieser limit of 12.8% (see 3). The nucleophilic character of the transition state for pertussis toxin is greater than the hydrolytic reactions, but less than a symmetric fully developed A_ND_N reaction. The transition state maintains much of its oxocarbenium character, as indicated by the relatively large secondary ^3H KIEs. The influence of the enzyme in stabilizing the transition state is suggested by the unchanged $1'$-^3H KIE and the small change in the β-secondary KIE from $2'$-^3H as nucleophilic participation increases. The enzyme stabilization of the reactive oxocarbenium permits the approach of a strong nucleophile. At this transition state, the reactive thiolate is brought near the reactive oxocarbenium ion, which accounts for the great rate acceleration for the enzymes.

Molecular Electrostatic Potential Surfaces for Transition State Structures. The source of transition state changes can be further investigated by comparing the molecular electrostatic potential surfaces of the reactant and transition states to see where large changes are occurring. This defines the regions of transition state recognition for the catalyst. By considering the atomic interactions which cause transition state stabilization in the catalytic site of the enzyme, it should eventually be possible to obtain information for the basis of enzymatic catalytic activity. Since the forces are determined by electronic distribution at or near the van der Waals surface, an appropriate form of analysis is a comparison of the molecular electrostatic potentials of the enzyme compared to those of the transition state.

Complementarity between the electronic structure of the enzyme and transition state has been proposed to be the single most important factor in the large rate accelerations achieved by enzymes (30). Even if it were possible to accurately predict non-enzymatic transition state structures by purely computational means, such predictions for enzymatic reactions would be more difficult, or impossible. Even in cases where the structure of the enzyme is well known, extrapolation to transition state information is speculative, since enzymatic conformational changes which stabilize the transition state are by necessity transient. Transition state analogues bound to enzymes and solved as co-crystals are also at best only approximations of the actual transition state, since the binding energies are only a fraction of the total transition state binding energy. The dissociation constant of the enzyme-transition state complex ($K_d^{\ddagger}$) is proportional to the catalytic rate enhancement and the relative binding of the Michaelis complex (31,32). Enzymes typically increase the reaction rate by 10^{10} to 10^{14}-fold relative to the same conditions in solution and exhibit enzyme-substrate dissociation equilibrium constants of 10^{-3} to 10^{-6} M, giving $K_d^{\ddagger}$ values of 10^{-13} to 10^{-20} M (32,33). Transition state inhibitors with dissociation constants below 10^{-11} M are rare, demonstrating the difficulty in exactly matching transition state geometries and charge with stable transition state analogues (33,34).

In contrast, electrostatic potential surfaces calculated from KIE-defined enzymatic transition state structures show what changes do occur between the reactant and transition state (Figure 7). In light of the electrostatic potential surface, the enzyme structure can be examined to identify interactions likely to be important in stabilizing these changes. Molecular electrostatic potential surfaces for a KIE-defined enzymatic transition state structure were first determined in 1993 for nucleoside hydrolase (35). This enzyme catalyzes the hydrolysis of the N-ribosidic bond of purine and pyrimidine nucleosides, and is therefore chemically related to the

Reactant   **Transition state**

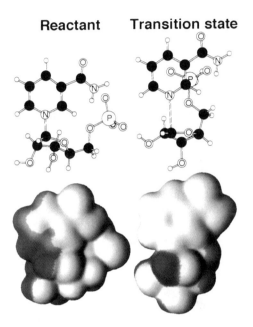

Figure 7.   Structures of reactant and transition state for diphtheria toxin-catalyzed NAD$^+$ hydrolysis. The transition state is shown without the water nucleophile to illustrate the change in electrostatic potential (6). **(top)** Molecular structures. **(bottom)** Location of positive electrostatic potentials projected onto the molecular surface. Electrostatic potential > 0.08 hartree/e appears shaded.

NAD+ N-ribohydrolase activity. The determination of molecular electrostatic potentials for KIE-defined enzymatic transition states has now been applied to several other enzymes (*19,36,37*).

The molecular electrostatic potential surfaces for diphtheria toxin-catalyzed hydrolysis of NAD+ are shown in Figure 7. The positive charge of the nicotinamide ring in reactant NAD+ is distributed over atoms of the pyridine ring and is shared by the ribose, so that nearly equal proportions of the charge reside in the ribosyl and pyridyl rings. This distribution is responsible for the weak N-ribosidic bond in the reactant (bond order = 0.77, see Figure 3). At the transition state, the positive charge migrates to the ribooxocarbenium ion, creating a neutral leaving group. The π-bonds of the transition state delocalize the positive charge across the O4'-C1'-C2' atoms (Figure 5), which become shortened and partially conjugated.

Effects of Catalysis on Transition State Structures, Implications for Mechanism

The diphtheria and pertussis toxin transition states share several characteristics. Both enzymes exhibit significant KIEs at sites remote from catalysis, at 4'-^{3}H and 5'-^{3}H, that are due to the enzymes using binding energy at remote sites to promote catalysis. Both transition states are dissociative; approach of the nucleophile significantly lags departure of the leaving group. In the case of pertussis toxin, however, the transition state is more concerted than for the hydrolytic reactions. If the oxocarbenium ion character of the transition state, OC_{TS}, is used as a gauge of the concertedness of each transition state, there appears to be a correlation with the relative rates of reaction ($k_{rel} = k_{cat}/k_{solvolysis}$): solvolysis, $k_{rel} = 1$, $OC_{TS} = 0.97$; diphtheria toxin hydrolysis, $k_{rel} = 6 \cdot 10^3$, $OC_{TS} = 0.94$; pertussis toxin ADP-ribosylation, $k_{rel} = 1.4 \cdot 10^6$, $OC_{TS} = 0.74$.

There is a relatively small increase in concertedness for the diphtheria toxin hydrolysis reaction, and a larger one for pertussis toxin, which has implications for the catalytic mechanisms of both enzymes. The complex of the transition state structure docked in the active site of the X-ray crystal structure of diphtheria toxin suggests several catalytic strategies (*6*). The oxocarbenium ion can be stabilized by apposition of the side chain carboxylate of Glu148 with the anomeric carbon of the ribosyl moiety. Glu148 is invariant in all bacterial ADP-ribosylating toxins. Further, Glu148 could increase the nucleophilicity of the attacking water by electrostatically polarizing the H-O bond, or even act as a general base catalyst. The nicotinamide ring binds in a deep hydrophobic pocket, becoming desolvated. The low dielectric environment can lower the energy difference between the reactant and transition state by increasing the energy of the positively charged nicotinamide ring in the reactant relative to the neutral form at the transition state, a case of ground state destabilization. His21 is located near the α-phosphate group of NAD+, where it could assist electrostatically in NAD+ binding. All of these features appear to stabilize formation of an oxocarbenium ion-like (that is, a dissociative) transition state. In the complex of the transition state structure docked in the active site cleft of pertussis toxin, all of these same features to stabilize a dissociative transition state appear to be operative (*10*). The additional 233-fold increase in relative rate and the more concerted transition state presumably arise from additional interactions between pertussis toxin and the second substrate, protein $G_{i\alpha1}$. The enzyme uses binding energy with the second substrate to increase its catalytic power by enforcing proximity of the thiolate nucleophile to the electrophilic carbon created as the leaving group departs.

Conclusions

The combination of KIE measurements and computational chemistry provide a fortuitous marriage in which the strengths of both partners contribute to the investigation of enzymatic transition state structure. Features of the enzymatic transition states differ from those in solution by the ability of the enzyme to stabilize reactive species and permit the near approach of attacking groups to form an exceptionally reactive transition state complex. The demonstration that these techniques can be used equally well for solution chemistry and complex enzymatic reactions involving covalent modification of proteins leads to an unprecedented ability to apply the principles of quantum chemistry to biological reactions.

Acknowledgements

This work was supported by NIH Research Grant AI34342 and a post doctoral fellowship from the Natural Sciences and Engineering Research Council (Canada) to P.J.B.

Literature cited

(1)   Bigeleisen, J.; Wolfsberg, M. *Adv. Chem. Phys.* **1958**, *1*, 15-76.
(2)   Suhnel, J.; Schowen, R. L. In *Enzyme Mechanism from Isotope Effects*, P. F. Cook, Ed.; CRC Press, Inc.: Boca Raton, FL., 1991; pp 3-35.
(3)   Huskey, W. P. In *Enzyme Mechanism from Isotope Effects*, P. F. Cook, Ed.; CRC Press, Inc.: Boca Raton, FL., 1991, pp 37-72.
(4)   Sims, L. B.; Lewis, D. E. In *Isotopes in Organic Chemistry,* E. Buncel and C. C. Lee.; Elsevier: New York, 1984; Vol. 6, pp 161-259.
(5)   Berti, P. J.; Schramm, V. L. *J. Am. Chem. Soc.* **1997**, *119*, 12069-12078.
(6)   Berti, P. J.; Blanke, S. R.; Schramm, V. L. *J. Am. Chem. Soc.* **1997**, *119*, 12079-12088.
(7)   Scheuring, J.; Schramm, V. L. *J. Am. Chem. Soc.* **1995**, *117*, 12653-12654.
(8)   Scheuring, J.; Schramm, V. L. *Biochemistry* **1997**, *36*, 4526-4534.
(9)   Scheuring, J.; Schramm, V. L. *Biochemistry* **1997**, *36*, 8215-8223.
(10) Scheuring, J.; Berti, P. J.; Schramm, V. L. *Biochemistry* **1998**, *37*, 2748-2758.
(11) Parkin, D. W. In *Enzyme Mechanism from Isotope Effects*, P. F. Cook, Ed.; CRC Press, Inc.: Boca Raton, 1991, pp 269-290.
(12) Kiick, D. M. In *Enzyme Mechanism from Isotope Effects*, P. F. Cook, Ed.; CRC Press, Inc.: Boca Raton, FL., 1991; pp 313-329
(13) Dalquist, F. W.; Rand-Meir, T.; Raftery, M.A. *Biochemistry* **1969**, *8*,4214-4221
(14) Weiss, P. M. In *Enzyme Mechanism from Isotope Effects*, P. F. Cook, Ed.; CRC Press, Inc.: Boca Raton, 1991, pp 291-311.
(15) Singleton, D. A.; Thomas, A. A. *J. Am. Chem. Soc.* **1995**, *117*, 9357-9358.
(16) Northrop, D. B. *Annu. Rev. Biochem.* **1981**, *50*, 103-131.
(17) Saunders, M.; Laidig, K. E.; Wolfsberg, M. *J. Am. Chem. Soc.* **1989**, *111*, 8989-8994.
(18) Rose, I. W. *Methods Enzymol.* **1980**, *64*, 47-59.
(19) Kline, P. C.; Schramm, V. L. *Biochemistry* **1995**, 34, 1153-1162.
(20) Cleland, W. W., In *Enzyme Mechanism from Isotope Effects*, P. F. Cook, Ed.; CRC Press, Inc.: Boca Raton, 1991, pp 247-265.
(21) McFarland, J. T., In *Enzyme Mechanism from Isotope Effects*, P. F. Cook, Ed.; CRC Press, Inc.: Boca Raton, 1991, pp 151-179.
(22) Rising, K. A.; Schramm, V. L. *J. Am. Chem. Soc.* **1994**, *116*, 6531-6536.
(23) Bernasconi, C. F. *Adv. Phys. Org. Chem.* **1992**, *27*, 119-238.

(24)   Richard, J. P. *Tetrahedron* **1995**, *51*, 1535-1573.

(25)   Sims, L. B.; Burton, G. W.; Lewis, D. E. *BEBOVIB-IV, QCPE No. 337*; Quantum Chemistry Program Exchange, Department of Chemistry, University of Indiana: Bloomington, IN, 1977.

(26)   Wong, M. W.; Wiberg, K. B.; Frisch,M. *J. Chem. Phys.* **1991**, *95*, 8991-8998

(27)   Becke, A. D. *Phys. Rev. A* **1988**, *38*, 3098-3100.

(28)   Perdew, J. P.; Wang, Y. *Phys. Rev. B* **1992**, *45*, 13244.

(29)   Markham, G. D.; Parkin, D. W.; Mentch, F., Schramm, V. L. *J. Biol. Chem.* **1987** *262*, 5609-5615

(30)   Warshel, A. *Computer modeling of chemical reactions in enzymes and solutions*; Wiley-Interscience: New York, 1991.

(31)   Wolfenden, R.; Frick, L. *The Royal Society of Chemistry,London* **1987**, 97-122.

(32)   Wolfenden, R.; Kati, W. M. *Acc. Chem. Res.* **1991**, *24*, 209-215.

(33)   Schramm, V. L.; Horenstein, B. A.; Kline, P. C. *J. Biol. Chem.* **1994**, *269*, 18259-18262.

(34)   Morrison, J. F.; Walsh, C. T. *Advances in Enzymology & Related Areas of Molecular Biology* **1988**, *61*, 201-301.

(35)   Horenstein, B. A.; Schramm, V. L. *Biochemistry* **1993**, *32*, 7089-7097.

(36)   Ehrlich, J. I.; Schramm, V. L. *Biochemistry* **1994**, *33*, 8890-8896.

(37)   Rising, K. A.; Schramm, V. L. *J. Am. Chem. Soc.* **1997**, *119*, 27-37.

Chapter 38

Energetics and Dynamics of Transition States of Reactions in Enzymes and Solutions

Arieh Warshel and Jörg Bentzien

Department of Chemistry, University of Southern California,
Los Angeles, CA 90089-1062

The validity of different proposals for the origin of the catalytic power of enzymes can be explored by computer simulation approaches. In this work we use such approaches to explore the catalytic role of dynamical and non-equilibrium effects. In agreement with our previous studies it is found that dynamical effects are not so different in enzymes and solutions and therefore do not contribute significantly to catalysis. It is also found that non-equilibrium solvation effects in proteins are somewhat smaller but similar to the corresponding effects in solution. The difference is due to the fact that the reorganization free energy is smaller in proteins and this reduces both the overall barrier and the non-equilibrium barrier. This effect, however, is a part of the previously proposed catalytic role of the preorganized enzyme active site and not a new "dynamical" effect. It is further found that the "solvent" coordinate is neither frozen in the transition state of the enzyme, nor in the transition state of the corresponding solution reaction.

1. Introduction

Reliable modeling of transition states of enzymatic reactions is crucial for a detailed understanding of these reactions. Such modeling can tell us which factors are important in enzyme catalysis and can help to establish the relative importance of different catalytic contributions.

Our previous studies[1,2] have demonstrated that enzyme catalysis is mainly due to the reduction of the relevant activation energy by electrostatic effects. However, there are still proposals that attribute catalytic powers to other factors and imply that enzyme reactions can be catalyzed by dynamical effects[3] and that non-equilibrium effects are fundametally different in enzymes and in solution.[4]

The validity of these assertions, as well as those of other proposals for ground state destabilization effects (see discussion in ref. 1), can, in principle, be verified by computer simulation approaches. This work briefly considers modeling of enzyme TSs and to use such modeling in examining the catalytic role of dynamical and non-equilibrium effects. This is done by comparing such effects in enzyme and solution reactions.

2. Analysis of Enzymatic Transition States

2.1 Rate Constants in Condensed Phase.

In order to establish the origin of the catalytic power of enzymes, it is crucial to gain a detailed understanding of the different contributions to the observed rate constant. This can be done by calculating rate constants in enzymes and solutions using microscopic simulation approaches.

Early on, it was assumed by many that the preexponential factor can change drastically between different reactions in condensed phases. However, it is clear now that once the barrier for the reaction is significant, the activation energy is the most important parameter. Most specifically, one can express the rate constant as[5-8]

$$k = \langle \dot{x} H(\dot{x}) \xi \rangle \exp\{-\Delta g(x^{\neq})\beta\} / Z_R \qquad (1)$$

where $\beta = 1/k_B T$ (k_B is the Boltzmann constant and T the absolute temperature), x is the reaction coordinate, $x^{\neq}$ is the value of x at the transition state, $Z_R = \int_{-\infty}^{x^{\neq}} \exp\{-g(x)\beta\}dx$, $H(\dot{x})$ is 1 and zero for positive and negative $\dot{x}$ respectively, ξ is a factor that counts each productive trajectory only once and is taken as $1/m$ for productive trajectories that cross the transition state m times.[7] Eq. (1) can be written as [9]

$$k = (\langle \dot{x}_+ \rangle / \Delta x^{\neq}) \exp\{-\Delta g(x^{\neq})\beta\} \qquad (2)$$

where $\Delta x^{\neq}$ is the width of the transition state region and $\langle \dot{x}_+ \rangle$ is the average of $\langle \dot{x} H(\dot{x}) \xi \rangle$. The pre-exponential factor includes all dynamical effects while the $\Delta g(x^{\neq})$ term reflects the non-dynamical probabilistic effects. Eq. (2) can also be written as:

$$k = F(k_b T / h) \exp\{-\Delta g^{\neq}\beta\} = F k_{TST} \qquad (3)$$

where k_{TST} is the transition state theory rate constant, which does not reflect dynamical effects. F is the so-called "transmission factor" that tells us how many times a trajectory that passes the transition state from the reactant state moves back and forth before falling to the product state. This recrossing process determines the dynamical correction to the rate constant. Note in this respect, that all chemical reactions involve fluctuations but the chance that these fluctuations will reach the TS is not a dynamical effect but simply a probability factor that can be determined by non-dynamical approaches (e.g. Monte Carlo simulations).

In order to explore the nature of enzymatic rate constants it is important to be able to estimate the activation barrier and the transmission factor. Approaches to calculate activation barriers will be outlined in section 2.2 while the evaluation of the transmission factor is considered below.

The transmission factor F of Eq. (3) is directly related to $\langle \dot{x}_+ \rangle$ of Eq. (2). This parameter can be evaluated by running downhill trajectories in the solute-solvent coordinate space. The same results can be obtained using the linear response approximation and two coupled equations for the solute and solvent coordinates.[9] The difference between the protein and solution case is entirely due to the difference in the corresponding environments of the solute. Thus, we can focus on the contribution of the environment (solvent) to $\langle \dot{x}_+ \rangle$. In doing this, we take advantage of the fact[9] that the solvent coordinate (S) is given by the electrostatic contribution to the energy gap

$$S(t) = \varepsilon_2^{elec}(t) - \varepsilon_1^{elec}(t) = \Delta\varepsilon'(t) \tag{4}$$

between the reactant and product valence bond states (the ε_1 and ε_2 of section 2.2). That is, using the linear response approximation, we can relate the effective solvent reaction coordinate to the dipole, μ, of the solute by[9,10]

$$\langle S(t) \rangle_+ = \langle \Delta\varepsilon'(t) \rangle_+ = (\Delta\varepsilon'_{max} / \mu_{max}) \text{ x}$$
$$\int_0^t (\langle \Delta\dot{\varepsilon}'(0)\Delta\varepsilon'(t') \rangle / \langle \Delta\dot{\varepsilon}'(0)\Delta\varepsilon'(0) \rangle) \langle \mu(t - t') \rangle_+ dt' \tag{5}$$

where μ_{max} is the maximum value of the solute dipole moment and is taken as the solute dipole at the product state (relative to the TS dipole). $\Delta\varepsilon'_{max}$ is the change in the electrostatic contribution to $\Delta\varepsilon$ upon moving from the TS to the product state.

Eq. (5) is found to give quite a reliable estimate of the corresponding results obtained by averaging many downhill trajectories.[9] The response function $\langle \Delta\dot{\varepsilon}'(0)\Delta\varepsilon'(t) \rangle$ is, of course, related to the autocorrelation of $\Delta\varepsilon'$ ($C(t) = \langle \Delta\varepsilon'(0)\Delta\varepsilon'(t) \rangle$) by

$$\frac{d}{dt}C(t) = \frac{d}{dt}\langle \Delta\varepsilon'(0)\Delta\varepsilon'(t) \rangle = -\langle \Delta\dot{\varepsilon}'(0)\Delta\varepsilon'(t) \rangle \tag{6}$$

Thus, the dynamical effects can be deduced from the autocorrelation of the electrostatic energy gap. Most importantly if $C(t)$ does not change significantly between two different environments we may conclude that the transmission factors are similar in these systems.

2.2 How to Evaluate Activation Barriers of Enzymatic Reactions.

In order to relate activation barriers of enzymatic reactions to the structure of the corresponding transition states it is essential to have a detailed model that captures the complexity of such systems. Several options are available for modeling enzymatic reactions. They range from hybrid QM/MM methods[11-13] to the EVB method.[1,14] At present, it is still very hard to evaluate by QM/MM approaches activation free energies, non-equilibrium effects, and dynamical effects. One problem is associated with the fact that the solute charges (or bond orders) cannot be used as mapping parameters in free energy perturbation studies, as these charges fluctuate when the solvent fluctuates. Without constraining the solute charges we cannot evelute non-equilibrium solvation effects (see below). Another well known problem is the need to use accurate QM methods which are of course very expensive. Some progress has been made recently by using the EVB potential surface as a reference for the ab initio calculations[15] but even these methods do not provide reliable free energies for the intramolecular contributions of the reacting fragments. Thus, at present, only the EVB method allows one to simulate all aspects of the rate constant of enzymatic reactions. This method can be parameterized by using ab initio results and forced to reproduce accurately the features of gas phase TSs. However, the most important feature of the EVB method is its ability to capture the physics of chemical reactions in condensed phases.[1,2] This is done by allowing the solvent (or proteins) to interact with the charge distribution of the reaction region.[9] The resulting electrostatic free energies were recently found to be very similar to the corresponding ab initio estimates.[15]

The EVB method has been shown to reproduce reasonable structures of TSs and accurate activation free energies of enzymatic reactions, e.g. refs. 16,17. Our focus here will be on dynamical and non-equilibrium effects. We start by noting that the EVB method provides an instructive picture of the relationship between the protein thermal fluctuations and the corresponding rate constant. For example, the proton transfer step from Glu35 to the sugar oxygen in lysozyme can be modeled by considering two resonance structures.

$$\psi_1 = (\ O_A - H \quad O_4\diagup\diagdown\ )$$

$$\psi_2 = (\ O_A^- \quad H - \overset{+}{O_4}\diagup\diagdown\ )$$

Here O_A and O_4 are the oxygens of Glu35 and the glycosidic bond, respectively. Figure 1 represents the reaction by considering the barrier for the proton transfer from O_A to O_4 as a function of time. We find that the barrier oscillates with the fluctuations of the surrounding polar environment. Once in a while the environment reaches a configuration where the barrier is small and the proton can be transferred. This figure clarifies that the proton transfer probability is strongly related to the fluctuations of the surrounding active site, and the resulting fluctuations of the potential energies ε_1 and ε_2 that correspond to ψ_1 and ψ_2, respectively. These environmental changes can be viewed as the fluctuations of the "reaction field" from the solvent/protein system. These fluctuations can be evaluated on a microscopic level by considering the solvent contributions to the gap between the energies ε_1 and ε_2 of ψ_1 and ψ_2, respectively. The relationship between the fluctuations of the "solvent" and the rate of the proton transfer (PT) was pointed out in our early works[18,19] and later elaborated by others.[20] However, the fact that we find an instructive relationship between the fluctuations of the reaction field and the chance for PT does not mean that we are dealing with a dynamical effect. That is, as mentioned above, any chemical process that reaches the transition state involves thermal fluctuations. If the probability for such fluctuations can be obtained from the corresponding Boltzmann probability, then we have a statistical rather than a dynamical effect. Thus, for example, if the chance of bringing the environment to the proper configuration can be determined by the corresponding Boltzmann probability, we do not have a dynamical environmental effect.

Another important factor that can be evaluated by the EVB method is the so-called "non-equilibrium" contribution to the activation barrier. This factor is missing when the solvent is allowed to completely relax along the solute reaction coordinate. In this way, one neglects the fact that the solvent must overcome some activation barrier in order to move from its reactant to its product configuration, when the solute is at its TS configuration. This barrier is associated with the solvent reorganization energy and is, in fact, the only barrier in cases of outer sphere electron transfer reactions. However, even in the case of regular chemical reactions the corresponding contribution is significant.[21] The non-equilibrium effect is entirely missing from standard QM/MM studies that use the solute coordinate as a mapping parameter[21] (see Figure 2). It is also missing from standard continuum models, although one can estimate the effect within the continuum formulation.[22] On the other hand, the EVB method captures this effect by using a mapping procedure that involves a combined free energy perturbation/umbrella sampling method.[9] This approach simply determines the probability of being at the solute-solvent TS without any need to divide the corresponding free energy to equilibrium and non-equilibrium contributions. At any rate, comparing the EVB mapping to the results obtained by relaxing the solvent at each point would give the non-equilibrium contribution.

2.3 The Transmission Factors in Enzymes and Water are Similar and not Much Different Than Unity.

Using Eqs. (5) and (6) we can examine the role of dynamical effects in enzymes by comparing $C(t)$ for enzymes to the corresponding $C(t)$ of the reference reaction in solution. *If these $C(t)$ are similar, then dynamical effects do not contribute to enzyme*

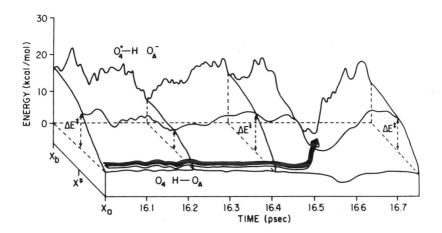

Figure 1. Time dependent changes in the barrier for the proton transfer, $\Delta E^{\neq}$, along a trajectory of the catalytic reaction of lysozyme. The trajectory involves a proton transfer from the oxygen of Glu 35 (O_A) to the oxygen of the glycosidic bond (O_4). The figure demonstrates that the effective barrier ($\Delta E^{\neq}$) is determined by the electrostatic fluctuations of the ionic state $O_4^+{-}H \quad O_A^-$ (see ref. 18 for more details).

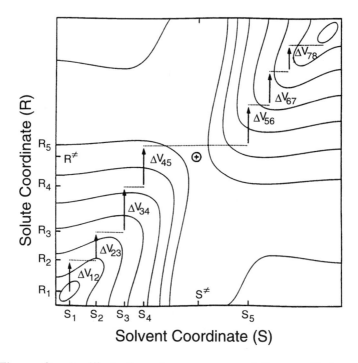

Solvent Coordinate (S)

Figure 2. An illustration of consequence of the so-called non-equilibrium solvation effects. The figure considers the potential surface of a charge transfer reaction (e.g. X^- H–X $\rightarrow$ X–H X^-) as a function of the solute (R) and solvent (S) coordinates. When we try to evaluate the activation free energy, $\Delta g^{\neq}$, by mapping along the solute (R) while allowing the solvent to equilibrate, we underestimate $\Delta g^{\neq}$. This underestimation is due to the neglect of the barrier associated with the reorganization of the solvent. The specific cases described in the figure consider the underestimation of $\Delta g^{\neq}$ in free energy perturbation approaches that use the solute coordinate as a mapping parameter. This is done by considering a low-temperature mapping, where the arrows represent the change in the solute coordinate for each mapping step and the dashed lines represent the change in the solvent coordinate between subsequent mapping steps. In each $i \rightarrow j$ mapping step, ΔG is approximated by $\langle \Delta V \rangle$, which is then approximated by ΔV (this is a reasonable assumption at low temperature) and the solvent is sampled near its equilibrium value for the given V_i. As is seen from the figure, the end of the third mapping step corresponds to S_3 whereas the beginning of the fourth mapping step corresponds to S_4, and thus the effect of changing the solvent coordinate from S_3 to S_4 is not reflected by the mapping procedure. Most importantly, since the equilibrated solvent in R_4 corresponds to S_4 and the equilibrated solvent at R_5 corresponds to S_5, the change of the solvent coordinate between S_4 and S_5 is not reflected by ΔV and the transition state for the solvent coordinate, $S^{\neq}$, is never sampled (see ref. 21 for more details).

catalysis. In Figure 3, we present a comparison of the $C(t)$ of the catalytic reaction of subtilisin to the corresponding $C(t)$ in solution. Both reactions were modeled using the Empirical Valance Bond (EVB) potential surfaces.[1] As seen from the figure, the two $C(t)$ are similar and therefore the dynamical effect for the enzyme and solution reaction is similar. The same conclusions were obtained by comparing the $C(t)$ of the native subtilisin and the Asn155→Ala mutant.[23] Here again the $C(t)$ are similar for the native and mutant enzymes. The conclusions that the enzyme does not use dynamical effects in catalysis can also be established by a direct calculation of the transmission factor running downhill trajectories[9] or by calculating the flux autocorrelation.[24] In fact, a recent study[4] that used our EVB method in simulating the catalytic reaction of triosephosphate isomerase (TIM) yielded a transmission factor of 1/2.8. This value is, of course, not far from unity and therefore reflects a minimal dynamical effect.

Interestingly the difference in activation free energy between the enzymatic reaction in TIM and the same reaction in water has been studied using the EVB method and these calculations reproduced the observed catalytic effect of the enzyme.[16] This indicates that the reaction rate is determined by the activation energy rather than by dynamical effect.

2.4 Non-equilibrium Effects have a Similar Nature in Water and in Enzymes.

While it starts to be clear now that the transmission factor is similar in enzymes and in water and also not very different than the TST limit, it has been proposed that non-equilibrium effects are fundamentally different in enzymes and water.[4] However, this proposal involved no attempt to calculate the relevant quantities in water but rather a comparison of the simulations of a reaction in a protein to the assumed result of the corresponding simulation in water. This was done by referring to an independent study of ref. 25 that considered a different reaction and did not take into account the large effect of the solute-solvent coupling (Ref. 25 used gas phase charges without considering the polarization of these charges by the solvent). On the other hand the proper solute-solvent coupling was considered in the enzyme simulation of ref. 4 since it is automatically reproduced by the EVB approach. Obviously, a consistent way of judging the catalytic role of non-equilibrium effects is to use the *same* model in studying the reaction in water and in the protein, rather than to compare studies that used two different approaches. Apparently the "frozen-solvent" picture assumed by Hynes and coworkers[25] does not hold when one uses the correct coupling between the solute charges and the reaction field from the solvent.[9]

In order to obtain a consistent analysis of the catalytic contributions of non-equilibrium effects, we consider the nucleophilic attack step in the catalytic reaction of subtilisin and the corresponding reaction in solution. We first considered the nature of the reactive fluctuations that reach the transition state. This is done, as usual, by running downhill trajectories from the transition state ($R^{\neq}$, $S^{\neq}$ in Figure 2). The time reversal of these trajectories gives us the proper uphill trajectories. Now, in a clear contrast to the frozen-solvent picture, we find that the solvent and the solute coordinates change at a similar rate. This means that the assumption that the solvent response is much slower than the solute[25] is not justified. As seen from Figure 4 the "solvent" moves fast in both the protein and solution reactions.

As far as dynamical effects in catalysis are concerned, we establish in Figure 4 that the overall reaction dynamics in the solute-solvent coordinate space is quite similar in proteins and solutions. Of course, the amplitude of the "solvent" coordinate is smaller in the protein, due to the smaller reorganization energy. Thus, the assertion that the non-equilibrium dynamics are different in proteins and solutions is not justified. Interestingly, the motion of the solvent coordinate can be separated to a very fast component followed by a slow diffusive motion. This feature can be, in fact, deduced from the slow and fast components of the autocorrelation $C(t)$ (see Figure 3 and related studies in refs. 26,27).

Next, we examined the actual "non-equilibrium" effect on the activation barriers. This was done by constraining the "solute" to its TS geometry ($R^{\neq}$ of Figure

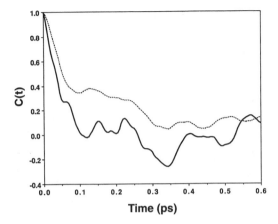

Figure 3. The autocorrelation, $C(t)$, of the time dependent energy gap for the nucleophilic attack step in the catalytic reaction of subtilisin (solid line) and for the corresponding reference reaction in water (dotted line). The similarity of these autocorrelations indicates that dynamical effects do not contribute significantly to catalysis.

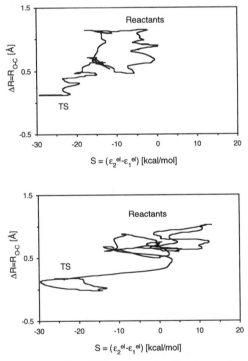

Figure 4. Comparing downhill trajectories for the nucleophilic attack step in subtilisin (top) and in water (bottom). The solvent reaction coordinate S is taken as the intermolecular electrostatic contribution to energy gap and the solute coordinate is the distance between the attacking O^- of Ser221 and the carbonyl carbon of the substrate.

2) and evaluating the free energy surface for moving the solvent from its equilibrium configuration on the left of $S^{\neq}$ (i.e. S_4) to its equilibrium configuration on the right of $S^{\neq}$ (i.e. S_5). Note that when the solute charges are coupled properly to the solvent reaction field, one finds that the solute charges are different than the corresponding gas phase value at $R^{\neq}$ and also different than the symmetric values at ($R^{\neq}$, $S^{\neq}$). That is, as demonstrated in ref. 21, the equilibrated solvent forces the solute charges to be different than their true transition state value. Now, using the solute charges $Q(S_4)$ and $Q(S_5)$, we evaluated the above mentioned activation barrier for $S_4 \rightarrow S_5$ at $R = R^{\neq}$. The results, which are presented in Figure 5, demonstrate that the non-equilibrium barriers in the enzyme and the solution reaction are similar in nature, although the barrier is lower in the enzyme due to its preorganized environment.

3. Concluding Remarks

Simulations of TSs of enzymatic reactions are crucial for the understanding of enzyme catalysis. Such simulations can be conducted by the EVB method that is capable of producing the relevant activation free energies, transmission factors, and non-equilibrium solvation effects. EVB simulations[1,4,23] have demonstrated that the transmission factor of reactions in enzymes and solutions are similar and thus dynamical effects are not likely to play a major role in enzyme catalysis. Furthermore, even the nature of the solvent thermal fluctuations that modulate the solute barrier are similar in enzymes and solution.[18,19] Here we find that the motions of the productive fluctuations that lead to the transition state involve concerted solute solvent motion so that the frozen solvent model is not justified for both the enzyme and solution reaction. As to non-equilibrium solvation effects, they appear to be associated with the reduction of the reorganization energy in the enzyme site rather than with dynamical effects. The reorganization energy is indeed reduced in enzyme active sites, whose preoriented polar environment is the key to the catalytic power of enzymes.[1]

Acknowledgement

This work was supported by NIH Grant GM-24492.

References

(1)   Warshel, A. *Computer Modeling of Chemical Reactions in Enzymes and Solutions*; John Wiley & Sons: New York, 1991.

(2)   Åqvist, J.; Warshel, A. *Chem. Rev.* **1993**, *93*, 2523.

(3)   Careri, G.; Fasella, P.; Gratton, E. *Ann. Rev. Biophys. Bioeng.* **1979**, *8*, 69.

(4)   Neria, E.; Karplus, M. *Chem. Phys. Lett.* **1997**, *267*, 23.

(5)   Keck, J. C. *Adv. Chem. Phys.* **1966**, *13*, 85.

(6)   Anderson, J. B. *J. Chem. Phys.* **1973**, *58*, 4684.

(7)   Bennett, C. H. In *Algorithms for Chemical Computations*; Christofferson, R. E., Ed.; ACS: Washington, D. C., 1977; pp 63.

(8)   Grimmelmann, E. K.; Tully, J. C.; Helfand, E. *J. Chem. Phys.* **1981**, *74*, 5300.

(9)   Hwang, J.-K.; King, G.; Creighton, S.; Warshel, A. *J. Am. Chem. Soc.* **1988**, 110, 5297.

(10)   Hwang, J.-K.; Creighton, S.; King, G.; Whitney, D.; Warshel, A. *J. Chem. Phys.* **1988**, *89*, 859.

(11)   Warshel, A.; Levitt, M. *J. Mol. Biol.* **1976**, *103*, 227.

(12)   Bash, P.A.; Field, M.J.; Davenport, R.C.; Petsko, G.A.; Ringe, D.; Karplus, M. *Biochemistry* **1991**, *30*, 5826.

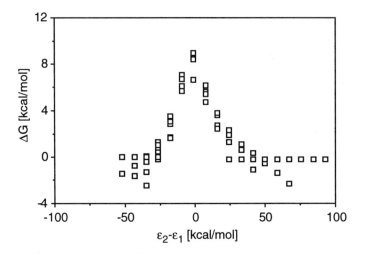

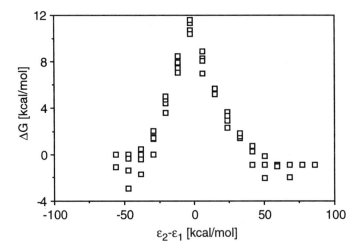

Figure 5. Comparing the non-equilibrium barrier for the enzyme (top) and solution (bottom) reaction. The calculated free energy profiles were obtained for the nucleophilic attack step in the catalytic reaction of subtilisin. This was done by keeping the "solute" in its transition state geometry ($R^{\neq}$ of Figure 2), and evaluating the free energy surface for moving the solvent across the transition state ($S_4 \rightarrow S_5$ in Figure 2). In the corresponding mapping procedure the solute charges change from their equilibrium value at $R^{\neq}$ and S_4 to their value at $R^{\neq}$ and S_5. Since the solvent is allowed to relax we do not obtain the solute charges that correspond to the actual transition state ($R^{\neq}$, $S^{\neq}$). Missing this charge distribution prevents one from obtaining the non-equilibrium barrier when using R as a mapping parameter. As seen from the figure, the trends in the non-equilibrium barriers are similar in the enzyme and solution cases, although the barrier is lower in the enzyme case due to the reduced reorganization energy.

(13) Mulholland, A.J.; Grant, G.H.; Richards, W.G. *Prot. Eng.* **1993**, *6*, 133.
(14) Warshel, A.; Weiss, R.M. *J. Am. Chem. Soc.* **1980**, *102*, 6218.
(15) Bentzien, J.; Muller, R.P.; Florián, J.; Warshel, A. *J. Phys. Chem. B* **1998**, *102*, 2293.
(16) Åqvist, J.; Fothergill, M. *J. Biol. Chem.* **1996**, *271*, 10010.
(17) Fuxreiter, M.; Warshel, A. *J. Am. Chem. Soc.* **1998**, *120*, 183.
(18) Warshel, A. *Proc. Natl. Acad. Sci. USA* **1984**, *81*, 444.
(19) Warshel, A. *J. Phys. Chem.* **1982**, *86*, 2218.
(20) Borgis, D.; Lee, S.; Hynes, J.T. *Chem. Phys. Lett.* **1989**, *162*, 19.
(21) Muller, R.P.; Warshel, A. *J. Phys. Chem.* **1995**, *99*, 17516.
(22) Truhlar, D.G.; Schenter, G.K.; Garret, B.C. *J. Chem. Phys* **1993**, *98*, 5756.
(23) Warshel, A.; Sussman, F.; Hwang, J.-K. *J. Mol. Biol.* **1988**, *201*, 139.
(24) Chandler, D. *J. Chem. Phys.* **1978**, *68*, 2959.
(25) Bergsma, J.P.; Gertner, B.J.; Wilson, K.R.; Hynes, J.T. *J. Chem. Phys.* **1987**, *86*, 1356.
(26) Warshel, A.; Hwang, J.-K. *J. Chem. Phys.* **1986**, *84*, 4938.
(27) Warshel, A.; Chu, Z. T.; Parson, W. W. *Science* **1989**, *246*, 112.

INDEXES

Author Index

Subject Index

A

Bestsellers from ACS Books

The ACS Style Guide: A Manual for Authors and Editors (2nd Edition)
Edited by Janet S. Dodd
470 pp; clothbound ISBN 0–8412–3461–2; paperback ISBN 0–8412–3462–0

Writing the Laboratory Notebook
By Howard M. Kanare
145 pp; clothbound ISBN 0–8412–0906–5; paperback ISBN 0–8412–0933–2

Career Transitions for Chemists
By Dorothy P. Rodmann, Donald D. Bly, Frederick H. Owens, and Anne-Claire Anderson
240 pp; clothbound ISBN 0–8412–3052–8; paperback ISBN 0–8412–3038–2

Chemical Activities (student and teacher editions)
By Christie L. Borgford and Lee R. Summerlin
330 pp; spiralbound ISBN 0–8412–1417–4; teacher edition, ISBN 0–8412–1416–6

Chemical Demonstrations: A Sourcebook for Teachers, Volumes 1 and 2, Second Edition
Volume 1 by Lee R. Summerlin and James L. Ealy, Jr.
198 pp; spiralbound ISBN 0–8412–1481–6
Volume 2 by Lee R. Summerlin, Christie L. Borgford, and Julie B. Ealy
234 pp; spiralbound ISBN 0–8412–1535–9

The Internet: A Guide for Chemists
Edited by Steven M. Bachrach
360 pp; clothbound ISBN 0–8412–3223–7; paperback ISBN 0–8412–3224–5

Laboratory Waste Management: A Guidebook
ACS Task Force on Laboratory Waste Management
250 pp; clothbound ISBN 0–8412–2735–7; paperback ISBN 0–8412–2849–3

Reagent Chemicals, Eighth Edition
700 pp; clothbound ISBN 0–8412–2502–8

Good Laboratory Practice Standards: Applications for Field and Laboratory Studies
Edited by Willa Y. Garner, Maureen S. Barge, and James P. Ussary
571 pp; clothbound ISBN 0–8412–2192–8

For further information contact:
Order Department
Oxford University Press
2001 Evans Road
Cary, NC 27513
Phone: 1-800-445-9714 or 919-677-0977

Highlights from ACS Books

Desk Reference of Functional Polymers: Syntheses and Applications
Reza Arshady, Editor
832 pages, clothbound, ISBN 0–8412–3469–8

Chemical Engineering for Chemists
Richard G. Griskey
352 pages, clothbound, ISBN 0–8412–2215–0

Controlled Drug Delivery: Challenges and Strategies
Kinam Park, Editor
720 pages, clothbound, ISBN 0–8412–3470–1

Chemistry Today and Tomorrow: The Central, Useful, and Creative Science
Ronald Breslow
144 pages, paperbound, ISBN 0–8412–3460–4

A Practical Guide to Combinatorial Chemistry
Anthony W. Czarnik and Sheila H. DeWitt
462 pages, clothbound, ISBN 0–8412–3485–X

Chiral Separations: Applications and Technology
Satinder Ahuja, Editor
368 pages, clothbound, ISBN 0–8412–3407–8

Molecular Diversity and Combinatorial Chemistry: Libraries and Drug Discovery
Irwin M. Chaiken and Kim D. Janda, Editors
336 pages, clothbound, ISBN 0–8412–3450–7

A Lifetime of Synergy with Theory and Experiment
Andrew Streitwieser, Jr.
320 pages, clothbound, ISBN 0–8412–1836–6

Chemical Research Faculties, An International Directory
1,300 pages, clothbound, ISBN 0–8412–3301–2

For further information contact:
Order Department
Oxford University Press
2001 Evans Road
Cary, NC 27513
Phone: 1-800-445-9714 or 919-677-0977
Fax: 919-677-1303